PLASTICITY THEORY

JACOB LUBLINER

Professor Emeritus of Engineering Science
Department of Civil and Environmental Engineering
University of California at Berkeley

DOVER PUBLICATIONS, INC.
MINEOLA, NEW YORK

Bibliographical Note

This Dover edition, first published in 2008, is a revised and corrected republication of the edition published by Macmillan Publishing Company (now Pearson Education), New York, in 1990.

Library of Congress Cataloging-in-Publication Data

Lubliner, Jacob.
Plasticity theory / Jacob Lubliner.
p. cm.
Originally published: New York : Macmillan, c1990.
ISBN-13: 978-0-486-46290-5
ISBN-10: 0-486-46290-0
1. Plasticity. I. Title.

QA931.L939 2008
531'.385—dc22

2007038484

Manufactured in the United States of America
Dover Publications, Inc., 31 East 2nd Street, Mineola, N.Y. 11501

Preface to the Dover Edition

Despite the proofreaders' efforts and mine, the original printed edition remained plagued with numerous errors. Most of them are typographic, but there are also some mathematical errors in equations and inline expressions, and graphic errors in the figures. It so happened that the time of proofreading coincided with my father's terminal illness, and I found myself unable to give the process the attention it required.

In the course of the following years I managed to catch — both on my own and with the help of others — lots and lots of those errors, eventually perhaps most if not all. Shortly after the book first came out I began to compile an errata file, which kept on growing, and which I made available (first in printed form, then as a PDF file) to anyone who inquired about the book, with updates as needed.

A few years ago I noticed that the errata file stopped growing. Any needles remaining in the haystack seemed to have become well hidden. It was then that I decided to compile a corrected — not revised, since I made no substantive changes — edition, in LaTeX (by way of WinEdt) rather than vi, which I had used for the original.

It also happened around that time that all the documents that I had kept in my office on the Berkeley campus, including the original drawings of the book's figures, were lost in the course of a move between buildings. I consequently had to redo the figures as well, and I made them, I believe, more readable than the original ones.

As the book gradually went out of print, I noticed that used copies were being sold at an exorbitant price. I also began to receive queries from students and teachers all over the world about how they might acquire the book. I therefore decided to place the entire book, first in chapter-by-chapter form and then in a single file, available for downloading from my academic Web site at no charge.

In 2006 my friend, colleague and former student Gautam Dasgupta of Columbia University suggested to me that I contact Dover Publications about publishing a paperback edition of the book that would be moderately priced. I did so early in 2007, and promptly received a favorable reply from John Grafton, Senior Reprint Editor. With his help and guidance, the project has now come to fruition, and I feel gratified that the work that I put into *Plasticity Theory* some twenty years ago is still bearing fruit, and not overpriced fruit at that.

Preface

When I first began to plan this book, I thought that I would begin the preface with the words "The purpose of this little book is..." While I never lost my belief that small is beautiful, I discovered that it is impossible to put together a treatment of a field as vast as plasticity theory between the covers of a truly "little" book and still hope that it will be reasonably comprehensive.

I have long felt that a modern book on the subject — one that would be useful as a primary reference and, more importantly, as a textbook in a graduate course (such as the one that my colleague Jim Kelly and I have been teaching) — should incorporate modern treatments of constitutive theory (including thermodynamics and internal variables), large-deformation plasticity, and dynamic plasticity. By no coincidence, it is precisely these topics — rather than the traditional study of elastic-plastic boundary-value problems, slip-line theory and limit analysis — that have been the subject of my own research in plasticity theory. I also feel that a basic treatment of plasticity theory should contain at least introductions to the physical foundations of plasticity (and not only that of metals) and to numerical methods — subjects in which I am not an expert.

I found it quite frustrating that no book in print came even close to adequately covering all these topics. Out of necessity, I began to prepare class notes to supplement readings from various available sources. With the aid of contemporary word-processing technology, the class notes came to resemble book chapters, prompting some students and colleagues to ask, "Why don't you write a book?" It was these queries that gave me the idea of composing a "little" book that would discuss both the topics that are omitted from most extant books and, for the sake of completeness, the conventional topics as well.

Almost two years have passed, and some 1.2 megabytes of disk space have been filled, resulting in over 400 pages of print. Naively perhaps, I still hope that the reader approaches this overgrown volume as though it were a little book: it must not be expected, despite my efforts to make it comprehensive, to be exhaustive, especially in the sections dealing with applications; I have preferred to discuss just enough problems to highlight various facets of any topic. Some oft-treated topics, such as rotating disks, are not touched at all, nor are such general areas of current interest as micromechanics (except on the elementary, qualitative level of dislocation theory), damage mechanics (except for a presentation of the general framework of internal-variable modeling), or fracture mechanics. I had to stop somewhere, didn't I?

The book is organized in eight chapters, covering major subject areas; the chapters are divided into sections, and the sections into topical subsections. Almost every section is followed by a number of exercises. The order of presentation of the areas is somewhat arbitrary. It is based on the order in which I have chosen to teach the field, and may easily be criticized by those partial to a different order. It may seem awkward, for example, that constitutive theory, both elastic and inelastic, is introduced in Chapter 1 (which is a general introduction to continuum thermomechanics), interrupted for a survey of the physics of plasticity as given in Chapter 2, and returned to with specific attention to viscoplasticity and (finally!) rate-independent plasticity in Chapter 3; this chapter contains the theory of yield criteria, flow rules, and hardening rules, as well as uniqueness theorems, extremum and variational principles, and limit-analysis and shakedown theorems. I believe that the book's structure and style are sufficiently loose to permit some juggling of the material; to continue the example, the material of Chapter 2 may be taken up at some other point, if at all.

The book may also be criticized for devoting too many pages to concepts of physics and constitutive theory that are far more general than the conventional constitutive models that are actually used in the chapters presenting applications. My defense against such criticisms is this: I believe that the physics of plasticity and constitutive modeling are in themselves highly interesting topics on which a great deal of contemporary research is done, and which deserve to be introduced for their own sake even if their applicability to the solution of problems (except by means of high-powered numerical methods) is limited by their complexity.

Another criticism that may, with some justification, be leveled is that the general formulation of continuum mechanics, valid for large as well as small deformations and rotations, is presented as a separate topic in Chapter 8, at the end of the book rather than at the beginning. It would indeed be more elegant to begin with the most general presentation and then to specialize. The choice I finally made was motivated by two factors. One is that most of the theory and applications that form the bulk of the book can be expressed quite adequately within the small-deformation framework. The other factor is pedagogical: it appears to me, on the basis of long experience, that most students feel overwhelmed if the new concepts appearing in large-deformation continuum mechanics were thrown at them too soon.

Much of the material of Chapter 1 — including the mathematical fundamentals, in particular tensor algebra and analysis — would normally be covered in a basic course in continuum mechanics at the advanced undergraduate or first-year graduate level of a North American university. I have included it in order to make the book more or less self-contained, and while I might have relegated this material to an appendix (as many authors have done), I chose to put it at the beginning, if only in order to establish a consistent set of notations at the outset. For more sophisticated students, this material may serve the purpose of review, and they may well study Section 8.1 along with Sections 1.2 and 1.3, and Section 8.2 along with Sections 1.4 and 1.5.

The core of the book, consisting of Chapters 4, 5, and 6, is devoted to classical quasi-static problems of rate-independent plasticity theory. Chapter 4 contains a selection of problems in contained plastic deformation (or elastic-plastic problems) for which analytical solutions have been found: some elementary problems, and those of torsion, the thick-walled sphere and cylinder, and bending. The last section, 4.5, is an introduction to numerical methods (although the underlying concepts of discretization are already introduced in Chapter 1). For the sake of completeness, numerical methods for both viscoplastic and (rate-independent) plastic solids are discussed, since numerical schemes based on viscoplasticity have been found effective in solving elastic-plastic problems. Those who are already familiar with the material of Sections 8.1 and 8.2 may study Section 8.3, which deals with numerical methods in large-deformation plasticity, immediately following Section 4.5.

Chapters 5 and 6 deal with problems in plastic flow and collapse. Chapter 5 contains some theory and some "exact" solutions: Section 5.1 covers the general theory of plane plastic flow and some of its applications, and Section 5.2 the general theory of plates and the collapse of axisymmetrically loaded circular plates. Section 5.3 deals with plastic buckling; its placement in this chapter may well be considered arbitrary, but it seems appropriate, since buckling may be regarded as another form of collapse. Chapter 6 contains applications of limit analysis to plane problems (including those of soil mechanics), beams and framed structures, and plates and shells.

Chapter 7 is an introduction to dynamic plasticity. It deals both with problems in the dynamic loading of elastic–perfectly plastic structures treated by an extension of limit analysis, and with wave-propagation problems, one-dimensional (with the significance of rate dependence explicitly discussed) and three-dimensional. The content of Chapter 8 has already been mentioned.

As the knowledgeable reader may see from the foregoing survey, a coherent course may be built in various ways by putting together selected portions of the book. Any recommendation on my part would only betray my own prejudices, and therefore I will refrain from making one. My hope is that those whose orientation and interests are different from mine will nonetheless find this would-be "little book" useful.

In shaping the book I was greatly helped by comments from some outstanding mechanicians who took the trouble to read the book in draft form, and to whom I owe a debt of thanks: Lallit Anand (M. I. T.), Satya Atluri (Georgia Tech), Maciej Bieniek (Columbia), Michael Ortiz (Brown), and Gerald Wempner (Georgia Tech).

An immeasurable amount of help, as well as most of the inspiration to write the book, came from my students, current and past. There are too many to cite by name — may they forgive me — but I cannot leave out Vassilis Panoskaltsis, who was especially helpful in the writing of the sections on numerical methods (including some sample computations) and who suggested useful improvements throughout the book, even the correct spelling of the classical Greek verb from which the word "plasticity" is derived.

Finally, I wish to acknowledge Barbara Zeiders, whose thoroughly professional copy editing helped unify the book's style, and Rachel Lerner and Harry Sices, whose meticulous proofreading found some needles in the haystack that might have stung the unwary. Needless to say, the ultimate responsibility for any remaining lapses is no one's but mine.

A note on cross-referencing: any reference to a number such as 3.2.1, without parentheses, is to a subsection; with parentheses, such as (4.3.4), it is to an equation.

Contents

Chapter 1

Introduction to Continuum Thermomechanics

Section 1.1 Mathematical Fundamentals

1.1.1. Notation

Solid mechanics, which includes the theories of elasticity and plasticity, is a broad discipline, with experimental, theoretical, and computational aspects, and with a twofold aim: on the one hand, it seeks to describe the mechanical behavior of solids under conditions as general as possible, regardless of shape, interaction with other bodies, field of application, or the like; on the other hand, it attempts to provide solutions to specific problems involving stressed solid bodies that arise in civil and mechanical engineering, geophysics, physiology, and other applied disciplines. These aims are not in conflict, but complementary: some important results in the general theory have been obtained in the course of solving specific problems, and practical solution methods have resulted from fundamental theoretical work. There are, however, differences in approach between workers who focus on one or the other of the two goals, and one of the most readily apparent differences is in the notation used.

Most of the physical concepts used in solid mechanics are modeled by mathematical entities known as *tensors*. Tensors have representations through components with respect to specific frames or coordinate systems (a vector is a kind of tensor), but a great deal can be said about them without reference to any particular frame. Workers who are chiefly interested in the solution of specific problems — including, notably, engineers — generally use a system of notation in which the various components of tensors appear explicitly. This system, which will here be called "engineering" notation, has as one of its advantages familiarity, since it is the one that is generally used in undergraduate "strength of materials" courses, but it is often cumbersome, requiring several lines of long equations where other notations permit one short line, and it sometimes obscures the mathematical nature of the objects and processes involved. Workers in constitutive theory tend to use either one of several systems of "direct" notation that in general use no indices (subscripts and superscripts), such as Gibbs' dyadic notation, matrix notation, and a combination of the two, or the so-called *indicial* notation in which the use of indices is basic. The indices are used to label components of tensors, but with respect to an *arbitrary* rather than a specific frame.

Indicial notation is the principal system used in this book, although other systems are used occasionally as seems appropriate. In particular, "engineering" notation is used when the solutions to certain specific problems are discussed, and the matrix-based direct notation is used in connection with the study of large deformation, in which matrix multiplication plays an important part.

Assuming the reader to be familiar with vectors as commonly taught in undergraduate engineering schools, we introduce indicial notation as follows: for Cartesian coordinates $(x,\ y,\ z)$ we write $(x_1,\ x_2,\ x_3)$; for unit vectors $(\mathbf{i},\ \mathbf{j},\ \mathbf{k})$ we write $(\mathbf{e}_1,\ \mathbf{e}_2,\ \mathbf{e}_3)$; for the components $(u_x,\ u_y,\ u_z)$ of a vector $\mathbf{u}$ we write $(u_1,\ u_2,\ u_3)$.

The **summation convention** is defined as follows: the symbol $\sum_i$ may be omitted (i.e., it is implied) if the summation (dummy) index (say i) appears exactly twice in each term of a sum. Example:

$$a_i b_i = a_1 b_1 + a_2 b_2 + a_3 b_3.$$

The *Kronecker delta* is defined as

$$\delta_{ij} = \left\{ \begin{array}{ll} 1 & if\ i = j \\ 0 & if\ i \neq j \end{array} \right\} = \delta_{ji}.$$

The *Levi-Civita "e" tensor* or *permutation tensor* is defined as

$$e_{ijk} = \begin{cases} 1 & \text{if } ijk = 123,\ 231,\ 312 \\ -1 & \text{if } ijk = 321,\ 213,\ 132 \\ 0 & \text{otherwise.} \end{cases}$$

There is a relation between the "e" tensor and the Kronecker delta known as the **e-delta identity**:

$$e_{ijk} e_{lmk} = \delta_{il}\delta_{jm} - \delta_{im}\delta_{jl}.$$

The fundamental operations of three-dimensional vector algebra, presented in indicial and, where appropriate, in direct notation, are as follows.

Decomposition: $\mathbf{u} = u_i \mathbf{e}_i$. (The *position vector* in $x_1 x_2 x_3$-space is denoted $\mathbf{x} = x_i \mathbf{e}_i$.)

Scalar (dot) product between unit vectors: $\mathbf{e}_i \cdot \mathbf{e}_j = \delta_{ij}$ (orthonormality).

Projection: $\mathbf{e}_i \cdot \mathbf{u} = \mathbf{e}_i \cdot \mathbf{e}_j u_j = \delta_{ij} u_j = u_i$.

Scalar (dot) product between any two vectors: $\mathbf{u} \cdot \mathbf{v} = u_i \mathbf{e}_i \cdot \mathbf{e}_j v_j = u_i v_i$.

Vector (cross) product between unit vectors: $\mathbf{e}_i \times \mathbf{e}_j = e_{ijk} \mathbf{e}_k$.

Vector (cross) product between any two vectors: $\mathbf{u} \times \mathbf{v} = \mathbf{e}_i e_{ijk} u_j v_k$.

Scalar triple product: $\mathbf{u} \cdot (\mathbf{v} \times \mathbf{w}) = (\mathbf{u} \times \mathbf{v}) \cdot \mathbf{w} = e_{ijk} u_i v_j w_k$.

Note that the parentheses in the direct notation for the scalar triple product can be omitted without ambiguity, since a product of the form $(\mathbf{u} \cdot \mathbf{v}) \times \mathbf{w}$ has no meaning.

The notation for *matrices* is as follows. A matrix with entries α_{ij}, where i is the row index and j is the column index, is denoted $[\alpha_{ij}]$ or $\underline{\alpha}$. The *transpose* of $\underline{\alpha}$ is the matrix $[\alpha_{ji}]$, also denoted $\underline{\alpha}^T$. The *determinant* of $\underline{\alpha}$ is denoted $\det \underline{\alpha}$, and the inverse of $\underline{\alpha}$ is $\underline{\alpha}^{-1}$, so that $\underline{\alpha}\,\underline{\alpha}^{-1} = \underline{\alpha}^{-1}\underline{\alpha} = \underline{I}$, where $\underline{I} = [\delta_{ij}]$ is the *unit matrix* or *identity matrix*.

1.1.2. Cartesian Tensors

Coordinate Transformation

Since our aim is to be able to make statements about physical behavior independently of any choice of coordinate axes, let us see what the relation is between two sets of axes. Limiting ourselves to Cartesian coordinate systems, let us consider a set of axes (x_i), with the corresponding set of unit vectors $(\mathbf{e}_i)$ (also known as the *basis* of the coordinate system), and another set (x_i^*) with the basis $(\mathbf{e}_i^*)$. If β_{ij} is the cosine of the angle between the x_i^*-axis and the x_j-axis, then

$$\mathbf{e}_i^* \cdot \mathbf{e}_j = \beta_{ij}.$$

According to this equation, β_{ij} is both the x_j-component of $\mathbf{e}_i^*$ and the x_i^*-component of $\mathbf{e}_j$, so that

$$\mathbf{e}_i^* = \beta_{ij} \mathbf{e}_j$$

and

$$\mathbf{e}_i = \beta_{ji} \mathbf{e}_j^*.$$

For any vector $\mathbf{u} = u_i \mathbf{e}_i = u_i^* \mathbf{e}_i^*$,

$$u_i^* = \beta_{ik} u_k, \qquad u_i = \beta_{ji} u_j^*.$$

If the free index i in the second equation is replaced by k and its right-hand side is substituted for u_k in the first equation, then

$$u_i^* = \beta_{ik}\beta_{jk}u_j^*.$$

Similarly,

$$u_i = \beta_{ki}\beta_{kj}u_j.$$

Since $u_i^* = \delta_{ij}u_j^*$ and $u_i = \delta_{ij}u_j$, and since the vector $\mathbf{u}$ is arbitrary, it follows that

$$\beta_{ik}\beta_{jk} = \beta_{ki}\beta_{kj} = \delta_{ij},$$

that is, the matrix $\underline{\beta} = [\beta_{ij}]$ is *orthogonal*. In matrix notation, $\underline{\beta}\,\underline{\beta}^T = \underline{\beta}^T\underline{\beta} = \underline{I}$. The determinant of a matrix equals the determinant of its transpose, that is, $\det\underline{\alpha} = \det\underline{\alpha}^T$, and the determinant of a product of matrices equals the product of the determinants, so that $\det(\underline{\alpha}\underline{\beta}) = \det\underline{\alpha}\det\underline{\beta}$. For an orthogonal matrix $\underline{\beta}$, therefore, $(\det\underline{\beta})^2 = \det\underline{I} = 1$, or $\det\underline{\beta} = \pm 1$. If the basis $(\mathbf{e}_i^*)$ is obtained from $(\mathbf{e}_i)$ by a pure rotation, then $\underline{\beta}$ is called *proper orthogonal* , and $\det\underline{\beta} = 1$.

An example of a proper orthogonal matrix is the matrix describing counterclockwise rotation by an angle θ about the x_3-axis, as shown in Figure 1.1.1.

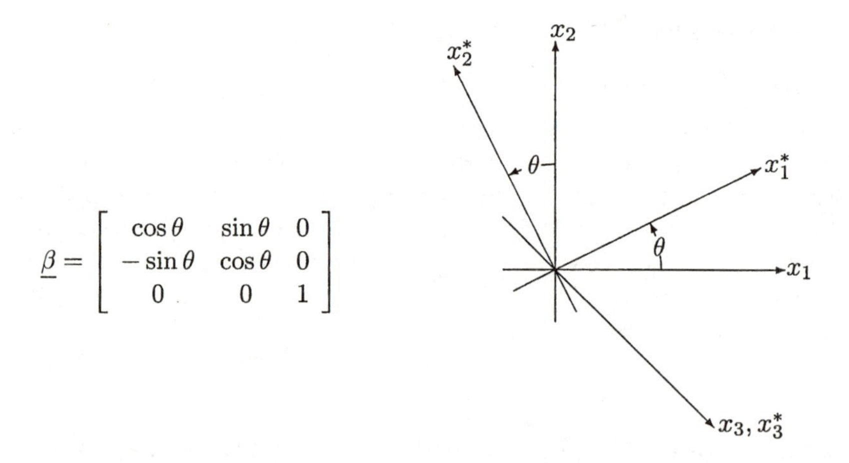

Figure 1.1.1. Example of a rotation represented by a proper orthogonal matrix.

Linear Operators

An *operator* $\boldsymbol{\lambda}$ on the space of three-dimensional vectors is simply a vector-valued function of a vector variable: given a vector $\mathbf{u}$, $\boldsymbol{\lambda}(\mathbf{u})$ uniquely defines another vector. $\boldsymbol{\lambda}$ is a *linear operator* if $\boldsymbol{\lambda}(a\mathbf{u}+b\mathbf{v}) = a\boldsymbol{\lambda}(\mathbf{u})+b\boldsymbol{\lambda}(\mathbf{v})$, where a and b are any real numbers, and $\mathbf{u}$ and $\mathbf{v}$ are any vectors.

The preceding definitions are independent of any decomposition of the vectors involved. If the vectors are to be represented with respect to a basis $(\mathbf{e}_i)$, then the linear operator must also be so represented. The Cartesian components of a linear operator are defined as follows: If $\mathbf{v} = \boldsymbol{\lambda}(\mathbf{u})$, then it follows from the definition of linearity that

$$v_i = \mathbf{e}_i \cdot \boldsymbol{\lambda}(\mathbf{e}_j u_j) = \mathbf{e}_i \cdot \boldsymbol{\lambda}(\mathbf{e}_j) u_j = \lambda_{ij} u_j,$$

where $\lambda_{ij} u_j \stackrel{\text{def}}{=} \mathbf{e}_i \cdot \boldsymbol{\lambda}(\mathbf{e}_j)$. Thus $\underline{\lambda} = [\lambda_{ij}]$ is the component matrix of $\boldsymbol{\lambda}$ with respect to the basis $(\mathbf{e}_i)$. In matrix notation we may write $\underline{v} = \underline{\lambda}\, \underline{u}$, where $\underline{u}$ ($\underline{v}$) is the column matrix whose entries are the components u_i (v_i) of the vector $\mathbf{u}$ ($\mathbf{v}$) with respect to the basis $(\mathbf{e}_i)$. In direct tensor notations it is also customary to omit parentheses: in Gibbs' notation we would write $\mathbf{v} = \boldsymbol{\lambda} \cdot \mathbf{u}$, and in the matrix-based direct notation, $\mathbf{v} = \boldsymbol{\lambda}\mathbf{u}$.

In a different basis $(\mathbf{e}_i^*)$, where $\mathbf{e}_i^* = \beta_{ij}\mathbf{e}_j$, the component matrix of $\boldsymbol{\lambda}$ is defined by

$$\lambda_{ij}^* = \beta_{ik}\beta_{jl}\lambda_{kl},$$

or, in matrix notation,

$$\underline{\lambda}^* = \underline{\beta}\,\underline{\lambda}\,\underline{\beta}^T.$$

If $\underline{\lambda}^* = \underline{\lambda}$, then $\boldsymbol{\lambda}$ is an *isotropic* or *spherical* operator. An example is the identity operator $\mathbf{I}$, whose component matrix is $\underline{I} = [\delta_{ij}]$. The most general isotropic operator is $c\mathbf{I}$, where c is any scalar.

If the component matrix of a linear operator has a property which is not changed by transformation to a different basis, that is, if the property is shared by $\underline{\lambda}$ and $\underline{\lambda}^*$ (for any $\underline{\beta}$), then the property is called *invariant*. An invariant property may be said to be a property of the linear operator $\boldsymbol{\lambda}$ itself rather than of its component matrix in a particular basis. An example is transposition: if $\underline{\lambda}^* = \underline{\beta}\,\underline{\lambda}\,\underline{\beta}^T$, then $\underline{\lambda}^{*T} = \underline{\beta}\,\underline{\lambda}^T\underline{\beta}^T$. Consequently we may speak of the transpose $\boldsymbol{\lambda}^T$ of the linear operator $\boldsymbol{\lambda}$, and we may define its symmetric and antisymmetric parts:

$$\boldsymbol{\lambda}^S = \tfrac{1}{2}(\boldsymbol{\lambda} + \boldsymbol{\lambda}^T) \quad \Leftrightarrow \quad \lambda_{ij}^S = \lambda_{(ij)} = \tfrac{1}{2}(\lambda_{ij} + \lambda_{ji}),$$

$$\boldsymbol{\lambda}^A = \tfrac{1}{2}(\boldsymbol{\lambda} - \boldsymbol{\lambda}^T) \quad \Leftrightarrow \quad \lambda_{ij}^A = \lambda_{[ij]} = \tfrac{1}{2}(\lambda_{ij} - \lambda_{ji}).$$

If $\boldsymbol{\lambda}^A = 0$ (i.e., $\lambda_{ij} = \lambda_{ji}$), then $\boldsymbol{\lambda}$ is a symmetric operator. If $\boldsymbol{\lambda}^S = 0$ (i.e., $\lambda_{ij} = -\lambda_{ji}$), then $\boldsymbol{\lambda}$ is an antisymmetric operator.

Tensors

A linear operator, as just defined, is also called a *tensor*. More generally, a *tensor of rank n* is a quantity $\mathbf{T}$ represented in a basis $(\mathbf{e}_i)$ by a component array $T_{i_1 \ldots i_n}$ (i_1, ..., $i_n = 1, 2, 3$) and in another basis $(\mathbf{e}_i^*)$ by the component array $T_{i_1 \ldots i_n}^*$, where

$$T_{i_1 \ldots i_n}^* = \beta_{i_1 k_1} \ldots \beta_{i_n k_n} T_{k_1 \ldots k_n}.$$

Thus a scalar quantity is a tensor of rank 0, a vector is a tensor of rank 1, and a linear operator is a tensor of rank 2. "Tensor" with rank unspecified is often used to mean a tensor or rank 2. The tensor whose component array is e_{ijk} is an isotropic tensor of rank 3. Tensors of rank 4 are found first in Section 1.4.

An array whose elements are products of tensor components of rank m and n, respectively, represents a tensor of rank $m + n$. An important example is furnished by the *tensor product* of two vectors $\mathbf{u}$ and $\mathbf{v}$ (a *dyad* in the terminology of Gibbs), the tensor of rank 2 represented by $\underline{u}\,\underline{v}^T = [u_i v_j]$ and denoted $\mathbf{u} \otimes \mathbf{v}$, or more simply $\mathbf{uv}$ in the Gibbs notation. Thus an arbitrary tensor $\boldsymbol{\lambda}$ of rank two, whose components with respect to a basis $(\mathbf{e}_i)$ are λ_{ij}, satisfies the equation

$$\boldsymbol{\lambda} = \lambda_{ij} \mathbf{e}_i \otimes \mathbf{e}_j.$$

Clearly, $(\mathbf{u} \otimes \mathbf{v})\mathbf{w} = \mathbf{u}(\mathbf{v} \cdot \mathbf{w})$; in the Gibbs notation both sides of this equation may be written as $\mathbf{uv} \cdot \mathbf{w}$.

An operation known as a *contraction* may be performed on a tensor of rank $n \geq 2$. It consists of setting any two indices in its component array equal to each other (with summation implied). The resulting array, indexed by the remaining indices, if any (the "free indices"), represents a tensor of rank $n-2$. For a tensor $\boldsymbol{\lambda}$ of rank 2, $\lambda_{ii} = \operatorname{tr} \boldsymbol{\lambda}$ is a scalar known as the *trace* of $\boldsymbol{\lambda}$. A standard example is $\mathbf{u} \cdot \mathbf{v} = u_i v_i = \operatorname{tr}(\mathbf{u} \otimes \mathbf{v})$. Note that if $n > 2$ then more than one contraction of the same tensor is possible, resulting in different contracted tensors; and, if $n \geq 4$, then we can have multiple contractions. For example, if $n = 4$ then we can have a double contraction resulting in a scalar, and three different scalars are possible: T_{iijj}, T_{ijij}, and T_{ijji}.

If $\mathbf{u}$ and $\mathbf{v}$ are vectors that are related by the equation

$$u_i = \alpha_{ij} v_j,$$

then $\underline{\alpha}$ necessarily represents a tensor $\boldsymbol{\alpha}$ of rank 2. Similarly, if $\boldsymbol{\alpha}$ and $\boldsymbol{\beta}$ are tensors of rank 2 related by

$$\alpha_{ij} = \rho_{ijkl} \beta_{kl},$$

then the array ρ_{ijkl} represents a tensor $\boldsymbol{\rho}$ of rank four. The generalization of these results is known as the **quotient rule.**

1.1.3. Vector and Tensor Calculus

A *tensor field* of rank n is a function (usually assumed continuously differentiable) whose values are tensors of rank n and whose domain is a region R in $x_1 x_2 x_3$ space. The boundary of R is a closed surface denoted ∂R, and the unit outward normal vector on ∂R will be denoted $\mathbf{n}$. The partial derivative operator $\partial/\partial x_i$ will be written more simply as ∂_i. A very common alternative notation, which is used extensively here, is $\partial_i \phi = \phi_{,i}$.

If ϕ is a tensor field of rank n, then the array of the partial derivatives of its components represents a tensor field of rank $n+1$.

The *del* operator is defined as $\nabla = \mathbf{e}_i \partial_i$, and the *Laplacian* operator as $\nabla^2 = \partial_i \partial_i = \sum_i \partial^2/\partial x_i^2$. For a scalar field ϕ, the *gradient* of ϕ is the vector field defined by

$$\nabla \phi = \text{grad}\phi = \mathbf{e}_i \phi_{,i}\,.$$

For a vector field $\mathbf{v}$, we use $\nabla \mathbf{v}$ to denote $(\nabla \otimes \mathbf{v})^T$, that is,

$$\nabla \mathbf{v} = v_{i,j}\, \mathbf{e}_i \otimes \mathbf{e}_j,$$

but this notation is not universal: many writers would call this $(\nabla \mathbf{v})^T$. There is no ambiguity, however, when only the symmetric part of $\nabla \mathbf{v}$ is used, or when the *divergence* of v is defined as the trace of $\nabla \mathbf{v}$:

$$\text{div}\, \mathbf{v} = \nabla \cdot \mathbf{v} = v_{i,i}\,.$$

Similarly, the *curl* of $\mathbf{v}$ is defined unambiguously as

$$\text{curl}\, \mathbf{v} = \nabla \times \mathbf{v} = \mathbf{e}_i e_{ijk} v_{k,j}\,.$$

For a tensor field $\boldsymbol{\phi}$ of rank 2, we define $\nabla \boldsymbol{\phi}$ as represented by $\phi_{jk,i}$, and

$$\nabla \cdot \boldsymbol{\phi} = \text{div}\, \boldsymbol{\phi} = \phi_{jk,j}\, \mathbf{e}_k.$$

These definitions are, again, not universal.

The three-dimensional equivalent of the fundamental theorem of calculus is **Gauss's theorem**:

$$\int_R \phi_{,i}\, dV = \int_{\partial R} n_i \phi\, dS, \tag{1.1.1}$$

where ϕ is any differentiable field. The particular case where ϕ is replaced (on both sides of the equation, of course) by v_i (the ith component of a vector field $\mathbf{v}$), with summation under the integral signs implied, is known as the **divergence theorem**. This is the case we use most often.

The two-dimensional Gauss's theorem refers to fields defined in an area A in the $x_1 x_2$-plane, bounded by a closed curve C on which an infinitesimal element of arc length is ds (positive when it is oriented counterclockwise). It is conventional to use Greek letters for indices whose range is 1, 2; thus the theorem reads

$$\int_A \phi_{,\alpha}\, dA = \oint_C n_\alpha \phi\, ds. \tag{1.1.2}$$

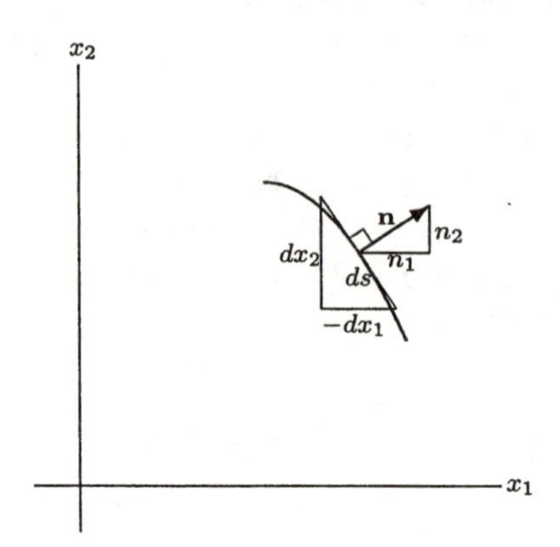

Figure 1.1.2. Normal vector to a plane curve

Now suppose that the curve C is described parametrically by $x_1 = x_1(s)$, $x_2 = x_2(s)$. Then, as can be seen from Figure 1.1.2,

$$n_1 = \frac{dx_2}{ds}, \quad n_2 = -\frac{dx_1}{ds}.$$

Thus, for any two functions $u_\alpha(x_1, x_2)$ ($\alpha = 1, 2$),

$$\oint_C u_1 \, dx_1 = -\oint_C n_2 u_1 \, ds = -\int_A u_{1,2} \, dA$$

and

$$\oint_C u_2 \, dx_2 = \oint_C n_1 u_2 \, ds = \int_A u_{2,1} \, dA.$$

Combining these two equations, we obtain **Green's lemma**:

$$\oint_C u_\alpha \, dx_\alpha = \int_A (u_{2,1} - u_{1,2}) \, dA. \tag{1.1.3}$$

If there exists a continuously differentiable function $\phi(x_1, x_2)$, defined in A, such that $u_\alpha = \phi_{,\alpha}$, then $\oint u_\alpha dx_\alpha = \oint d\phi = 0$ around *any* closed contour, so that $u_{2,1} = u_{1,2}$. The converse is also true (i.e., the last equality implies the existence of ϕ), provided that the area A is *simply connected* (i.e., contains no holes); otherwise additional conditions are required.

The preceding result is known as the two-dimensional **integrability theorem** and will be used repeatedly.

There exists an extension of Green's lemma to a curved surface S in $x_1x_2x_3$ space, bounded by a (not necessarily plane) closed curve C parametrized by $x_i = x_i(s)$, $i = 1, 2, 3$. This extension (derived from Gauss's theorem) is known as **Stokes' theorem** and takes the form

$$\oint_C u_i\, dx_i = \int_S n_i e_{ijk} u_{k,j}\, dS. \tag{1.1.4}$$

Clearly, Green's lemma represents the special case of Stokes' theorem when $\mathbf{n} = \mathbf{e}_3$. From Stokes' theorem follows the three-dimensional integrability theorem: a field $\phi(\mathbf{x})$ such that $\mathbf{u} = \nabla\phi$ in a region R exists only if $\nabla\times\mathbf{u} = 0$, or, equivalently, $u_{j,i} = u_{i,j}$. As in the two-dimensional version, this last condition is also sufficient if R is simply connected.

1.1.4. Curvilinear Coordinates

The study of tensor fields in curvilinear coordinates is intimately tied to differential geometry, and in many books and courses of study dealing with continuum mechanics it is undertaken at the outset. The traditional methodology is as follows: with a set of curvilinear coordinates ξ^i ($i = 1, 2, 3$) such that the position of a point in three-dimensional space is defined by $\mathbf{x}(\xi^1, \xi^2, \xi^3)$, the *natural basis* is defined as the ordered triple of vectors $\mathbf{g}_i = \partial\mathbf{x}/\partial\xi^i$, so that $d\mathbf{x} = \mathbf{g}_i\, d\xi^i$; the summation convention here applies whenever the pair of repeated indices consists of one subscript and one superscript. The basis vectors are not, in general, unit vectors, nor are they necessarily mutually perpendicular, although it is usual for them to have the latter property. One can find, however, the *dual basis* $(\mathbf{g}^i)$ such that $\mathbf{g}_i \cdot \mathbf{g}^j = \delta_i^j$. A vector $\mathbf{v}$ may be represented as $v_i\mathbf{g}^i$ or as $v^i\mathbf{g}_i$, where $v_i = \mathbf{v}\cdot\mathbf{g}_i$ and $v^i = \mathbf{v}\cdot\mathbf{g}^i$ are respectively the *covariant* and *contravariant* components of v. For tensors of higher rank one can similarly define covariant, contravariant, and several kinds of *mixed* components. The gradient of a tensor field is defined in terms of the so-called *covariant derivatives* of its components, which, except in the case of a scalar field, differ from the partial derivatives with respect to the ξ^i because the basis vectors themselves vary. A central role is played by the *metric tensor* with components $g_{ij} = \mathbf{g}_i \cdot \mathbf{g}_j$, having the property that $d\mathbf{x}\cdot d\mathbf{x} = g_{ij}d\xi^i d\xi^j$.

An alternative approach is based on the theory of differentiable manifolds (see, e.g., Marsden and Hughes [1983]).

Curvilinear tensor analysis is especially useful for studying the mechanics of curved surfaces, such as shells; when this topic does not play an important part, a simpler approach is available, based on the so-called "physical" components of the tensors involved. In this approach mutually perpendicular unit vectors (forming an *orthonormal basis*) are used, rather than the natural and dual bases. We conclude this section by examining cylindrical and spherical coordinates in the light of this methodology.

Cylindrical Coordinates

In the cylindrical coordinates (r, θ, z), where $r = \sqrt{x_1^2 + x_2^2}$, $\theta = \tan^{-1}(x_2/x_1)$, and $z = x_3$, the unit vectors are

$$\mathbf{e}_r = \mathbf{e}_1 \cos\theta + \mathbf{e}_2 \sin\theta, \qquad \mathbf{e}_\theta = -\mathbf{e}_1 \sin\theta + \mathbf{e}_2 \cos\theta, \qquad \mathbf{e}_z = \mathbf{e}_3,$$

so that

$$\frac{d}{d\theta}\mathbf{e}_r = \mathbf{e}_\theta, \qquad \frac{d}{d\theta}\mathbf{e}_\theta = -\mathbf{e}_r.$$

Using the chain rule for partial derivatives, we may show the ∇ operator to be given by

$$\nabla = \mathbf{e}_r \frac{\partial}{\partial r} + \mathbf{e}_\theta \frac{1}{r}\frac{\partial}{\partial \theta} + \mathbf{e}_z \frac{\partial}{\partial z}. \tag{1.1.5}$$

When this operator is applied to a vector field $\mathbf{v}$, represented as

$$\mathbf{v} = v_r \mathbf{e}_r + v_\theta \mathbf{e}_\theta + v_z \mathbf{e}_z,$$

the result may be written as

$$\begin{aligned} \nabla \otimes \mathbf{v} &= \left(\mathbf{e}_r \frac{\partial}{\partial r} + \mathbf{e}_\theta \frac{1}{r}\frac{\partial}{\partial \theta} + \mathbf{e}_z \frac{\partial}{\partial z}\right) \otimes (v_r \mathbf{e}_r + v_\theta \mathbf{e}_\theta + v_z \mathbf{e}_z) \\ &= \mathbf{e}_r \otimes \left(\mathbf{e}_r \frac{\partial v_r}{\partial r} + \mathbf{e}_\theta \frac{\partial v_\theta}{\partial r} + \mathbf{e}_z \frac{\partial v_z}{\partial r}\right) \\ &\quad + \mathbf{e}_\theta \otimes \frac{1}{r}\left(\mathbf{e}_r \frac{\partial v_r}{\partial \theta} + \mathbf{e}_\theta v_r + \mathbf{e}_\theta \frac{\partial v_\theta}{\partial \theta} - \mathbf{e}_r v_\theta + \mathbf{e}_z \frac{\partial v_z}{\partial \theta}\right) \\ &\quad + \mathbf{e}_z \otimes \left(\mathbf{e}_r \frac{\partial v_r}{\partial z} + \mathbf{e}_\theta \frac{\partial v_\theta}{\partial z} + \mathbf{e}_z \frac{\partial v_z}{\partial z}\right). \end{aligned} \tag{1.1.6}$$

The trace of this second-rank tensor is the divergence of $\mathbf{v}$:

$$\nabla \cdot \mathbf{v} = \frac{\partial v_r}{\partial r} + \frac{1}{r}\left(v_r + \frac{\partial v_\theta}{\partial \theta}\right) + \frac{\partial v_z}{\partial z};$$

and when $\mathbf{v} = \nabla u$, the gradient of a scalar field, then this is

$$\nabla^2 u = \left(\frac{\partial^2}{\partial r^2} + \frac{1}{r}\frac{\partial}{\partial r} + \frac{1}{r^2}\frac{\partial^2}{\partial \theta^2} + \frac{\partial^2}{\partial z^2}\right) u.$$

Lastly, let us consider a *symmetric* second-rank tensor field

$$\begin{aligned} \boldsymbol{\lambda} = \ & \mathbf{e}_r \otimes \mathbf{e}_r \lambda_{rr} + (\mathbf{e}_r \otimes \mathbf{e}_\theta + \mathbf{e}_\theta \otimes \mathbf{e}_r)\lambda_{r\theta} + \mathbf{e}_\theta \otimes \mathbf{e}_\theta \lambda_{\theta\theta} \\ & + (\mathbf{e}_r \otimes \mathbf{e}_z + \mathbf{e}_z \otimes \mathbf{e}_r)\lambda_{rz} + (\mathbf{e}_\theta \otimes \mathbf{e}_z + \mathbf{e}_z \otimes \mathbf{e}_\theta)\lambda_{\theta z} + \mathbf{e}_z \otimes \mathbf{e}_z \lambda_{zz}; \end{aligned}$$

its divergence — which we have occasion to use in Section 1.3 — is

$$\begin{aligned} \nabla \cdot \boldsymbol{\lambda} = \ & \mathbf{e}_r \left(\frac{\partial \lambda_{rr}}{\partial r} + \frac{\lambda_{rr} - \lambda_{\theta\theta}}{r} + \frac{1}{r}\frac{\partial \lambda_{r\theta}}{\partial \theta} + \frac{\partial \lambda_{rz}}{\partial z}\right) \\ & + \mathbf{e}_\theta \left(\frac{\partial \lambda_{r\theta}}{\partial r} + \frac{2\lambda_{r\theta}}{r} + \frac{1}{r}\frac{\partial \lambda_{\theta\theta}}{\partial \theta} + \frac{\partial \lambda_{\theta z}}{\partial z}\right) \\ & + \mathbf{e}_z \left(\frac{\partial \lambda_{rz}}{\partial r} + \frac{\lambda_{rz}}{r} + \frac{1}{r}\frac{\partial \lambda_{\theta z}}{\partial \theta} + \frac{\partial \lambda_{zz}}{\partial z}\right). \end{aligned} \tag{1.1.7}$$

Spherical Coordinates

The spherical coordinates (r, θ, ϕ) are defined by $r = \sqrt{x_1^2 + x_2^2 + x_3^2}$, $\theta = \tan^{-1}(x_2/x_1)$, and $\phi = \cot^{-1}\left(x_3/\sqrt{x_1^2 + x_2^2}\right)$. The unit vectors are

$$\mathbf{e}_r = (\mathbf{e}_1 \cos\theta + \mathbf{e}_2 \sin\theta)\sin\phi + \mathbf{e}_3\cos\phi, \quad \mathbf{e}_\phi = (\mathbf{e}_1\cos\theta + \mathbf{e}_2\sin\theta)\cos\phi - \mathbf{e}_3\sin\phi,$$

$$\mathbf{e}_\theta = -\mathbf{e}_1\sin\theta + \mathbf{e}_2\cos\theta,$$

so that

$$\frac{\partial}{\partial\phi}\mathbf{e}_r = \mathbf{e}_\phi, \quad \frac{\partial}{\partial\phi}\mathbf{e}_\phi = -\mathbf{e}_r, \quad \frac{\partial}{\partial\phi}\mathbf{e}_\theta = 0,$$

$$\frac{\partial}{\partial\theta}\mathbf{e}_r = \mathbf{e}_\theta\sin\phi, \quad \frac{\partial}{\partial\theta}\mathbf{e}_\phi = \mathbf{e}_\theta\cos\phi, \quad \frac{\partial}{\partial\theta}\mathbf{e}_\theta = -\mathbf{e}_r\sin\phi - \mathbf{e}_\phi\cos\phi.$$

The ∇ operator is given by

$$\nabla = \mathbf{e}_r\frac{\partial}{\partial r} + \mathbf{e}_\phi\frac{1}{r}\frac{\partial}{\partial\phi} + \mathbf{e}_\theta\frac{1}{r\sin\phi}\frac{\partial}{\partial\theta}.$$

For a vector field $\mathbf{v}$ we accordingly have

$$\begin{aligned}
\nabla\otimes\mathbf{v} = {} & \mathbf{e}_r\otimes\left(\mathbf{e}_r\frac{\partial v_r}{\partial r} + \mathbf{e}_\phi\frac{\partial v_\phi}{\partial r} + \mathbf{e}_\theta\frac{\partial v_\theta}{\partial r}\right) \\
& + \mathbf{e}_\phi\otimes\left[\mathbf{e}_r\left(\frac{1}{r}\frac{\partial v_r}{}\partial\phi - \frac{v_\phi}{r}\right) + \mathbf{e}_\phi\left(\frac{1}{r}\frac{\partial v_\phi}{\partial\phi} + \frac{v_r}{r}\right) + \mathbf{e}_\theta\frac{1}{r}\frac{\partial v_\theta}{\partial\phi}\right] \\
& + \mathbf{e}_\theta\otimes\left[\mathbf{e}_r\left(\frac{1}{r\sin\phi}\frac{\partial v_r}{\partial\theta} - \frac{v_\theta}{r}\right) + \mathbf{e}_\phi\left(\frac{1}{r\sin\phi}\frac{\partial v_\phi}{\partial\theta} - \frac{\cot\phi}{r}v_\theta\right)\right. \\
& \left. + \mathbf{e}_\theta\left(\frac{1}{r\sin\phi}\frac{\partial v_\theta}{\partial\theta} + \frac{v_r}{r} + \frac{\cot\phi}{r}v_\phi\right)\right],
\end{aligned} \tag{1.1.8}$$

so that

$$\nabla\cdot\mathbf{v} = \frac{\partial v_r}{\partial r} + 2\frac{v_r}{r} + \frac{1}{r}\frac{\partial v_\phi}{\partial\phi} + \frac{\cot\phi}{r}v_\phi + \frac{1}{r\sin\phi}\frac{\partial v_\theta}{\partial\theta},$$

$$\nabla^2 u = \frac{\partial^2 u}{\partial r^2} + \frac{2}{r}\frac{\partial u}{\partial r} + \frac{1}{r^2}\frac{\partial^2 u}{\partial\phi^2} + \frac{\cot\phi}{r^2}\frac{\partial u}{\partial\phi} + \frac{1}{r^2\sin^2\phi}\frac{\partial^2 u}{\partial\theta^2},$$

and, for a symmetric second-rank tensor field $\boldsymbol{\lambda}$,

$$\begin{aligned}
\nabla\cdot\boldsymbol{\lambda} = {} & \mathbf{e}_r\left(\frac{\partial\lambda_{rr}}{\partial r} + \frac{1}{r}\frac{\partial\lambda_{r\phi}}{\partial\phi} + \frac{1}{r\sin\phi}\frac{\partial\lambda_{r\theta}}{\partial\theta} + \frac{2\lambda_{rr} - \lambda_{\phi\phi} - \lambda_{\theta\theta} + \lambda_{r\phi}\cot\phi}{r}\right) \\
& + \mathbf{e}_\phi\left(\frac{\partial\lambda_{r\phi}}{\partial r} + \frac{1}{r}\frac{\partial\lambda_{\phi\phi}}{\partial\phi} + \frac{1}{r\sin\phi}\frac{\partial\lambda_{\phi\theta}}{\partial\theta}\right. \\
& \qquad \left. + \frac{\lambda_{\phi\phi}\cot\phi - \lambda_{\theta\theta}\cot\phi + 3\lambda_{r\phi}}{r}\right) \\
& + \mathbf{e}_\theta\left(\frac{\partial\lambda_{r\theta}}{\partial r} + \frac{1}{r}\frac{\partial\lambda_{\phi\theta}}{\partial\phi} + \frac{1}{r\sin\phi}\frac{\partial\lambda_{\theta\theta}}{\partial\theta} + \frac{3\lambda_{r\theta} + 2\lambda_{\phi\theta}\cot\phi}{r}\right).
\end{aligned} \tag{1.1.9}$$

Exercises: Section 1.1

1. Show that

   (a) $\delta_{ii} = 3$

   (b) $\delta_{ij}\delta_{ij} = 3$

   (c) $e_{ijk}e_{jki} = 6$

   (d) $e_{ijk}A_jA_k = 0$

   (e) $\delta_{ij}\delta_{jk} = \delta_{ik}$

   (f) $\delta_{ij}e_{ijk} = 0$

2. Using indicial notation and the summation convention, prove that

$$(\mathbf{s} \times \mathbf{t}) \cdot (\mathbf{u} \times \mathbf{v}) = (\mathbf{s} \cdot \mathbf{u})(\mathbf{t} \cdot \mathbf{v}) - (\mathbf{s} \cdot \mathbf{v})(\mathbf{t} \cdot \mathbf{u}).$$

3. For the matrix

$$[a_{ij}] = \begin{bmatrix} 1 & 1 & 0 \\ 1 & 2 & 2 \\ 0 & 2 & 3 \end{bmatrix},$$

   calculate the values of

   (a) a_{ii},

   (b) $a_{ij}a_{ij}$,

   (c) $a_{ij}a_{jk}$ when $i = 1$, $k = 1$ and when $i = 1$, $k = 2$.

4. Show that the matrix

$$\underline{\beta} = \begin{bmatrix} \frac{12}{25} & -\frac{9}{25} & \frac{4}{5} \\ \frac{3}{5} & -\frac{4}{5} & 0 \\ \frac{16}{25} & \frac{12}{25} & \frac{3}{5} \end{bmatrix}$$

   is proper orthogonal, that is, $\underline{\beta}\,\underline{\beta}^T = \underline{\beta}^T\underline{\beta} = \underline{I}$, and $\det \underline{\beta} = 1$.

5. Find the rotation matrix $\underline{\beta}$ describing the transformation composed of, first, a 90° rotation about the x_1-axis, and second, a 45° rotation about the *rotated* x_3-axis.

6. Two Cartesian bases, $(\mathbf{e}_i)$, and $(\mathbf{e}_i^*)$ are given, with $\mathbf{e}_1^* = (2\mathbf{e}_1 + 2\mathbf{e}_2 + \mathbf{e}_3)/3$ and $\mathbf{e}_2^* = (\mathbf{e}_1 - \mathbf{e}_2)/\sqrt{2}$.

   (a) Express $\mathbf{e}_3^*$ in terms of the $\mathbf{e}_i$.

   (b) Express the $\mathbf{e}_i$ in terms of the $\mathbf{e}_i^*$.

(c) If $\mathbf{v} = 6\mathbf{e}_1 - 6\mathbf{e}_2 + 12\mathbf{e}_3$, find the v_i^*.

7. The following table shows the angles between the original axes x_i and the transformed axes x_i^*.

	x_1	x_2	x_3
x_1^*	135°	60°	120°
x_2^*	90°	45°	45°
x_3^*	45°	60°	120°

(a) Find the transformation matrix $\underset{\sim}{\beta}$, and verify that it describes a rotation.

(b) If a second-rank tensor $\boldsymbol{\lambda}$ has the following component matrix with the respect to the original axes,

$$\underset{\sim}{\lambda} = \begin{bmatrix} 3 & -4 & 2 \\ -4 & 0 & 1 \\ 2 & 1 & 3 \end{bmatrix},$$

find its component matrix $\underset{\sim}{\lambda}^*$ with respect to the rotated axes.

8. (a) Use the chain rule of calculus to prove that if ϕ is a scalar field, then $\nabla\phi$ is a vector field.

(b) Use the quotient rule to prove the same result.

9. Using indicial notation, prove that (a) $\nabla\times\nabla\phi = 0$ and (b) $\nabla\cdot\nabla\times\mathbf{v} = 0$.

10. If $\mathbf{x} = x_i\mathbf{e}_i$ and $r = |\mathbf{x}|$, prove that

$$\nabla^2(r^n) = n(n+1)r^{n-2}.$$

11. If $\phi(x_1, x_2, x_3) = a_{ij}x_ix_j$, with a_{ij} constant, show that $\phi_{,i} = (a_{ij} + a_{ji})x_j$ and $\phi_{,ij} = a_{ij} + a_{ji}$.

12. Show that $\nabla^2(\phi\psi) = \phi\nabla^2\psi + 2(\nabla\phi)\cdot(\nabla\psi) + \psi\nabla^2\phi$.

13. Use Gauss's theorem to prove that, if V is the volume of a three-dimensional region R, then $V = \frac{1}{3}\int_{\partial R} x_i n_i \, dS$.

14. Verify Green's lemma for the area A bounded by the square with corners at $(0, 0)$, $(a, 0)$, (a, a), $(0, a)$, of $u_1(x_1, x_2) = 0$ and $u_2(x_1, x_2) = bx_1$, where b is a constant.

15. Find the natural basis $(\mathbf{g}_i)$ and the dual basis $(\mathbf{g}^i)$ (a) for cylindrical coordinates, with $\xi^1 = r$, $\xi^2 = \theta$, and $\xi^3 = z$, and (b) for spherical coordinates, with $\xi^1 = r$, $\xi^2 = \phi$, and $\xi^3 = \theta$.

16. Starting with the expression in Cartesian coordinates for the gradient operator ∇ and using the chain rule for partial derivatives, derive Equation (1.1.5).

Section 1.2 Continuum Deformation

1.2.1. Displacement

The first application of the mathematical concepts introduced in Section 1.1 will now be to the description of the deformation of bodies that can be modeled as continua. A body is said to be modeled as a continuum if to any configuration of the body there corresponds a region R in three-dimensional space such that every point of the region is occupied by a particle (material point) of the body.

Any one configuration may be taken as the *reference configuration*. Consider a particle that in this configuration occupies the point defined by the vector $\mathbf{r} = x_i \mathbf{e}_i$. When the body is displaced, the same particle will occupy the point $\mathbf{r}^* = x_i^* \mathbf{e}_i$. (Note that here the x_i^* no longer mean the coordinates of the same point with respect to a rotated basis, as in the Section 1.1, but the coordinates of a different point with respect to the same basis.) The difference $\mathbf{r}^* - \mathbf{r}$ is called the *displacement* of the particle and will be denoted $\mathbf{u}$. The reference position vector $\mathbf{r}$ will be used to label the given particle; the coordinates x_i are then called *Lagrangian* coordinates. Consequently the displacement may be given as a function of $\mathbf{r}$, $\mathbf{u}(\mathbf{r})$, and it forms a vector field defined in the region occupied by the body in the reference configuration.

Now consider a neighboring particle labeled by $\mathbf{r} + \Delta\mathbf{r}$. In the displaced configuration, the position of this point will be

$$\mathbf{r}^* + \Delta\mathbf{r}^* = \mathbf{r} + \Delta\mathbf{r} + \mathbf{u}(\mathbf{r} + \Delta\mathbf{r})$$

(see Figure 1.2.1), so that

$$\Delta\mathbf{r}^* = \Delta\mathbf{r} + \mathbf{u}(\mathbf{r} + \Delta\mathbf{r}) - \mathbf{u}(\mathbf{r}),$$

or, in indicial notation,

$$\Delta x_i^* = \Delta x_i + u_i(\mathbf{r} + \Delta\mathbf{r}) - u_i(\mathbf{r}).$$

But if $\Delta\mathbf{r}$ is sufficiently small, then $u_i(\mathbf{r} + \Delta\mathbf{r}) - u_i(\mathbf{r}) \doteq u_{i,j}(\mathbf{r})\Delta x_j$, the error in the approximation being such that it tends to zero faster than $|\Delta\mathbf{r}|$. It is conventional to replace $\Delta\mathbf{r}$ by the *infinitesimal* $d\mathbf{r}$, and to write the approximation as an equality. Defining the displacement-gradient matrix $\underline{\alpha}$ by $\alpha_{ij} \overset{\text{def}}{=} u_{i,j}$, we may write in matrix notation

$$d\underline{x}^* = (\underline{I} + \underline{\alpha})d\underline{x}.$$

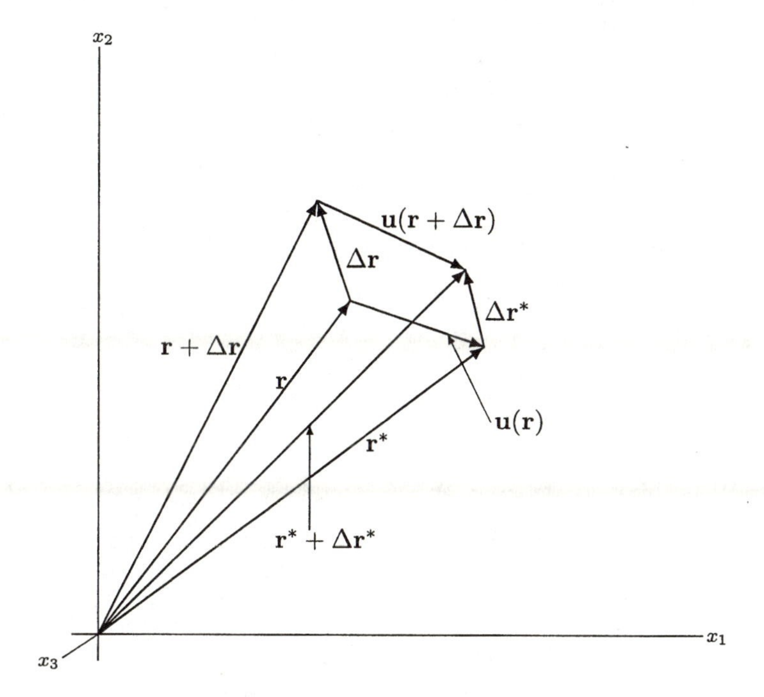

Figure 1.2.1. Displacement

1.2.2. Strain

A body is said to undergo a *rigid-body displacement* if the distances between all particles remain unchanged; otherwise the body is said to be *deformed.*

Let us limit ourselves, for the moment, to an infinitesimal neighborhood of the particle labeled by $\mathbf{r}$; the deformation of the neighborhood may be measured by the extent to which the lengths of the infinitesimal vectors $d\mathbf{r}$ emanating from $\mathbf{r}$ change in the course of the displacement. The square of the length of $d\mathbf{r}^*$ is

$$|d\mathbf{r}^*|^2 = d\mathbf{r}^* \cdot d\mathbf{r}^* = d\underline{x}^{*T} d\underline{x}^* = d\underline{x}^T (\underline{I} + \underline{\alpha}^T)(\underline{I} + \underline{\alpha})\, \underline{x} = d\underline{x}^T (\underline{I} + 2\underline{E})\, \underline{x},$$

where $\underline{E} = \frac{1}{2}(\underline{\alpha}^T + \underline{\alpha} + \underline{\alpha}^T \underline{\alpha})$, or, in indicial notation,

$$E_{ij} = \frac{1}{2}(u_{j,i} + u_{i,j} + u_{k,i}\, u_{k,j}),$$

which defines the symmetric second-rank tensor $\mathbf{E}$, known as the *Green–Saint-Venant strain tensor* and sometimes called the *Lagrangian strain ten-*

sor.[1] Clearly, $\mathbf{E}(\mathbf{r})$ describes the deformation of the infinitesimal neighborhood of $\mathbf{r}$, and the tensor field $\mathbf{E}$ that of the whole body; $\mathbf{E}(\mathbf{r}) = 0$ for all $\mathbf{r}$ in R if and only if the displacement is a rigid-body one.

The deformation of a region R is called *homogeneous* if $\mathbf{E}$ is constant. It is obvious that a necessary and sufficient condition for the deformation to be homogeneous is that the $u_{i,j}$ are constant, or equivalently, that $\mathbf{u}$ varies linearly with $\mathbf{r}$,

Infinitesimal Strain and Rotation

We further define the tensor $\boldsymbol{\varepsilon}$ and $\boldsymbol{\omega}$, respectively symmetric and antisymmetric, by

$$\varepsilon_{ij} = \frac{1}{2}(u_{j,i} + u_{i,j}), \quad \omega_{ij} = \frac{1}{2}(u_{i,j} - u_{j,i}),$$

so that $u_{i,j} = \varepsilon_{ij} + \omega_{ij}$, and

$$E_{ij} = \varepsilon_{ij} + \frac{1}{2}(\varepsilon_{ik}\varepsilon_{jk} - \varepsilon_{ik}\omega_{jk} - \omega_{ik}\varepsilon_{jk} + \omega_{ik}\omega_{jk}).$$

If $|\varepsilon_{ij}| \ll 1$ and $|\omega_{ij}| \ll 1$ for all i, j, then $\boldsymbol{\varepsilon}$ is an approximation to $\mathbf{E}$ and is known as the *infinitesimal strain tensor*. The displacement field is then called *small* or *infinitesimal*. Moreover, $\boldsymbol{\omega}$ can then be defined as the *infinitesimal rotation tensor*: if $\boldsymbol{\varepsilon} = 0$, then $\underline{\alpha} = \underline{\omega}$, and therefore $d\underline{x}^* = (\underline{I} + \underline{\omega})\,\underline{x}$. Now

$$(\underline{I} + \underline{\omega})^T(\underline{I} + \underline{\omega}) = \underline{I} + \underline{\omega}^T + \underline{\omega} + \underline{\omega}^T\underline{\omega} \doteq \underline{I} + \underline{\omega}^T + \underline{\omega} = \underline{I};$$

In other words, a matrix of the form $\underline{I} + \underline{\omega}$, where $\underline{\omega}$ is any antisymmetric matrix whose elements are small, is *approximately orthogonal*.

The tensor $\boldsymbol{\omega}$, because of its antisymmetry, has only three independent components: $\omega_{32} = -\omega_{23}$, $\omega_{13} = -\omega_{31}$, and $\omega_{21} = -\omega_{12}$. Let these components be denoted θ_1, θ_2, and θ_3, respectively. Then it is easy to show the two reciprocal relations

$$\theta_i = \frac{1}{2}e_{ijk}\omega_{kj}, \qquad \omega_{ik} = e_{ijk}\theta_j.$$

Since, moreover, $e_{ijk}\varepsilon_{jk} = 0$ because of the symmetry of $\boldsymbol{\varepsilon}$, the first relation implies that

$$\theta_i = \frac{1}{2}e_{ijk}u_{k,j},$$

or, in vector notation,

$$\boldsymbol{\theta} = \frac{1}{2}\nabla \times \mathbf{u}.$$

[1] As will be seen in Chapter 8, $\mathbf{E}$ is only one of several tensors describing finite deformation.

In a rigid-body displacement, then, $du_i = e_{ijk}\theta_j dx_k$, or $d\mathbf{u} = \boldsymbol{\theta} \times d\mathbf{r}$. That is, $\boldsymbol{\theta}$ is the infinitesimal rotation *vector*: its magnitude is the angle of rotation and its direction gives the axis of rotation. Note that $\nabla \cdot \boldsymbol{\theta} = 0$; a vector field with this property is called *solenoidal.*

It must be remembered that a *finite* rotation is described by an orthogonal, not an antisymmetric, matrix. Since the orthogonality conditions are six in number, such a matrix is likewise determined by only three independent numbers, but it is not equivalent to a vector, since the relations among the matrix elements are not linear.

Significance of Infinitesimal Strain Components

The study of finite deformation is postponed until Chapter 8. For now, let us explore the meaning of the components of the infinitesimal strain tensor $\boldsymbol{\varepsilon}$. Consider, first, the unit vector $\mathbf{n}$ such that $d\mathbf{r} = \mathbf{n}dr$, where $dr = |d\mathbf{r}|$; we find that

$$dr^{*2} = d\mathbf{r}^* \cdot d\mathbf{r}^* = (1 + 2E_{ij}n_i n_j)dr^2 \doteq (1 + 2\varepsilon_{ij}n_i n_j)\, r^2.$$

But for α small, $\sqrt{1+2\alpha} \doteq 1 + \alpha$, so that $dr^* \doteq (1 + \varepsilon_{ij}n_i n_j)\, r$. Hence

$$\frac{dr^* - dr}{dr} \doteq \varepsilon_{ij}n_i n_j$$

is the *longitudinal strain* along the direction $\mathbf{n}$ (note that the left-hand side is just the "engineering" definition of strain).

Next, consider two infinitesimal vectors, $d\mathbf{r}^{(1)} = \mathbf{e}_1 dr^{(1)}$ and $d\mathbf{r}^{(2)} = \mathbf{e}_2 dr^{(2)}$. In indicial notation, we have $dx_i^{(1)} = \delta_{i1}dr^{(1)}$ and $dx_i^{(2)} = \delta_{i2}dr^{(2)}$. Obviously, $d\mathbf{r}^{(1)} \cdot d\mathbf{r}^{(2)} = 0$. The displacement changes $d\mathbf{r}^{(1)}$ to $d\mathbf{r}^{(1)*}$ and $d\mathbf{r}^{(2)}$ to $d\mathbf{r}^{(2)*}$, where

$$dx_i^{(1)*} = (\delta_{ij} + u_{i,j}\,)\, x_j^{(1)} = (\delta_{i1} + u_{i,1}\,)\, r^{(1)},$$

$$dx_i^{(2)*} = (\delta_{ij} + u_{i,j}\,)\, x_j^{(2)} = (\delta_{i2} + u_{i,2}\,)\, r^{(2)},$$

so that

$$d\mathbf{r}^{(1)*} \cdot d\mathbf{r}^{(2)*} = dx_i^{(1)*}dx_i^{(2)*} \doteq (u_{2,1} + u_{1,2}\,)\, r^{(1)}dr^{(2)} = 2\varepsilon_{12}dr^{(1)}dr^{(2)}.$$

But

$$d\mathbf{r}^{(1)*} \cdot d\mathbf{r}^{(2)*} = \left|d\mathbf{r}^{(1)*}\right|\left|d\mathbf{r}^{(2)*}\right| \cos\left(\frac{\pi}{2} - \gamma_{12}\right),$$

where γ_{12} is the *shear angle* in the x_1x_2-plane (Figure 1.2.2). Consequently,

$$2\varepsilon_{12} = (1 + \varepsilon_{11})(1 + \varepsilon_{22})\gamma_{12} \doteq \gamma_{12}$$

for infinitesimal strains.

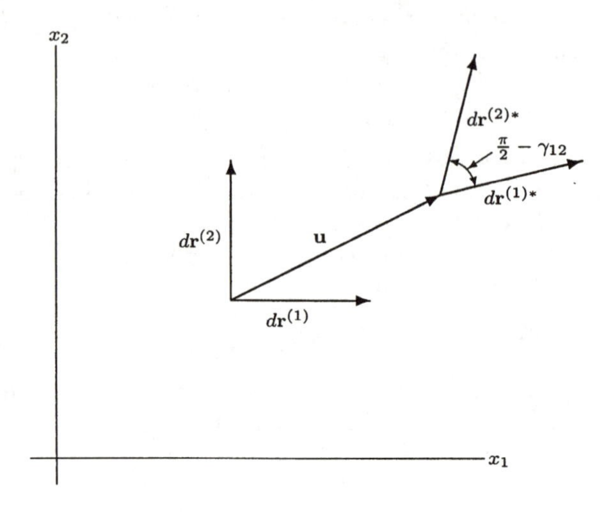

Figure 1.2.2. Shear angle

Since the labeling of the axes is arbitrary, we may say in general that, for $i \neq j$, $\varepsilon_{ij} = \frac{1}{2}\gamma_{ij}$. Both the ε_{ij} and γ_{ij}, for $i \neq j$, are referred to as *shear strains;* more specifically, the former are the *tensorial* and the latter are the *conventional* shear strains. A state of strain that can, with respect to *some* axes, be represented by the matrix

$$\underline{\varepsilon}^* = \begin{bmatrix} 0 & \frac{1}{2}\gamma & 0 \\ \frac{1}{2}\gamma & 0 & 0 \\ 0 & 0 & 0 \end{bmatrix}$$

is called a state of *simple shear* with respect to those axes, and *pure shear* in general.

It cannot be emphasized strongly enough that the tensor $\boldsymbol{\varepsilon}$ is an approximation to $\mathbf{E}$, and therefore deserves to be called the infinitesimal strain tensor, *only* if both the deformation and rotation are infinitesimal, that is, if both $\boldsymbol{\varepsilon}$ *and* $\boldsymbol{\omega}$ (or $\boldsymbol{\theta}$) are small compared to unity. If the rotation is finite, then the strain must be described by $\mathbf{E}$ (or by some equivalent finite deformation tensor, discussed further in Chapter 8) even if the deformation per se is infinitesimal.

As an illustration, we consider a homogeneous deformation in which the x_1x_3-plane is rotated counterclockwise about the x_3-axis by a finite angle θ, while the the x_2x_3-plane is rotated counterclockwise about the x_3-axis by the slightly different angle $\theta - \gamma$, with $|\gamma| \ll 1$. Since all planes perpendicular to the x_3 axis deform in the same way, it is sufficient to study the deformation of the x_1x_2-plane, as shown in Figure 1.2.3. It is clear that, with respect to axes rotated by the angle θ, the deformation is just one of simple shear,

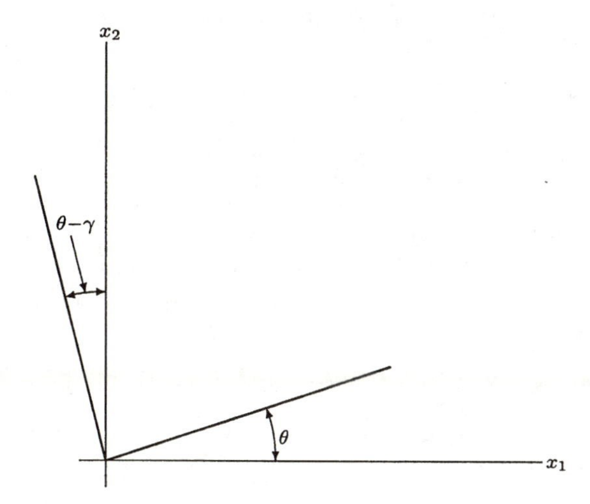

Figure 1.2.3. Infinitesimal shear strain with finite rotation

and can be described by the infinitesimal strain matrix given above. With respect to the reference axes, we determine first the displacement of points originally on the x_1-axis,

$$u_1(x_1, 0, x_3) = -(1 - \cos\theta)x_1, \quad u_2(x_1, 0, x_3) = \sin\theta x_1,$$

and that of points originally on the x_2-axis,

$$u_1(0, x_2, x_3) = -\sin(\theta - \gamma)x_2, \quad u_2(0, x_2, x_3) = -[1 - \cos(\theta - \gamma)]x_2.$$

The latter can be linearized with respect to γ:

$$u_1(0, x_2, x_3) = -(\sin\theta - \gamma\cos\theta)x_2, \quad u_2(0, x_2, x_3) = -(1 - \cos\theta - \gamma\sin\theta)x_2.$$

Since the deformation is homogeneous, the displacement must be linear in x_1 and x_2 and therefore can be obtained by superposition:

$$u_1(x_1, x_2, x_3) = -(1 - \cos\theta)x_1 - (\sin\theta - \gamma\cos\theta)x_2,$$

$$u_2(x_1, x_2, x_3) = \sin\theta x_1 - (1 - \cos\theta - \gamma\sin\theta)x_2.$$

Knowing that $u_3 = 0$, we can now determine the Green–Saint-Venant strain tensor, and find, with terms of order γ^2 neglected,

$$\underline{E} = \begin{bmatrix} 0 & \frac{1}{2}\gamma & 0 \\ \frac{1}{2}\gamma & 0 & 0 \\ 0 & 0 & 0 \end{bmatrix},$$

that is, precisely the same form as obtained for the infinitesimal strain with respect to rotated axes. Moreover, the result is independent of θ. The reason is that $\mathbf{E}$ measures strain with respect to axes that are, in effect, fixed in the body.

Further discussion of the description of finite deformation is postponed until Chapter 8.

Alternative Notations and Coordinate Systems

The "engineering" notations for the Cartesian strain components are ε_x, ε_y, and ε_z for ε_{11}, ε_{22}, and ε_{33}, respectively, and γ_{xy} for γ_{12}, and so on.

In cylindrical coordinates we find the strain components by taking the symmetric part of $\nabla \otimes \mathbf{u}$ as given by Equation (1.1.6):

$$\varepsilon_r = \frac{\partial u_r}{\partial r}, \quad \varepsilon_\theta = \frac{u_r}{r} + \frac{1}{r}\frac{\partial u_\theta}{\partial \theta}, \quad \varepsilon_z = \frac{\partial u_z}{\partial z},$$
$$\gamma_{r\theta} = \frac{\partial u_\theta}{\partial r} + \frac{1}{r}\frac{\partial u_r}{\partial \theta} - \frac{u_\theta}{r}, \quad \gamma_{rz} = \frac{\partial u_z}{\partial r} + \frac{\partial u_r}{\partial z}, \quad \gamma_{\theta z} = \frac{1}{r}\frac{\partial u_z}{\partial \theta} + \frac{\partial u_\theta}{\partial z}. \tag{1.2.1}$$

In spherical coordinates we similarly find, from Equation (1.1.8),

$$\varepsilon_r = \frac{\partial u_r}{\partial r}, \quad \varepsilon_\phi = \frac{1}{r}\frac{\partial u_\phi}{\partial \phi} + \frac{u_r}{r}, \quad \varepsilon_\theta = \frac{1}{r\sin\phi}\frac{\partial u_\theta}{\partial \theta} + \frac{u_r}{r} + \frac{\cot\phi}{r}u_\phi,$$

$$\gamma_{r\phi} = \frac{\partial u_\phi}{\partial r} + \frac{1}{r}\frac{\partial u_r}{\partial \phi} - \frac{u_\phi}{r}, \quad \gamma_{r\theta} = \frac{\partial u_\theta}{\partial r} + \frac{1}{r\sin\phi}\frac{\partial u_r}{\partial \theta} - \frac{u_\theta}{r}, \tag{1.2.2}$$

$$\gamma_{\phi\theta} = \frac{1}{r}\frac{\partial u_\theta}{\partial \phi} + \frac{1}{r\sin\phi}\frac{\partial u_\phi}{\partial \theta} - \frac{\cot\phi}{r}u_\theta.$$

Volumetric and Deviatoric Strain

The trace of the strain tensor, $\varepsilon_{kk} = \nabla \cdot \mathbf{u}$, has a special geometric significance: it is the (infinitesimal) *volumetric strain*, defined as $\Delta V/V_0$, where ΔV is the volume change and V_0 the initial volume of a small neighborhood. An easy way to show this is to look at a unit cube ($V_0 = 1$) whose edges parallel the coordinate axes. When the cube is infinitesimally deformed, the lengths of the edges change to $1 + \varepsilon_{11}$, $1 + \varepsilon_{22}$, and $1 + \varepsilon_{33}$, respectively, making the volume $(1 + \varepsilon_{11})(1 + \varepsilon_{22})(1 + \varepsilon_{33}) \doteq 1 + \varepsilon_{kk}$. The volumetric strain is also known as the *dilatation*.

The total strain tensor may now be decomposed as

$$\varepsilon_{ij} = \tfrac{1}{3}\varepsilon_{kk}\delta_{ij} + e_{ij}.$$

The *deviatoric strain* or *strain deviator* tensor $\mathbf{e}$ is defined by this equation. Its significance is that it describes *distortion*, that is, deformation without volume change. A state of strain with $\mathbf{e} = \mathbf{0}$ is called *spherical* or *hydrostatic*.

1.2.3. Principal Strains

It is possible to describe any state of infinitesimal strain as a superposition of three uniaxial extensions or contractions along mutually perpendicular axes, that is, to find a set of axes x_i^* such that, with respect to these axes, the strain tensor is described by

$$\underline{\varepsilon}^* = \begin{bmatrix} \varepsilon_1 & 0 & 0 \\ 0 & \varepsilon_2 & 0 \\ 0 & 0 & \varepsilon_3 \end{bmatrix}.$$

Let $\mathbf{n}$ be unit vector parallel to such an axis; then the longitudinal strain along $\mathbf{n}$ is $\varepsilon_{ij}n_in_j$. By hypothesis, the shear strains $\varepsilon_{ij}n_i\bar{n}_j$, where $\bar{\mathbf{n}}$ is any unit vector perpendicular to $\mathbf{n}$, are zero. Consequently the vector whose components are $\varepsilon_{ij}n_j$ is parallel to $\mathbf{n}$, and if its magnitude is ε, then $\varepsilon_{ij}n_j = \varepsilon n_i$, or

$$\varepsilon_{ij}n_j - \varepsilon n_i = (\varepsilon_{ij} - \varepsilon\delta_{ij})n_j = 0.$$

Then, in order that $\mathbf{n} \neq 0$, it is necessary that

$$\det(\underline{\varepsilon} - \varepsilon\underline{I}) = -\varepsilon^3 + K_1\varepsilon^2 + K_2\varepsilon + K_3 = 0, \tag{1.2.3}$$

where $K_1 = \varepsilon_{kk}$, $K_2 = \frac{1}{2}(\varepsilon_{ij}\varepsilon_{ij} - \varepsilon_{ii}\varepsilon_{kk})$, and $K_3 = \det\underline{\varepsilon}$ are the so-called *principal invariants* of the tensor $\boldsymbol{\varepsilon}$. Since Equation (1.2.3) is a cubic equation, it has three roots, which are the values of ε for which the assumption holds, namely ε_1, ε_2, and ε_3. Such roots are known in general as the *eigenvalues* of the matrix $\underline{\varepsilon}$, and in the particular case of strain as the *principal strains*.

The principal invariants have simple expressions in terms of the principal strains:

$$\begin{aligned} K_1 &= \varepsilon_1 + \varepsilon_2 + \varepsilon_3, \\ K_2 &= -(\varepsilon_1\varepsilon_2 + \varepsilon_2\varepsilon_3 + \varepsilon_3\varepsilon_1), \\ K_3 &= \varepsilon_1\varepsilon_2\varepsilon_3. \end{aligned}$$

To each ε_I ($I = 1, 2, 3$) there corresponds an *eigenvector* $\mathbf{n}^{(I)}$; an axis directed along an eigenvector is called a *principal axis* of strain.

We must remember, however, that a cubic equation with real coefficients need not have roots that are all real: it may be that one root is real and the other two are complex conjugates. It is important to show that the eigenvalues of a *symmetric* second-rank tensor — and hence the principal strains — are real (if they were not real, their physical meaning would be dubious). We can also show that the principal axes are mutually perpendicular.

Theorem 1. If $\boldsymbol{\varepsilon}$ is symmetric then the ε_I are real.

Proof. Let $\varepsilon_1 = \varepsilon, \varepsilon_2 = \bar{\varepsilon}$, where the bar denotes the complex conjugate. Now, if $n_i^{(1)} = n_i$, then $n_i^{(2)} = \bar{n}_i$. Since $\varepsilon_{ij}n_j = \varepsilon n_i$, we have $\bar{n}_i\varepsilon_{ij}n_j =$

$\varepsilon\bar{n}_i n_i$; similarly, $\varepsilon_{ij}\bar{n}_j = \bar{\varepsilon}\bar{n}_i$, so that $n_i\varepsilon_{ij}\bar{n}_j = \bar{\varepsilon}\bar{n}_i n_i$. However, $\varepsilon_{ij} = \varepsilon_{ji}$; consequently $n_i\varepsilon_{ij}\bar{n}_j = \bar{n}_i\varepsilon_{ij}n_j$, so that $(\varepsilon - \bar{\varepsilon})\bar{n}_i n_i = 0$. Since $\bar{n}_i n_i$ is, for any nonzero vector $\mathbf{n}$, a positive real number, it follows that $\varepsilon = \bar{\varepsilon}$ (i.e., ε is real).

Theorem 2. If $\boldsymbol{\varepsilon}$ is symmetric then the $\mathbf{n}^{(I)}$ are mutually perpendicular.

Proof. Assume that $\varepsilon_1 \neq \varepsilon_2$; then $\varepsilon_{ij}n_j^{(1)} = \varepsilon_1 n_i^{(1)}$ and $\varepsilon_{ij}n_j^{(2)} = \varepsilon_2 n_i^{(2)}$. But $n_i^{(2)}\varepsilon_{ij}n_j^{(1)} - n_i^{(1)}\varepsilon_{ij}n_j^{(2)} = 0 = (\varepsilon_1 - \varepsilon_2)n_i^{(1)}n_i^{(2)}$. Hence $\mathbf{n}^{(1)} \cdot \mathbf{n}^{(2)} = 0$.

If $\varepsilon_1 = \varepsilon_2 \neq \varepsilon_3$, then any vector perpendicular to $\mathbf{n}^{(3)}$ is an eigenvector, so that we can choose two that are perpendicular to each other. If $\varepsilon_1 = \varepsilon_2 = \varepsilon_3$ (hydrostatic strain), then every nonzero vector is an eigenvector; hence we can always find three mutually perpendicular eigenvectors. Q.E.D.

If the eigenvectors $\mathbf{n}^{(i)}$ are normalized (i.e., if their magnitudes are defined as unity), then we can always choose from among them or their negatives a right-handed triad, say $\boldsymbol{l}^{(1)}$, $\boldsymbol{l}^{(2)}$, $\boldsymbol{l}^{(3)}$, and we define Cartesian coordinates x_i^* ($i = 1, 2, 3$) along them, then the direction cosines β_{ij} are given by $\boldsymbol{l}^{(i)} \cdot \mathbf{e}_j$, so that the strain components with respect to the new axes are given by

$$\varepsilon_{ij*} = l_k^{(i)} l_l^{(j)} \varepsilon_{kl}.$$

But from the definition of the $\mathbf{n}^{(i)}$, we have

$$l_l^{(i)}\varepsilon_{kl} = \varepsilon_i \delta_{kl} l_l^{(i)} = \varepsilon_i l_k^{(i)} \quad \text{(no sum on } i\text{)},$$

so that

$$\varepsilon_{ij}^* = \varepsilon_i l_k^{(i)} l_k^{(j)} = \varepsilon_i \delta_{ij} \quad \text{(no sum on } i\text{)},$$

or

$$\underline{\varepsilon}^* = \begin{bmatrix} \varepsilon_1 & 0 & 0 \\ 0 & \varepsilon_2 & 0 \\ 0 & 0 & \varepsilon_3 \end{bmatrix} \stackrel{\text{def}}{=} \underline{\Lambda}.$$

The same manipulations can be carried out in direct matrix notation. With the matrix $\underline{\Lambda}$ as just defined, and with $\underline{L}$ defined by $L_{ij} = l_i^{(j)}$ (so that $\underline{L} = \underline{\beta}^T$), the equations defining the eigenvectors can be written as

$$\underline{\varepsilon}\underline{L} = \underline{L}\underline{\Lambda},$$

and therefore

$$\underline{\varepsilon}^* = \underline{\beta}\underline{\varepsilon}\underline{\beta}^T = \underline{L}^T\underline{\varepsilon}\underline{L} = \underline{L}^T\underline{L}\underline{\Lambda} = \underline{\Lambda}.$$

If one of the basis vectors $(\mathbf{e}_i)$, say $\mathbf{e}_3$, is already an eigenvector of $\boldsymbol{\varepsilon}$, then $\varepsilon_{13} = \varepsilon_{23} = 0$, and ε_{33} is a principal strain, say ε_3. The remaining principal strains, ε_1 and ε_2, are governed by the quadratic equation

$$\varepsilon^2 - (\varepsilon_{11} + \varepsilon_{22})\varepsilon + \varepsilon_{11}\varepsilon_{22} - \varepsilon_{12}^2.$$

This equation can be solved explicitly, yielding

$$\varepsilon_{1,2} = \frac{1}{2}(\varepsilon_{11} + \varepsilon_{22}) \pm \frac{1}{2}\sqrt{(\varepsilon_{11} - \varepsilon_{22})^2 + 4\varepsilon_{12}{}^2}.$$

In the special case of simple shear, we have $\varepsilon_{11} = \varepsilon_{22} = 0$ and $\varepsilon_{12} = \frac{1}{2}\gamma$. Consequently $\varepsilon_{1,2} = \pm\frac{1}{2}\gamma$. With respect to principal axes, the strain tensor is represented by

$$\underset{\sim}{\varepsilon}^* = \begin{bmatrix} \frac{1}{2}\gamma & 0 & 0 \\ 0 & -\frac{1}{2}\gamma & 0 \\ 0 & 0 & 0 \end{bmatrix},$$

that is, the strain can be regarded as the superposition of a uniaxial extension and a uniaxial contraction of equal magnitudes and along mutually perpendicular directions. Conversely, any strain state that can be so represented is one of pure shear.

1.2.4. Compatibility Conditions

If a second-rank tensor field $\boldsymbol{\varepsilon}(x_1, x_2, x_3)$ is given, it does not automatically follow that such a field is indeed a strain field, that is, that there exists a displacement field $\mathbf{u}(x_1, x_2, x_3)$ such that $\varepsilon_{ij} = \frac{1}{2}(u_{j,i} + u_{i,j})$; if it does, then the strain field is said to be *compatible*.

The determination of a necessary condition for the compatibility of a presumed strain field is closely related to the integrability theorem. Indeed, if there were given a second-rank tensor field $\boldsymbol{\alpha}$ such that $\alpha_{ji} = u_{j,i}$, then the condition would be just $e_{ikm}\alpha_{ji,k} = 0$. Note, however, that if there exists a displacement field $\mathbf{u}$, then there also exists a rotation field $\boldsymbol{\theta}$ such that $\varepsilon_{ij} + e_{ijl}\theta_l = u_{j,i}$. Consequently, the condition may also be written as $e_{ikm}(\varepsilon_{ij} + e_{ijl}\theta_l)_{,k} = 0$. But

$$e_{ikm}e_{ijl}\theta_{l,k} = (\delta_{jk}\delta_{lm} - \delta_{jm}\delta_{kl})\theta_{l,k} = \theta_{m,j}\,,$$

since $\theta_{k,k} = 0$. Therefore the condition reduces to

$$e_{ikm}\varepsilon_{ij,k} = -\theta_{m,j}\,.$$

The condition for a $\boldsymbol{\theta}$ field to exist such that the last equation is satisfied may be found by again invoking the integrability theorem, namely,

$$e_{ikm}e_{jln}\varepsilon_{ij,kl} = 0. \tag{1.2.4}$$

The left-hand side of Equation (1.2.4) represents a symmetric second-rank tensor, called the *incompatibility tensor*, and therefore the equation represents six distinct component equations, known as the **compatibility conditions**. If the region R is simply connected, then the compatibility conditions are also sufficient for the existence of a displacement field from which the strain field can be derived. In a multiply connected region (i.e., a region with holes), additional conditions along the boundaries of the holes are required.

Other methods of derivation of the compatibility conditions lead to the fourth-rank tensor equation

$$\varepsilon_{ij,kl} + \varepsilon_{kl,ij} - \varepsilon_{ik,jl} - \varepsilon_{jl,ik} = 0;$$

the sufficiency proof due to Cesaro (see, e.g., Sokolnikoff [1956]) is based on this form. It can easily be shown, however, that only six of the 81 equations are algebraically independent, and that these six are equivalent to (1.2.4). A sufficiency proof based directly on (1.2.4) is due to Tran-Cong [1985].

The algebraic independence of the six equations does not imply that they represent six independent conditions. Let the incompatibility tensor, whose components are defined by the left-hand side of (1.2.4), be denoted **R**. Then

$$R_{mn,n} = e_{ikm}e_{jln}\varepsilon_{ij,kln} = \frac{1}{2}e_{ikm}e_{jln}(u_{i,jkln} + u_{j,ikln}) = 0,$$

regardless of whether (1.2.4) is satisfied, because $e_{ikm}u_{j,ikln} = e_{jln}u_{i,jkln} = 0$. The identity $R_{mn,n} = 0$ is known as the **Bianchi formula** (see Washizu [1958] for a discussion).

Compatibility in Plane Strain

Plane strain in the x_1x_2-plane is defined by the conditions $\varepsilon_{i3} = 0$ and $\varepsilon_{ij,3} = 0$ for all i, j. The strain tensor is thus determined by the two-dimensional components $\varepsilon_{\alpha\beta}(x_1, x_2)$ (α, β = 1, 2), and the only nontrivial compatibility condition is the one corresponding to $m = n = 3$ in Equation (1.2.4), namely,

$$e_{\alpha\gamma 3}e_{\beta\delta 3}\varepsilon_{\alpha\beta,\gamma\delta} = 0.$$

In terms of strain components, this equation reads

$$\varepsilon_{11,22} + \varepsilon_{22,11} - 2\varepsilon_{12,12} = 0,$$

or, in engineering notation,

$$\frac{\partial^2 \varepsilon_x}{\partial y^2} + \frac{\partial^2 \varepsilon_y}{\partial x^2} = \frac{\partial^2 \gamma}{\partial x \partial y}, \tag{1.2.5}$$

where $\gamma = \gamma_{xy}$. Lastly, an alternative form in indicial notation is

$$\varepsilon_{\alpha\alpha,\beta\beta} - \varepsilon_{\alpha\beta,\alpha\beta} = 0. \tag{1.2.6}$$

It can easily be shown that, to within a rigid-body displacement, the only displacement field that is consistent with a compatible field of plane strain is one of *plane displacement*, in which u_1 and u_2 are functions of x_1 and x_2 only, and u_3 vanishes identically. The conditions $\varepsilon_{i3} = 0$ are, in terms of displacement components,

$$u_{3,3} = 0, \qquad u_{1,3} + u_{3,1} = 0, \qquad u_{2,3} + u_{3,2} = 0,$$

leading to

$$u_3 = w(x_1, x_2), \qquad u_\alpha = u^0_\alpha(x_1, x_2) - x_3 w_{,\alpha}.$$

The strain components are now

$$\varepsilon_{\alpha\beta} = \varepsilon^0_{\alpha\beta}(x_1, x_2) - x_3 w_{,\alpha\beta},$$

where $\boldsymbol{\varepsilon}^0$ is the strain derived from the plane displacement field $\mathbf{u}^0 = u^0_\alpha \mathbf{e}_\alpha$. The conditions $\varepsilon_{\alpha\beta,3} = 0$ require that $w_{,\alpha\beta} = 0$, that is, $w(x_1, x_2) = ax_1 + bx_2 + c$, where a, b and c are constants. The displacement field is thus the superposition of $\mathbf{u}^0$ and of $-ax_3\mathbf{e}_1 - bx_3\mathbf{e}_2 + (ax_1 + bx_2 + c)\mathbf{e}_3$, the latter being obviously a rigid-body displacement. In practice, "plane strain" is synonymous with plane displacement.

Exercises: Section 1.2

1. For each of the following displacement fields, with $\gamma \ll 1$, sketch the displaced positions in the x_1x_2-plane of the points initially on the sides of the square bounded by $x_1 = 0$, $x_1 = 1$, $x_2 = 0$, $x_2 = 1$.

   (a) $\mathbf{u} = \frac{1}{2}\gamma x_2\mathbf{e}_1 + \frac{1}{2}\gamma x_1\mathbf{e}_2$

   (b) $\mathbf{u} = -\frac{1}{2}\gamma x_2\mathbf{e}_1 + \frac{1}{2}\gamma x_1\mathbf{e}_2$

   (c) $\mathbf{u} = \gamma x_1\mathbf{e}_2$

2. For each of the displacement fields in the preceding exercise, determine the matrices representing the finite (Green–Saint-Venant) and infinitesimal strain tensors and the infinitesimal rotation tensor, as well as the infinitesimal rotation vector.

3. For the displacement field (a) of Exercise 1, determine the longitudinal strain along the direction $(\mathbf{e}_1 + \mathbf{e}_2)/\sqrt{2}$.

4. For the displacement field given in cylindrical coordinates by

$$\mathbf{u} = ar\,\mathbf{e}_r + brz\,\mathbf{e}_\theta + c\sin\theta\,\mathbf{e}_z,$$

   where a, b and c are constants, determine the infinitesimal strain components as functions of position in cylindrical coordinates.

5. Determine the infinitesimal strain and rotation fields for the displacement field $\mathbf{u} = -w'(x_1)x_3\mathbf{e}_1 + w(x_1)\mathbf{e}_3$, where w is an arbitrary continuously differentiable function. If $w(x) = kx^2$, find a condition on k in order that the deformation be infinitesimal in the region $-h < x_3 < h$, $0 < x_1 < l$.

6. For the displacement-gradient matrix

$$\underset{\sim}{\alpha} = \begin{bmatrix} 4 & -1 & 0 \\ 1 & -4 & 2 \\ 4 & 0 & 6 \end{bmatrix} \times 10^{-3},$$

determine

(a) the strain and rotation matrices,

(b) the volume strain and the deviatoric strain matrix,

(c) the principal strain invariants K_1, K_2, K_3,

(d) the principal strains and their directions.

7. For the displacement field $\mathbf{u} = \alpha(-x_2x_3\mathbf{e}_1 + x_1x_3\mathbf{e}_2)$, determine (a) the strain and rotation fields, (b) the principal strains and their directions as functions of position.

8. For the plane strain field

$$\varepsilon_x = Bxy, \quad \varepsilon_y = -\nu Bxy, \quad \gamma_{xy} = (1+\nu)B(h^2 - y^2),$$

where B, ν and h are constants,

(a) check if the compatibility condition is satisfied;

(b) if it is, determine the displacement field $u(x, y)$, $v(x, y)$ in $0 < x < L$, $-h < y < h$ such that $u(L, 0) = 0$, $v(L, 0) = 0$, and $\left(\frac{\partial v}{\partial x} - \frac{\partial u}{\partial y}\right)(L, 0) = 0$.

Section 1.3 Mechanics of Continuous Bodies

1.3.1. Introduction

Global Equations of Motion

Mechanics has been defined as the study of forces and motions. It is easy enough to define motion as the change in position of a body, in time, with respect to some frame of reference. The definition of force is more elusive, and has been the subject of much controversy among theoreticians, especially with regard to whether force can be defined independently of Newton's second law of motion. An interesting method of definition is based on a thought experiment due to Mach, in which two particles, A and B, are close to each other but so far away from all other bodies that the motion of each one can be influenced only by the other. It is then found that there exist numbers m_A, m_B (the *masses* of the particles) such that the motions of the particles obey the relation $m_A \mathbf{a}_A = -m_B \mathbf{a}_B$, where $\mathbf{a}$ denotes acceleration. The force exerted by A on B can now be defined as $\mathbf{F}_{AB} = m_B \mathbf{a}_B$, and $\mathbf{F}_{BA}$ is defined analogously. If B, rather than being a single particle, is a set of several particles, then $\mathbf{F}_{AB}$ is the sum of the forces exerted by A on all the particles contained in B, and if A is also a set of particles, then $\mathbf{F}_{AB}$ is the sum of the forces exerted on B by all the particles in A.

The total force $\mathbf{F}$ on a body B is thus the vector sum of all the forces exerted on it by all the other bodies in the universe. In reality these forces are of two kinds: long-range and short-range. If B is modeled as a continuum occupying a region R, then the effect of the long-range forces is felt throughout R, while the short-range forces act as contact forces on the boundary surface ∂R. Any volume element dV experiences a long-range force $\rho \mathbf{b}\, dV$, where ρ is the density (mass per unit volume) and $\mathbf{b}$ is a vector field (with dimensions of force per unit mass) called the *body force*. Any oriented surface element $d\mathbf{S} = \mathbf{n}\, dS$ experiences a contact force $\mathbf{t}(\mathbf{n})\, dS$, where $\mathbf{t}(\mathbf{n})$ is called the *surface traction*; it is not a vector field because it depends not only on position but also on the local orientation of the surface element as defined by the local value (direction) of $\mathbf{n}$.

If $\mathbf{a}$ denotes the acceleration field, then the global force equation of motion (balance of linear momentum) is

$$\int_R \rho \mathbf{b}\, dV + \int_{\partial R} \mathbf{t}(\mathbf{n})\, dS = \int_R \rho \mathbf{a}\, dV. \tag{1.3.1}$$

When all moments are due to forces (i.e. when there are no distributed couples, as there might be in an electromagnetic field), then the global **moment** equation of motion (balance of angular momentum) is

$$\int_R \rho \mathbf{x} \times \mathbf{b}\, dV + \int_{\partial R} \mathbf{x} \times \mathbf{t}(\mathbf{n})\, dS = \int_R \rho \mathbf{x} \times \mathbf{a}\, dV, \tag{1.3.2}$$

where $\mathbf{x}$ is the position vector.

Equations (1.3.1)–(1.3.2) are known as **Euler's equations of motion**, applied by him to the study of the motion of rigid bodies. If a body is represented as an assemblage of discrete particles, each governed by Newton's laws of motion, then Euler's equations can be derived from Newton's laws. Euler's equations can, however, be taken as axioms describing the laws of motion for extended bodies, independently of any particle structure. They are therefore the natural starting point for the mechanics of bodies modeled as continua.

Lagrangian and Eulerian Approaches

The existence of an acceleration field means, of course, that the displacement field is time-dependent. If we write $\mathbf{u} = \mathbf{u}(x_1, x_2, x_3, t)$ and interpret the x_i as Lagrangian coordinates, as defined in Section 1.2, then we have simply $\mathbf{a} = \dot{\mathbf{v}} = \ddot{\mathbf{u}}$; here $\mathbf{v} = \dot{\mathbf{u}}$ is the velocity field, and the superposed dot denotes partial differentiation with respect to time at constant x_i (called *material* time differentiation). With this interpretation, however, it must be agreed that R is the region occupied by B in the reference configuration, and similarly that dV and dS denote volume and surface elements measured in the reference configuration, ρ is the mass per unit reference volume, and $\mathbf{t}$ is force per unit reference surface. This convention constitutes the so-called *Lagrangian* approach (though Lagrange did not have much to do with it) to continuum mechanics, and the quantities associated with it are called Lagrangian, referential, or material (since a point (x_1, x_2, x_3) denotes a fixed particle or material point). It is, by and large, the preferred approach in solid mechanics. In problems of flow, however — not only fluid flow, but also plastic flow of solids — it is usually more instructive to describe the motion of particles with respect to coordinates that are fixed in space — *Eulerian* or spatial coordinates. In this *Eulerian* approach the motion is described not by the displacement field $\mathbf{u}$ but by the velocity field $\mathbf{v}$. If the x_i are spatial coordinates, then the *material time derivative* of a function $\phi(x_1, x_2, x_3, t)$, defined as its time derivative with the Lagrangian coordinates held fixed, can be found by applying the chain rule to be

$$\dot{\phi} = \frac{\partial \phi}{\partial t} + v_i \phi_{,i}\,.$$

The material time derivative of ϕ is also known as its *Eulerian derivative* and denoted $\frac{D}{Dt}\phi$.

If the displacement field is infinitesimal, as defined in the preceding section, then the distinction between Lagrangian and Eulerian coordinates can usually be neglected, and this will generally be done here until finite deformations are studied in Chapter 8. The fundamental approach is Lagrangian, except when problems of plastic *flow* are studied; but many of the equations derived are not exact for the Lagrangian formulation. Note, however, one point: because of the postulated constancy of mass of any fixed part of B, the product $\rho\, dV$ is time-independent regardless of whether ρ and dV are given Lagrangian or Eulerian readings; thus the relation

$$\frac{d}{dt}\int_R \rho\psi\, dV = \int_R \rho\dot{\psi}\, dV$$

is exact.

1.3.2. Stress

To determine how $\mathbf{t}$ depends on $\mathbf{n}$, we employ the *Cauchy tetrahedron* illustrated in Figure 1.3.1.Assuming $\mathbf{b}$, ρ, $\mathbf{a}$ and $\mathbf{t}(\mathbf{n})$ to depend continuously on $\mathbf{x}$, we have, if the tetrahedron is sufficiently small,

$$\int_R \rho\mathbf{b}\,dV \doteq \rho\mathbf{b}\,\Delta V;$$
$$\int_R \rho\mathbf{a}\,dV \doteq \rho\mathbf{a}\,\Delta V;$$
$$\begin{aligned}\int_{\partial R} \mathbf{t}(\mathbf{n})\,dS &\doteq \mathbf{t}(\mathbf{n})\,\Delta A \\ &\quad + \sum_j \mathbf{t}(-\mathbf{e}_j)\,\Delta A_j \\ &= [\mathbf{t}(\mathbf{n}) - \mathbf{t}_j n_j]\,\Delta A,\end{aligned}$$

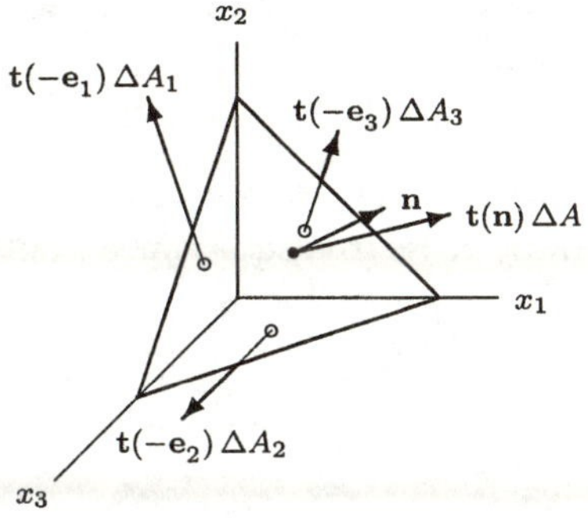

Figure 1.3.1. Cauchy tetrahedron

where $\mathbf{t}_j \stackrel{\text{def}}{=} -\mathbf{t}(-\mathbf{e}_j)$. Thus we have, approximately,

$$\mathbf{t}(\mathbf{n}) - \mathbf{t}_j n_j + \rho(\mathbf{b} - \mathbf{a})\frac{\Delta V}{\Delta A} \doteq 0.$$

This becomes exact in the limit as the tetrahedron shrinks to a point, i.e. $\Delta V/\Delta A \to 0$, so that

$$\mathbf{t}(\mathbf{n}) = \mathbf{t}_j n_j,$$

that is, $\mathbf{t}(\cdot)$ is a *linear* function of its argument. If we define $\sigma_{ij} \stackrel{\text{def}}{=} \mathbf{e}_i \cdot \mathbf{t}_j$, then

$$t_i(\mathbf{n}) = \sigma_{ij} n_j,$$

so that $\underline{\sigma} = [\sigma_{ij}]$ represents a second-rank tensor field called the *stress* tensor. Denoting this tensor by $\boldsymbol{\sigma}$, the preceding equation may be rewritten in direct tensor notation as

$$\mathbf{t}(\mathbf{n}) = \boldsymbol{\sigma}\mathbf{n}.$$

The force equation of motion (1.3.1) can now be written in indicial notation as

$$\int_R \rho b_i\,dV + \int_{\partial R} \sigma_{ij} n_j\,dS = \int_R \rho a_i\,dV.$$

By Gauss's theorem we have

$$\int_{\partial R} \sigma_{ij} n_j\,dS = \int_R \sigma_{ij,j}\,dV;$$

therefore,

$$\int_R (\sigma_{ij},_j + \rho b_i - \rho a_i)\, dV = 0.$$

This equation, since it embodies a fundamental physical law, must be independent of how we define a given body and therefore it must be valid for any region R, including very small regions. Consequently, the integrand must be zero, and thus we obtain the **local force equations of motion** (due to Cauchy):

$$\sigma_{ij},_j + \rho b_i = \rho a_i. \tag{1.3.3}$$

When the relation between traction and stress is introduced into Equation (1.3.2), this equation becomes, in indicial notation,

$$\int_R \rho e_{ijk} x_j b_k\, dV + \int_{\partial R} e_{ijk} x_j \sigma_{kl} n_l\, dS = \int_R \rho e_{ijk} x_j a_k\, dV.$$

By Gauss's theorem,

$$\begin{aligned} \int_{\partial R} x_j \sigma_{kl} n_l\, dS &= \int_R (x_j \sigma_{kl}),_l\, dV \\ &= \int_R (\delta_{ij} \sigma_{kl} + x_j \sigma_{kl},_l)\, dV \\ &= \int_R (\sigma_{kj} + x_j \sigma_{kl},_l)\, dV; \end{aligned}$$

therefore

$$\int_R e_{ijk} [x_j (\rho b_k + \sigma_{kl},_l - \rho a_k) + \sigma_{kj}] dV = 0,$$

which, as a result of (1.3.3), reduces to

$$\int_R e_{ijk} \sigma_{kj}\, dV = 0.$$

Since this result, again, must be valid for any region R, it follows that

$$e_{ijk} \sigma_{kj} = 0,$$

or equivalently,

$$\sigma_{ij} = \sigma_{ji}. \tag{1.3.4}$$

In words: **the stress tensor is symmetric.**

In the usual "engineering" notation, the normal stresses σ_{11}, σ_{22}, and σ_{33} are designated σ_x, σ_y, and σ_z, respectively, while the shear stresses are written as τ_{xy} in place of σ_{12}, and so on. This notation is invariably used in conjunction with the use of x, y, z for the Cartesian coordinates and of u_x, u_y, and u_z for the Cartesian components of a vector $\mathbf{u}$, except that the components of the displacement vector are usually written u, v, w, and the body-force vector components are commonly written X, Y, Z rather than

ρb_x, and so on. Thus the local force equations of motion are written in engineering notation as

$$\frac{\partial \sigma_x}{\partial x} + \frac{\partial \tau_{xy}}{\partial y} + \frac{\partial \tau_{xz}}{\partial z} + X = \rho a_x,$$

and two similar equations.

The equations in cylindrical coordinates are obtained from Equation (1.1.7), with the changes of notation self-explanatory:

$$\frac{\partial \sigma_r}{\partial r} + \frac{\sigma_r - \sigma_\theta}{r} + \frac{1}{r}\frac{\partial \tau_{r\theta}}{\partial \theta} + \frac{\partial \tau_{rz}}{\partial z} + R = \rho a_r,$$
$$\frac{\partial \tau_{r\theta}}{\partial r} + \frac{2\tau_{r\theta}}{r} + \frac{1}{r}\frac{\partial \sigma_\theta}{\partial \theta} + \frac{\partial \tau_{\theta z}}{\partial z} + \Theta = \rho a_\theta,$$
$$\frac{\partial \tau_{rz}}{\partial r} + \frac{\tau_{rz}}{r} + \frac{1}{r}\frac{\partial \tau_{\theta z}}{\partial \theta} + \frac{\partial \sigma_z}{\partial z} + Z = \rho a_z.$$

The corresponding equations in spherical coordinates are obtained from Equation (1.1.9):

$$\frac{\partial \sigma_r}{\partial r} + \frac{1}{r}\frac{\partial \tau_{r\phi}}{\partial \phi} + \frac{1}{r\sin\phi}\frac{\partial \tau_{r\theta}}{\partial \theta} + \frac{2\sigma_r - \sigma_\phi - \sigma_\theta + \tau_{r\phi}\cot\phi}{r} + R = \rho a_r,$$
$$\frac{\partial \sigma_\phi}{\partial \phi} + \frac{1}{r\sin\phi}\frac{\partial \tau_{\phi\theta}}{\partial \theta} + \frac{\sigma_\phi\cot\phi - \sigma_\theta\cot\phi + 3\tau_{r\phi}}{r} + \Phi = \rho a_\phi,$$
$$\frac{\partial \tau_{r\theta}}{\partial r} + \frac{1}{r}\frac{\partial \tau_{\phi\theta}}{\partial \phi} + \frac{1}{r\sin\phi}\frac{\partial \sigma_\theta}{\partial \theta} + \frac{3\tau_{r\theta} + 2\tau_{\phi\theta}\cot\phi}{r} + \Theta = \rho a_\theta.$$

Projected Stresses

If $\mathbf{n}$ is an arbitrary unit vector, the traction $\mathbf{t} = \boldsymbol{\sigma}\mathbf{n}$ has, in general, a component parallel to $\mathbf{n}$ and one perpendicular to $\mathbf{n}$. These are the *projected stresses*, namely,

$$\textit{normal}\text{ stress}: \quad \sigma(\mathbf{n}) = \mathbf{n}\cdot\mathbf{t}(\mathbf{n}) = \sigma_{ij}n_i n_j,$$
$$\textit{shear}\text{ stress}: \quad \tau(\mathbf{n}) = \sqrt{|\mathbf{t}(\mathbf{n})|^2 - [\sigma(\mathbf{n})]^2};$$

note that this is the magnitude of the shear-stress *vector*

$$\boldsymbol{\tau}(\mathbf{n}) = \mathbf{t}(\mathbf{n}) - \mathbf{n}\sigma(\mathbf{n}) = \mathbf{n}\times(\mathbf{t}\times\mathbf{n}).$$

Principal Stresses

As in the case of strains, it is possible to find directions $\mathbf{n}$ such that $\tau(\mathbf{n}) = 0$, so that $\mathbf{t}(\mathbf{n}) = \sigma(\mathbf{n})\mathbf{n}$, or

$$\sigma_{ij}n_j - \sigma n_i = (\sigma_{ij} - \sigma\delta_{ij})n_j = 0.$$

Then, for $\mathbf{n} \neq 0$, we need

$$\det(\underline{\sigma} - \sigma \underline{I}) = -\sigma^3 + I_1\sigma^2 + I_2\sigma + I_3 = 0,$$

where $I_1 = \sigma_{kk}$, $I_2 = \frac{1}{2}(\sigma_{ij}\sigma_{ij} - \sigma_{ii}\sigma_{kk})$, and $I_3 = \det \underline{\sigma}$ are the principal invariants of $\boldsymbol{\sigma}$. The *principal stresses* can now be defined exactly like the principal strains; by Theorem 1 of 1.2.3[1] they are real, and by Theorem 2 the *principal axes of stress* are mutually perpendicular. The principal invariants of stress can be expressed in the form

$$\begin{aligned} I_1 &= \sigma_1 + \sigma_2 + \sigma_3, \\ I_2 &= -(\sigma_1\sigma_2 + \sigma_2\sigma_3 + \sigma_3\sigma_1), \\ I_3 &= \sigma_1\sigma_2\sigma_3. \end{aligned}$$

The *mean stress* or *hydrostatic stress* is defined as $\sigma_m = \frac{1}{3}I_1$.

Stress Deviator

The *stress deviator* or *deviatoric stress* tensor $\mathbf{s}$ is defined by $s_{ij} \stackrel{\text{def}}{=} \sigma_{ij} - \sigma_m\delta_{ij}$. The principal invariants of the stress deviator are denoted $J_1 = s_{kk}$, which vanishes identically, J_2 $(= \frac{1}{2}s_{ij}s_{ij})$, and J_3. The principal axes of $\mathbf{s}$ are the same as those of $\boldsymbol{\sigma}$, and the principal deviatoric stresses are $s_I = \sigma_I - \frac{1}{3}I_1$. J_2 and J_3 may be expressed in terms of the principal stresses through the principal-stress differences $\sigma_1 - \sigma_2$ etc., namely,

$$J_2 = \frac{1}{6}[(\sigma_1 - \sigma_2)^2 + (\sigma_2 - \sigma_3)^2 + (\sigma_3 - \sigma_1)^2],$$

$$\begin{aligned} J_3 = \frac{1}{27}[&(\sigma_1 - \sigma_2)^2(\sigma_1 - \sigma_3 + \sigma_2 - \sigma_3) + (\sigma_2 - \sigma_3)^2(\sigma_2 - \sigma_1 + \sigma_3 - \sigma_1) \\ &+ (\sigma_3 - \sigma_1)^2(\sigma_3 - \sigma_2 + \sigma_1 - \sigma_2)]. \end{aligned}$$

Octahedral Stresses

Let the basis vectors $(\mathbf{e}_i)$ be directed along the principal axes, and suppose that $\mathbf{n}$ is one of the eight vectors

$$\mathbf{n} = \frac{1}{\sqrt{3}}(\pm\mathbf{e}_1 \pm \mathbf{e}_2 \pm \mathbf{e}_3);$$

a regular octahedron can be formed with planes perpendicular to these vectors. The traction on such a plane (called an *octahedral plane*) is

$$\mathbf{t}(\mathbf{n}) = \frac{1}{\sqrt{3}}(\pm\sigma_1\mathbf{e}_1 \pm \sigma_2\mathbf{e}_2 \pm \sigma_3\mathbf{e}_3).$$

---

[1] Any cross-reference such as 1.2.3, not enclosed in parentheses, refers to a subsection, unless it is specified as a figure or a table. With parentheses, for example (1.2.3), the reference is to an equation.

The normal stress is

$$\sigma(\mathbf{n}) = \frac{1}{3}(\sigma_1 + \sigma_2 + \sigma_3) = \frac{1}{3}I_1 = \sigma_m \quad \text{(mean stress)},$$

and the shear stress is given by

$$\begin{aligned}
[\tau(\mathbf{n})]^2 &= \frac{1}{3}(\sigma_1^2 + \sigma_2^2 + \sigma_3^2) - \frac{1}{9}(\sigma_1 + \sigma_2 + \sigma_3)^2 \\
&= \frac{2}{9}(\sigma_1^2 + \sigma_2^2 + \sigma_3^2 - \sigma_1\sigma_2 - \sigma_2\sigma_3 - \sigma_3\sigma_1) \\
&\stackrel{\text{def}}{=} \tau_{oct}^2,
\end{aligned}$$

where τ_{oct} is called the *octahedral shear stress*. By comparing the just-derived result with the previously obtained expression for J_2 in terms of the principal stresses, it can be shown that

$$\tau_{oct}^2 = \frac{2}{3}J_2. \tag{1.3.5}$$

1.3.3. Mohr's Circle

Let the x_3-axis coincide with the principal axis defined by σ_3, that is, let $\mathbf{e}_3 = \mathbf{n}^{(3)}$. Now any unit vector $\mathbf{n}$ that is perpendicular to this axis may be written as $\mathbf{n} = \mathbf{e}_1 \cos\theta + \mathbf{e}_2 \sin\theta$. It follows that

$$\mathbf{t}(\mathbf{n}) = \mathbf{e}_1(\sigma_{11}\cos\theta + \sigma_{12}\sin\theta) + \mathbf{e}_2(\sigma_{12}\cos\theta + \sigma_{22}\sin\theta),$$

since $\sigma_{13} = \sigma_{23} = 0$, and

$$\sigma(\mathbf{n}) = \sigma_{11}\cos^2\theta + \sigma_{22}\sin^2\theta + 2\sigma_{12}\sin\theta\cos\theta,$$

which will be designated σ_θ. By means of the trigonometric identities $\cos^2\theta = \frac{1}{2}(1 + \cos 2\theta)$, $\sin^2\theta = \frac{1}{2}(1 - \cos 2\theta)$, and $2\sin\theta\cos\theta = \sin 2\theta$, σ_θ may be rewritten as

$$\sigma_\theta = \frac{1}{2}(\sigma_{11} + \sigma_{22}) + \frac{1}{2}(\sigma_{11} - \sigma_{22})\cos 2\theta + \sigma_{12}\sin 2\theta. \tag{1.3.6}$$

The projected shear-stress vector is

$$\boldsymbol{\tau}(\mathbf{n}) = [\sigma_{12}(\cos^2\theta - \sin^2\theta) + (\sigma_{22} - \sigma_{11})\sin 2\theta\cos 2\theta](-\mathbf{e}_1\sin\theta + \mathbf{e}_2\cos\theta);$$

the quantity in brackets will be designated τ_θ, and clearly $|\tau_\theta| = \tau(\mathbf{n})$, since the vector in parentheses is a unit vector. With the help of the trigonometric identities $\cos 2\theta = \cos^2\theta - \sin^2\theta$, $\sin 2\theta = 2\sin 2\theta\cos 2\theta$, we may write

$$\tau_\theta = \sigma_{12}\cos 2\theta + \frac{1}{2}(\sigma_{22} - \sigma_{11})\sin 2\theta. \tag{1.3.7}$$

From Equation (1.3.7) we may obtain the principal directions $\mathbf{n}^{(1)}$ and $\mathbf{n}^{(2)}$ directly by finding the values of θ for which τ_θ vanishes, namely, those that satisfy

$$\tan 2\theta = \frac{2\sigma_{12}}{\sigma_{11} - \sigma_{22}}, \qquad (1.3.8)$$

unless σ_{12} and $\sigma_{11} - \sigma_{22}$ are both zero, in which case $\tau_\theta = 0$ for all θ. In the general case, if θ_1 is a solution of Equation (1.3.8) (so that $\mathbf{n}^{(1)} = \mathbf{e}_1 \cos\theta_1 + \mathbf{e}_2 \sin\theta_1$), then so is $\theta_1 \pm \frac{1}{2}\pi$, showing the perpendicularity of nondegenerate principal directions. As is readily seen, however, $d\sigma_\theta/d\theta = 2\tau_\theta$. It follows that the principal stresses σ_1 and σ_2 are the extrema of σ_θ, one being the maximum and the other the minimum, again with the exception of the degenerate case in which σ_θ is constant. The principal stresses, with the numbering convention $\sigma_1 \geq \sigma_2$, are given by

$$\sigma_1, 2 = \frac{1}{2}(\sigma_{11} + \sigma_{22}) \ \pm \ \sqrt{\tfrac{1}{4}(\sigma_{11} - \sigma_{22})^2 + \sigma_{12}^2};$$

this convention is consistent with defining θ_1 in such a way that

$$\cos 2\theta_1 = \frac{\frac{1}{2}(\sigma_{11} - \sigma_{22})}{\sqrt{\frac{1}{4}(\sigma_{11} - \sigma_{22})^2 + \sigma_{12}^2}}, \qquad \sin 2\theta_1 = \frac{\sigma_{12}}{\sqrt{\frac{1}{4}(\sigma_{11} - \sigma_{22})^2 + \sigma_{12}^2}},$$

so that $\sigma_{\theta_1} = \sigma_1$.

Using the just-derived expressions for $\cos 2\theta_1$ and $\sin 2\theta_1$ and the trigonometric identities

$$\cos 2(\theta - \theta_1) = \cos 2\theta \cos 2\theta_1 + \sin 2\theta \sin 2\theta_1,$$

$$\sin 2(\theta - \theta_1) = \sin 2\theta \cos 2\theta_1 - \sin 2\theta_1 \cos 2\theta,$$

we may rewrite Equations (1.3.6)–(1.3.7) as

$$\sigma_\theta = \frac{1}{2}(\sigma_1 + \sigma_2) + \frac{1}{2}(\sigma_1 - \sigma_2)\cos 2(\theta - \theta_1), \qquad (1.3.9)$$

$$\tau_\theta = -\frac{1}{2}(\sigma_1 - \sigma_2)\sin 2(\theta - \theta_1). \qquad (1.3.10)$$

Equations (1.3.9)–(1.3.10) are easily seen to be the parametric representation of a circle, known as **Mohr's circle**, in the σ_θ-τ_θ plane, with its center at $(\frac{1}{2}(\sigma_1 + \sigma_2), 0)$ and with radius $\frac{1}{2}(\sigma_1 - \sigma_2)$, a value (necessarily positive in view of the numbering convention) equal to the maximum of $|\tau_\theta|$. Note that this maximum occurs when $\sin 2(\theta - \theta_1) = \pm 1$, that is, when $\theta = \theta_1 \pm \frac{1}{4}\pi$. The significance of the angle θ_1, and other aspects of Mohr's circle, can be seen from Figure 1.3.2.

It can be shown that the maximum over all $\mathbf{n}$ (in three dimensions) of the projected shear stress $\tau(\mathbf{n})$ is just the largest of the three maxima of τ_θ found in the planes perpendicular to each of the principal directions. With no regard for any numbering convention for the principal stresses, we have

$$\tau_{max} = \max_{\mathbf{n}} \tau(\mathbf{n}) = \frac{1}{2}\max\{|\sigma_1 - \sigma_2|, |\sigma_2 - \sigma_3|, |\sigma_1 - \sigma_3|\}. \qquad (1.3.11)$$

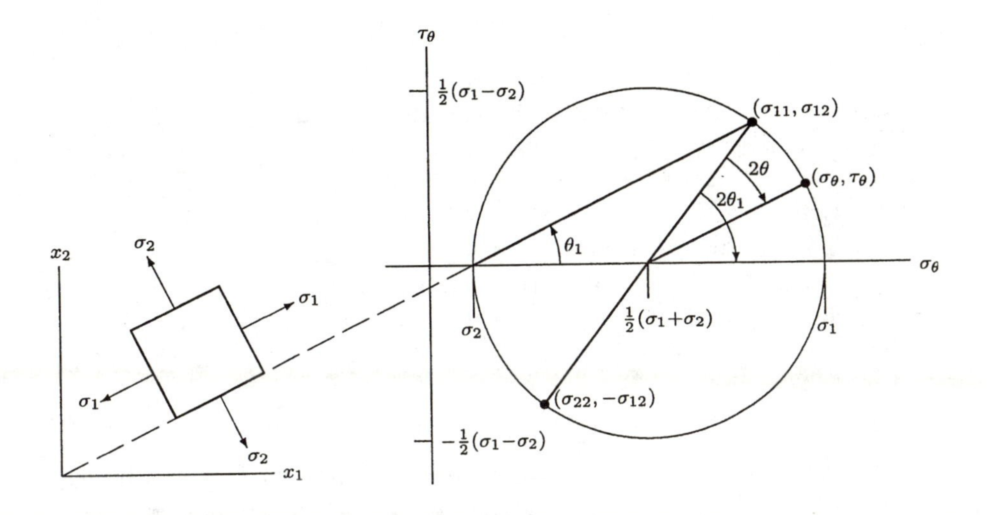

Figure 1.3.2. Mohr's circle.

1.3.4. Plane Stress

Plane stress is defined by the conditions $\sigma_i3 = 0$ $(i = 1, 2, 3)$ and $\sigma_{ij},_3 = 0$, so that the stress field is given by $\sigma_{\alpha\beta}(x_1, x_2)$ $(\alpha, \beta = 1, 2)$. The summation convention applies to Greek indices ranging over 1, 2. Consequently, the equilibrium equations *without body force* are

$$\sigma_{\alpha\beta},_\beta = 0,$$

or (1) $\sigma_{11},_1 + \sigma_{12},_2 = 0$ and (2) $\sigma_{12},_1 + \sigma_{22},_2 = 0$, so that there exist functions $\phi_\alpha(x_1, x_2)$ $(\alpha = 1, 2)$ such that (1) $\sigma_{11} = \phi_1,_2$, $\sigma_{12} = -\phi_1,_1$, and (2) $\sigma_{22} = \phi_2,_1$, $\sigma_{12} = -\phi_2,_2$. Hence $\phi_1,_1 = \phi_2,_2$, and therefore there exists a function $\Phi(x_1, x_2)$ such that $\phi_1 = \Phi,_2$, $\phi_2 = \Phi,_1$. Thus

$$\sigma_{11} = \Phi,_{22}\,, \quad \sigma_{22} = \Phi,_{11}\,, \quad \sigma_{12} = -\Phi,_{12}\,,$$

or, in two-dimensional indicial notation,

$$\sigma_{\alpha\beta} = \delta_{\alpha\beta}\Phi,_{\gamma\gamma} - \Phi,_{\alpha\beta}\,.$$

The function Φ is known as the *Airy stress function.*

In plane-stress problems described in "engineering" notation, the shear stress $\tau_{xy} = \tau_{yx}$ is sometimes written simply as τ, and the relation between the stress components and the Airy stress function is accordingly written as

$$\sigma_x = \frac{\partial^2\Phi}{\partial y^2}, \quad \sigma_y = \frac{\partial^2\Phi}{\partial x^2}, \quad \tau = -\frac{\partial^2\Phi}{\partial x \partial y}.$$

1.3.5. Boundary-Value Problems

The standard boundary-value problem of solid mechanics is the following: find the fields $\mathbf{u}$ and $\boldsymbol{\sigma}$ throughout R if the given information consists of, first, the body-force field $\mathbf{b}$ throughout R, and second, *boundary conditions* on ∂R, namely, that at every point of ∂R there is a local basis $(\mathbf{e}_i)$ such that *either* $t_i(\mathbf{n})$ *or* u_i is prescribed for each i, $i = 1, 2, 3$.

It may be that ∂R consists of two parts, ∂R_u and ∂R_t, such that $\mathbf{u}$ is prescribed on the former and $\mathbf{t}$ on the latter. This is not the most general case, but it is often cited for convenience. The prescribed t_i and u_i will be denoted t_i^a and u_i^a, respectively. The boundary conditions may be written as

$$u_i = u_i^a \text{ on } \partial R_u, \quad n_j\sigma_{ij} = t_i^a \text{ on } \partial R_t, \tag{1.3.12}$$

even if the boundary is not strictly divided into two parts; points at which both displacement and traction components are prescribed may be regarded as belonging to both ∂R_u and ∂R_t — in other words, ∂R_u and ∂R_t may be thought of as overlapping.

The prescribed body-force field $\mathbf{b}$ and surface tractions t_i^a are together known as the *loads*, while the generally unknown surface tractions t_i at the points where the displacements u_i are prescribed are called *reactions*. Whenever displacements are prescribed, the body is said to be subject to *external constraints*.[1] There may, in addition, exist *internal constraints* which restrict the displacement field in the interior of R; an example is *incompressibility*, that is, the inability of a body or any part thereof to change its volume, expressed by $\varepsilon_{kk} = \nabla \cdot \mathbf{u} = 0$.

A displacement field is called *kinematically admissible* if it is mathematically well-behaved (for example, continuous and piecewise continuously differentiable) and obeys the external and internal constraints, if any.

The boundary-value problem is called *static* if the data are independent of time and the acceleration is assumed to be zero. It is called *quasi-static* if the acceleration is neglected even though the data depend on time. In static or quasi-static boundary-value problems, the equations of motion (1.3.3) may be replaced by the **equilibrium equations**

$$\sigma_{ij},_j + \rho b_i = 0. \tag{1.3.13}$$

A stress field $\boldsymbol{\sigma}$ that obeys the equilibrium equations (1.3.13) and the traction boundary conditions $(1.3.12)_2$ is called *statically admissible*. If the acceleration is not assumed to vanish, then the problem is *dynamic*, and then

[1] More specifically, these are *holonomic* external constraints. A constraint is nonholonomic if, for example, it is given by an inequality — a displacement component may be required to have less than (or greater than) a specified value. Such a constraint is called unilateral.

additional data are required, namely, the *initial conditions* consisting of the displacement and velocity fields throughout R at the initial time.

Virtual Displacements

A *virtual displacement* field is defined as the difference between two neighboring kinematically admissible displacement fields. In other words, it is a vector field $\delta\mathbf{u}$ which is such that, if $\mathbf{u}$ is a kinematically admissible displacement field, then so is $\mathbf{u} + \delta\mathbf{u}$. It is furthermore assumed that the virtual displacement field is infinitesimal, that is, $|\delta u_{i,j}| \ll 1$.

Corresponding to a virtual displacement field $\delta\mathbf{u}$ we may define the virtual strain field $\delta\boldsymbol{\varepsilon}$ by $\delta\varepsilon_{ij} = \frac{1}{2}(\delta u_{j,i} + \delta u_{i,j})$. Note that the operator δ when applied to a field represents taking the difference between two possible fields and is therefore a linear operator which commutes with partial differentiation. Note further that if any displacement components are prescribed on any part of the boundary then the corresponding virtual displacement components vanish there, that is, $\delta\mathbf{u} = 0$ on ∂R_u.

Virtual Work

Given a set of loads and a virtual displacement field $\delta\mathbf{u}$, we define the *external virtual work* as[1]

$$\overline{\delta W}_{ext} = \int_R \rho b_i \,\delta u_i \, dV + \int_{\partial R_t} t_i^a \,\delta u_i \, dS.$$

The *internal virtual work* is defined as

$$\overline{\delta W}_{int} = \int_R \sigma_{ij} \,\delta\varepsilon_{ij} \, dV.$$

Since $\sigma_{ij} = \sigma_{ji}$, we have $\sigma_{ij}\,\delta\varepsilon_{ij} = \sigma_{ij}\delta u_{i,j}$, and therefore

$$\sigma_{ij}\,\delta\varepsilon_{ij} = (\sigma_{ij}\,\delta u_i)_{,j} - \sigma_{ij,j}\,\delta u_i.$$

Using Gauss's theorem, we obtain

$$\overline{\delta W}_{int} = \int_{\partial R} n_j\sigma_{ij}\,\delta u_i \, dS - \int_R \sigma_{ij,j}\,\delta u_i \, dV.$$

Since, however, $\delta u_i = 0$ on ∂R_u, the surface integral may be restricted to ∂R_t. It follows that

$$\overline{\delta W}_{ext} - \overline{\delta W}_{int} = \int_R (\sigma_{ij,j} + \rho b_i)\,\delta u_i \, dV - \int_{\partial R_t} (n_j\sigma_{ij} - t_i^a)\,\delta u_i \, dS.$$

The right-hand side vanishes for *all* virtual displacement fields $\delta\mathbf{u}$ if and only if the quantities multiplying δu_i in both integrals vanish identically, that

[1] A note on notation: we write $\overline{\delta W}$ instead of the more usual δW in order to indicate that this is not a case of an operator δ applied to a quantity W.

is, if and only if the equilibrium equations (1.3.13) and the traction boundary conditions $(1.3.12)_2$ are satisfied. Thus the body is in equilibrium under the applied loads if and only if the **principle of virtual work**, namely,

$$\overline{\delta W}_{ext} = \overline{\delta W}_{int}, \tag{1.3.14}$$

is obeyed.

The principle of virtual work, also known as the **principle of virtual displacements**, may be interpreted as an application of the **method of weighted residuals**, whose essential idea is as follows. Suppose that a certain stress field $\boldsymbol{\sigma}$ is not exactly statically admissible, and therefore the equations obeyed by it have the form

$$\sigma_{ij},_j + \rho b_i + \rho\,\Delta b_i = 0 \quad \text{in } R, \qquad n_j\sigma_{ij} = t_i^a + \Delta t_i^a \quad \text{on } \partial R_t;$$

we may think of $\Delta\mathbf{b}$ and $\Delta\mathbf{t}^a$ as being the *residuals* of the body force and applied surface traction, respectively. If we cannot make these residuals vanish everywhere (which would make the stress field obey the equations exactly), then we can try to make them vanish in some average sense, namely, by multiplying them with a vector-valued *weighting function*, say $\mathbf{w}$, belonging to a suitable family (say W) of such functions, such that

$$\int_R \rho\,\Delta\mathbf{b}\cdot\mathbf{w}\,dV + \int_{\partial R_t} \Delta t_i^a w_i\,dS = 0$$

for every $\mathbf{w}$ belonging to W. If we identify W with the set of all virtual displacement fields, the principle of virtual work results. An advantage of this point of view is that it permits the application of the principle to dynamic problems as well: since the weighting functions $\mathbf{w}$ are not a priori identified with the virtual displacement fields, the inertia force $-\rho\mathbf{a}$ may be added to $\mathbf{b}$, and $\Delta\mathbf{b}$ may be interpreted as the residual of $\rho(\mathbf{b}-\mathbf{a})$.

Viewed in the light of the method of weighted residuals, the principle of virtual work may be represented by the equation

$$\int_R (\sigma_{ij},_j + \rho b_i) w_i\,dV - \int_{\partial R_t} (n_j\sigma_{ij} - t_i^a) w_i\,dS = 0 \tag{1.3.15}$$

for any $\mathbf{w}$ that obeys the same conditions as a virtual displacement field, that is, $w_i = 0$ on ∂R_u, in addition to any internal constraints. Equation (1.3.15) is also known as the **weak form** of the equilibrium equations with the traction boundary conditions, and forms the foundation for many approximate methods of solution, based on different choices of the family W to which the weighting functions $\mathbf{w}$ belong. If W contains only a finite number of linearly independent functions, then the body is said to be *discretized*. Discretization is discussed below. When $\mathbf{w}$ is interpreted as a *virtual velocity field*, then Equation (1.3.15) is called the **principle of virtual velocities**, and, when, in addition, b_i is replaced by $b_i - a_i$, as the **dynamic principle of virtual velocities**.

A principle of the virtual-work type but different from the one just discussed is the **principle of virtual forces** (also called the **principle of complementary virtual work**). This principle is based on the notion of a *virtual stress field* $\delta\sigma$, defined (by analogy with the definition of a virtual displacement field) as the *difference between two statically admissible stress fields.* A virtual stress field therefore obeys

$$\delta\sigma_{ij,j} = 0 \text{ in } R \textit{ and } n_j\,\delta\sigma_{ij} = 0 \text{ on } \partial R_t.$$

The external and internal complementary virtual work are defined respectively by

$$\overline{\delta W}^c_{ext} = \int_{\partial R_u} u_i^a n_j\,\delta\sigma_{ij}\,dS \quad \textit{and} \quad \overline{\delta W}^c_{int} = \int_R \varepsilon_{ij}\,\delta\sigma_{ij}\,dV.$$

An analysis similar to the one for virtual displacements leads to

$$\overline{\delta W}^c_{int} - \overline{\delta W}^c_{ext} = \int_{\partial R_u} (u_i - u_i^a) n_j \delta\sigma_{ij}\,dS + \int_R [\varepsilon_{ij} - \tfrac{1}{2}(u_{i,j} + u_{j,i})]\delta\sigma_{ij}\,dV.$$

Consequently

$$\overline{\delta W}^c_{int} - \overline{\delta W}^c_{ext} = 0 \tag{1.3.16}$$

for all virtual stress fields $\delta\boldsymbol{\sigma}$ if and only if the strain field $\boldsymbol{\varepsilon}$ is compatible with a kinematically admissible displacement field $\mathbf{u}$.

Discretization

Approximate treatments of continuum mechanics are often based on a procedure known as **discretization**, and virtual work provides a consistent framework for the procedure. A displacement-based discretized model of the body may be formulated as follows. Let $q_1, ..., q_N, q_{N+1}, ..., q_{N+K}$ denote an ordered set of $N + K$ scalars, called *generalized coordinates.* The displacement field $\mathbf{u}(\mathbf{x})$ is *assumed* to be given by

$$\mathbf{u}(\mathbf{x}) = \sum_{n=1}^{N+K} \boldsymbol{\phi}_n(\mathbf{x}) q_n, \tag{1.3.17}$$

where $\boldsymbol{\phi}_n$ $(n = 1, ..., N + K)$ is a given set of vector-valued functions of position. For *given* values of the q_n, $n = N+1, ..., N+K$, the displacement field given by (1.3.17) is kinematically admissible for *all* values of the q_n, $n = 1, ..., N$. The q_n for $n = 1, ..., N$ are called the *free* generalized coordinates, and those for $n = N+1, ..., N+K$, to be denoted q_n^a $(n = 1, ..., K)$, are called the *constrained* generalized coordinates. The former will be assembled in the $1 \times N$ matrix $\underline{q}$, and the latter in the $1 \times K$ matrix $\underline{q}^a$. The integer N

is called the *number of degrees of freedom*, and K the *number of constraints*. In particular, any or all of the q_n^a may be zero.

The strain-displacement relation for infinitesimal deformation may be written in direct vector notation as $\boldsymbol{\varepsilon} = (\nabla \mathbf{u})^S$, and consequently the strain tensor at $\mathbf{x}$ is

$$\boldsymbol{\varepsilon}(\mathbf{x}) = \sum_{n=1}^{N+K} (\nabla \boldsymbol{\phi}_n)^S(\mathbf{x}) q_n. \tag{1.3.18}$$

Before inserting the discretized displacement and strain fields into the principle of virtual work, it is convenient to express them in matrix notation. Matrix notation for vector-valued quantities such as displacement and force is obvious. The column matrices representing stress and strain are, respectively,

$$\underline{\sigma} = \begin{Bmatrix} \sigma_{11} \\ \sigma_{22} \\ \sigma_{33} \\ \sigma_{23} \\ \sigma_{13} \\ \sigma_{12} \end{Bmatrix}, \quad \underline{\varepsilon} = \begin{Bmatrix} \varepsilon_{11} \\ \varepsilon_{22} \\ \varepsilon_{33} \\ 2\varepsilon_{23} \\ 2\varepsilon_{13} \\ 2\varepsilon_{12} \end{Bmatrix}, \tag{1.3.19}$$

so that ε_4, ε_5, and ε_6 represent conventional shear strains, and, most important,

$$\underline{\sigma}^T \underline{\varepsilon} = \sigma_{ij}\varepsilon_{ij}.$$

In certain simpler problems, the dimension of the column matrices $\underline{\sigma}$, $\underline{\varepsilon}$ may be less than six. In problems of plane stress *or* plane strain in the xy-plane, with the component ε_z or σ_z ignored (because it does not participate in any virtual work), the matrices are

$$\underline{\sigma} = \begin{Bmatrix} \sigma_x \\ \sigma_y \\ \tau_{xy} \end{Bmatrix}, \quad \underline{\varepsilon} = \begin{Bmatrix} \varepsilon_x \\ \varepsilon_y \\ \gamma_{xy} \end{Bmatrix}. \tag{1.3.20}$$

In problems with only one component each of normal stress and shear stress (or longitudinal strain and shear strain) they are

$$\underline{\sigma} = \begin{Bmatrix} \sigma \\ \tau \end{Bmatrix}, \quad \underline{\varepsilon} = \begin{Bmatrix} \varepsilon \\ \gamma \end{Bmatrix}. \tag{1.3.21}$$

In every case, the stress and strain matrices are conjugate in the sense that the internal virtual work is given by

$$\overline{\delta W}_{int} = \int_R \underline{\sigma}^T \delta\underline{\varepsilon}\, dV. \tag{1.3.22}$$

Equations (1.3.17) and (1.3.18) may now be rewritten in matrix notation as

$$\underline{u}(\underline{x}) = \underline{\Phi}(\underline{x})\underline{q} + \underline{\Phi}^a(\underline{x})\underline{q}^a, \qquad \underline{\varepsilon}(\underline{x}) = \underline{B}(\underline{x})\underline{q} + \underline{B}^a(\underline{x})\underline{q}^a, \tag{1.3.23}$$

where $\underline{x}$ and $\underline{u}$ are the column-matrix representations of position and displacement, respectively, that is, for three-dimensional continua,

$$\underline{x} = \begin{Bmatrix} x_1 \\ x_2 \\ x_3 \end{Bmatrix}, \quad \underline{u} = \begin{Bmatrix} u_1 \\ u_2 \\ u_3 \end{Bmatrix}.$$

Since $\underline{q}^a$ is prescribed, virtual displacement and strain fields are defined only in terms of the variations of $\underline{q}$:

$$\delta\underline{u}(\underline{x}) = \underline{\Phi}(\underline{x})\,\delta\underline{q}, \qquad \delta\underline{\varepsilon}(\underline{x}) = \underline{B}(\underline{x})\,\delta\underline{q}. \tag{1.3.24}$$

We may now apply the principle of virtual work to these fields. Since

$$\underline{\sigma}^T\delta\underline{\varepsilon} = \underline{\sigma}^T\underline{\Phi}\,\delta\underline{q} = (\underline{\Phi}^T\underline{\sigma})^T\delta\underline{q},$$

we obtain, using (1.3.24), the internal virtual work in discrete form:

$$\overline{\delta W}_{int} = \underline{Q}^T\delta\underline{q},$$

where

$$\underline{Q} = \int_R \underline{B}^T\underline{\sigma}\,dV$$

is the internal generalized force matrix conjugate to $\underline{q}$. Similarly, with the body force per unit volume $\mathbf{f} = \rho\mathbf{b}$ and the prescribed surface traction $\mathbf{t}^a$ represented by matrices $\underline{f}$ and $\underline{t}^a$, the external virtual work is

$$\overline{\delta W}_{ext} = \underline{F}^T\delta\underline{q},$$

where

$$\underline{F} = \int_R \underline{\Phi}^T\underline{f}\,dV + \int_{\partial R_t} \underline{\Phi}^T\underline{t}^a\,dS$$

is the (known) external generalized force matrix conjugate to $\underline{q}$. If the *residual force* matrix is defined by $\underline{R} = \underline{Q} - \underline{F}$, then the equilibrium condition may be written as

$$\underline{R} = 0, \tag{1.3.25}$$

a matrix equation representing N independent scalar equations.

Exercises: Section 1.3

1. If the acceleration field $\mathbf{a}$ is the material derivative $\dot{\mathbf{v}}$ of the velocity field $\mathbf{v}$, find the acceleration at the point (1, 1, 0) at the time $t = 0$ if the velocity field in the Eulerian description is

$$\mathbf{v} = C[(x_1^3 + x_1x_2^2)\mathbf{e}_1 - (x_1^2x_2 + x_2^3)\mathbf{e}_2]e^{-at},$$

where C and a are constants.

2. With respect to a basis $(\mathbf{e}_i)$, a stress tensor is represented by the matrix

$$\underline{\sigma} = \begin{bmatrix} 0.1 & 0.6 & 0.0 \\ 0.6 & 1.2 & 0.0 \\ 0.0 & 0.0 & 0.3 \end{bmatrix} \quad \text{MPa}$$

   (a) Find the traction vector on an element of the plane $2x_1 - 2x_2 + x_3 = 1$.

   (b) Find the magnitude of the traction vector in (a), and the normal stress and the shear stress on the plane given there.

   (c) Find the matrix representing the stress tensor with respect to a basis $\mathbf{e}_i^*$, where $\mathbf{e}_i^* = \beta_{ij}\mathbf{e}_j$ with $\underline{\beta}$ given in Exercise 4 of Section 1.1.

3. For the stress tensor in Exercise 2, find

   (a) the principal invariants I_1, I_2, I_3,

   (b) the principal stresses and principal directions,

   (c) the octahedral shear stress,

   (d) the component matrix of the stress deviator $\mathbf{s}$ with respect to the basis $(\mathbf{e}_i)$,

   (e) the principal deviatoric invariants J_2 and J_3.

4. Draw the three Mohr's circles for each of the following states of stress (units are MPa; stress components not given are zero).

   (a) Uniaxial tension, $\sigma_{11} = 150$

   (b) Uniaxial compression, $\sigma_{22} = -100$

   (c) Biaxial stress, $\sigma_{11} = 50$, $\sigma_{22} = 100$

   (d) Biaxial stress, $\sigma_{11} = 50$, $\sigma_{22} = -50$

   (e) Triaxial stress, $\sigma_{11} = 80$, $\sigma_{22} = \sigma_{33} = -40$

   (f) $\sigma_{11} = 50$, $\sigma_{22} = -10$, $\sigma_{12} = \sigma_{21} = 40$, $\sigma_{33} = 30$

5. Find the maximum shear stress τ_{max} for each of the stress states of the preceding exercise.

6. Derive Equation (1.3.11) for τ_{max} as follows: fix a Cartesian basis $(\mathbf{e}_i)$ coinciding with the principal stress directions, and form $[\tau(\mathbf{n})]^2$ for all possible directions $\mathbf{n}$ by identifying $\mathbf{n}$ with the spherical unit vector $\mathbf{e}_r$, so that $[\tau(\mathbf{n})]^2$ is a function of the spherical surface coordinates ϕ and θ. Lastly, show that this function can become stationary only in the planes formed by the principal stress directions.

7. For the plane stress field given in engineering notation by

$$\sigma_x = Axy, \quad \tau = \frac{A}{2}(h^2 - y^2), \quad \sigma_y = 0,$$

where A and h are constants,

(a) show that it is in equilibrium under a zero body force,

(b) find an Airy stress function $\Phi(x, y)$ corresponding to it.

8. Show that the equilibrium equations in the absence of body force (i.e. Equation (1.3.13) with $b_i = 0$) are satisfied if $\sigma_{ij} = e_{ikm}e_{jln}\phi_{kl,mn}$, where ϕ is a symmetric second-rank tensor field.

9. If a stress field is given by the matrix

$$A \begin{bmatrix} x_1^2 x_2 & (a^2 - x_2^2)x_1 & 0 \\ (a^2 - x_2^2)x_1 & \frac{1}{3}(x_2^3 - 3a^2 x_2) & 0 \\ 0 & 0 & 2ax_3^2 \end{bmatrix},$$

where A and a are constants, find the body-force field necessary for the stress field to be in equilibrium.

10. The following displacement field is assumed in a prismatic bar with the x_1 axis being the longitudinal axis, and the cross-section defined by a closed curve C enclosing an area A in the x_2x_3 plane, such that the x_1-axis intersects the cross-sections at their centroids, so that $\int_A x_2\, dA = \int_A x_3\, dA = 0$.

$$u_1 = u(x_1) - v'(x_1)x_2, \quad u_2 = v(x_1), \quad u_3 = 0.$$

Show that the internal virtual work is

$$\overline{\delta W}_{int} = \int_0^L (P\,\delta u' + M\delta v'')\, dx_1,$$

where $P = \int_A \sigma_{11}\, dA$ and $M = -\int_A x_2\sigma_{11}\, dA$.

11. If the bar described in Exercise 10 is subject to a body force and a surface traction distributed along its cylindrical surface, show that the external virtual work can be written as

$$\overline{\delta W}_{ext} = \int_0^L (p\,\delta u + q\,\delta v + m\,\delta v')\, dx_1.$$

Find expressions for p, q, and m.

Section 1.4 Constitutive Relations: Elastic

1.4.1. Energy and Thermoelasticity

Energy

As discussed in Section 1.3, the solution of a boundary-value problem in solid mechanics requires finding the displacement field $\mathbf{u}$ and the stress field $\boldsymbol{\sigma}$, that is, in the most general case, nine component fields. Thus far, the only field equations we have available are the three equations of motion (or of equilibrium, in static and quasi-static problems). In certain particularly simple problems, the number of *unknown* stress components equals that of nontrivial equilibrium equations, and these may be solved to give the stress field, though not the displacement field; such problems are called *statically determinate*, and are discussed further in Section 4.1. In general, however, *constitutive relations* that relate stress to displacement (more particularly, to the strain that is derived from the displacement) are needed. Such relations are characteristic of the material or materials of which the body is made, and are therefore also called simply *material properties*. In this section a simple class of constitutive relations, in which the current value of the strain at a point depends only on the stress at that point, is studied; a body described by such relations is called *elastic*. The influence of the thermal state (as given, for example, by the temperature) on the stress-strain relations cannot, however, be ignored. In order to maintain generality, instead of introducing elasticity directly a more basic concept is first presented: that of *energy*.

The concept of energy is fundamental in all physical science. It makes it possible to relate different physical phenomena to one another, as well as to evaluate their relative significance in a given process. Here we focus only on those forms of energy that are relevant to solid mechanics.

The *kinetic energy* of a body occupying a region R is

$$K = \frac{1}{2}\int_R \rho\mathbf{v}\cdot\mathbf{v}\,dV$$

($\mathbf{v} = \dot{\mathbf{u}}$ = velocity field), so that

$$\dot{K} = \frac{d}{dt}K = \int_R \rho\mathbf{v}\cdot\mathbf{a}\,dV$$

($\mathbf{a} = \dot{\mathbf{v}}$ = acceleration field), because, even though the mass density ρ and the volume element dV may vary in time, their product, which is a mass element, does not (**conservation of mass**).

The *external power* acting on the body is

$$P = \int_R \rho\mathbf{b}\cdot\mathbf{v}\,dV + \int_{\partial R}\mathbf{t}(\mathbf{n})\cdot\mathbf{v}\,dS = \int_R \rho b_i v_i\,dV + \int_{\partial R} t_i(\mathbf{n})v_i\,dS.$$

But $t_i(\mathbf{n}) = \sigma_{ij}n_j$, and the divergence theorem leads to

$$P = \int_R [(\sigma_{ij},_j + \rho b_i)v_i + \sigma_{ij}v_{i},_j]\, dV.$$

In infinitesimal deformations, the difference between Eulerian and Lagrangian coordinates can be ignored, so that $\frac{1}{2}(v_{i},_j + v_{j},_i) \doteq \dot{\varepsilon}_{ij}$, and hence $\sigma_{ij}v_{i},_j \doteq \sigma_{ij}\dot{\varepsilon}_{ij}$. In addition, $\sigma_{ij},_j + \rho b_i = \rho a_i$, so that

$$P - \dot{K} = \int_R \sigma_{ij}\dot{\varepsilon}_{ij}\, dV \stackrel{\text{def}}{=} P_d \quad \textit{(deformation power)}.$$

If $\dot{\boldsymbol{\varepsilon}} \equiv 0$ (rigid-body motion), then $P = \dot{K}$. More generally, if P_d can be neglected in comparison with P and $\dot{K}$, then the body may approximately be treated as rigid. If, on the other hand, it is $\dot{K}$ that is negligible, then the problem is approximately quasi-static. A problem of *free vibrations* may arise if the external power P is neglected.

The *heat flow* into the body is

$$Q = \int_R \rho r\, dV - \int_{\partial R} h(\mathbf{n})\, dS$$

($r =$ body heating or radiation, $h(\mathbf{n}) =$ heat outflow per unit time per unit surface area with orientation $\mathbf{n}$).

The **first law of thermodynamics** or *principle of energy balance* asserts that there exists a state variable u (internal-energy density) such that

$$\frac{d}{dt}\int_R \rho u\, dV = Q + P_d.$$

If we apply this law to the Cauchy tetrahedron, we obtain $h(\mathbf{n}) = \mathbf{h}\cdot\mathbf{n}$ ($\mathbf{h} =$ heat flux vector), so that

$$Q = \int_R \rho r\, dV - \int_{\partial R} \mathbf{h}\cdot\mathbf{n}\, dS = \int_R (\rho r - \text{div}\, \mathbf{h})\, dV$$

by Gauss's theorem, and we obtain the *local energy-balance equation*,

$$\rho\dot{u} = \sigma_{ij}\dot{\varepsilon}_{ij} + \rho r - \text{div}\, \mathbf{h}. \tag{1.4.1}$$

Thermoelasticity

The local deformation, as defined by the strain tensor $\boldsymbol{\varepsilon}$, may be assumed to depend on the stress tensor $\boldsymbol{\sigma}$, on the internal-energy density u, and on additional variables $\xi_1, \ldots, \xi_N$ (*internal variables*). By definition, a body is called *thermoelastic* if $\boldsymbol{\varepsilon}$ everywhere depends only on $\boldsymbol{\sigma}$ and u. The dependence is not arbitrary; it must be consistent with the **second law of thermodynamics**.

We assume for the time being that at a fixed value of u, the relation between strain and stress is invertible, so that $\boldsymbol{\sigma}$ may be regarded as a function of $\boldsymbol{\varepsilon}$ and u. Possible restrictions on this assumption, resulting from internal constraints, are discussed later.

The second law of thermodynamics *for a thermoelastic body* can be stated as follows: there exists a state function $\eta = \bar{\eta}(u, \boldsymbol{\varepsilon})$ (*entropy density*) such that $\dot{\eta} = 0$ in an adiabatic process, that is, in a process in which $\rho r - \text{div}\,\mathbf{h} = 0$.

As a consequence, the two equations

$$\frac{\partial \bar{\eta}}{\partial u}\dot{u} + \frac{\partial \bar{\eta}}{\partial \varepsilon_{ij}}\dot{\varepsilon}_{ij} = 0 \quad \textit{and} \quad \rho\dot{u} - \sigma_{ij}\dot{\varepsilon}_{ij} = 0$$

must be satisfied simultaneously. If we now define the *absolute temperature* T by

$$T^{-1} = \frac{\partial \bar{\eta}}{\partial u}, \tag{1.4.2}$$

then, upon eliminating $\dot{u}$ between the two equations,

$$\left(\sigma_{ij} + T\rho\frac{\partial \bar{\eta}}{\partial \varepsilon_{ij}}\right)\dot{\varepsilon}_{ij} = 0.$$

If, moreover, the $\dot{\varepsilon}_{ij}$ are independent (i.e. if there are no internal constraints as discussed in Section 1.3), then their coefficients must vanish, since any five of the six independent components $\dot{\varepsilon}_{ij}$ may be arbitrarily taken as zero. The vanishing of the coefficients yields

$$\sigma_{ij} = -T\rho\frac{\partial \bar{\eta}}{\partial \varepsilon_{ij}},$$

an equation that gives an explicit form to the assumed dependence of $\boldsymbol{\sigma}$ on u and $\boldsymbol{\varepsilon}$. The fact that this relation was derived on the basis of an adiabatic process is irrelevant, since, by hypothesis, the relation is only among the current values of $\boldsymbol{\sigma}$, $\boldsymbol{\varepsilon}$, and u and is independent of process.

It follows further that

$$T\rho\dot{\eta} = \rho\dot{u} - \sigma_{ij}\dot{\varepsilon}_{ij} = \rho r - \text{div}\,\mathbf{h}. \tag{1.4.3}$$

If the total entropy of the body is defined by $S \stackrel{\text{def}}{=} \int_R \rho\eta\, dV$, then

$$\begin{aligned}
\dot{S} &= \int_R \rho\dot{\eta}\, dV = \int_R T^{-1}(\rho r - \text{div}\,\mathbf{h})\, dV \\
&= \int_R \left[\rho\frac{r}{T} - \text{div}\left(\frac{\mathbf{h}}{T}\right) + \mathbf{h}\cdot\nabla T^{-1}\right] dV \\
&= \int_R \rho\frac{r}{T} dV - \int_{\partial R} \frac{h(\mathbf{n})}{T} dS + \int_R \mathbf{h}\cdot\nabla T^{-1}\, dV.
\end{aligned}$$

We can now bring in the experimental fact that heat flows from the hotter to the colder part of a body. The mathematical expression of this fact is $\mathbf{h} \cdot \nabla T^{-1} \geq 0$, so that

$$\dot{S} - \left(\int_R \rho \frac{r}{T} dV - \int_{\partial R} \frac{h(\mathbf{n})}{T} dS \right) \stackrel{\text{def}}{=} \Gamma \geq 0. \tag{1.4.4}$$

Inequality (1.4.4), known as the **global Clausius–Duhem inequality**, is usually taken to be the general form of the second law of thermodynamics for continua, whether thermoelastic or not, though its physical foundation in the general case is not so firm as in the thermoelastic case (see, e.g., Woods [1981]). The quantity in parentheses is called the external entropy supply, and Γ is the internal entropy production. In the thermoelastic case, clearly,

$$\Gamma = \int_R \mathbf{h} \cdot \nabla T^{-1} \, dV.$$

More generally, Γ may be assumed to be given by $\int_R \rho\gamma \, dV$, where γ is the internal entropy production per unit mass, and $\rho\gamma$ contains, besides the term $\mathbf{h} \cdot \nabla T^{-1}$, additional terms representing energy dissipation. These are discussed in Section 1.5. With the help of the divergence theorem, (1.4.4) may be transformed into the equation

$$\int_R \left[\rho\dot{\eta} - \rho\frac{r}{T} + \nabla \cdot \left(\frac{\mathbf{h}}{T}\right) - \rho\gamma \right] dV = 0.$$

The *local* Clausius–Duhem inequality is obtained when it is assumed, as usual, that the equation must apply to any region of integration R, however small, and, in addition, that γ is nonnegative:

$$\rho\dot{\eta} - \rho\frac{r}{T} + \nabla \cdot \left(\frac{\mathbf{h}}{T}\right) = \rho\gamma \geq 0. \tag{1.4.5}$$

The assumptions underlying the derivation of (1.4.5) have been severely criticized by Woods [1981]. The criticisms do not, however, apply to thermoelastic bodies, to which we now return.

Given the definition (1.4.2) of the absolute temperature T and the fact that T is always positive, it follows from the implicit function theorem of advanced calculus that we can solve for u as a function of η and ε [i.e. $u = \tilde{u}(\eta, \varepsilon)$], and we can use η as a state variable. Since, by Equation (1.4.3), $\dot{u} = T\dot{\eta} + \rho^{-1}\sigma_{ij}\dot{\varepsilon}_{ij}$, we have $T = \partial\tilde{u}/\partial\eta$ and

$$\left(\sigma_{ij} - \rho\frac{\partial\tilde{u}}{\partial\varepsilon_{ij}} \right) \dot{\varepsilon}_{ij} = 0. \tag{1.4.6}$$

Again, in the absence of internal constraints, the $\dot{\varepsilon}_{ij}$ can be specified independently. The coefficients in Equation (1.4.6) must therefore all vanish, yielding the relation

$$\sigma_{ij} = \rho\frac{\partial\tilde{u}}{\partial\varepsilon_{ij}},$$

in which the specific entropy η is the other variable. The equation thus gives the stress-strain relation at constant specific entropy and is accordingly known as the *isentropic* stress-strain relation. It is also called the *adiabatic* stress-strain relation, since in a thermoelastic body isentropic processes are also adiabatic. We must remember, however, that the relation as such is independent of process.

We now drop the assumption that the body is free of internal constraints. In the presence of an internal constraint of the form $c_{ij}\dot{\varepsilon}_{ij} = 0$, a term pc_{ij}, with p an arbitrary scalar, may be added to the the stress without invalidating Equation (1.4.6). This equation is therefore satisfied whenever

$$\sigma_{ij} = \rho\frac{\partial\tilde{u}}{\partial\varepsilon_{ij}} - pc_{ij}.$$

The stress, as can be seen, is not completely determined by the strain and the entropy density.

As noted in 1.3.5, a commonly encountered internal constraint is incompressibility. In an incompressible body the volume does not change, so that $\delta_{ij}\dot{\varepsilon}_{ij} = 0$ and thus c_{ij} may be identified with δ_{ij}. The stress is, accordingly, determinate only to within a term given by $p\delta_{ij}$, where p is, in this case, a pressure.

The use of the entropy density η as an independent state variable is not convenient. A far more convenient thermal variable is, of course, the temperature, since it is fairly easy to measure and to control. If the *Helmholtz free energy* per unit mass is defined as

$$\psi \stackrel{\text{def}}{=} u - T\eta = \psi(T, \varepsilon),$$

then $\dot{\psi} = -\eta\dot{T} + \rho^{-1}\sigma_{ij}\dot{\varepsilon}_{ij}$, so that $\eta = -\partial\psi/\partial T$ and, in the absence of internal constraints,

$$\sigma_{ij} = \rho\frac{\partial\psi}{\partial\varepsilon_{ij}}. \tag{1.4.7}$$

Equation (1.4.7) embodies the *isothermal* stress-strain relation.

In addition to the stress-strain relations, the properties of a thermoelastic continuum include the *thermal stress coefficients* and the *specific heat*. The former are the increases in the stress components per unit decrease in temperature with no change in the strain, that is,

$$\beta_{ij} = -\left.\frac{\partial\sigma_{ij}}{\partial T}\right|_{\varepsilon=\text{const}} = -\rho\frac{\partial^2\psi}{\partial T\partial\varepsilon_{ij}} = \rho\frac{\partial\eta}{\partial\varepsilon_{ij}}.$$

Clearly, $\boldsymbol{\beta}$ is a second-rank tensor. The specific heat (per unit mass) at constant strain is

$$C = \left.\frac{\partial u}{\partial T}\right|_{\varepsilon=\text{const}} = \frac{\partial}{\partial T}(\psi + \eta T) = \frac{\partial\psi}{\partial T} + \eta + T\frac{\partial\eta}{\partial T} = T\frac{\partial\eta}{\partial T} = -T\frac{\partial^2\psi}{\partial T^2}.$$

Linearization

For sufficiently small deviations in strain and temperature from a given reference state, the stress-strain relation, if it is smooth, can be approximated by a linear one. Let us consider a reference state "0" at zero strain and temperature T_0, and let us expand $\psi(T, \varepsilon)$ in a Taylor series about this state:

$$\psi(T, \varepsilon) = \psi_0 + \left.\frac{\partial \psi}{\partial T}\right|_0 (T - T_0) + \left.\frac{\partial \psi}{\partial \varepsilon_{ij}}\right|_0 \varepsilon_{ij} + \left.\frac{1}{2}\frac{\partial^2 \psi}{\partial T^2}\right|_0 (T - T_0)^2$$
$$+ \left.\frac{\partial^2 \psi}{\partial T \partial \varepsilon_{ij}}\right|_0 (T - T_0)\varepsilon_{ij} + \left.\frac{1}{2}\frac{\partial^2 \psi}{\partial \varepsilon_{ij} \partial \varepsilon_{kl}}\right|_0 \varepsilon_{ij}\varepsilon_{kl} + \ldots$$

Now, $\partial\psi/\partial T|_0 = -\eta_0$ ($\eta_0 =$ specific entropy in the reference state) and $\left.\frac{\partial \psi}{\partial \varepsilon_{ij}}\right|_0 = \rho^{-1}\sigma_{ij}^0$ ($\boldsymbol{\sigma}^0 =$ initial stress). Furthermore,

$$\left.\frac{\partial^2 \psi}{\partial T^2}\right|_0 = -\frac{1}{T_0}C_0, \quad \left.\frac{\partial^2 \psi}{\partial T \partial \varepsilon_{ij}}\right|_0 = -\rho^{-1}\beta_{ij}^0,$$

and

$$\rho \left.\frac{\partial^2 \psi}{\partial \varepsilon_{ij} \partial \varepsilon_{kl}}\right|_0 = \left.\frac{\partial \sigma_{ij}}{\partial \varepsilon_{kl}}\right|_0 = C_{ijkl}^0 = C_{klij}^0,$$

the isothermal elastic modulus tensor (of rank 4) at the temperature T_0. Thus

$$\eta = \eta_0 + \frac{1}{T_0}C_0(T - T_0) + \rho^{-1}\beta_{ij}^0\varepsilon_{ij}$$

and

$$\sigma_{ij} = \sigma_{ij}^0 - \beta_{ij}^0(T - T_0) + C_{ijkl}^0\varepsilon_{kl}$$

are the constitutive relations of linear thermoelasticity.

In an isentropic process, $\eta \equiv \eta_0$, so that $T - T_0 = -(T_0/\rho C_0)\beta_{ij}^0\varepsilon_{ij}$. Consequently,

$$\sigma_{ij} = \sigma_{ij}^0 + \left(C_{ijkl}^0 + \frac{T_0}{\rho C_0}\beta_{ij}^0\beta_{kl}^0\right)\varepsilon_{kl},$$

which defines the isentropic or adiabatic elastic modulus tensor.

1.4.2. Linear Elasticity

Elasticity

The dependence of the stress-strain relation on the thermal state is often ignored, and the simple relation $\boldsymbol{\sigma} = \boldsymbol{\sigma}(\boldsymbol{\varepsilon})$ is assumed. It is then that a body is called simply, in the traditional sense, *elastic*. The internal-energy density or the free-energy density, as the case may be, may be replaced by the *strain-energy function* $W(\boldsymbol{\varepsilon})$ (per unit volume), so that the stress-strain relation, again in the absence of internal constraints, is

$$\sigma_{ij} = \frac{\partial W}{\partial \varepsilon_{ij}}.$$

This relation is exact in an isentropic process, with η at a fixed value, if $W(\boldsymbol{\varepsilon})$ is identified with $\tilde{u}(\eta, \boldsymbol{\varepsilon})$, and in an isothermal process, with T at a fixed value, if $W(\boldsymbol{\varepsilon})$ is identified with $\psi(T, \boldsymbol{\varepsilon})$.

The introduction of the strain-energy function into elasticity is due to Green, and elastic solids for which such a function is assumed to exist are called *Green-elastic* or *hyperelastic*. Elasticity without an underlying strain-energy function is called *Cauchy elasticity*. Throughout this book, "elasticity" means hyperelasticity as a matter of course.

Generalized Hooke's Law

Linearization for an elastic continuum will be carried out with respect to a reference configuration which is stress-free at the reference temperature T_0, so that $\boldsymbol{\sigma}^0 = 0$. We may now let C denote either the isothermal or the isentropic modulus tensor, as appropriate, and under isothermal or isentropic conditions we obtain the **generalized Hooke's law**

$$\sigma_{ij} = C_{ijkl}\varepsilon_{kl}.$$

The C_{ijkl}, called the *elastic constants* (recall that they depend on the temperature), are components of a tensor of rank 4, likewise symmetric with respect to the index pairs ij and kl; this symmetry reduces the number of independent components from 81 to 36. But there is an additional symmetry: since

$$C_{ijkl} = \left.\frac{\partial^2 W}{\partial \varepsilon_{ij}\partial \varepsilon_{kl}}\right|_{\boldsymbol{\varepsilon}=0} = \left.\frac{\partial^2 W}{\partial \varepsilon_{kl}\partial \varepsilon_{ij}}\right|_{\boldsymbol{\varepsilon}=0}, \tag{1.4.8}$$

we also have $C_{ijkl} = C_{klij}$, and thus the number of independent components is further reduced to 21. This number may be reduced even more by material symmetries. Of these, only isotropy is considered here.

If stress and strain are represented in matrix notation as given by Equation (1.3.19), then the stress-strain relation may be written as

$$\underline{\sigma} = \underline{C}\,\underline{\varepsilon}, \tag{1.4.9}$$

where the symmetric square matrix $\underline{C} = [C_{IJ}]$ (we use capital letters as indices in the six-dimensional space of stress and strain components) is defined as follows:

$$C_{11} = C_{1111},\ C_{12} = C_{1212},\ C_{14} = C_{1123},\ C_{44} = C_{2323},\ \text{etc.}$$

Assuming the matrix $\underline{C}$ to be invertible, we also have the strain-stress relations

$$\underline{\varepsilon} = \underline{C}^{-1}\underline{\sigma}.$$

Reverting to tensor component notation, we may write these relations as

$$\varepsilon_{ij} = C^{-1}_{ijkl}\sigma_{kl},$$

where the *compliance tensor* C^{-1} is given as follows:

$$C^{-1}_{1111} = C^{-1}_{11},\ C^{-1}_{1122} = C^{-1}_{12},\ C^{-1}_{1123} = \tfrac{1}{2}C^{-1}_{14},\ C^{-1}_{1212} = \tfrac{1}{4}C^{-1}_{66},\ \text{etc.}$$

The *complementary-energy function* is defined for a general elastic material as

$$W^c = \sigma_{ij}\varepsilon_{ij} - W.$$

Note that if we assume $W^c = W^c(\varepsilon, \boldsymbol{\sigma})$, then

$$\frac{\partial W^c}{\partial \varepsilon_{ij}} = \sigma_{ij} - \frac{\partial W}{\partial \varepsilon_{ij}} = 0,$$

so that W^c is in fact a function of $\boldsymbol{\sigma}$ only, and

$$\varepsilon_{ij} = \frac{\partial W^c}{\partial \sigma_{ij}}.$$

For the linear material we have

$$W^c = \frac{1}{2}C^{-1}_{ijkl}\sigma_{ij}\sigma_{kl} = \frac{1}{2}\underline{\sigma}^T\underline{C}^{-1}\underline{\sigma},$$

and therefore

$$\varepsilon_I = \frac{\partial W^c}{\partial \sigma_I}.$$

Isotropic Linear Elasticity

The most general isotropic tensor of rank 4 has the representation

$$\lambda\delta_{ij}\delta_{kl} + \mu\delta_{ik}\delta_{jl} + \nu\delta_{il}\delta_{jk}.$$

If C_{ijkl} has this form, then, in order to satisfy the symmetry condition $C_{ijkl} = C_{jikl}$ (symmetry of the stress tensor) we must have $\mu = \nu$. The symmetry condition $C_{ijkl} = C_{klij}$ (existence of strain-energy function) is then automatically satisfied. Thus

$$C_{ijkl} = \lambda\delta_{ij}\delta_{kl} + \mu(\delta_{ik}\delta_{jl} + \delta_{il}\delta_{jk}), \tag{1.4.10}$$

so that the isotropic linear elastic stress-strain relation is

$$\sigma_{ij} = \lambda\delta_{ij}\varepsilon_{kk} + 2\mu\varepsilon_{ij}.$$

λ and μ are known as the *Lamé coefficients*. In particular, $\mu = G$, the shear modulus. The matrix $\underline{C}$ takes the form

$$\underline{C} = \begin{bmatrix} \lambda+2\mu & \lambda & \lambda & 0 & 0 & 0 \\ \lambda & \lambda+2\mu & \lambda & 0 & 0 & 0 \\ \lambda & \lambda & \lambda+2\mu & 0 & 0 & 0 \\ 0 & 0 & 0 & \mu & 0 & 0 \\ 0 & 0 & 0 & 0 & \mu & 0 \\ 0 & 0 & 0 & 0 & 0 & \mu \end{bmatrix}.$$

Inverting this matrix, we obtain

$$\underline{C}^{-1} = \frac{1}{\mu(3\lambda+2\mu)} \begin{bmatrix} \lambda+\mu & -\lambda/2 & -\lambda/2 & 0 & 0 & 0 \\ -\lambda/2 & \lambda+\mu & -\lambda/2 & 0 & 0 & 0 \\ -\lambda/2 & -\lambda/2 & \lambda+\mu & 0 & 0 & 0 \\ 0 & 0 & 0 & 3\lambda+2\mu & 0 & 0 \\ 0 & 0 & 0 & 0 & 3\lambda+2\mu & 0 \\ 0 & 0 & 0 & 0 & 0 & 3\lambda+2\mu \end{bmatrix}.$$

If we define the *Young's modulus* as $E = \mu(3\lambda + 2\mu)/(\lambda + \mu)$ and the *Poisson's ratio* as $\nu = \frac{1}{2}\lambda/(\lambda + \mu)$, then

$$\underline{C}^{-1} = \frac{1}{E} \begin{bmatrix} 1 & -\nu & -\nu & 0 & 0 & 0 \\ -\nu & 1 & -\nu & 0 & 0 & 0 \\ -\nu & -\nu & 1 & 0 & 0 & 0 \\ 0 & 0 & 0 & 2(1+\nu) & 0 & 0 \\ 0 & 0 & 0 & 0 & 2(1+\nu) & 0 \\ 0 & 0 & 0 & 0 & 0 & 2(1+\nu) \end{bmatrix}.$$

Note that the determinant of $\underline{C}^{-1}$ is $8(1 + \nu)^5(1 - 2\nu)E^{-6}$, so that when $\nu = \frac{1}{2}$ (incompressible material), the compliance matrix is singular. In that case there exists no matrix $\underline{C}$, that is, stress cannot be given as a function of strain, as we already know.

From the nonzero elements of $\underline{C}^{-1}$ we can obtain the corresponding components of the compliance tensor C^{-1}:

$$C_{11}^{-1} = \frac{1}{E} = C_{1111}^{-1}, \text{ etc.,}$$

$$C_{12}^{-1} = -\frac{\nu}{E} = C_{1122}^{-1}, \text{ etc.,}$$

$$C_{44}^{-1} = \frac{2(1+\nu)}{E} = \frac{1}{\mu} = \frac{1}{G} = 4C_{1212}^{-1}, \text{ etc.}$$

Consequently the isotropic linearly elastic strain-stress relation in indicial notation is

$$\varepsilon_{ij} = \frac{1}{E}[(1+\nu)\sigma_{ij} - \nu\sigma_{kk}\delta_{ij}]. \tag{1.4.11}$$

If the only nonzero stress component is $\sigma_{11} = \sigma$, and if we denote ε_{11} by ε, then Equation (1.4.11) gives the uniaxial stress-strain relation

$$\sigma = E\varepsilon. \tag{1.4.12}$$

When $\sigma_{33} = 0$, then Equation (1.4.11), with the indices limited to the values 1 and 2, reads

$$\varepsilon_{\alpha\beta} = \frac{1}{E}[(1+\nu)\sigma_{\alpha\beta} - \nu\sigma_{\gamma\gamma}\delta_{\alpha\beta}],$$

which may be inverted to yield

$$\sigma_{\alpha\beta} = \frac{E}{1-\nu^2}[(1-\nu)\varepsilon_{\alpha\beta} + \nu\varepsilon_{\gamma\gamma}\delta_{\alpha\beta}]. \tag{1.4.13}$$

In plane problems, with $\underline{\sigma}$ and $\underline{\varepsilon}$ defined by Equations (1.3.20), $\underline{C}$ is given by

$$\underline{C} = \frac{E}{1-\nu^2}\begin{bmatrix} 1 & \nu & 0 \\ \nu & 1 & 0 \\ 0 & 0 & \frac{1}{2}(1-\nu) \end{bmatrix}$$

for plane stress and by

$$\underline{C} = \frac{E}{(1+\nu)(1-2\nu)}\begin{bmatrix} 1-\nu & \nu & 0 \\ \nu & 1-\nu & 0 \\ 0 & 0 & \frac{1}{2}(1-2\nu) \end{bmatrix}$$

for plane strain. In a stress state composed of uniaxial stress and simple shear, with stress and strain matrices given by (1.3.21), we have

$$\underline{C} = \begin{bmatrix} E & 0 \\ 0 & G \end{bmatrix}.$$

If E, ν are interpreted as representing the isothermal elastic stiffness, then for small changes in temperature from the reference temperature T_0 we may add to the elastic strain given by Equation (1.4.11) the *thermal strain* $\alpha(T - T_0)\delta_{ij}$, where α is the familiar coefficient of thermal expansion, obtaining

$$\varepsilon_{ij} = \frac{1}{E}[(1+\nu)\sigma_{ij} - \nu\sigma_{kk}\delta_{ij}] + \alpha(T - T_0)\delta_{ij}. \tag{1.4.14}$$

The thermal stress coefficients β_{ij} are accordingly given by $3K\alpha\delta_{ij}$, where

$$K = \lambda + \frac{2}{3}\mu = \frac{1}{3}\frac{E}{1-2\nu}$$

is the *bulk modulus*, which appears in the *volumetric* constitutive relation

$$\sigma_{kk} = 3K[\varepsilon_{kk} - 3\alpha(T - T_0)].$$

We may also derive the *deviatoric* or *distortional* constitutive relation

$$s_{ij} = 2\mu e_{ij},$$

which includes, in particular, **Hooke's law in shear**:

$$\tau = G\gamma. \tag{1.4.15}$$

It is characteristic of an isotropic material that the volumetric and deviatoric stress-strain relations are *uncoupled*. The uncoupling can also be seen from the strain-energy function,

$$W = \frac{1}{2}\lambda(\varepsilon_{kk})^2 + \mu\varepsilon_{ij}\varepsilon_{ij} = \frac{1}{2}K(\varepsilon_{kk})^2 + \mu e_{ij}e_{ij}, \tag{1.4.16}$$

and of the complementary-energy function,

$$W^c = \frac{1}{18K}(\sigma_{kk})^2 + \frac{1}{4G}s_{ij}s_{ij}. \tag{1.4.17}$$

1.4.3. Energy Principles

Internal Potential Energy, Variations

Let $\Pi_{int} = \int_R W\, dV$ be the *total strain energy* or *internal potential energy* of the body. Given a virtual displacement field $\delta\mathbf{u}$ and the corresponding virtual strain field $\delta\varepsilon$, the *first variation* of Π_{int}, denoted $\delta\Pi_{int}$, is defined as follows. We let Π_{int} denote, more specifically, the internal potential energy evaluated at the displacement field $\mathbf{u}$, while $\Pi_{int}+\Delta\Pi_{int}$ denotes the internal potential energy evaluated at the varied displacement field $\mathbf{u}+\delta\mathbf{u}$. Assuming the dependence of Π_{int} on $\mathbf{u}$ to be smooth, we can write

$$\Delta\Pi_{int} = \delta\Pi_{int} + \frac{1}{2}\delta^2\Pi_{int} + \ldots,$$

where $\delta\Pi_{int}$ is linear in $\delta\mathbf{u}$ (and/or in its derivatives, and therefore also in $\delta\varepsilon$), $\delta^2\Pi_{int}$ (the *second variation* of Π_{int}) is quadratic, and so on. From the definition of Π_{int},

$$\begin{aligned}\Delta\Pi_{int} &= \int_R [W(\varepsilon + \delta\varepsilon) - W(\varepsilon)]dV \\ &= \int_R \frac{\partial W}{\partial\varepsilon_{ij}}\delta\varepsilon_{ij}\, dV + \frac{1}{2}\int_R \frac{\partial^2 W}{\partial\varepsilon_{ij}\partial\varepsilon_{kl}}\delta\varepsilon_{ij}\delta\varepsilon_{kl}\, dV + \ldots.\end{aligned}$$

According to our definition, the first integral in the last expression is $\delta\Pi_{int}$, while the second integral is $\delta^2\Pi_{int}$.

Even in the presence of internal constraints,

$$\sigma_{ij}\delta\varepsilon_{ij} = \frac{\partial W}{\partial\varepsilon_{ij}}\delta\varepsilon_{ij}.$$

In view of the definition of internal virtual work as given in 1.3.5, therefore,

$$\overline{\delta W}_{int} = \delta \Pi_{int}.$$

Total Potential Energy, Minimum Principle

For a fixed set of loads $\mathbf{b}$, $\mathbf{t}^a$, let the *external potential energy* be defined as

$$\Pi_{ext} = -\int_R \rho b_i u_i \, dV - \int_{\partial R_t} t_i^a u_i \, dS;$$

then the external virtual work over a virtual displacement field $\delta \mathbf{u}$ is

$$\overline{\delta W}_{ext} = -\delta \Pi_{ext}.$$

If we now define the *total potential energy* as $\Pi = \Pi_{int} + \Pi_{ext}$, then we have $\overline{\delta W}_{int} - \overline{\delta W}_{ext} = \delta \Pi$, the first variation of Π. The principle of virtual work, (1.3.15), then tells us that *an elastic body is in equilibrium if and only if*

$$\delta \Pi = 0; \tag{1.4.18}$$

that is, *at equilibrium the displacement field makes the total potential energy stationary with respect to virtual displacements.*

It can further be shown that for the equilibrium to be stable, the total potential energy must be a minimum, a result known as the **principle of minimum potential energy**. If Π is a local minimum, then, for any nonzero virtual displacement field $\delta \mathbf{u}$, $\Delta \Pi$ must be positive. Since the first variation $\delta \Pi$ vanishes, the second variation $\delta^2 \Pi$ must be positive. Π_{ext} is, by definition, linear in $\mathbf{u}$, and therefore, *in the absence of significant changes in geometry,*[1] $\delta^2 \Pi_{ext} = 0$. Hence

$$\delta^2 \Pi = \delta^2 \Pi_{int} = \int_R C_{ijkl} \delta \varepsilon_{ij} \delta \varepsilon_{kl} \, dV.$$

An elastic body is thus stable under fixed loads if $C_{ijkl} \varepsilon_{ij} \varepsilon_{kl} > 0$, or, in matrix notation, $\underline{\varepsilon}^T \underline{C} \, \underline{\varepsilon} > 0$, for all $\underline{\varepsilon} \neq \underline{0}$. A matrix having this property is known as *positive definite*, and the definition can be naturally extended to the fourth-rank tensor C. The positive-definiteness of C is assumed henceforth.

If the loads $\mathbf{b}$ and $\mathbf{t}^a$ are not fixed but depend on the displacement, then the just-derived principle is still valid, provided that the loads are derivable from potentials, that is, that there exist functions $\phi(\mathbf{u}, \mathbf{x})$ and $\psi(\mathbf{u}, \mathbf{x})$ defined on R and ∂R, respectively, such that $\rho b_i = -\partial \phi / \partial u_i$ and $t_i^a = -\partial \psi / \partial u_i$ (examples: spring support, elastic foundation). In that case,

$$\Pi_{ext} = \int_R \phi \, dV + \int_{\partial R} \psi \, dS,$$

[1] It is the changes in geometry that are responsible for unstable phenomena such as buckling, discussed in Section 5.3.

and the second variation $\delta^2\Pi_{ext}$ does not, in general, vanish, even in the absence of significant geometry changes.

Complementary Potential Energy

From the principle of virtual forces we may derive a complementary energy principle for elastic bodies. If the prescribed displacements are u_i^a (assumed independent of traction), and if the external and internal complementary potential energies for a given stress field $\boldsymbol{\sigma}$ are defined as

$$\Pi_{ext}^c = -\int_{\partial R} u_i^a n_j \sigma_{ij}\, dS$$

and

$$\Pi_{int}^c = \int_{\partial R} W^c(\boldsymbol{\sigma})\, dV,$$

respectively, then, analogously,

$$\overline{\delta W}_{int}^c - \overline{\delta W}_{ext}^c = \delta\Pi^c,$$

where $\Pi^c = \Pi_{ext}^c + \Pi_{int}^c$ is the total complementary potential energy. According to the virtual-force principle, then, Π^c is stationary ($\delta\Pi^c = 0$) if and only if the stress field $\boldsymbol{\sigma}$ is related by the stress-strain relations to a strain field that is compatible — both with internal constraints (if any) and with the prescribed displacements or external constraints. Since Π^c can also be shown to be a minimum at stable equilibrium, this principle is the **principle of minimum complementary energy**.

Discretized Elastic Body

If an elastic continuum is discretized as in 1.3.5, then the stress-strain relation (1.4.9) implies

$$\underline{\sigma}(\underline{x}) = \underline{C}[\underline{B}(\underline{x})\underline{q} + \underline{B}^a(\underline{x})\underline{q}^a],$$

and therefore

$$\underline{Q} = \underline{K}\underline{q} + \underline{K}^a\underline{q}^a,$$

where

$$\underline{K} = \int_R \underline{B}^T \underline{C}\, \underline{B}\, dV \tag{1.4.19}$$

and

$$\underline{K}^a = \int_R \underline{B}^T \underline{C}\, \underline{B}^a\, dV.$$

The symmetric matrix $\underline{K}$ is generally referred to as the *stiffness matrix* of the discretized model. The equilibrium equation $\underline{Q} = \underline{F}$ can now be solved for the q_n in terms of prescribed data as

$$\underline{q} = \underline{K}^{-1}(\underline{F} - \underline{K}^a\underline{q}^a),$$

provided that the stiffness matrix is invertible.

It is easy to extend the preceding result to include the effects of initial and thermal stresses. The new result is

$$\underline{q} = \underline{K}^{-1}(\underline{F} - \underline{K}^a \underline{q}^a - \underline{Q}^0 + \underline{F}_T),$$

where

$$\underline{Q}^0 = \int_R \underline{B}^T \underline{\sigma}^0 \, dV, \quad \underline{F}_T = \int_R \underline{B}^T \underline{\beta}(T - T^0) \, dV,$$

$\underline{\sigma}^0$ denoting the initial stress field, T^0 and T respectively the initial and current temperature fields, and $\underline{\beta}$ the 6×1 matrix form of the the thermal-stress coefficient tensor $\boldsymbol{\beta}$ defined in 1.4.1.

The invertibility requirement on $\underline{K}$ is equivalent to the nonvanishing of its eigenvalues. Note that

$$\delta \underline{q}^T \underline{K} \, \delta \underline{q} = \int_R \delta \underline{\varepsilon}^T \underline{C} \, \delta \underline{\varepsilon} \, dV,$$

so that since the elastic strain energy cannot be negative, the left-hand side of the preceding equation also cannot be negative — that is, $\underline{K}$ must be *positive semidefinite*. The invertibility requirement therefore translates into one of the *positive definiteness* of the stiffness matrix. This condition means that any variation in $\underline{q}$ implies deformation, that is, that the model has no rigid-body degrees of freedom.

Exercises: Section 1.4

1. Assume that the internal-energy density can be given as $u = \bar{u}(\varepsilon, T)$, that the heat flux is governed by the Fourier law $\mathbf{h} = -k(T)\nabla T$, and that $r = 0$. Defining the specific heat $C = \partial \bar{u}/\partial T$, write the equation resulting from combining these assumptions with the local energy-balance equation (1.4.1).

2. Let the complementary internal-energy density be defined by $\kappa = \sigma_{ij}\varepsilon_{ij}/\rho - u$. Neglecting density changes, assume that the entropy density can be given as $\eta = \hat{\eta}(\boldsymbol{\sigma}, \kappa)$.

   (a) Show that $T^{-1} = -\partial \hat{\eta}/\partial \kappa$.

   (b) Assuming $\varepsilon = \varepsilon(\boldsymbol{\sigma}, \eta)$ and $\kappa = \hat{\kappa}(\boldsymbol{\sigma}, \eta)$, show that $T = -\partial \hat{\kappa}/\partial \eta$ and $\varepsilon_{ij} = \rho \partial \hat{\kappa}/\partial \sigma_{ij}$.

   (c) Defining the complementary free-energy density $\chi(\boldsymbol{\sigma}, T) = \kappa + T\eta$, show that $\eta = \partial \chi/\partial T$ and the isothermal strain-stress relation is $\varepsilon_{ij} = \rho \partial \chi/\partial \sigma_{ij}$.

3. Expand the complementary free energy χ (Exercise 2) in powers of $\boldsymbol{\sigma}$ and $T - T_0$, and find the linearized expressions for η and ε.

4. A solid is called *inextensible* in a direction $\mathbf{n}$ if the longitudinal strain along that direction is always zero. Find an isothermal relation for the stress in an inextensible thermoelastic solid, and explain the meaning of any undetermined quantity that may appear in it.

5. Show that, in an isotropic linearly elastic solid, the principal stress and principal strain directions coincide.

6. Write the elastic modulus matrix $\underline{C}$ for an isotropic linearly elastic solid in terms of the Young's modulus E and the Poisson's ratio ν.

7. Combine the stress-strain relations for an isotropic linearly elastic solid with the equations of motion and the strain-displacement relations in order to derive the equations of motion for such a solid entirely in terms of displacement, using (a) λ and μ, (b) G and ν.

8. Derive the forms given in the text for $\underline{C}$ in plane stress and plane strain.

9. Find the relation between the isothermal and adiabatic values of the Young's modulus E in an isotropic linearly elastic solid, in terms of the linear-expansion coefficient α and any other quantities that may be necessary.

10. Derive an explicit expression for the strain-energy function $W(\varepsilon)$ in terms of the "engineering" components of strain $\varepsilon_x, \ldots, \gamma_{xy}, \ldots$, using E and ν.

11. Show that, if the bar described in Exercise 10 of Section 1.3 is made of a homogeneous, isotropic, linearly elastic material, then its internal potential energy is

$$\Pi_{int} = \frac{1}{2}\bar{E}\int_0^L (Au'^2 + Iv''^2)\,dx_1,$$

where A is the cross-sectional area, $I = \int_A x_2^2\,dA$, and

$$\bar{E} = \frac{E(1-\nu)}{(1-2\nu)(1+\nu)}.$$

12. Give a justification for replacing $\bar{E}$ in Exercise 11 by E.

13. Combining the results of Exercise 11, as modified by Exercise 12, with those of Exercise 10 of Section 1.3, show that $P = EAu'$ and $M = EIv''$.

14. For the quantity Ξ defined by

$$\Xi = \int_R \{W(\boldsymbol{\varepsilon}) - \rho b_i u_i - \sigma_{ij}[\varepsilon_{ij} - \tfrac{1}{2}(u_{i,j} + u_{j,i})]\}\, dV - \int_{\partial R_t} t_i^a u_i\, dS - \int_{\partial R_u} t_i(u_i - u_i^a)\, dS,$$

show that, if $\mathbf{u}$, $\boldsymbol{\varepsilon}$, $\boldsymbol{\sigma}$ and (on ∂R_u) $\mathbf{t}$ can be varied independently of one another, then $\delta\Xi = 0$ leads to six sets of equations describing a static boundary-value problem in linear elasticity. The result is known as the **Hu–Washizu principle**.

Section 1.5 Constitutive Relations: Inelastic

In this section a theoretical framework for the description of inelastic materials is presented, building on the discussion of elasticity in the preceding section. The concepts introduced here are applied to the development of the constitutive theory of plasticity in Chapter 3, following a brief survey of the physics of plasticity in Chapter 2.

1.5.1. Inelasticity

Introduction

If we return to the weak (Cauchy) definition of an elastic body as one in which the strain at any point of the body is completely determined by the current stress and temperature there, then an obvious definition of an *inelastic* body is as one in which there is something else, besides the current stress and temperature, that determines the strain. That "something else" may be thought of, for example, as the *past history* of the stress and temperature at the point. While the term "past history" seems vague, it can be defined quite precisely by means of concepts from functional analysis, and a highly mathematical theory, known as the theory of **materials with memory**, has been created since about 1960 in order to deal with it. The dependence of the current strain on the history of the stress (and its converse dependence of stress on strain history, whenever that can be justified) can be expressed explicitly when the behavior is linear. The relevant theory is known as the theory of **linear viscoelasticity** and is briefly reviewed later in this section.

One way in which history affects the relation between strain and stress is through *rate sensitivity*: the deformation produced by slow stressing is different — almost invariably greater — than that produced by rapid stressing.

A particular manifestation of rate sensitivity is the fact that deformation will in general increase in time at constant stress, except possibly under hydrostatic stress; this phenomenon is called *flow* in fluids and *creep* in solids. Rate sensitivity, as a rule, increases with temperature, so that materials which appear to behave elastically over a typical range of times at ordinary temperatures (at least within a certain range of stresses) become markedly inelastic at higher temperatures. For this reason creep is an important design factor in metals used at elevated temperatures, while it may be ignored at ordinary temperatures.

If strain and stress can be interchanged in the preceding discussion, then, since a slower rate implies a greater deformation at a given stress, it accordingly implies a lower stress at a given strain. Consequently, the stress will in general decrease in time at a fixed strain, a phenomenon known as *relaxation*.

The rate sensitivity of many materials, including polymers, asphalt, and concrete, can often be adequately described, within limits, by means of the linear theory. The inelasticity of metals, however, tends to be highly nonlinear in that their behavior is very nearly elastic within a certain range of stresses (the *elastic range*) but strongly history-dependent outside that range. When the limit of the elastic range (the *elastic limit*) is attained as the stress is increased, the metal is said to *yield*. When the elastic range forms a region in the space of the stress components, then it is usually called the *elastic region* and its boundary is called the *yield surface*.

Internal Variables

An alternative way of representing the "something else" is through an array of variables, $\xi_1, \ldots, \xi_n$, such that the strain depends on these variables in addition to the stress and the temperature. These variables are called *internal* (or *hidden*) *variables*, and are usually assumed to take on scalar or second-rank tensor values. The array of the internal variables, when the tensorial ones, if any, are expressed in invariant form, will be denoted $\boldsymbol{\xi}$. The strain is accordingly assumed as given by

$$\boldsymbol{\varepsilon} = \boldsymbol{\varepsilon}(\boldsymbol{\sigma}, T, \boldsymbol{\xi}).$$

The presence of additional variables in the constitutive relations requires additional constitutive equations. The additional equations that are postulated for a *rate-sensitive* or *rate-dependent* inelastic body reflect the hypothesis that, if the *local state* that determines the strain is defined by $\boldsymbol{\sigma}$, T, $\boldsymbol{\xi}$, then the *rate of evolution* of the internal variables is also determined by the local state:

$$\dot{\xi}_\alpha = g_\alpha(\boldsymbol{\sigma}, T, \boldsymbol{\xi}). \tag{1.5.1}$$

Equations (1.5.1) are known as the **equations of evolution** or **rate equations** for the internal variables ξ_α.

The relation $\boldsymbol{\varepsilon} = \boldsymbol{\varepsilon}(\boldsymbol{\sigma}, T, \boldsymbol{\xi})$ cannot always be inverted to give $\boldsymbol{\sigma} = \boldsymbol{\sigma}(\boldsymbol{\varepsilon}, T, \boldsymbol{\xi})$, even when there are no internal constraints governing the strain. As an illustration of this **principle of nonduality**, as it was called by Mandel [1967], let us consider the classical model known as *Newtonian viscosity*. When limited to infinitesimal deformation, the model can be described by the equation

$$\varepsilon_{ij} = \frac{1}{9K}\sigma_{kk}\delta_{ij} + \varepsilon_{ij}^v,$$

where $\boldsymbol{\varepsilon}^v$ is a tensor-valued internal variable, called the *viscous strain*, whose rate equation is

$$\dot{\boldsymbol{\varepsilon}}^v = \frac{1}{2\eta}\mathbf{s},$$

where $\mathbf{s}$ is the stress deviator. The elastic bulk modulus K and the viscosity η are functions of temperature; if necessary, a thermal strain $\alpha(T - T_0)\delta_{ij}$ may be added to the expression for ε_{ij}. It is clearly not possible to express $\boldsymbol{\sigma}$ as a function of $\boldsymbol{\varepsilon}$, T, and a set of internal variables governed by rate equations. Instead, the stress is given by

$$\sigma_{ij} = K\varepsilon_{kk}\delta_{ij} + 2\eta\dot{e}_{ij}.$$

If it is possible to express the stress as $\boldsymbol{\sigma}(\boldsymbol{\varepsilon}, T, \boldsymbol{\xi})$, then this expression may be substituted in the right-hand side of (1.5.1), resulting in an alternative form of the rate equations:

$$\dot{\xi}_\alpha = g_\alpha(\boldsymbol{\sigma}(\boldsymbol{\varepsilon}, T, \boldsymbol{\xi}), T, \boldsymbol{\xi}) = \bar{g}_\alpha(\boldsymbol{\varepsilon}, T, \boldsymbol{\xi}).$$

Inelastic Strain

For inelastic bodies undergoing infinitesimal deformation, it is almost universally assumed that the strain tensor can be decomposed additively into an elastic strain $\boldsymbol{\varepsilon}^e$ and an inelastic strain $\boldsymbol{\varepsilon}^i$:

$$\varepsilon_{ij} = \varepsilon_{ij}^e + \varepsilon_{ij}^i,$$

where $\varepsilon_{ij}^e = C_{ijkl}^{-1}\sigma_{kl}$ (with thermal strain added if needed). Newtonian viscosity, as discussed above, is an example of this decomposition, with $\boldsymbol{\varepsilon}^i = \boldsymbol{\varepsilon}^v$.

1.5.2. Linear Viscoelasticity

The aforementioned theory of linear viscoelasticity provides additional examples of the additive decomposition. For simplicity, we limit ourselves to states that can be described by a single stress component σ with the conjugate strain ε; the extension to arbitrary states of multiaxial stress and strain is made later. The temperature will be assumed constant and will not be explicitly shown.

"Standard Solid" Model

Suppose that the behavior of a material element can be represented by the mechanical model of Figure 1.5.1(a) (page 63), with force representing stress and displacement representing strain. Each of the two springs models elastic response (with moduli E_0 and E_1), and the dashpot models viscosity. The displacement of the spring on the left represents the elastic strain ε^e, and the displacement of the spring-dashpot combination on the right represents the inelastic strain ε^i. Equilibrium requires that the force in the left-hand spring be the same as the sum of the forces in the other two elements, and therefore we have two equations for the stress σ:

$$\sigma = E_0\varepsilon^e, \qquad \sigma = E_1\varepsilon^i + \eta\dot{\varepsilon}^i,$$

where η is the viscosity of the dashpot element. For the total strain ε we may write

$$\varepsilon = \frac{\sigma}{E_0} + \varepsilon^i,$$

$$\dot{\varepsilon}^i = \frac{1}{\eta}\sigma - \frac{E_1}{\eta}\varepsilon^i.$$

The inelastic strain may consequently be regarded as an internal variable, the last equation being its rate equation (E_1 and η are, of course, functions of the temperature).

Given an input of stress as a function of time, the rate equation for ε^i is a differential equation that can be solved for $\varepsilon^i(t)$:

$$\varepsilon^i(t) = \frac{1}{\eta}\int_{-\infty}^{t} e^{-(t-t')/\tau}\sigma(t')\,dt',$$

where the reference time (at which $\varepsilon^i = 0$) is chosen as $-\infty$ for convenience, and $\tau = \eta/E_1$ is a material property having the dimension of time. In particular, if $\sigma(t) = 0$ for $t < 0$ and $\sigma(t) = \sigma_0$ (constant) for $t > 0$, then $\varepsilon^i(t) = (1/E_1)(1 - e^{-t/\tau})\sigma_0$; this result demonstrates a form of creep known as *delayed elasticity*.

The limiting case represented by $E_1 = 0$ is known as the *Maxwell model* of viscoelasticity. Note that in this case τ is infinite, so that the factor $\exp[-(t - t')/\tau]$ inside the integral becomes unity, and the creep solution is just $\varepsilon^i(t) = (t/\eta)\sigma_0$, displaying *steady creep*.

Generalized Kelvin Model

If models of viscoelasticity with more than one dashpot per strain component are used — say a number of parallel spring-dashpot combinations in

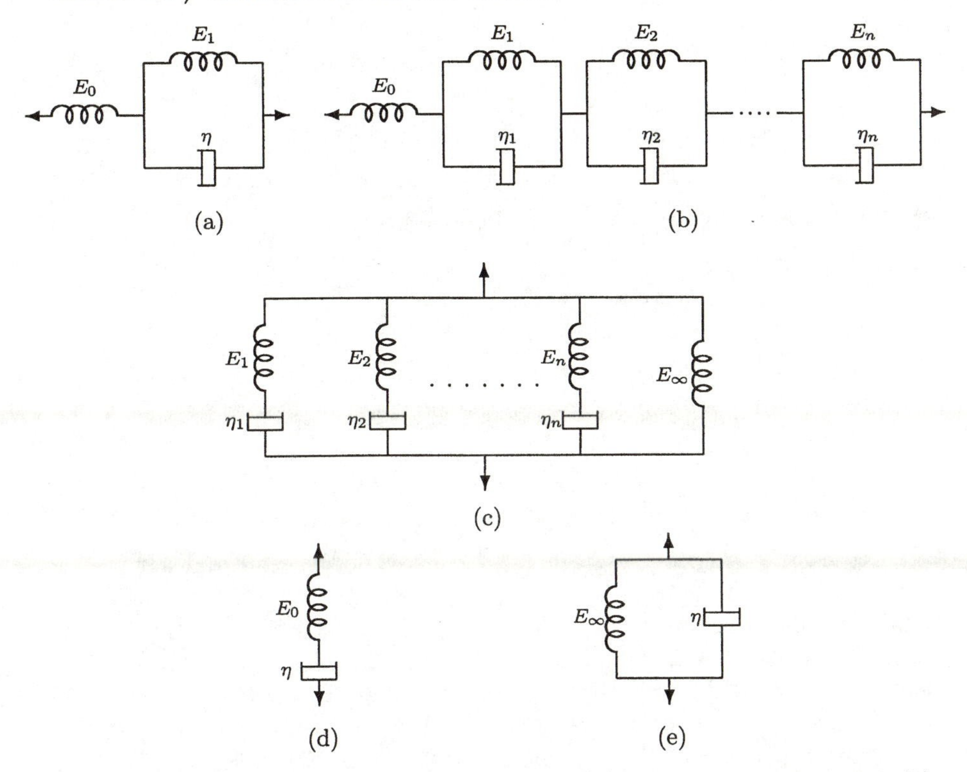

Figure 1.5.1. Models of linear viscoelasticity: (a) "standard solid" model; (b) generalized Kelvin model; (c) generalized Maxwell model; (d) Maxwell model; (e) Kelvin model.

series with a spring, the so-called *generalized Kelvin model* shown in Figure 1.5.1(b) — then the inelastic strain is represented by the sum of the dashpot displacements, and every dashpot displacement constitutes an internal variable. Designating these by ξ_α, we have

$$\varepsilon^i = \sum_{\alpha=1}^{n} \xi_\alpha.$$

By analogy with the previous derivation, the rate equations for the ξ_α are, if $\tau_\alpha = \eta_\alpha / E_\alpha$,

$$\dot{\xi}_\alpha = \frac{\sigma}{\eta_\alpha} - \frac{\xi_\alpha}{\tau_\alpha}.$$

As before, for a given stress history $\sigma(t')$, $-\infty < t' < t$ (where t is the current time), the rate equations are ordinary linear differential equations for the ξ_α that can be integrated explicitly:

$$\xi_\alpha(t) = \int_{-\infty}^{t} \frac{1}{\eta_\alpha} e^{-(t-t')/\tau_\alpha} \sigma(t')\, dt'.$$

The current strain $\varepsilon(t)$ can therefore be expressed as

$$\varepsilon(t) = \frac{1}{E_0}\sigma(t) + \int_{-\infty}^{t} \left(\sum_{\alpha=1}^{n} \frac{1}{\eta_\alpha} e^{-(t-t')/\tau_\alpha} \right) \sigma(t')\, dt'.$$

With the uniaxial *creep function* $J(t)$ defined by

$$J(t) = \frac{1}{E_0} + \sum_{\alpha=1}^{n} \frac{1}{E_\alpha}\left(1 - e^{-t/\tau_\alpha}\right),$$

the current strain can, with the help of integration by parts [and the assumption $\sigma(-\infty) = 0$], also be expressed as

$$\varepsilon(t) = \int_{-\infty}^{t} J(t-t')\frac{d\sigma}{dt'}dt'.$$

If the stress history is given by $\sigma(t') = 0$ for $t' < 0$ and $\sigma(t') = \sigma$ (constant) for $t' > 0$, then, for $t > 0$, the strain is $\varepsilon(t) = \sigma J(t)$. The creep function can therefore be determined experimentally from a single creep test, independently of an internal-variable model (unless the material is one whose properties change in time, such as concrete). Similarly, in a relaxation test in which $\varepsilon(t') = 0$ for $t' < 0$ and $\varepsilon(t') = \varepsilon$ (constant) for $t' > 0$, the measured stress for $t > 0$ has the form $\sigma(t) = \varepsilon R(t)$, where $R(t)$ is the uniaxial *relaxation function*. Under an arbitrary strain history $\varepsilon(t')$, the stress at time t is then

$$\sigma(t) = \int_{-\infty}^{t} R(t-t')\frac{d\varepsilon}{dt'}dt'.$$

An explicit form of the relaxation function in terms of internal variables can be obtained by means of the *generalized Maxwell model* [Figure 1.5.1(c)]. It can be shown that, in general, $R(0) = 1/J(0) = E_0$ (the instantaneous elastic modulus), and $R(\infty) = 1/J(\infty) = E_\infty$ (the asymptotic elastic modulus), with $E_0 > E_\infty$ except in the case of an elastic material. In particular, E_∞ may be zero, as in the Maxwell model [Figure 1.5.1(d)], while E_0 may be infinite, as in the Kelvin model [Figure 1.5.1(e)]. The relaxation function of the Kelvin model is a singular function given by $R(t) = E_\infty + \eta\delta(t)$, where $\delta(t)$ is the Dirac delta function.

A generalization of the preceding description from uniaxial to multiaxial behavior is readily accomplished by treating $J(t)$ and $R(t)$ as functions whose values are fourth-rank tensors. For isotropic materials, the tensorial form of these functions is analogous to that of the elastic moduli as given in 1.4.2:

$$J_{ijkl}(t) = J_0(t)\delta_{ij}\delta_{kl} + J_1(t)(\delta_{ik}\delta_{jl} + \delta_{ik}\delta_{jl}),$$

$$R_{ijkl}(t) = R_0(t)\delta_{ij}\delta_{kl} + R_1(t)(\delta_{ik}\delta_{jl} + \delta_{ik}\delta_{jl}).$$

By analogy with Newtonian viscosity, it is frequently assumed (not always with physical justification) that the volumetric strain is purely elastic, that is, $\varepsilon_{kk} = \sigma_{kk}/3E$. It follows that the creep and relaxation functions obeying this assumption must satisfy the relations

$$3J_0(t) + 2J_1(t) = \frac{1}{3K}, \qquad 3R_0(t) + 2R_1(t) = 3K.$$

The use of creep and relaxation functions makes it possible to represent the strain explicitly in terms of the history of stress, and vice versa, with no reference to internal variables. Indeed, the concept of internal variables is not necessary in linear viscoelasticity theory: once the creep function is known, the assumption of linear response is by itself sufficient to determine the strain for any stress history (the **Boltzmann superposition principle**).

As was mentioned before, a mathematical theory of materials with memory, without internal variables, exists for nonlinear behavior as well. The theory, however, has proved too abstract for application to the description of real materials. Virtually all constitutive models that are used for nonlinear inelastic materials rely on internal variables.

1.5.3. Internal Variables: General Theory

Internal Variables and Thermomechanics

An *equilibrium state* of a system is a state that has no tendency to change with no change in the external controls. In the thermomechanics of an inelastic continuum with internal variables, a local state $(\boldsymbol{\sigma}, T, \boldsymbol{\xi})$ may be called a *local equilibrium state* if the internal variables remain constant at constant stress and temperature, or, in view of Equation (1.5.1), if

$$g_\alpha(\boldsymbol{\sigma}, T, \boldsymbol{\xi}) = 0, \quad \alpha = 1, ..., n.$$

In an elastic continuum, every local state is an equilibrium state, though the continuum need not be globally in equilibrium: a nonuniform temperature field will cause heat conduction and hence changes in the temperature. The existence of nonequilibrium states is an essential feature of rate-dependent inelastic continua; such states evolve in time by means of *irreversible* processes, of which creep and relaxation are examples.[1]

The thermomechanics of inelastic continua consequently belongs to the domain of the thermodynamics of irreversible processes (also known as *nonequilibrium thermomechanics*). The fundamental laws of thermodynamics

[1]A process is reversible if the equations governing it are unaffected when the time t is replaced by $-t$; otherwise it is irreversible. In the rate-independent plastic continuum, which forms the main subject of this book, irreversible processes occur even in the absence of nonequilibrium states, as is seen in Chapter 3.

discussed in 1.4.1 are assumed to be valid in this domain as well, but there is no full agreement in the scientific community about the meaning of such variables as the entropy and the temperature, which appear in the second law, or about the range of validity of this law (there is no comparable controversy about the first law).

Entropy and temperature were defined in 1.4.1 for thermoelastic continua only; in other words, they are intrinsically associated with equilibrium states. The assumption that these variables are uniquely defined at nonequilibrium states as well, and obey the Clausius–Duhem inequality at all states, represents the point of view of the "rational thermodynamics" school, most forcefully expounded by Truesdell [1984] and severely criticized by Woods [1981].

For continua with internal variables, another point of view, articulated by Kestin and Rice [1970] and Bataille and Kestin [1975] (see also Germain, Nguyen and Suquet [1983]), may be taken. According to this school of thought, entropy and temperature may be defined at a nonequilibrium state if one can associate with this state a fictitious "accompanying equilibrium state" at which the internal variables are somehow "frozen" so as to have the same values as at the actual (nonequilibrium) state. If we follow this point of view, then we are allowed to assume the existence of a free-energy density given by $\psi(\boldsymbol{\varepsilon}, T, \boldsymbol{\xi})$ such that the entropy density and the stress may be derived from it in the same way as for elastic continua, with the internal variables as parameters. Consequently the stress is given by Equation (1.4.7) *if* it is entirely determined by $\boldsymbol{\varepsilon}$, T, $\boldsymbol{\xi}$, and the entropy density is given by $\eta = -\partial\psi/\partial T$.

In any statement of the second law of thermodynamics, however, the internal variables must be "unfrozen," since this law governs irreversible processes. Let us assume that the second law is expressed by the local Clausius–Duhem inequality (1.4.5), and let us rewrite this inequality as

$$\rho\dot{\eta} - T^{-1}(\rho r - \nabla\cdot\mathbf{h}) + \mathbf{h}\cdot\nabla T^{-1} = \rho\gamma \geq 0.$$

With the help of the local energy-balance equation (1.4.1), the expression in parentheses may be replaced by $\rho\dot{u} - \sigma_{ij}\dot{\varepsilon}_{ij}$, and, since the definition of the free-energy density ψ leads to $T\dot{\eta} - \dot{u} = -(\eta\dot{T} + \dot{\psi})$, the left-hand side of the inequality becomes

$$\mathbf{h}\cdot\nabla T^{-1} - \rho T^{-1}(\dot{\psi} + \eta\dot{T} - \frac{1}{\rho}\sigma_{ij}\dot{\varepsilon}_{ij}).$$

Furthermore,

$$\dot{\psi} = \frac{\partial\psi}{\partial\varepsilon_{ij}}\dot{\varepsilon}_{ij} + \frac{\partial\psi}{\partial T}\dot{T} + \sum_{\alpha}\frac{\partial\psi}{\partial\xi_{\alpha}}\dot{\xi}_{\alpha}.$$

With η and $\boldsymbol{\sigma}$ expressed in terms of ψ, the left-hand side of (1.4.5) — that is, the internal entropy production — becomes

$$\mathbf{h}\cdot\nabla T^{-1} - \rho T^{-1}\sum_\alpha \frac{\partial\psi}{\partial\xi_\alpha}\dot{\xi}_\alpha. \tag{1.5.2}$$

This quantity must be nonnegative in any process and at any state, and in particular when the temperature gradient vanishes. The Clausius–Duhem inequality is therefore obeyed if and only if, in addition to the already mentioned heat-conduction inequality $\mathbf{h}\cdot\nabla T^{-1} \ge 0$, the material also obeys the **dissipation inequality** (Kelvin inequality)

$$D = \sum_\alpha p_\alpha \dot{\xi}_\alpha \ge 0, \tag{1.5.3}$$

where

$$p_\alpha = -\rho\frac{\partial\psi}{\partial\xi_\alpha} \tag{1.5.4}$$

is the "thermodynamic force" conjugate to ξ_α.

The preceding results must be generalized somewhat if the material possesses viscosity in the sense that generalizes the Newtonian model: the stress depends continuously[1] on the strain-rate tensor $\dot{\boldsymbol{\varepsilon}}$ in addition to the thermodynamic state variables $\boldsymbol{\varepsilon}$, T, $\boldsymbol{\xi}$. The definition of a local equilibrium state now requires the additional condition $\dot{\boldsymbol{\varepsilon}} = 0$. The stress in the accompanying equilibrium state is still given (by definition) by $\rho\,\partial\psi/\partial\varepsilon_{ij}$, but it is not equal to the actual stress $\boldsymbol{\sigma}(\boldsymbol{\varepsilon}, T, \boldsymbol{\xi}; \dot{\boldsymbol{\varepsilon}})$. Instead, it equals

$$\sigma(\boldsymbol{\varepsilon}, T, \boldsymbol{\xi}; \mathbf{0}) \stackrel{\text{def}}{=} \boldsymbol{\sigma}^e(\boldsymbol{\varepsilon}, T, \boldsymbol{\xi})$$

(*equilibrium stress* or *elastic stress*). If the *viscous stress* is defined as $\boldsymbol{\sigma}^v = \boldsymbol{\sigma} - \boldsymbol{\sigma}^e$, then the additional term $T^{-1}\sigma^v_{ij}\dot{\varepsilon}_{ij}$ must be added to the internal entropy production as expressed by (1.5.2). The quantity $\sigma^v_{ij}\dot{\varepsilon}_{ij}$ is the *viscous dissipation*, and must also be nonnegative.

If the decomposition of the strain into elastic and inelastic parts is assumed to take the form

$$\boldsymbol{\varepsilon} = \boldsymbol{\varepsilon}^e(\boldsymbol{\sigma}, T) + \boldsymbol{\varepsilon}^i(\boldsymbol{\xi}),$$

then, as was shown by Lubliner [1972], such a decomposition is compatible with the existence of a free-energy density $\psi(\boldsymbol{\varepsilon}, \boldsymbol{\xi}, T)$ if and only if ψ can be decomposed as

$$\psi(\boldsymbol{\varepsilon}, T, \boldsymbol{\xi}) = \psi^e(\boldsymbol{\varepsilon} - \boldsymbol{\varepsilon}^i(\boldsymbol{\xi}), T) + \psi^i(\boldsymbol{\xi}, T). \tag{1.5.5}$$

[1]The *continuity* of the dependence of the stress on the strain rate is what distinguishes viscosity from, say, dry friction.

The Kelvin inequality then takes the special form

$$D = D_i - \rho \sum_{\alpha} \frac{\partial \psi^i}{\partial \xi_\alpha} \dot{\xi}_\alpha \geq 0,$$

where $D_i \stackrel{\text{def}}{=} \sigma_{ij}\dot{\varepsilon}^i_{ij}$ is the *inelastic work rate* per unit volume. Note that this rate may be negative, without violating the second law of thermodynamics, if ψ^i decreases fast enough, that is, if enough stored inelastic energy is liberated. The contraction of muscle under a tensile force, driven by chemical energy, is an example of such a process.

The Nature of Internal Variables

What are internal variables in general? In principle, they may be any variables which, in addition to the strain (or stress) and temperature, define the local state in a small[2] neighborhood of a continuum. As we have seen, the components of ε^i themselves may or may not be included among the internal variables.

As a general rule, internal variables may be said to be of two types. On the one hand, they may be "physical" variables describing aspects of the local physico-chemical structure which may change spontaneously; for example, if the material can undergo a chemical reaction or a change of phase, then a quantity describing locally the extent of the reaction or the relative density of the two phases may serve as an internal variable. Other internal variables of this type include densities of structural defects, as discussed in Chapter 2. On the other hand, internal variables may be mathematical constructs; they are then called *phenomenological* variables. The inelastic strain ε^i itself is of this type, as are the dashpot displacements in viscoelastic models. Here the *form* of the functional dependence of the stress (or strain) on the internal variables, and of their rate equations, is assumed a priori.

In the simplest constitutive models describing nonlinear inelastic materials, the internal variables are often assumed to consist of the ε^i_{ij} and an additional variable κ, called a *hardening variable*. The rate equation for κ is further assumed to be related to the rate equations for the ε^i_{ij} in such a way $\dot{\kappa} = 0$ whenever $\dot{\varepsilon}^i = 0$, but, in a cyclic process at the end of which the ε^i_{ij} return to their original values (see Figure 1.5.2), κ will have changed. Usually, κ is defined so that $\dot{\kappa} > 0$ whenever $\dot{\varepsilon}^i \neq 0$. Two commonly used definitions of κ are, first, the *inelastic work*, defined as

$$\kappa = \int D_i \, dt \stackrel{\text{def}}{=} W_i, \tag{1.5.6}$$

[2] "Small" means small enough so that the state may be regarded as uniform, but large enough for the continuum viewpoint to be valid.

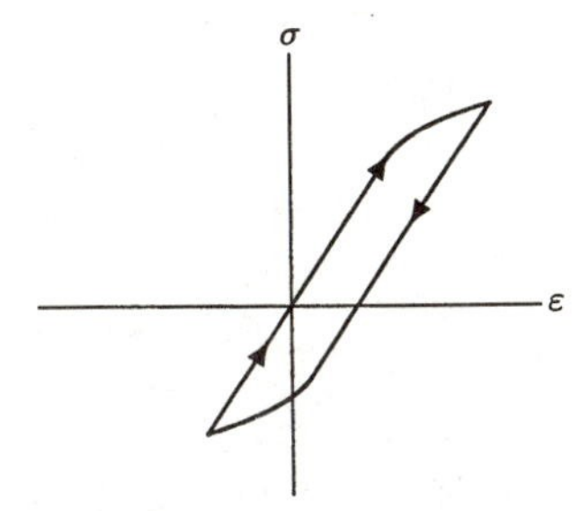

Figure 1.5.2. Closed stress-strain cycle with inelastic deformation: the inelastic strain returns to zero at the end of the cycle, but the internal state may be different, so that internal variables other than the inelastic strain may be necessary.

and second, the *equivalent* (or *effective*) *inelastic strain*,

$$\kappa = \int \sqrt{\frac{2}{3}\dot{\varepsilon}^i_{ij}\dot{\varepsilon}^i_{ij}}\, dt \stackrel{\text{def}}{=} \bar{\varepsilon}^i. \tag{1.5.7}$$

The reason for the traditional factor of $\frac{2}{3}$ in the latter definition (which is due to Odqvist [1933]) is the following: if a specimen of a material that is (a) isotropic and (b) characterized by inelastic incompressibility (so that $\dot{\varepsilon}^i_{kk} = 0$) is subjected to a uniaxial tensile or compressive stress, then the inelastic strain-rate tensor must be given by

$$\dot{\boldsymbol{\varepsilon}}^i = \begin{bmatrix} \dot{\varepsilon}^i & 0 & 0 \\ 0 & -\frac{1}{2}\dot{\varepsilon}^i & 0 \\ 0 & 0 & -\frac{1}{2}\dot{\varepsilon}^i \end{bmatrix},$$

so that $\sqrt{\frac{2}{3}\dot{\varepsilon}^i_{ij}\dot{\varepsilon}^i_{ij}}$ is just equal to $|\dot{\varepsilon}^i|$.

In practice there is little difference in the way the two types of internal variables are used.[1] Whether the functions involved are provided by physical theory or by hypothesis, they contain parameters that must be evaluated by comparison of theoretical predictions with experimental results. It is seen in Chapter 2 that in the case of metal plasticity, physical theory has been remarkably successful in providing a qualitative understanding of the phenomena, but attempts to generate constitutive equations in terms of physical variables have not met with success.

1.5.4. Flow Law and Flow Potential

Regardless of whether the inelastic strain components are directly included among the internal variables, it is always possible to define a *flow law*, that

[1] As is seen in Section 3.1, phenomenological variables are sometimes given physical-sounding names.

is, a rate equation for $\boldsymbol{\varepsilon}^i$, by applying the chain rule to the basic assumption $\boldsymbol{\varepsilon}^i = \boldsymbol{\varepsilon}^i(\boldsymbol{\xi})$. The result is

$$\dot{\varepsilon}^i_{ij} = g_{ij}(\boldsymbol{\sigma}, T, \boldsymbol{\xi}),$$

where

$$g_{ij} = \sum_\alpha \frac{\partial \varepsilon^i_{ij}}{\partial \xi_\alpha} g_\alpha,$$

the g_α being the right-hand sides of Equation (1.5.1).

Mainly for convenience, it is often assumed that the g_{ij} can be derived from a scalar function $g(\boldsymbol{\sigma}, T, \boldsymbol{\xi})$, called a *flow potential*, by means of

$$g_{ij} = \phi \frac{\partial g}{\partial \sigma_{ij}},$$

$\phi(\boldsymbol{\sigma}, T, \boldsymbol{\xi})$ being a positive scalar function.

The flow potential g is commonly assumed to be a function of the stress alone, the most frequently used form being $g(\boldsymbol{\sigma}, T, \boldsymbol{\xi}) = J_2$, where J_2 is the second stress-deviator invariant defined in 1.3.2. Since

$$\frac{\partial}{\partial \sigma_{ij}} J_2 = \frac{\partial s_{kl}}{\partial \sigma_{ij}} \frac{\partial}{\partial s_{kl}} \left(\frac{1}{2} s_{mn} s_{mn} \right) = \left(\delta_{ik}\delta_{jl} - \frac{1}{3}\delta_{ij}\delta_{kl} \right) s_{kl} = s_{ij},$$

it follows that the flow law has the form

$$\dot{\varepsilon}^i_{ij} = \phi(\boldsymbol{\sigma}, T, \boldsymbol{\xi}) s_{ij}. \tag{1.5.8}$$

One consequence of this flow law is that inelastic deformation is volume-preserving, or, equivalently, that volume deformation is purely elastic — a result that is frequently observed in real materials.

Generalized Potential and Generalized Normality

A stronger concept of the flow potential is due to Moreau [1970] and Rice [1970, 1971] (see also Halphen and Nguyen [1975]). A function Ω of $\boldsymbol{\sigma}, T, \boldsymbol{\xi}$ is assumed to depend on stress only through the thermodynamic forces p_α, defined by (1.5.4), conjugate to the internal variables ξ_α, that is, $\Omega = \Omega(\mathbf{p}, T, \boldsymbol{\xi})$r, where $\mathbf{p} \stackrel{\text{def}}{=} \{p_1, ..., p_n\}$. It is further assumed that the rate equations are

$$\dot{\xi}_\alpha = \frac{\partial \Omega}{\partial p_\alpha}. \tag{1.5.9}$$

Equations (1.5.9) represent the hypothesis of **generalized normality**, and Ω is called a *generalized potential*.

The thermodynamic forces p_α can be obtained as functions of $\boldsymbol{\sigma}$ by means of the *complementary free-energy density* $\chi(\boldsymbol{\sigma}, T, \boldsymbol{\xi})$ (also called the free-enthalpy density or Gibbs function), defined by

$$\chi = \rho^{-1} \sigma_{ij} \varepsilon_{ij} - \psi,$$

where ψ is the Helmholtz free-energy density. It can easily be shown that

$$p_\alpha = \rho \frac{\partial \chi}{\partial \xi_\alpha}$$

and

$$\varepsilon_{ij} = \rho \frac{\partial \chi}{\partial \sigma_{ij}}.$$

It follows that

$$\dot{\varepsilon}^i_{ij} = \sum_\alpha \frac{\partial \varepsilon_{ij}}{\partial \xi_\alpha} \dot{\xi}_\alpha = \rho \sum_\alpha \frac{\partial^2 \chi}{\partial \sigma_{ij} \partial \xi_\alpha} \dot{\xi}_\alpha = \sum_\alpha \frac{\partial p_\alpha}{\partial \sigma_{ij}} \dot{\xi}_\alpha.$$

Combining with (1.5.9), we find

$$\dot{\varepsilon}^i_{ij} = \sum_\alpha \frac{\partial \Omega}{\partial p_\alpha} \frac{\partial p_\alpha}{\partial \sigma_{ij}},$$

or

$$\dot{\varepsilon}^i_{ij} = \frac{\partial \Omega}{\partial \sigma_{ij}}. \tag{1.5.10}$$

A sufficient condition for the existence of a generalized potential was found by Rice [1971]. The condition is that each of the rate functions $g_\alpha(\boldsymbol{\sigma}, T, \boldsymbol{\xi})$ depends on the stress only through its own conjugate thermodynamic force p_α:

$$\dot{\xi}_\alpha = \hat{g}_\alpha(p_\alpha, T, \boldsymbol{\xi}) = \frac{\partial \Omega_\alpha}{\partial p_\alpha},$$

where, by definition,

$$\Omega_\alpha(p_\alpha, T, \boldsymbol{\xi}) = \int_0^{p_\alpha} \hat{g}_\alpha(p_\alpha, T, \boldsymbol{\xi})\, dp_\alpha.$$

If we now define

$$\Omega(\mathbf{p}, T, \boldsymbol{\xi}) = \sum_\alpha \Omega_\alpha(p_\alpha, T, \boldsymbol{\xi}),$$

then Equations (1.5.9)–(1.5.10) follow.

The preceding results are independent of whether the inelastic strain components ε^i_{ij} are themselves included among the internal variables.

For mathematical reasons, the generalized potential Ω is usually assumed to be a *convex* function of $\mathbf{p}$, that is, with the remaining variables not shown,

$$\Omega(t\mathbf{p} + (1-t)\mathbf{p}^*) \le t\Omega(\mathbf{p}) + (1-t)\Omega(\mathbf{p}^*)$$

for any admissible $\mathbf{p}$, $\mathbf{p}^*$ and any t such that $0 \le t \le 1$. It follows from this definition that

$$(\partial\Omega/\partial\mathbf{p}) \cdot (\mathbf{p} - \mathbf{p}^*) \ge \Omega(\mathbf{p}) - \Omega(\mathbf{p}^*), \tag{1.5.11}$$

where $\partial\Omega/\partial\mathbf{p}$ is evaluated at $\mathbf{p}$, and the dot defines the scalar product in n-dimensional space. Thus, for any $\mathbf{p}^*$ such that $\Omega(\mathbf{p}^*) \le \Omega(\mathbf{p})$, $(\mathbf{p}-\mathbf{p}^*)\cdot\dot{\boldsymbol{\xi}} \ge 0$.

The hypothesis of generalized normality has often been invoked in constitutive models formulated by French researchers, though not always consistently. An example is presented when constitutive theories of plasticity and viscoplasticity are discussed in detail. Before such a discussion, however, it is worth our while to devote a chapter to the physical bases underlying the theories. We return to theory in Chapter 3.

Exercises: Section 1.5

1. Consider a model of linear viscoelasticity made up of a spring of modulus E_∞ *in parallel* with a Maxwell model consisting of a spring of modulus E' and a dashpot of viscosity η'.

   (a) If the dashpot displacement constitutes the internal variable, find the relation among the stress σ, the strain ε, and ξ.

   (b) Find the rate equation for ξ (*i*) in terms of ε and ξ and (*ii*) in terms of σ and ξ.

   (c) Show that a model of the type shown in Figure 1.5.1(a) can be found that is fully equivalent to the present one, and find the relations between the parameters E_0, E_1, η of that model and those of the present one.

2. For a standard solid model with creep function given by

$$J(t) = \frac{1}{E_\infty} - \left(\frac{1}{E_\infty} - \frac{1}{E_0}\right) e^{-t/\tau},$$

   find an expression for the strain *as a function of stress* when the stress history is such that $\sigma = 0$ for $t < 0$ and $\sigma = \alpha\sigma_0 t/\tau$, α being a dimensionless constant and σ_0 a reference stress. Assuming $E_0/E_\infty = 1.5$, sketch plots of σ/σ_0 against $E_\infty\varepsilon/\sigma_0$ for $\alpha = 0.1$, 1.0, and 10.0.

3. Show that the relation between the uniaxial creep function $J(t)$ and relaxation function $R(t)$ of a linearly viscoelastic material is

$$\int_0^t J(t')R(t-t')\,dt' = t.$$

4. Show that if the free-energy density is as given by Equation (1.5.5), and if the inelastic strain components ε^i_{ij} are themselves used as internal variables, then the conjugate thermodynamic forces are just the stress components σ_{ij}.

5. Show that if the free-energy density is as given by Equation (1.5.5), and if in addition the specific heat C is independent of the internal variables, then the free-energy density reduces to the form

$$\psi(\varepsilon, T, \boldsymbol{\xi}) = \psi^e(\varepsilon - \varepsilon^i(\boldsymbol{\xi})) + u^i(\boldsymbol{\xi}) - T\eta^i(\boldsymbol{\xi}).$$

Discuss the possible meaning of u^i and η^i.

6. Find the flow law derived from a flow potential given by $g(\sigma, T, \boldsymbol{\xi}) = f(I_1, J_2, J_3)$.

7. Show that if $\chi(\sigma, T, \boldsymbol{\xi})$ is the complementary free-energy density and p_α is the thermodynamic force conjugate to ξ_α, then $p_\alpha = \rho\, \partial\chi/\partial\xi_\alpha$.

Chapter 2

The Physics of Plasticity

Section 2.1 Phenomenology of Plastic Deformation

The adjective "plastic" comes from the classical Greek verb πλάσσειν, meaning "to shape"; it thus describes materials, such as ductile metals, clay, or putty, which have the property that bodies made from them can have their shape easily changed by the application of appropriately directed forces, and retain their new shape upon removal of such forces. The shaping forces must, of course, be of sufficient intensity — otherwise a mere breath could deform the object — but often such intensity is quite easy to attain, and for the object to have a useful value it may need to be hardened, for example through exposure to air or the application of heat, as is done with ceramics and thermosetting polymers. Other materials — above all metals — are quite hard at ordinary temperatures and may need to be softened by heating in order to be worked.

It is generally observed that the considerable deformations which occur in the plastic shaping process are often accompanied by very slight, if any, volume changes. Consequently plastic deformation is primarily a *distortion*, and of the stresses produced in the interior of the object by the shaping forces applied to the surface, it is their deviators that do most of the work. A direct test of the plasticity of the material could thus be provided by producing a state of simple shearing deformation in a specimen through the application of forces that result in a state of shear stress. In a soft, semi-fluid material such as clay, or soil in general, this may be accomplished by means of a direct shear test such as the shear-box test, which is discussed in Section 2.3. In hard solids such as metals, the only experiment in which uniform simple shear is produced is the twisting of a thin-walled tube, and this is not always a simple experiment to perform. A much simpler test is the *tension test*.

2.1.1 Experimental Stress-Strain Relations

Tension Tests

Of all mechanical tests for structural materials, the tension test is the most common. This is true primarily because it is a relatively rapid test and requires simple apparatus. It is not as simple to interpret the data it gives, however, as might appear at first sight. J. J. Gilman [1969]

The tensile test [is] very easily and quickly performed but it is not possible to do much with its results, because one does not know what they really mean. They are the outcome of a number of very complicated physical processes. . . . The extension of a piece of metal [is] in a sense more complicated than the working of a pocket watch and to hope to derive information about its mechanism from two or three data derived from measurement during the tensile test [is] perhaps as optimistic as would be an attempt to learn about the working of a pocket watch by determining its compressive strength. E. Orowan [1944]

Despite these caveats, the tension test remains the preferred method of determining the material properties of metals and other solids on which it is easily performed: glass, hard plastics, textile fibers, biological tissues, and many others.

Stress-Strain Diagrams

The immediate result of a tension test is a relation between the axial force and either the change in length (elongation) of a gage portion of the specimen or a representative value of longitudinal strain as measured by one or more strain gages. This relation is usually changed to one between the stress σ (force F divided by cross-sectional area) and the strain ε (elongation divided by gage length or strain-gage output), and is plotted as the *stress-strain diagram*. Parameters that remain constant in the course of a test include the temperature and the rate of loading or of elongation. If significant changes in length and area are attained, then it is important to specify whether the area used in calculating the stress is the original area A_0 (*nominal* or "engineering" stress, here to be denoted simply σ_e) or the current area A (*true* or *Cauchy* stress, σ_t) — in other words, whether the Lagrangian or the Eulerian definition is used — and whether the strain plotted is the change in length Δl divided by the original length l_0 (*conventional* or "engineering" strain, ε_e) or the natural logarithm of the ratio of the current length l ($= l_0 + \Delta l$) to l_0 (*logarithmic* or *natural* strain, ε_l).

Examples of stress-strain diagrams, both as σ_e versus ε_e and as σ_t versus ε_l, are shown in Figure 2.1.1. It is clear that the Cauchy stress, since it does not depend on the initial configuration, reflects the actual state in the specimen better than the nominal stress, and while both definitions of strain involve the initial length, the rates (time derivatives) of conventional and logarithmic strain are respectively $\dot{l}/l_0$ and $\dot{l}/l$, so that it is the latter that is independent of initial configuration. In particular, in materials in which it is possible to perform a compression test analogous to a tension test, it is often found that the stress-strain diagrams in tension and compression coincide to a remarkable degree when they are plots of Cauchy stress against logarithmic strain [see Figure 2.1.1(b)].

The rate of work done by the force is $F\dot{l} = \sigma_e A_0 l_0 \dot{\varepsilon}_e = \sigma_t A l \dot{\varepsilon}_l$, so that $\sigma_e \dot{\varepsilon}_e$ and $\sigma_t \dot{\varepsilon}_l$ are the rates of work per unit original and current volume, respectively. While the calculation of Cauchy stress requires, strictly speaking, measurement of cross-sectional area in the course of the test, in practice this is not necessary if the material is a typical one in which the volume does not change significantly, since the current area may be computed from the volume constancy relation $Al = A_0 l_0$.

As is shown in Chapter 8, the logarithmic strain rate $\dot{\varepsilon}_l$ has a natural and easily determined extension to general states of deformation, but the logarithmic strain itself does not, except in situations (such as the tension test) in which the principal strain axes are known and remain fixed. The use of the logarithmic strain in large-deformation problems with rotating principal strain axes may lead to erroneous results.

Compression Tests

As seen in Figure 2.1.1(b), the results of a simple compression test on a specimen of ductile metal are virtually identical with those of a tensile test if Cauchy stress is plotted against logarithmic strain. The problem is that a "simple compression test" is difficult to achieve, because of the friction that is necessarily present between the ends of the specimen and the pressure plates. Compression of the material causes an increase in area, and therefore a tendency to slide outward over the plates, against the shear stress on the interfaces due to the frictional resistance. Thus the state of stress cannot be one of simple compression. Lubrication of the interface helps the problem, as does the use of specimens that are reasonably slender — though not so slender as to cause buckling — so that, at least in the middle portion, a state of simple compressive stress is approached.

Unlike ductile metals, brittle solids behave quite differently in tension and compression, the highest attainable stress in compression being many times that in tension. Classically brittle solids, such as cast iron or glass, fracture almost immediately after the proportional limit is attained, as in Figure 2.1.1(c). Others, however, such as concrete and many rocks, produce stress-strain diagrams that are qualitatively similar to those of many ductile materials, as in Figure 2.1.1(d). Of course, the strain scale is quite different: in brittle materials the largest strains attained rarely exceed 1%. The stress peak represents the onset of fracture, while the decrease in slope of the stress-strain curve represents a loss in stiffness due to progressive crack-

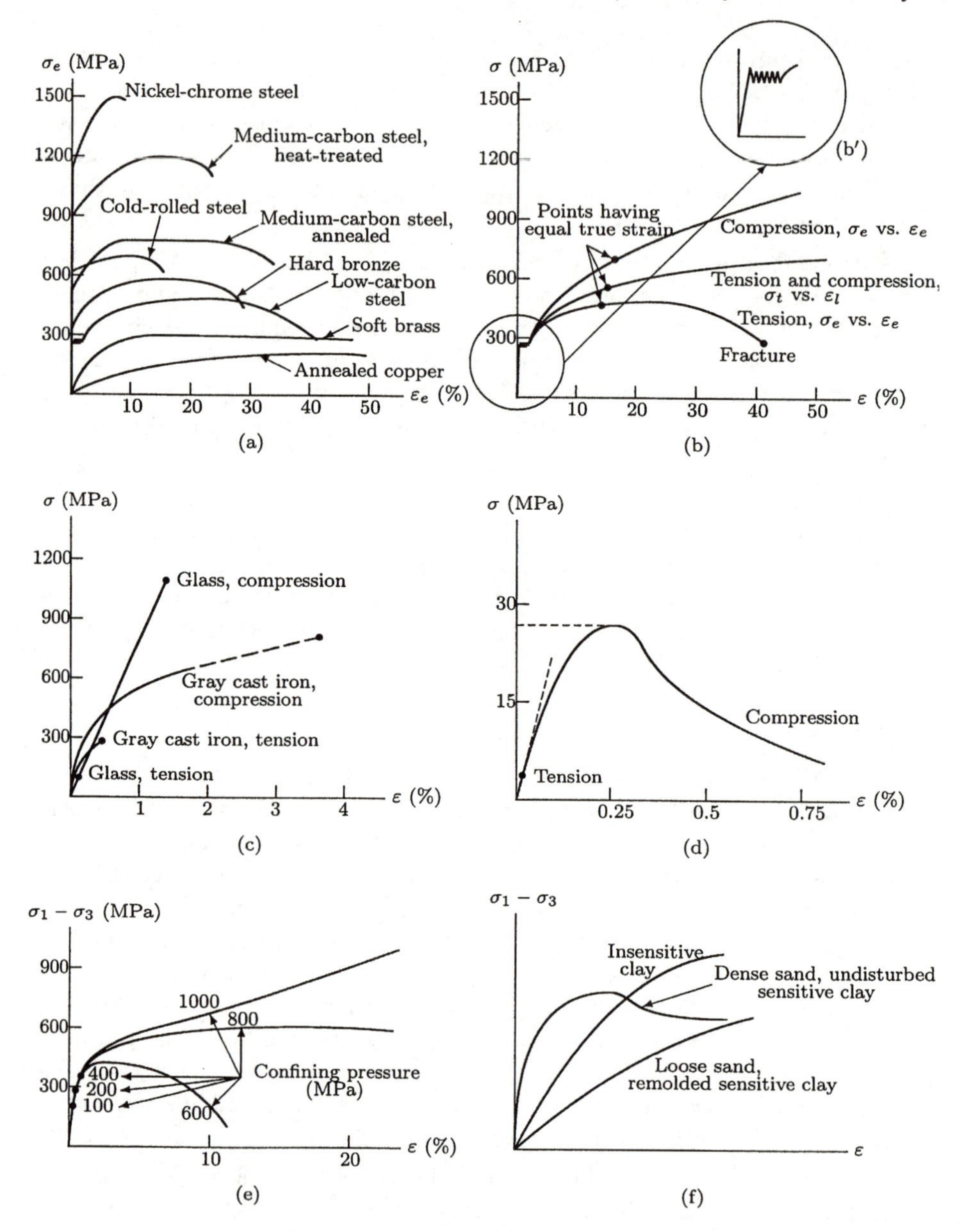

Figure 2.1.1. Stress-strain diagrams: (a) ductile metals, simple tension; (b) ductile metal (low-carbon steel), simple tension and compression; (b') yield-point phenomenon; (c) cast iron and glass, simple compression and tension; (d) typical concrete or rock, simple compression and tension; (e) rock (limestone), triaxial compression; (f) soils, triaxial compression.

ing. The post-peak portion of the curve is highly sensitive to test conditions and specimen dimensions, and therefore it cannot be regarded as a material property. Moreover, it is not compression per se that brings about fracture, but the accompanying shear stresses and secondary tensile stresses. Nevertheless, the superficial resemblance between the curves makes it possible to apply some concepts of plasticity to these materials, as discussed further in Section 2.3.

Unless the test is performed very quickly, soils are usually too soft to allow the use of a compression specimen without the application of a *confining pressure* to its sides through air or water pressure. This *confined compression test* or *triaxial shear test* is frequently applied to rock and concrete as well, for reasons discussed in Section 2.3. The specimen in this test is in an axisymmetric, three-dimensional stress state, the principal stresses being the longitudinal stress σ_1 and the confining pressure $\sigma_2 = \sigma_3$, both taken conventionally as **positive in compression**, in contrast to the usual convention of solid mechanics. The results are usually plotted as $\sigma_1 - \sigma_3$ (which, when positive — as it usually is — equals $2\tau_{\max}$) against the compressive longitudinal strain ε_1; typical curves are shown in Figure 2.1.1(e) and (f).

Elastic and Proportional Limits, Yield Strength

Some of the characteristic features of tensile stress-strain diagrams for ductile solids when rate sensitivity may be neglected will now be described. Such diagrams are characterized by a range of stress, extending from zero to a limiting stress (say σ_o) in which the stress is proportional to strain (the corresponding strains are normally so small that it does not matter which definitions of stress and strain are used); σ_o is called the *proportional limit*. Also, it is found that the same proportionality obtains when the stress is decreased, so that the material in this range is linearly elastic, described by the uniaxial Hooke's law given by Equation (1.4.12), that is, $\sigma = E\varepsilon$. The range of stress-strain proportionality is thus also essentially the *elastic range*, and the proportional limit is also the *elastic limit* as defined in Section 1.5.

When the specimen is stressed slightly past the elastic limit and the stress is then reduced to zero, the strain attained at the end of the process is, as a rule, different from zero. The material has thus acquired a *permanent strain*.

Rate effects, which are more or less present in all solids, can distort the results. The "standard solid" model of viscoelasticity discussed in 1.5.1, for example, predicts that in a test carried out at a constant rate of stressing or of straining, the stress-strain diagram will be curved, but no permanent strain will be present after stress removal; the complete loading-unloading diagram presents a hysteresis loop. The curvature depends on the test rate; it is negligible if the time taken for the test is *either very long or very short* compared with the characteristic time τ of the model.

Even in the absence of significant rate effects, it is not always easy to determine an accurate value for the elastic or proportional limit. Some materials, such as soft copper, present stress-strain curves that contain no discernible straight portions. In others, such as aluminum, the transition from the straight to the curved portion of the diagram is so gradual that the determination of the limit depends on the sensitivity of the strain-measuring apparatus. For design purposes, it has become conventional to define as the "yield strength" of such materials the value of the stress that produces a *specified* value of the "offset" or conventional permanent strain, obtained by drawing through a given point on the stress-strain curve a straight line whose slope is the elastic modulus E (in a material such as soft copper, this would be the slope of the stress-strain curve at the origin). Typically used values of the offset are 0.1%, 0.2% and 0.5%. When this definition is used, it is necessary to specify the offset, and thus we would speak of "0.2% offset yield strength."

2.1.2 Plastic Deformation

Plastic Strain, Work-Hardening

The strain defined by the offset may be identified with the inelastic strain as defined in 1.5.1. In the context in which rate sensitivity is neglected, this strain is usually called the *plastic strain*, and therefore, if it is denoted ε^p, it is given by

$$\varepsilon^p = \varepsilon - \frac{\sigma}{E}. \tag{2.1.1}$$

The plastic strain at a given value of the stress is often somewhat different from the permanent strain observed when the specimen is unloaded from this stress, because the stress-strain relation in unloading is not always ideally elastic, whether as a result of rate effects or other phenomena (one of which, the **Bauschinger effect**, is discussed below).

Additional plastic deformation results as the stress is increased. The stress-strain curve resulting from the initial loading into the plastic range is called the *virgin curve* or *flow curve*. If the specimen is unloaded after some plastic deformation has taken place and then reloaded, the reloading portion of the stress-strain diagram is, like the unloading portion, approximately a straight line of slope E, more or less up to the highest previously attained stress (that is, the stress at which unloading began). The reloading then follows the virgin curve. Similar results occur with additional unloadings and reloadings. The highest stress attained before unloading is therefore a new yield stress, and the material may be regarded as having been *strengthened* or *hardened* by the plastic deformation (or *cold-working*). The increase of stress with plastic deformation is consequently called *work-hardening* or *strain-hardening*.

The virgin curve of work-hardening solids, especially ones without a sharply defined yield stress, is frequently approximated by the **Ramberg–Osgood formula**

$$\varepsilon = \frac{\sigma}{E} + \alpha\frac{\sigma_R}{E}\left(\frac{\sigma}{\sigma_R}\right)^m, \tag{2.1.2}$$

where α and m are dimensionless constants,[1] and σ_R is a reference stress. If m is very large, then ε^p remains small until σ approaches σ_R, and increases rapidly when σ exceeds σ_R, so that σ_R may be regarded as an approximate yield stress. In the limit as m becomes infinite, the plastic strain is zero when $\sigma < \sigma_R$, and is indeterminate when $\sigma = \sigma_R$, while $\sigma > \sigma_R$ would produce an infinite plastic strain and is therefore impossible. This limiting case accordingly describes a *perfectly plastic* solid with yield stress σ_R.

If the deformation is sufficiently large for the elastic strain to be neglected, then Equation (2.1.2) can be solved for σ in terms of ε:

$$\sigma = C\varepsilon^n, \tag{2.1.3}$$

where $C = \sigma_R(E/\alpha\sigma_R)^n$, and $n = 1/m$ is often called the *work-hardening exponent*. Equation (2.1.3), proposed by Ludwik [1909], is frequently used in applications where an explicit expression for stress as a function of strain is needed. Note that the stress-strain curve representing (2.1.3) has an infinite initial slope. In order to accommodate an elastic range with an initial yield stress σ_E, Equation (2.1.3) is sometimes modified to read

$$\sigma = \begin{cases} E\varepsilon, & \varepsilon \le \dfrac{\sigma_E}{E}, \\ \sigma_E\left(\dfrac{E\varepsilon}{\sigma_E}\right)^n, & \varepsilon \ge \dfrac{\sigma_E}{E}. \end{cases} \tag{2.1.4}$$

Ultimate Tensile Strength

It must be emphasized that when the strain is greater than a few percent, the distinction between the different types of stress and strain must be taken into account. The decomposition (2.1.1) applies, strictly speaking, to the logarithmic strain. The nature of the stress-strain curve at larger strains is, as discussed above, also highly dependent on whether the stress plotted is nominal or true stress [see Figure 2.1.1(b)]. True stress is, in general, an increasing function of strain until fracture occurs. Since the cross-sectional area of the specimen decreases with elongation, the nominal stress increases more slowly, and at a certain point in the test it begins to decrease. Since, very nearly, $\sigma_e = \sigma_t exp(-\varepsilon_l)$, it follows that

$$d\sigma_e = (d\sigma_t - \sigma_t d\varepsilon_l)exp(-\varepsilon_l),$$

[1] If m is a number other than an odd integer, then $size - 2(\sigma/\sigma_R)^m$ may be replaced by $size - 2|\sigma/\sigma_R|^{m-1}(\sigma/\sigma_R)$ if the curve is the same for negative as for positive stress and strain.

and therefore the nominal stress (and hence the load) is maximum when

$$\frac{d\sigma_t}{d\varepsilon_l} = \sigma_t.$$

If Equation (2.1.3) is assumed to describe the flow curve in terms of Cauchy stress and logarithmic strain, then the maximum nominal stress can easily be seen to occur when $\varepsilon_l = n$.

The maximum value of nominal stress attained in a tensile test is called the *ultimate tensile strength* or simply the *tensile strength*. When the specimen is extended beyond the strain corresponding to this stress, its weakest portion begins to elongate — and therefore also to thin — faster than the remainder, and so a *neck* will form. Further elongation and thinning of the neck — or *necking* — proceeds at decreasing load, until fracture.

Discontinuous Yielding

The stress-strain curves of certain impurity-containing metals, such as mild steel and nitrogen-containing brass, present a phenomenon known as *discontinuous yielding*. When the initial elastic limit is reached, suddenly a significant amount of stretching (on the order of 1 or 2%, and thus considerably larger than the elastic strain achieved up to that point) occurs at essentially constant stress, of a value equal to or somewhat lower than the initial elastic limit. If the value is the same, then it is called the *yield point* of the material. If it is lower, then it is called the *lower yield point*, while the higher value is called the *upper yield point*. The portion of the stress-strain diagram represented by the constant stress value is called the *yield plateau*, and the drop in stress, if any, that precedes it is called the *yield drop*. Following the plateau, work-hardening sets in, as described above. Figure 2.1.1(b') shows a typical stress-strain diagram for a material with a yield point.

As shown in the figure, the stress on the plateau is not really constant but shows small, irregular fluctuations. They are due to the fact that plastic deformation in this stage is not a homogeneous process but concentrated in discrete narrow zones known as *Lüders bands*, which propagate along the specimen as it is stretched, giving rise to surface marks called *stretcher strains*.

When a specimen of a material with a yield point is loaded into the work-hardening range, unloaded, and reloaded soon after unloading, the virgin curve is regained smoothly as described previously. If, however, some time — of the order of hours — is allowed to elapse before reloading, the yield point recurs at a higher stress level (see Figure 2.1.2). This phenomenon is called *strain aging*.

Bauschinger Effect, Anisotropy

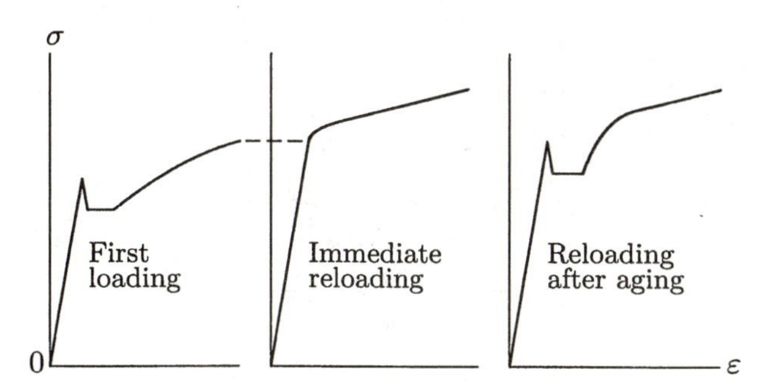

Figure 2.1.2. Strain aging

specimen of a ductile material that has been subjected to increasing tensile stress and then unloaded ("cold-worked") is different from a virgin specimen. We already know that it has a higher tensile yield stress. If, however, it is now subjected to increasing *compressive* stress, it is found that the yield stress in compression is *lower* than before. This observation is known as the **Bauschinger effect** [see Figure 2.1.3(a)].

The Bauschinger effect can be observed whenever the direction of straining is reversed, as, for example, compression followed by tension, or shearing (as in a torsion test on a thin-walled tube) followed by shearing in the opposite direction. More generally, the term "Bauschinger effect" can be used to describe the lowering of the yield stress upon reloading that follows unloading, even if the reloading is in the same direction as the original loading (Lubahn and Felgar [1961]) [see Figure 2.1.3(b)]. Note the hysteresis loop which appears with large strains, even at very slow rates of straining at which the viscoelastic effects mentioned above may be neglected.

Another result of plastic deformation is the loss of isotropy. Following cold-working in a given direction, differences appear between the values of the tensile yield strength in that direction and in a direction normal to it. These differences may be of the order of 10%, but are usually neglected in practice.

Annealing, Recovery

The term "cold-working" used in the foregoing discussions refers to plastic deformation carried out at temperatures below the so-called *recrystallization temperature* of the metal, typically equal, in terms of absolute temperature, to some 35 to 50% of the melting point (although, unlike the melting point, it is not sharply defined); the reason for the name is explained in the next section. The effects of cold-working, such as work-hardening, the Bauschinger effect, and induced anisotropy, can largely be removed by a process called *annealing*, consisting of heating the metal to a relatively high tem-

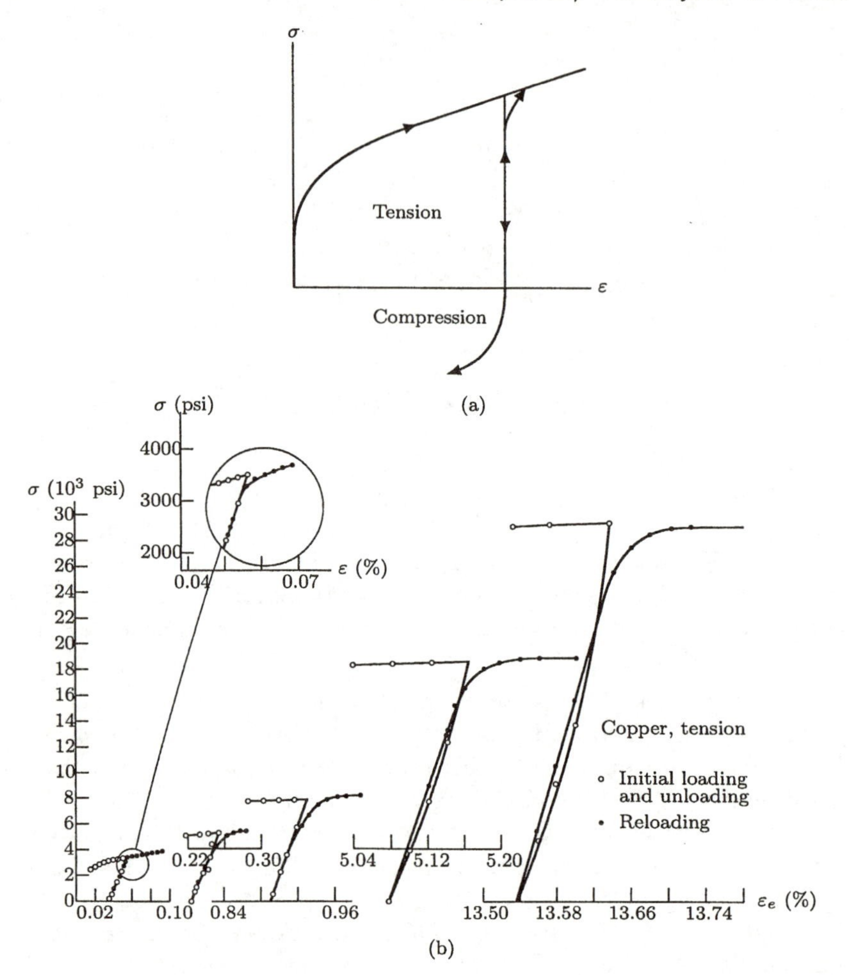

Figure 2.1.3. Bauschinger effect: (a) classical; (b) generalized (from Lubahn and Felgar [1961]).

perature (above the recrystallization temperature) and holding it there for a certain length of time before slowly cooling it. The length of time necessary for the process decreases with the annealing temperature and with the amount of cold work.

Plastic deformation that takes place at temperatures in the annealing range (i.e., above the recrystallization temperature) is known as *hot-working*, and does not produce work-hardening, anisotropy, or the Bauschinger effect. For metals with low melting points, such as lead and tin, the recrystallization temperature is about 0°C and therefore deformation at room temperature must be regarded as hot-working. Conversely, metals with very high melting points, such as molybdenum and tungsten (with recrystallization temperatures of 1100 to 1200°C can be "cold-worked" at temperatures at which the metal is red-hot.

The recrystallization temperature provides a qualitative demarcation between stress-strain diagrams that show work-hardening and those that do not. Within each of the two ranges, however, the stress needed to achieve a given plastic deformation at a given strain rate also depends on the temperature. In particular, it decreases with increasing temperature (see Figure 2.1.4).

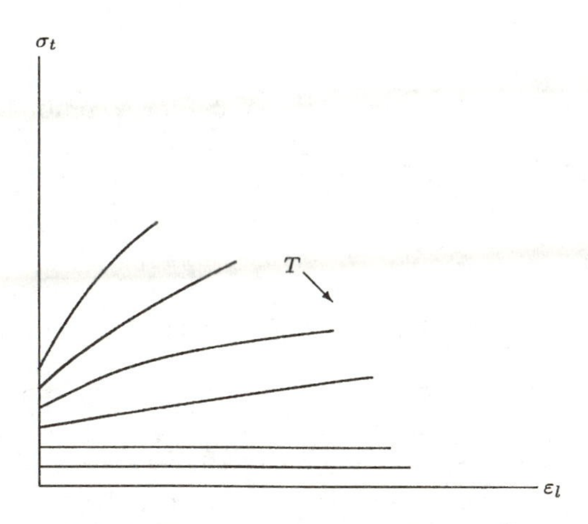

Figure 2.1.4. Temperature dependence of flow stress

A characteristic of some metals (including mild steel), with important implications for design, is a change of behavior from ductile to brittle when the temperature falls below the so-called *transition temperature*.

Softening (that is, a spontaneous decrease in yield strength) of work-hardened metals also occurs at temperatures below recrystallization. This process, whose rate is considerably slower than that of annealing, is called *recovery*. The rate of recovery decreases with decreasing temperature, and is negligible at room temperature for such metals as aluminum, copper and steel. These metals may accordingly be regarded for practical purposes as work-hardening permanently.

2.1.3 Temperature and Rate Dependence

The preceding discussion of the rates of annealing and recovery shows the close relationship between temperature and rate. A great many physico-chemical rate processes — specifically, those that are *thermally activated* — are governed by the **Arrhenius equation**, which has the general form

$$\text{rate} \propto e^{-\Delta E/kT}, \tag{2.1.5}$$

where k is Boltzmann's constant (1.38×10^{-23} J/K), T is the absolute temperature, and ΔE is the *activation energy* of the process. The rate sensitivity of the work-hardening stress-strain curve itself increases with increasing temperature. In a good many metals, the dependence on the plastic strain rate of the stress required to achieve a given plastic strain can be approximated quite well by $\dot{\varepsilon}^r$, where the exponent r (sometimes called simply the rate sensitivity) depends on the plastic strain and the temperature, increasing with both. Some typical results for r, obtained from tests at strain rates between 1 and 40 per second, are shown in Table 2.1.1.

Table 2.1.1

Metal	Temperature	Value of r for a compression of		
	(°C)	10%	30%	50%
Aluminum	18	0.013	0.018	0.020
	350	0.055	0.073	0.088
	550	0.130	0.141	0.155
Copper	18	0.001	0.002	0.010
	450	0.001	0.008	0.031
	900	0.134	0.154	0.190
Mild steel	930	0.088	0.094	0.105
	1200	0.116	0.141	0.196

Source: Johnson and Mellor [1973].

The Arrhenius equation (2.1.5) permits, in principle, the simultaneous representation of the rate sensitivity and temperature sensitivity of the stress-strain relation by means of the parameter $\dot{\varepsilon} exp(\Delta E/RT)$, or, more generally, $\dot{\varepsilon} f(T)$, where $f(T)$ is an experimentally determined function, since the activation energy ΔE may itself be a function of the temperature.

Creep

The preceding results were obtained from tests carried out at constant strain rate (since the strains are large, total and plastic strain need not be distinguished). Following Ludwik [1909], it is frequently assumed that at a given temperature, a relation exists among stress, plastic (or total) strain, and plastic (or total) strain rate, independently of the process, and therefore this relation also describes *creep*, that is, continuing deformation at constant stress. Such a relation is reminiscent of the "standard solid" model of viscoelasticity, in which this relation is linear. It will be recalled that this model describes both the rate dependence of the stress-strain relation (discussed above in this section) and the increasing deformation at constant stress known as creep, which in this case asymptotically attains a finite value (*bounded creep*), though in the limiting case of the Maxwell model it becomes steady creep. In fact, all linear spring-dashpot models of viscoelasticity lead

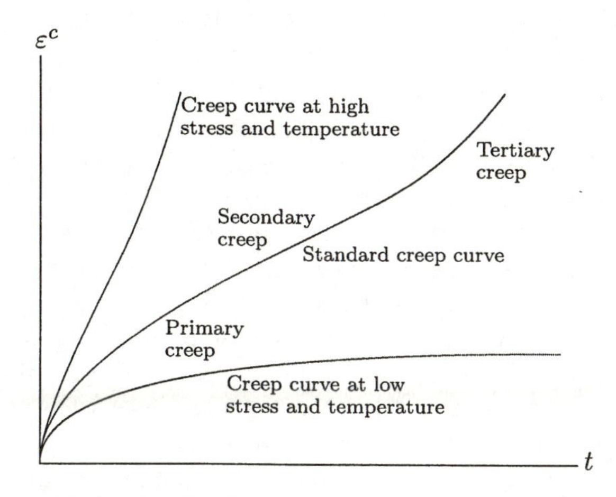

Figure 2.1.5. Typical creep curves for metals.

to creep that is either bounded or steady.

For metals, the relation, if it exists, is nonlinear — many different forms have been proposed — and therefore the resulting creep need not belong to one of the two types predicted by the linear models. Typical creep curves for a metal, showing the creep strain ε^c (equal to the total strain less the initial strain) as a function of time at constant stress and temperature, are shown in Figure 2.1.5. The standard curve is conventionally regarded as consisting of three stages, known respectively as primary (or transient), secondary (or steady), and tertiary (or accelerating) creep, though not all creep curves need contain all three stages. At low stresses and temperatures, the primary creep resembles the bounded creep of linear viscoelasticity, with a limiting value attained asymptotically, and secondary and tertiary creep never appear. At higher stress or temperature, however, the primary creep shows a logarithmic or a power dependence on time:

$$\varepsilon^c \propto \ln t \quad \text{or} \quad \varepsilon^c \propto t^\alpha,$$

where α is between 0 and 1, a frequently observed value being $\frac{1}{3}$ (**Andrade's creep law**). The logarithmic form is usually found to prevail below, and the power form above, the recrystallization temperature.

Creep described by the power law can be derived from a formula relating stress, creep strain and creep-strain rate that has the form (due to Nadai [1950])

$$\sigma = C(\varepsilon^c)^n(\dot{\varepsilon}^c)^r, \tag{2.1.6}$$

where C, n, and r depend on the temperature; this formula reduces to the Ludwik equation (2.1.3) at constant strain rate, and implies a rate sensitivity that is independent of the strain. At constant stress, the equation can be integrated, resulting in a power law with $\alpha = r/(n+r)$.

Tertiary (accelerating) creep is generally regarded as resulting from structural changes leading to a loss of strength and, eventually, fracture. Whether secondary (steady) creep really takes place over a significant time interval, or is merely an approximate description of creep behavior near an inflection point of the creep curve, is not certain (see Lubahn and Felgar [1961], pp. 136–141). In either case, however, one may speak of a *minimum creep rate* characteristic of the metal at a given stress and temperature, if these are sufficiently high. At a given stress, the temperature dependence of this minimum creep rate is usually found to be given fairly closely by the Arrhenius equation. Its dependence on stress at a given temperature can be approximated by an exponential function at higher stresses, and by a power function of the form $\dot{\varepsilon}^c_{\min} \propto \sigma^q$, where q is an exponent greater than 1 (the frequently used **Bailey–Norton–Nadai law**), at lower stresses. (Note that Equation (2.1.6) describes the Bailey–Norton law if $n = 0$ and $r = 1/q$.) A commonly used approximation for the creep strain as a function of time, at a given stress and temperature, is

$$\varepsilon^c(t) = \varepsilon^c_0 + \dot{\varepsilon}^c_{\min} t,$$

where $\dot{\varepsilon}^c_{\min}$ is the minimum creep rate, and ε^c_0 is a fictitious initial value defined by the ε^c-intercept of the straight line tangent to the actual creep curve at the point of inflection or in the steady-creep portion.

In many materials at ordinary temperatures, rate-dependent inelastic deformation is insignificant when the stress is below a yield stress. A simple model describing this effect is the **Bingham model**:

$$\dot{\varepsilon}^i = \begin{cases} 0, & |\sigma| < \sigma_Y, \\ \left(1 - \dfrac{\sigma_Y}{|\sigma|}\right)\dfrac{\sigma}{\eta}, & |\sigma| \geq \sigma_Y, \end{cases} \tag{2.1.7}$$

where η is a viscosity, and the yield stress σ_Y may depend on strain. The Bingham model is the simplest model of **viscoplasticity**. Its generalizations are discussed in Section 3.1.

Exercises: Section 2.1

1. Show that the relation between the conventional strain ε_e and the logarithmic strain ε_l is $\varepsilon_l = \ln(1 + \varepsilon_e)$.

2. It is assumed that the stress-strain relations of isotropic linear elasticity, with Young's modulus E and Poisson's ratio ν, are exact in terms of true stress and logarithmic strain. For uniaxial stress, find the relation (parametric if necessary) between the conventional stress and the

conventional strain. Show that the second-order approximation to the relation is $\sigma_e = E[\varepsilon_e - (\frac{1}{2} + 2\nu)\varepsilon_e^2]$.

3. A uniaxial tension test produces a curve of true stress against logarithmic strain that is fitted by $\sigma_t = 2 \times 10^5 \varepsilon_l$ in the elastic range and $\sigma_t = 635\varepsilon_l^{1/6}$ in the plastic range, with stresses in MPa. Determine (a) the elastic-limit stress, (b) the logarithmic and conventional strains at maximum load, and (c) the true and conventional stresses at maximum load, assuming negligible volume change.

4. If the reference stress σ_R in the Ramberg–Osgood formula (2.1.2) is the offset yield strength for a given permanent strain ε_R, find α in terms of σ_R, ε_R, and E.

5. Find a formula describing a stress-strain relation that (a) is linear for $\sigma < \sigma_E$, (b) asymptotically tends to $\varepsilon \propto \sigma^m$, and (c) is smooth at $\sigma = \sigma_E$.

6. Suppose that in Equation (2.1.6) only C depends on the temperature. Show that, for a given stress, the creep curves corresponding to different temperatures are parallel if they are plotted as creep strain against the logarithm of time.

7. Determine the form of the creep law resulting from Equation (2.1.6).

8. Assuming $\varepsilon = \sigma/E + \varepsilon^c$, and letting $n = 0$ in Equation (2.1.6), determine the resulting relaxation law, i. e. σ as a function of t when a strain ε is suddenly imposed at $t = 0$ and maintained thereafter.

9. To what does the Bingham model described by Equation (2.1.7) reduce when $\sigma_Y = 0$? When $\eta = 0$?

Section 2.2 Crystal Plasticity

2.2.1 Crystals and Slip

Crystal Structure

Plasticity theory was developed primarily in order to describe the behavior of ductile metals. Metals in their usual form are *polycrystalline aggregates*, that is, they are composed of large numbers of grains, each of which has the structure of a simple crystal.

A crystal is a three-dimensional array of atoms forming a regular lattice; it may be regarded as a molecule of indefinite extent. The atoms vibrate about fixed points in the lattice but, by and large, do not move away from them, being held more or less in place by the forces exerted by neighboring atoms. The forces may be due to ionic, covalent, or metallic bonding. Ionic bonds result from electron transfer from electropositive to electronegative atoms, and therefore can occur only in compounds of unlike elements. Ionic crystal structures range from very simple, such as the sodium chloride structure in which Na^+ and Cl^- alternate in a simple cubic array, to the very complex structures found in ceramics. Covalent bonds are due to the sharing of electrons, and are found in diamond and, to some extent, in crystalline polymers.

In a metallic crystal, the outer or valence electrons move fairly freely through the lattice, while the "cores" (consisting of the nucleus and the filled shells of electrons) vibrate about the equilibrium positions. The metallic bond is the result of a rather complex interaction among the cores and the "free" electrons. It is the free electrons that are responsible for the electrical and thermal conductivity of metals.

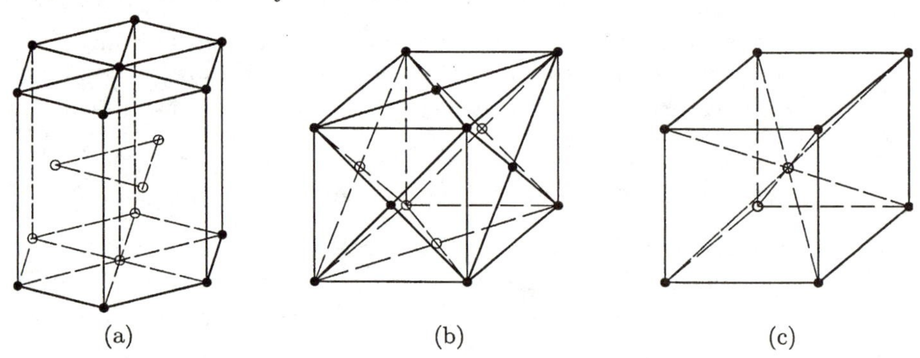

Figure 2.2.1. Crystal structures: (a) hexagonal close-packed (hcp); (b) face-centered cubic (fcc); (c) body-centered cubic (bcc).

The most common crystal structures in metals are the hexagonal close-packed (hcp), face-centered cubic (fcc) and body-centered cubic (bcc), shown in Figure 2.2.1. Because of the random orientation of individual grains in a typical metallic body, the overall behavior of the aggregate is largely isotropic, but such phenomena as the Bauschinger effect and preferred orientation, which occur as a result of different plastic deformation of grains with different orientations, demonstrate the effect of crystal structure on plastic behavior. It is possible, however, to produce specimens of crystalline solids — not only metals — in the form of *single crystals* of sufficiently large size to permit mechanical testing.

Crystal Elasticity

The linear elastic behavior of a solid is described by the elastic modulus matrix $\underline{C}$ defined in 1.4.2. The most general anisotropic solid has 21 independent elements of $\underline{C}$. For the isotropic solid, on the other hand, the only nonzero elements of $\underline{C}$ are (a) $C_{11} = C_{22} = C_{33}$, (b) $C_{44} = C_{55} = C_{66}$, and (c) $C_{12} = C_{13} = C_{23}$ (the symmetry $C_{IJ} = C_{JI}$ is not explicitly shown). But only two of the three values are independent, since $C_{11} = \lambda + 2\mu$, $C_{44} = \mu$, and $C_{12} = \lambda$, so that

$$C_{44} = \frac{1}{2}(C_{11} - C_{12}).$$

In a crystal with cubic symmetry (such as simple cubic, fcc or bcc), with the Cartesian axes oriented along the cube edges, the nonzero elements of $\underline{C}$ are the same ones as for the isotropic solid, but the three values C_{11}, C_{12} and C_{44} are independent. It may, of course, happen fortuitously that the isotropy condition expressed by the preceding equation is satisfied for a given cubic crystal; this is the case for tungsten.

A crystal with hexagonal symmetry is isotropic in the basal plane. Thus, if the basal planes are parallel to the x_1x_2-plane, $C_{66} = \frac{1}{2}(C_{11} - C_{12})$. The following elements of $\underline{C}$ are independent: (a) $C_{11} = C_{22}$, (b) C_{33}, (c) C_{12}, (d) $C_{13} = C_{23}$, and (e) $C_{44} = C_{55}$.

The anisotropy of crystals is often studied by performing tension tests on specimens with different orientations, resulting in orientation-dependent values of the Young's modulus E. If the maximum and minimum values are denoted E_{max} and E_{min}, respectively, while E_{ave} denotes the polycrystalline average, the *anisotropy index* may be defined as $(E_{\text{max}} - E_{\text{min}})/E_{\text{ave}}$. Values range widely: 1.13 for copper, 0.73 for α-iron, 0.2 for aluminum, and, as indicated above, zero for tungsten.

Crystal Plasticity

Experiments show that plastic deformation is the result of relative motion, or *slip*, on specific crystallographic planes, in response to shear stress along these planes. It is found that the *slip planes* are most often those that are parallel to the planes of closest packing; a simple explanation for this is that the separation between such planes is the greatest, and therefore slip between them is the easiest, since the resistance to slip as a result of interatomic forces decreases rapidly with interatomic distance. Within each slip plane there are in turn preferred *slip directions*, which once more are those of the atomic rows with the greatest density, for the same reason. A slip plane and a slip direction together are said to form a *slip system*.

In hcp crystals, which include zinc and magnesium, the planes of closest packing are those containing the hexagons, and the slip directions in those planes are parallel to the diagonals. Hexagonal close-packed crystals therefore have three primary slip systems, although at higher temperatures other, secondary, slip systems may become operative.

Face-centered cubic crystals, by contrast, have twelve primary slip systems: the close-packed planes are the four octahedral planes, and each contains three face diagonals as the closest-packed lines. As a result, fcc metals, such as aluminum, copper, and gold, exhibit considerably more ductility than do hcp metals.

In body-centered cubic crystals there are six planes of closest packing and two slip directions in each, for a total of twelve primary slip systems. However, the difference in packing density between the closest-packed planes and certain other planes is not great, so that additional slip systems become available even at ordinary temperatures. Consequently, metals having a bcc structure, such as α-iron (the form of iron found at ordinary temperatures), tungsten, and molybdenum, have a ductility similar to that of fcc metals.

The preceding correlation between ductility and lattice type is valid in very broad terms. Real metal crystals almost never form perfect lattices containing one type of atom only; they contain imperfections such as geometric lattice defects and impurity atoms, besides the grain boundaries contained in polycrystals. In fact, these imperfections are the primary determinants of crystal plasticity. Ductility must therefore be regarded as a *structure-sensitive* property, as are other inelastic properties. It is only the thermoelastic properties discussed in 1.4.1 — the elastic moduli, thermal stress (or strain) coefficients, and specific heat — that are primarily influenced by the ideal lattice structure, and are therefore called *structure-insensitive*.

Slip Bands

In principle, slip in a single crystal can occur on every potential slip plane when the necessary shear stress is acting. Observations, however, show slip to be confined to discrete planes.[1] When a slip plane intersects the outer surface, an observable *slip line* is formed, and slip lines form clusters called *slip bands*. In a given slip band, typically, a new slip line forms at a distance of the order of 100 interatomic spacings from the preceding one when the amount of slip on the latter has reached something of the order of 1,000 interatomic spacings. It follows from these observations that slip does not take place by a uniform relative displacement of adjacent atomic planes.

Critical Resolved Shear Stress

It was said above that slip along a slip plane occurs in response to shear stress on that plane. In particular, in a tensile specimen of monocrystalline metal in which the tensile stress σ acts along an axis forming an angle ϕ with the normal to the slip plane and an angle λ with the slip direction, then the relation between σ and the resolved shear stress on the slip plane and in the slip direction, τ, is

$$\sigma = (\cos\phi\cos\lambda)^{-1}\tau. \tag{2.2.1}$$

It was found by Schmid [1924], and has been confirmed by many experiments,

[1]Or, more generally, surfaces (*slip surfaces*), since slip may transfer from one slip plane to another which intersects it in the interior of the crystal, especially in bcc metals.

that slip in a single crystal is initiated when the resolved shear stress on some slip system reaches a critical value τ_c, which is a constant for a given material at a given temperature and is known as the *critical resolved shear stress*. This result is called **Schmid's law.** The critical resolved shear stress is, as a rule, very much higher for bcc metals (iron, tungsten) than for fcc metals (aluminum, copper) or hcp metals (zinc, magnesium).

Theoretical Shear Strength

A value of the shear stress necessary to produce slip may be calculated by assuming that slip takes place by the uniform displacement of adjacent atomic planes. Consider the two-dimensional picture in Figure 2.2.2: two

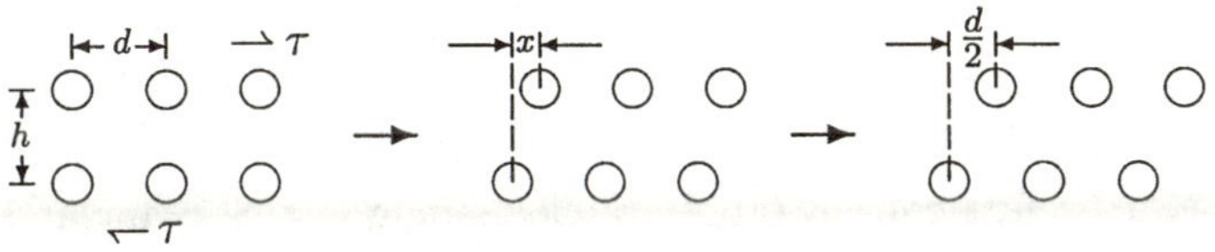

Figure 2.2.2. Slip between two neighboring rows of atoms

neighboring rows of atoms, the distance between the centers of adjacent atoms in each row being d, and the distance between the center lines of the two rows being h. Suppose the two rows to be in a stable equilibrium configuration under zero stress. If one row is displaced by a distance d relative to the other, a new configuration is achieved that is indistinguishable from the first. A displacement of $d/2$, on the other hand, would lead to an unstable equilibrium configuration at zero stress. As a first approximation, then, the shear stress necessary to produce a relative displacement x may be assumed to be given by

$$\tau = \tau_{\max} \sin \frac{2\pi x}{d}, \tag{2.2.2}$$

and slip would proceed when $\tau = \tau_{\max}$. When the displacement x is small, the stress-displacement relation is approximately linear: $\tau = 2\pi\tau_{\max}x/d$. But a small displacement x between rows a distance h apart corresponds to a lattice shear of $\gamma = x/h$, and Hooke's law in shear reads $\tau = G\gamma$ [Equation (1.4.15)]. Consequently,

$$\tau_{\max} = \frac{Gd}{2\pi h}.$$

Since $h \equiv d$, the value $G/6$ is a first, structure-insensitive approximation to the so-called *theoretical shear strength* of a crystal.

More refined calculations that take actual crystal structures into account reduce the value of the theoretical shear strength to about $G/30$. In reality, however, the shear strength of single crystals is less than this by one to three orders of magnitude, that is, it is of order $10^{-3}G$ to $10^{-5}G$. Only in so-called *whiskers*, virtually perfect crystals about 1 μm in diameter, is a shear strength of the theoretical order of magnitude observed.

2.2.2. Dislocations and Crystal Plasticity

The discrepancy between theoretical and observed shear strength, as well as the observation of slip bands, have led to the inevitable conclusion that slip in ordinary crystals must take place by some mechanism other than the movement of whole planes of atoms past one another, and that it is somehow associated with lattice defects. A mechanism based on a specific defect called a *dislocation* was proposed independently by G. I. Taylor [1934] and E. Orowan [1934].

Defects in Crystals

All real crystals contain defects, that is, deviations from the ideal crystal structure. A defect concentrated about a single lattice point and involving only a few atoms is called a *point defect*; if it extends along a row of many atoms, it is called a *line defect*; and if it covers a whole plane of atoms, a *planar defect*.

Point defects are shown in Figure 2.2.3. They may be purely structural, such as (a) a vacancy or (b) an interstitial atom, or they may involve foreign atoms (*impurities*): (c) a substitutional impurity, (d) an interstitial impurity. As shown in the figure, point defects distort the crystal lattice locally,

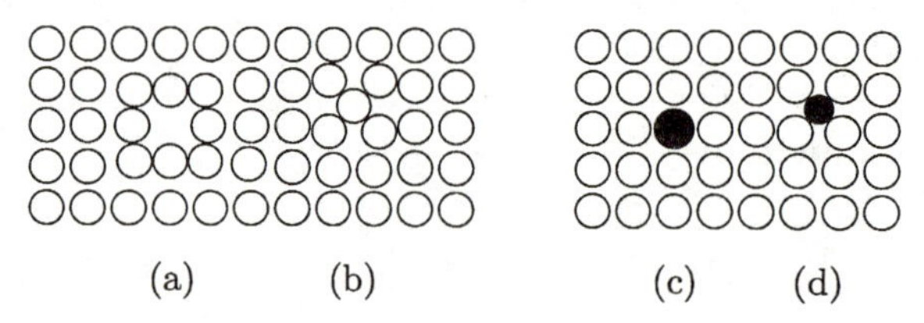

Figure 2.2.3. Point defects: (a) vacancy; (b) interstitial atom; (c) substitutional impurity; (d) interstitial impurity.

the distortion being significant over a few atomic distances but negligible farther away. Planar defects, illustrated in Figure 2.2.4, include (a) grain boundaries in polycrystals, and within single crystals, (b) twin boundaries and (c) stacking faults.

Dislocations

The most important line defects in crystals are *dislocations*. The concept of a dislocation has its origin in continuum mechanics, where it was introduced by V. Volterra. Consider a hollow thick-walled circular cylinder in which a radial cut, extending through the wall, is made [see Figure

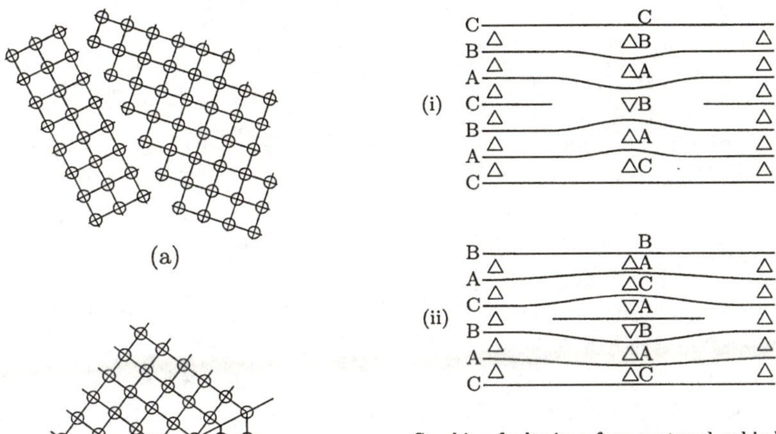

Stacking faults in a face-centered cubic lattice. The normal stacking sequence of (111) planes is denoted by ABCA... Planes in normal relation to one another are separated by Δ, those with a stacking error by ∇; (i) intrinsic stacking fault, (ii) extrinsic stacking fault.

(b) (c) (from Hull and Bacon [1984])

Figure 2.2.4. Planar defects: (a) grain boundary; (b) twin boundary; (c) stacking fault.

2.2.5(a)]. The two faces of the cut may be displaced relative to each other by a distance b, either in the (b) radial or (c) axial direction, and then reattached. The result is a *Volterra dislocation*, with Figures 2.2.5(b) and (c)

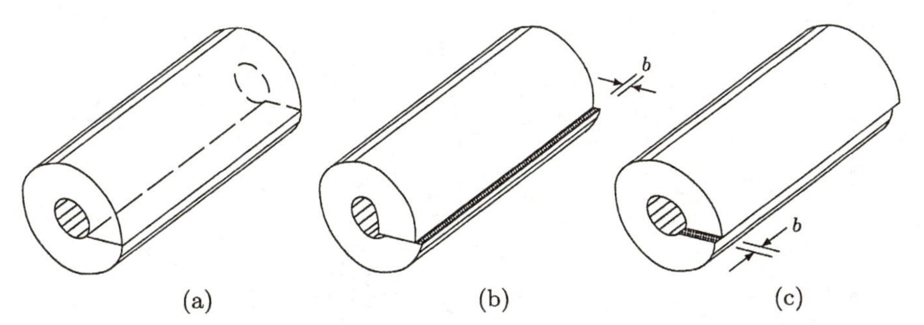

Figure 2.2.5. Volterra dislocation: (a) Volterra cut; (b) edge dislocation; (c) screw dislocation.

representing respectively an *edge* and a *screw* dislocation. When the rough edges are smoothed, the result is a cylinder looking much as it did before the operation, but containing a self-equilibrating internal stress field. If the material is isotropic and linearly elastic, then the stress and displacement fields can be calculated by means of the theory of elasticity. In particular, the strain energy per unit length of cylinder is found to be

$$W' = \frac{Gb^2}{4\pi(1-\nu)}\left(\ln\frac{R}{a} - 1\right) \tag{2.2.3a}$$

for an edge dislocation and

$$W' = \frac{Gb^2}{4\pi}\left(\ln\frac{R}{a} - 1\right) \tag{2.2.3b}$$

for a screw dislocation, where G is the shear modulus, ν is the Poisson's ratio, and R and a are respectively the outer and inner radii of the cylinder.

An edge dislocation in a crystal can be visualized as a line on one side of which an extra half-plane of atoms has been introduced, as illustrated in Figure 2.2.6(a) for a simple cubic lattice. At a sufficient number of atomic distances away from the dislocation line, the lattice is virtually undisturbed. Consider, now, a path through this "good" crystal which would be closed if the lattice were perfect. If such a path, consisting of the same number of atom-to-atom steps in each direction, encloses a dislocation, then, as shown in the figure, it is not closed; the vector **b** needed to close it is called the *Burgers vector* of the dislocation, and the path defining it is called the *Burgers circuit*.

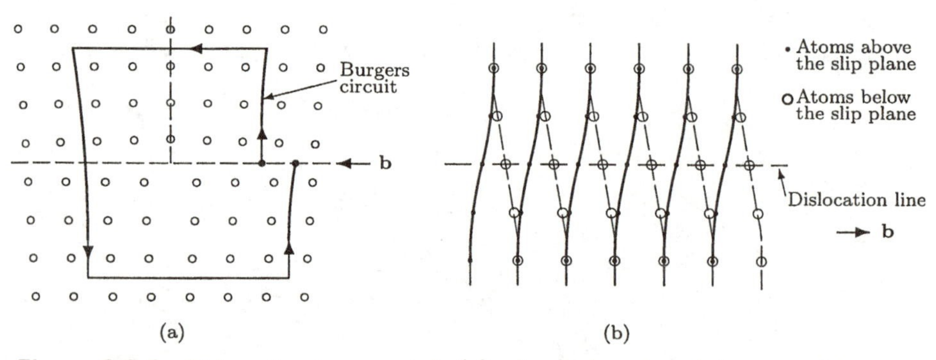

Figure 2.2.6. Dislocation in a crystal: (a) edge dislocation; (b) screw dislocation.

Note that, for an edge dislocation, the Burgers vector is necessarily perpendicular to the dislocation line. Indeed, this can be used as the defining property of an edge dislocation. Similarly, a screw dislocation can be defined as one whose Burgers vector is parallel to the dislocation line [see Figure 2.2.6(b)].

A dislocation in a crystal need not be a straight line. However, the Burgers vector must remain constant. Thus, a dislocation can change from edge to screw, or vice versa, if it makes a right-angle turn. It cannot, moreover, terminate inside the crystal, but only at the surface of a crystal or at a grain boundary. It can form a closed loop, or branch into other dislocations (at points called *nodes*), subject to the **conservation of the Burgers vectors**: the sum of the Burgers vectors of the dislocations meeting at a node must vanish if each dislocation is considered to go into the node (Frank [1951]).

Dislocations and Slip

It is now universally accepted that plastic deformation in crystals results from the movement of dislocations. As can be seen from Figure 2.2.7, in order

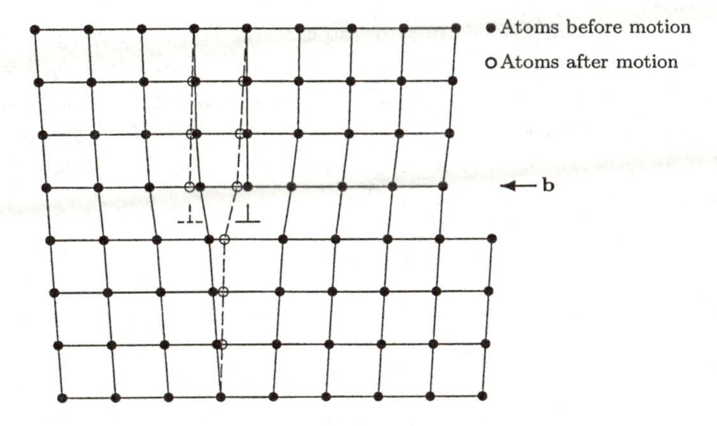

Figure 2.2.7. Slip by means of an edge dislocation.

for an edge dislocation to move one atomic distance in the plane containing it and its Burgers vector (the slip plane), each atom need move only a small fraction of an atomic distance. Consequently, the stress required to move the dislocation is only a small fraction of the theoretical shear strength discussed in 2.2.1. An approximate value of this stress is given by the *Peierls–Nabarro stress*,

$$\tau_{\text{PN}} = \frac{2G}{1-\nu} \exp\left[-\frac{2\pi h}{d(1-\nu)}\right],$$

where h and d denote, as before, the distances between adjacent planes of atoms and between atoms in each plane, respectively. The Peierls–Nabarro stress is clearly much smaller than the theoretical shear strength. Its value, moreover, depends on h/d, and the smallest value is achieved when h/d is largest, that is, for close-packed planes that are as far apart as possible; this result explains why such planes are the likeliest slip planes. When $h = \sqrt{2}d$, τ_{PN} is of the order $10^{-5}G$, consistent with the observed shear strength of pure single crystals.

If the stress is maintained, the dislocation can move to the next position, and the next, and so on. As the dislocation moves in its slip plane, the portion of the plane that it leaves behind can be regarded as having experienced slip of the amount of one Burgers-vector magnitude $b = |\mathbf{b}|$. When the dislocation reaches the crystal boundary, slip will have occurred on the entire slip plane. Suppose that the length of the slip plane is s, and that an edge dislocation moves a distance x in the slip plane; then it contributes a displacement bx/s, so that n dislocations moving an average distance $\bar{x}$ produce a displacement $u = nb\bar{x}/s$. If the average spacing between slip planes is l, then the plastic shear strain is

$$\gamma^p = \frac{u}{l} = \frac{nb\bar{x}}{ls}.$$

However, n/ls is just the average number of dislocation lines per unit perpendicular area, or, equivalently, the total length of dislocation lines of the given family per unit crystal volume — a quantity known as the *density* of dislocations, usually denoted ρ. Since only the mobile dislocations contribute to plastic strain, it is their density, denoted ρ_m, that must appear in the equation for the plastic strain, that is,

$$\gamma^p = \rho_m b\bar{x},$$

and the plastic shear-strain rate is

$$\dot{\gamma}^p = \rho_m b\bar{v},$$

where $\bar{v}$ is the average dislocation velocity.

Forces on and Between Dislocations

A shear stress τ acting on the slip plane and in the direction of the Burgers vector produces a force per unit length of dislocation that is perpendicular to the dislocation line and equal to τb. To prove this result, we consider an infinitesimal dislocation segment of length dl; as this segment moves by a distance ds, slip of an amount b occurs over an area $dl\,ds$, and therefore the work done by the shear stress is $(\tau\,dl\,ds)b = (\tau b)\,dl\,ds$, equivalent to that done by a force $(\tau b)dl$, or τb per unit length of dislocation.

Equations (2.2.3) for the strain energy per unit length of a dislocation in an isotropic elastic continuum may be used to give an order-of-magnitude estimate for the strain energy per unit length of a dislocation in a crystal, namely,

$$W' = \alpha G b^2, \tag{2.2.4}$$

where α is a numerical factor between 0.5 and 1.

Two parallel edge dislocations having the same slip plane have, when they are far apart, a combined energy equal to the sum of their individual energies, that is, $2\alpha G b^2$ per unit length, since any interaction between them is negligible. When they are very close together, then, if they are unlike (that is, if their Burgers vectors are equal and opposite), they will annihilate each other and the resulting energy will be zero; thus they attract each other in order to minimize the total energy. Like dislocations, on the other hand, when close together are equivalent to a single dislocation of Burgers vector $2\mathbf{b}$, so that the energy per unit length is $\alpha G(2b)^2$, and therefore they repel each other in order to reduce the energy.

Frank–Read Source

The number of dislocations typically present in an unstressed, annealed crystal is not sufficient to produce plastic strains greater than a few percent. In order to account for the large plastic strains that are actually produced, it is necessary for large numbers of dislocations to be created, and on a relatively small number of slip planes, in order to account for slip bands. The *Frank–Read source* is a mechanism whereby a single segment of an edge dislocation, anchored at two interior points of its slip plane, can produce a large number of dislocation loops. The anchor points can be point defects, or points at which the dislocation joins other dislocations in unfavorable planes.

If α in Equation (2.2.4) is constant along the dislocation, independently of its orientation, then an increase ΔL in dislocation length requires an energy increment $W'\,\Delta L$, that is, work in that amount must be done on it. This is equivalent to assuming that a line tension T equal to W' is acting along the dislocation. In order to deform an initially straight dislocation segment into a circular arc subtending an angle 2θ, equilibrium requires a restoring force $F = 2T\sin\theta$ perpendicular to the original dislocation segment. If the length of the segment is L, then the force per unit length is F/L and can be produced by a shear stress $\tau = F/bL$, or

$$\tau = \frac{2\alpha Gb}{L} r \sin\theta.$$

When $\theta = \pi/2$, that is, when the dislocation segment forms a semicircle, the shear stress is maximum and equal to

$$\tau_{\max} = \frac{Gb}{L}$$

if $\alpha = 0.5$, as it is frequently taken.

If the maximum necessary shear stress is acting on a dislocation segment pinned at two points, as in Figure 2.2.8, the semicircular form is soon attained, whereupon the dislocation becomes unstable and expands indefinitely. The expanding loop doubles back on itself, as in (c) and (d), until two sections meet and annihilate each other, since they have the same Burgers vector but opposite line sense, forming a closed outer loop that continues to expand and a new dislocation segment that will repeat the process.

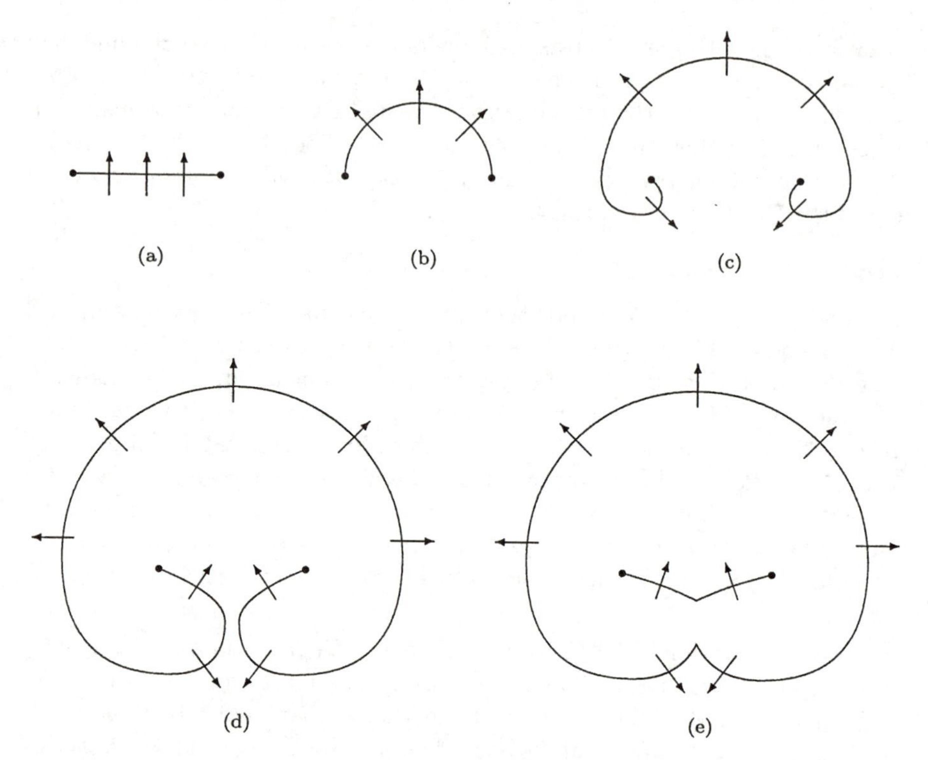

Figure 2.2.8. Frank–Read source (after Read [1953]).

Other mechanisms for the multiplication of dislocations that are similar to the Frank–Read source involve screw dislocations and include *cross-slip* and the *Bardeen–Herring source* (see, e.g., Hull and Bacon [1984]).

2.2.3. Dislocation Models of Plastic Phenomena

W. T. Read, Jr., in his classic *Dislocations in Crystals* (Read [1953]), offered the following caution: "Little is gained by trying to explain any and all experimental results by dislocation theory; the number of possible explanations is limited only by the ingenuity, energy, and personal preference of the theorist."

Indeed, much theoretical work has been expended in the past half-century in attempts to explain the phenomena of metal plasticity, discussed in Section 2.1, by means of dislocation theory. No comprehensive theory has been achieved, but numerous qualitative or semi-quantitative explanations have been offered, and some of these are now generally accepted. A few are described in what follows.

Yield Stress

If the loops generated by Frank–Read sources or similar mechanisms could all pass out of the crystal, then an indefinite amount of slip could be produced under constant stress. In reality, obstacles to dislocation movement are present. These may be scattered obstacles such as impurity atoms or precipitates, extended barriers such as grain boundaries, or other dislocations that a moving dislocation has to intersect ("forest dislocations"). In addition, if a dislocation is stopped at a barrier, then successive dislocations emanating from the same Frank–Read source pile up behind it, stopped from further movement by the repulsive forces that like dislocations exert on one another.

The yield stress is essentially the applied shear stress necessary to provide the dislocations with enough energy to overcome the short-range forces exerted by the obstacles as well as the long-range forces due to other dislocations. The mechanisms are many and complex, and therefore there is no single dislocation theory of the yield strength but numerous theories attempting to explain specific phenomena of metal plasticity. This is especially true for alloys, in which the impurity atoms may present various kinds of obstacles, depending on the form they take in the host lattice — for example, whether as solutes or precipitates (for a general review, see, e.g., Nabarro [1975]).

Yield Point

Under some conditions, solute atoms tend to segregate in the vicinity of a dislocation at a much greater density than elsewhere in the lattice, forming so-called *Cottrell atmospheres*. In order to move the dislocation, an extra stress is required to overcome the anchoring force exerted on it by the solutes. Once the dislocation is dislodged from the atmosphere, however, the extra stress is no longer necessary, and the dislocation can move under a stress that is lower than that required to initiate the motion. This is the explanation, due to Cottrell and Bilby [1949], of the yield-point phenomenon discussed in 2.1.2 [see Figure 2.1.1(b')]. Strain-aging (Figure 2.1.2) is explained by the fact that the formation of atmospheres takes place by diffusion and is therefore a rate process. Thus if a specimen is unloaded and immediately reloaded, not enough time will have passed for the atmospheres to form anew. After a sufficient time, whose length decreases with increasing temperature, the solutes segregate once more and the upper yield point returns.

Work-Hardening

As plastic deformation proceeds, dislocations multiply and eventually get stuck. The stress field of these dislocations acts as a *back stress* on mobile dislocations, whose movement accordingly becomes progressively more difficult, and an ever greater applied stress is necessary to produce additional plastic deformation. This is the phenomenon of work-hardening.

In a first stage, when only the most favorably oriented slip systems are active, the back stress is primarily due to interaction between dislocations on parallel slip planes and to the pile-up mechanism. In this stage work-hardening is usually slight, and the stage is therefore often called *easy glide*. Later, as other slip systems become activated, the intersection mechanism becomes predominant, resulting in much greater work-hardening. In a final stage, screw dislocations may come into play.

Since the number of possible mechanisms producing forces on dislocations is great, there is as yet no comprehensive theory of work-hardening that would permit the formulation of a stress-strain relation from dislocation theory. For reviews of work-hardening models, see Basinski and Basinski [1979] or Hirsch [1975].

Yield Strength of Polycrystals

The plastic deformation of polycrystals differs from that of single crystals in that, in the former, individual crystals have different orientations and therefore, under a given applied stress, the resolved shear stress varies from grain to grain. The critical value of this stress is therefore attained in the different grains at different values of the applied stress, so that the grains yield progressively. Furthermore, the grain boundaries present strong barriers to dislocation motion, and therefore the yield stress is in general a decreasing function of grain size, other factors being the same; the dependence is often found to be described by the **Hall–Petch relation**,

$$\sigma_Y = \sigma_{Y\infty} + \frac{k_Y}{\sqrt{d}},$$

where d is the grain diameter, and $\sigma_{Y\infty}$ and k_Y are temperature-dependent material constants.

The stress $\sigma_{Y\infty}$, corresponding (theoretically) to infinite grain size, may be interpreted as the yield stress when the effects of grain boundaries can be neglected. As such it should be determinable, in principle, from the single-crystal yield stress by a suitable averaging process, on the assumption of random orientation of the grains. Such a determination was made by Taylor [1938], who obtained the result that, if the stress-strain curve for a single crystal in shear on an active slip system is given by $\tau = f(\gamma^p)$, then for the polycrystal it is given by

$$\sigma = \bar{m} f(\bar{m}\varepsilon^p),$$

where $\bar{m}$ is the average value of the factor $(\cos\phi\cos\lambda)^{-1}$ in Equation (2.2.1), a value that Taylor calculated to be about 3.1 for fcc metals.

Bauschinger Effect

A fairly simple explanation of the Bauschinger effect is due to Nabarro [1950]. The dislocations in a pile-up are in equilibrium under the applied stress σ, the internal stress σ_i due to various obstacles, and the back stress σ_b due to the other dislocations in the pile-up; σ_i may be identified with the elastic limit. When the applied stress is reduced, the dislocations back off somewhat, with very little plastic deformation, in order to reduce the internal stress acting on them. They can do so until they are in positions in which the internal stress on them is $-\sigma_i$. When this occurs, they can move freely backward, resulting in reverse plastic flow when the applied stress has been reduced by $2\sigma_i$.

Exercises: Section 2.2

1. For a crystal with cubic symmetry, find the Young's modulus E in terms of C_{11}, C_{12}, and C_{44} for tension (a) parallel to a cube edge, (b) perpendicular to a cube edge and at 45° to the other two edges.

2. Show the close-packed planes and slip directions in a face-centered cubic crystals.

3. Derive Equation (2.2.1).

4. For what range of R/a do Equations (2.2.3) give Equation (2.2.4) with the values of α given in the text?

Section 2.3 Plasticity of Soils, Rocks, and Concrete

In recent years the term "geomaterials" has become current as one encompassing soils, rocks, and concrete. What these materials have in common, and in contrast to metals, is the great sensitivity of their mechanical behavior to pressure, resulting in very different strengths in tension and compression. Beyond this common trait, however, the differences between soils on the one hand and rocks and concrete on the other are striking. Soils can usually undergo very large shearing deformations, and thus can be regarded as plastic in the usual sense, although soil mechanicians usually label as "plastic" only cohesive, claylike soils that can be easily molded without crumbling. Rock and concrete, on the other hand, are brittle, except under high triaxial compression. Nevertheless, unlike classically brittle solids, which fracture shortly after the elastic limit is attained, concrete and many rocks can undergo inelastic deformations that may be significantly greater than the elastic strains, and their stress-strain curves superficially resemble those of plastic solids.

2.3.1. Plasticity of Soil

The Nature of Soil

The essential property of soils is that they are *particulate*, that is, they are composed of many small solid particles, ranging in size from less than 0.001 mm (in clays) to a few millimeters (in coarse sand and gravel). Permanent shearing deformation of a soil mass occurs when particles slide over one another. Beyond this defining feature, however, there are fundamental differences among various types of soils, differences that are strongly reflected in their mechanical behavior.

The voids between the particles are filled with air and water; the ratio of the void (air and water) volume to the solid volume is known as the *void ratio* of the soil. While much of the water may be in the usual liquid form (*free water*), and will evaporate on drying, some of the water is attached to the particle surfaces in the form of adsorbed layers, and does not evaporate unless the solid is heated to a temperature well above the boiling point of water. A soil is called *saturated* if all the voids are filled with water. If both water and air are present, the soil is called *partially saturated*, and if no free water is present, the soil is called *dry*.

Clay was mentioned at the beginning of this chapter as a prototype of a plastic material. Clays are fine-grained soils whose particles contain a significant proportion of minerals known as *clay minerals*. The chemistry of these minerals permits the formation of an adsorbed water film that is many times thicker than the grain size. This film permits the grains to move past one another, with no disintegration of the matrix, when stress is applied. It is this property that soil mechanicians label as plasticity. Claylike soils are also generally known as *cohesive soils*.

In cohesionless soils, such as gravels, sands, and silts, the movement of grains past one another is resisted by dry friction, resulting in shear stresses that depend strongly on the compression. Materials of this type are sometimes called *frictional materials*.

Soil Compressibility

If soil that is prevented from expanding laterally is loaded in compression between layers, at least one of which is permeable to water, an irreversible decrease in void ratio occurs, a result of the seepage of water from the voids. The process, known as *consolidation*, takes time, and sometimes goes on indefinitely, though at an ever-diminishing rate, much like creep. As a rule, though, something very near the ultimate compression is attained in a finite time which depends on the properties of the soil layer. A typical compression curve is shown in Figure 2.3.1(a). The figure shows both the virgin curve and the hysteresis loop resulting from decompression followed by recompression. A soil that has been decompressed is called *overconsolidated*. The curves

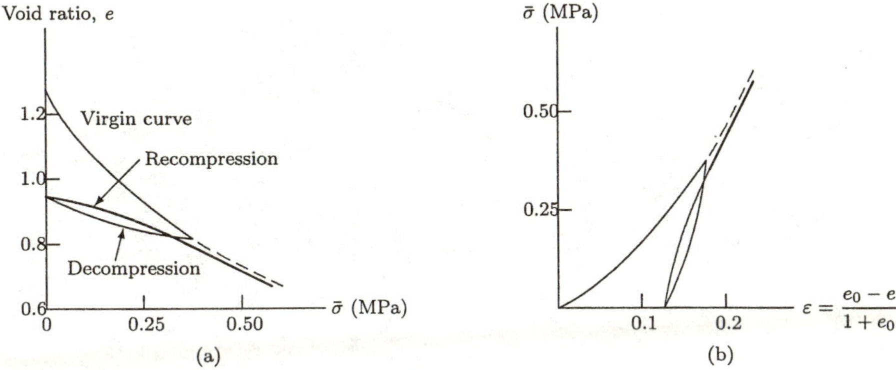

Figure 2.3.1. Compression curve for soil: (a) consolidation curve; (b) compressive stress-strain diagram [(b) is (a) replotted].

are replotted in Figure 2.3.1(b) as a compressive stress-strain diagram. It is seen that except for the upward convexity of the virgin curve, the diagram resembles that of work-hardening metals.

Shearing Behavior

As in ductile metals, failure in soils occurs primarily in shear. Unlike metals, the shear strength of soils is, in most circumstances, strongly influenced by the compressive normal stress acting on the shear plane and therefore by the hydrostatic pressure. Since soils have little or no tensile strength, the tension test cannot be applied to them. Other means are necessary in order to determine their shear strength.

Direct Shear Test. A traditional test of the shear strength of soft clays and of dry sands and gravels is the *direct shear test* or *shear-box test.* A sample of soil is placed in a rectangular box whose top half can slide over the bottom half and whose lid can move vertically, as shown in Figure 2.3.2(a). A normal load is applied to the lid, and a shear force is applied to the top half of the box, shearing the soil sample.

Simple Shear Test. In this test, developed by Roscoe [1953], it is the strain that is maintained as one of simple shear [see Figure 2.3.2(b)].

The two tests just described, along with others like them, provide simple means of estimating the shear strength. However, the stress distribution in the sample is far from uniform, so that these tests do not actually measure stress, and no stress-strain diagrams can result from them.

Triaxial Test. This is generally regarded as the most reliable test of the shearing behavior of soils. As we shall see, it is used to test rock and concrete as well. This test was discussed in 2.2.1; a normal compressive stress σ_3 ($= \sigma_2$) is applied to the sides of a cylindrical sample by means of air or

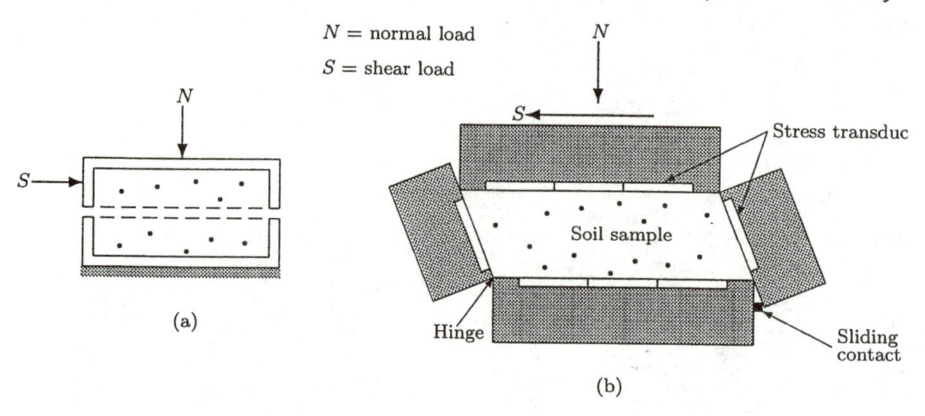

Figure 2.3.2. Shear tests: (a) direct shear test; (b) simple shear test (after Roscoe [1953]).

water pressure, and an axial compressive stress σ_1, numerically greater than σ_3, is applied at the ends (Figure 2.3.3). The results are commonly plotted

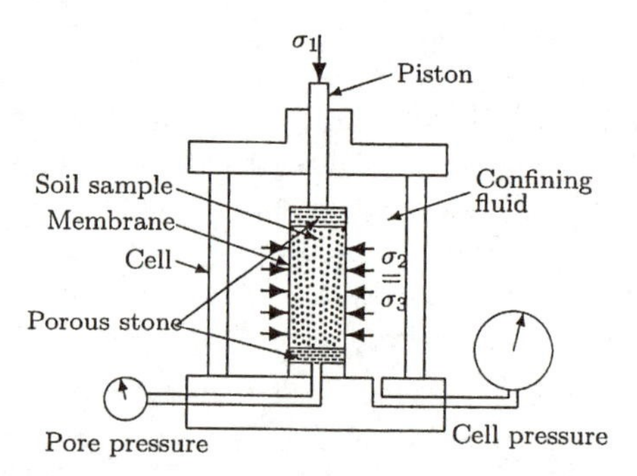

Figure 2.3.3. Triaxial test apparatus.

as graphs of $\sigma_1 - \sigma_3$ against the axial shortening strain ε_1, with σ_3 as a parameter. (Alternatively, the mean stress $(\sigma_1 + 2\sigma_3)/3$ or the normal stress on the maximum-shear plane $(\sigma_1 + \sigma_3)/2$ may be used as a parameter.) Note that $\sigma_1 - \sigma_3$ is a measure both of the maximum shear stress given by Equation (1.3.11), namely, $\tau_{\max} = \frac{1}{2}|\sigma_1 - \sigma_3|$, and of the octahedral shear stress, given in accordance with Equation (1.3.5) as $\tau_{\text{oct}} = (\sqrt{2}/3)|\sigma_1 - \sigma_3|$. If $\sigma_3 = 0$ then the test is called an *unconfined compression test*, used most commonly on hard materials such as rock and concrete, but occasionally on clay if it is performed fast enough ("quick test"). Some typical stress-strain curves for soils are shown in Figure 2.1.1(f) (page 78).

The dependence of the shear strength of soils on the normal stress acting on the shearing plane varies with the type and condition of the soil. It is simplest in dry cohesionless soils (gravels, sands, and silts), in which resistance to shear is essentially due to dry friction between the grains, and therefore is governed by the Coulomb law of friction:

$$\tau = \sigma \tan\phi, \tag{2.3.1}$$

where τ and σ are respectively the shear and normal stresses on the shearing plane, and ϕ is the *angle of internal friction*, a material property.

In wet cohesionless soils, the applied stress is the sum of the *effective stress* in the grains and the *neutral stress* due to water pressure and possibly capillary tension. If the latter stress is denoted σ_w (like σ, positive in compression), then the Coulomb law is expressed by

$$\tau = (\sigma - \sigma_w)\tan\phi, \tag{2.3.2}$$

since the water pressure provides a counterthrust on potential sliding surfaces, and therefore it is only the effective stress that governs frictional resistance. The concept of effective stress is due to Terzaghi.

Cohesionless soils also undergo significant volume changes when sheared. They tend to swell if they are dense, since closely packed grains must climb over one another in the course of shearing, and shrink if they are loose, since grains fall into the initially large voids and thus reduce the void volume. A granular soil thus has a *critical density* which remains essentially constant as shearing proceeds, and the soil is termed *dense* or *loose*, respectively, if its density is above or below critical.

In a sample of fine sand or silt that is dense and saturated, and which has no source of additional water, the swelling that accompanies shearing produces surface tension on the water which acts as a negative neutral stress. Consequently, in accord with Equation (2.3.2), such a sample has shear strength under zero applied stress.

In clays, the stresses in the adsorbed water layers play an important role in determining strength, and in partially saturated clays this role is predominant. The shear strength of such clays is given approximately by

$$\tau = c + \sigma \tan\phi, \tag{2.3.3}$$

where ϕ is the angle of internal friction and c is a material constant called the *cohesion*, representing the shear strength under zero normal stress.

The shear response of a saturated clay depends on whether it is in a *drained* or *undrained* condition. The former condition is achieved in a slow application of the stresses, so that the neutral stresses are not changed during the loading and therefore play little part in determining the shear strength. Equation (2.3.1) is consequently a good approximation to the relation between shear stress and normal stress in this condition. In the undrained condition, on the other hand, the loading is quick and the applied stress is carried by the neutral stress. In this condition the shear strength is independent of the applied normal stress, and is therefore given by Equation (2.3.3) with $\phi = 0$; the cohesion c is then called the *undrained strength* and denoted c_u. Volume changes accompanying shearing are negligible in saturated clays. The shear-strength response of undrained clays thus resembles that of metals. Much of soil engineering practice is based on this model, though it is not universally accepted; see Bolton [1979], Section 5.1, for a survey of the criticisms.

2.3.2. "Plasticity" of Rock and Concrete

Unlike soils, materials such as rock, mortar and concrete are generally not plastic in the sense of being capable of considerable deformation before failure. Instead, in most tests they fracture through crack propagation when fairly small strains (on the order of 1% or less) are attained, and must therefore be regarded as brittle. However, concrete, mortar, and many rocks (such as marble and sandstone) are also unlike such characteristically brittle solids as glass and cast iron, which fracture shortly after the elastic limit is attained. Instead, they attain their ultimate strength after developing permanent strains that, while small in absolute terms, are significantly greater than the elastic strains. The permanent deformation is due to several mechanisms, the foremost of which is the opening and closing of cracks.

Strain-Softening

Following the attainment of the ultimate strength, concrete and many rocks exhibit *strain-softening*, that is, a gradual decrease in strength with additional deformation. The nature of this decrease, however, depends on factors associated with the testing procedure, including sample dimensions and the stiffness of the testing machine.

The effect of machine stiffness can be described as follows. Let P denote the load applied by the machine to the sample, and u the sample displacement. In the course of a small change Δu in the displacement, the sample absorbs energy in the amount $P\Delta u$. If the machine acts like an elastic spring with stiffness k, then a change ΔP in the load implies a change $P\,\Delta P/k$ in the energy stored in the machine. This change represents *release* of energy if $P\,\Delta P < 0$, that is, once softening takes place. The energy released by the machine is greater than that which can be absorbed by the sample if $k < |\Delta P/\Delta u|$, resulting in an unstable machine-sample system in the case of a "soft" machine; the sample breaks violently shortly after the ultimate strength is passed. A "stiff" machine, on the other hand, makes for a system that is stable under displacement control. It is only with a stiff machine, therefore, that a complete load-displacement (or stress-displacement) curve can be traced.

It is not certain, however, whether the stress-displacement curve may legitimately be converted into a stress-strain curve, such as is shown in Figure 2.1.1(d) (page 78), that reflects material properties, since specimen deformation is often far from homogeneous. Experiments by Hudson, Brown and Fairhurst [1971] show a considerable effect of both the size and the shape of the specimens on the compressive stress-strain curve of marble, including as a particular result the virtual disappearance of strain-softening in squat specimens. Read and Hegemier [1984] conclude that no strain-softening occurs in specimens of soil, rock and concrete that are homogeneously deformed. A similar conclusion was reached by Kotsovos and Cheong [1984] for concrete. It should be remarked that some rocks, such as limestone, exhibit classically brittle behavior in unconfined compression tests even with stiff testing machines — that is, they fracture shortly after the elastic limit is reached.

The Effect of Pressure

An important feature of the triaxial behavior of concrete, mortar and rocks (including those which are classically brittle in unconfined tests) is that, if the confining pressure σ_3 is sufficiently great, then crack propagation is prevented, so that brittle behavior disappears altogether and is replaced by ductility with work-hardening. Extensive tests were performed on marble and limestone by von Kármán [1911] and by Griggs [1936]; some results are shown in Figure 2.1.1(e). Note that the strains attained in these tests can become quite large.

The relation between hydrostatic pressure and volumetric strain also exhibits ductility with work-hardening; the curves resemble those of Figure 2.3.1(b). It can be said, in general, that rocks and concrete behave in a ductile manner if all three principal stresses are compressive and close to one another.

Dilatancy

If the transverse strain $\varepsilon_2 = \varepsilon_3$ is measured in uniaxial compression tests of rock and concrete specimens in addition to the axial strain ε_1, then, as discussed in 1.2.2, the volumetric strain ε_V equals $\varepsilon_1+\varepsilon_2+\varepsilon_3$. If the stress σ_1 is plotted against ε_V (positive in compression), it is found that ε_V begins to decrease from its elastic value at stresses greater than about half the ultimate strength, reaches zero at a stress near the ultimate strength, and becomes negative (signifying an *increase* in volume) in the strain-softening range (see Figure 2.3.4, showing both a σ_1-ε_1 and a σ_1-ε_V diagram). Similar results are obtained in triaxial tests under low confining pressures. This volume increase, which results from the formation and growth of cracks parallel to the direction of the greatest compressive stress, is known as *dilatancy*. This term is sometimes also applied to the swelling of dense granular soils,

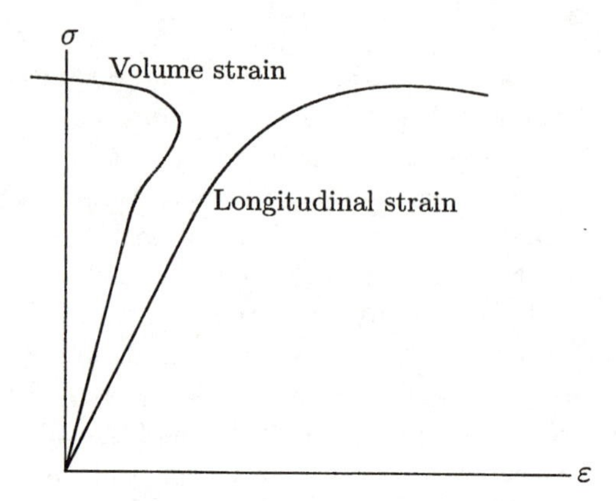

Figure 2.3.4. Compression tests on concrete or rock: stress against longitudinal strain and volume strain.

although the mechanism causing it is unrelated.

Tensile Behavior

Uniaxial tension tests are difficult to perform on rock and concrete, and the results of such tests vary considerably. The most reliable direct tension tests are those in which the ends of the specimen are cemented with epoxy resin to steel plates having the same cross-section as the specimen, with the tensile force applied through cables in order to minimize bending effects. The uniaxial tensile strength of rock and concrete is typically between 6 and 12% the uniaxial compressive strength. Strain-softening, associated with the opening of cracks perpendicular to the direction of tension, is observed in tests performed in stiff machines.

Chapter 3

Constitutive Theory

Section 3.1 Viscoplasticity

We saw in the preceding chapter that while "yielding" is the most striking feature of plastic behavior, the existence of a well-defined yield stress is the exception rather than the rule. It so happens, however, that mild steel, which belongs to this exceptional class, is one of the most commonly used of metals, and attempts at a theoretical description of its behavior preceded those for other metals; such attempts naturally incorporated a criterion as an essential feature of what came to be known as plasticity theory, as well as of a later development, known as viscoplasticity theory, which takes rate sensitivity into account.

It should be pointed out that while most workers in solid mechanics use "viscoplasticity" in its classical meaning (see Prager [1961]), that is, to denote the description of rate-dependent behavior with a well-defined yield criterion, this usage is not universal. Others, following Bodner [1968], use the term for models of highly nonlinear viscoelastic behavior, without any elastic range, that is characteristic of metals, especially at higher temperatures. Such models are discussed in 3.1.3. 3.1.1 is limited to models of classical viscoplasticity. Both classes of models are subclasses of the internal-variable models presented in Section 1.5. In 3.1.2, rate-independent plasticity, the foundation for most of the remainder of this book, is derived as a limiting case of classical viscoplasticity.

3.1.1. Internal-Variable Theory of Viscoplasticity

Yield Surface

As in Section 1.5, let $\boldsymbol{\xi}$ denote the array of internal variables $\xi_1, \ldots, \xi_n$. If there is a continuous function $f(\boldsymbol{\sigma}, T, \boldsymbol{\xi})$ such that there exists a region in the space of the stress components in which (at given values of T, $\boldsymbol{\xi}$)

$f(\boldsymbol{\sigma}, T, \boldsymbol{\xi}) < 0$, and such that the inelastic strain-rate tensor $\dot{\boldsymbol{\varepsilon}}^i$ vanishes in that region but not outside it, then this region constitutes the aforementioned elastic range, and $f(\boldsymbol{\sigma}, T, \boldsymbol{\xi}) = 0$ defines the yield surface in stress space; the orientation of the yield surface is defined in such a way that the elastic range forms its interior. A material having such a *yield function* $f(\cdot)$ is viscoplastic in the stricter sense. This definition, it should be noted, does not entail the simultaneous vanishing of *all* the internal-variable rates $\dot{\xi}_\alpha$ in the elastic region; if such were the case, strain-aging as described in the preceding chapter would not be possible, since it requires an evolution of the local structure while the material is stress-free. However, this proviso is of significance only for processes whose time scale is of the order of magnitude of the relaxation time for strain-aging, which for mild steel at ordinary temperatures is of the order of hours. Thus, for a process lasting a few minutes or less, the internal variables governing strain-aging are essentially constant and their rates may be ignored. For the sake of simplicity, we adopt a somewhat more restricted definition of viscoplasticity, according to which all the internal-variable rates vanish in the elastic region, that is, the functions $g_\alpha(\boldsymbol{\sigma}, T, \boldsymbol{\xi})$ constituting the right-hand sides of the rate equations (1.5.1) are assumed to vanish whenever $f(\boldsymbol{\sigma}, T, \boldsymbol{\xi}) \le 0$. In particular, this definition includes all those models (such as that of Perzyna [1971]) in which the rates of the internal variables depend linearly on $\dot{\boldsymbol{\varepsilon}}^i$.

In view of this definition it now becomes convenient to redefine the g_α as $g_\alpha = \phi h_\alpha$, where ϕ is a scalar function that embodies the rate and yielding characteristics of the material, with the property that $\phi = 0$ when $f \le 0$ and $\phi > 0$ when $f > 0$. Such a function was introduced by Perzyna [1963] in the form $\gamma(T)<\Phi(f)>$, where $\gamma(T)$ is a temperature-dependent "viscosity coefficient" (actually an inverse viscosity, or fluidity), and the notation $<\Phi(f)>$ is defined — somewhat misleadingly — as

$$<\Phi(f)> = \begin{cases} 0 & \text{for } f \le 0 \\ \Phi(f) & \text{for } f > 0 \end{cases}$$

(the more usual definition of the operator $< \cdot >$ is given below). Note that our definition of ϕ is determinate only to within a multiplicative scalar; that is, if λ is a positive continuous function of the state variables, then ϕ may be replaced by ϕ/λ and the h_α by λh_α without changing the rate equations.

Hardening

The dependence of the yield function f on the internal variables ξ_α describes what are usually called the *hardening* properties of the material. The relationship between this dependence and the behavior of the material can be understood by considering a stress $\boldsymbol{\sigma}$ that is close to the yield surface but outside it, that is, $f(\boldsymbol{\sigma}, T, \boldsymbol{\xi}) > 0$. In particular, let us look at a case of uniaxial stress in a specimen of a material whose static stress-strain curve is given by the solid curve of Figure 3.1.1, which shows both rising ("hardening") and falling ("softening") portions. If the material is viscoplastic, then its behavior is elastic at points below the curve, and viscoelastic at points above the curve — that is, the curve represents the yield surface.

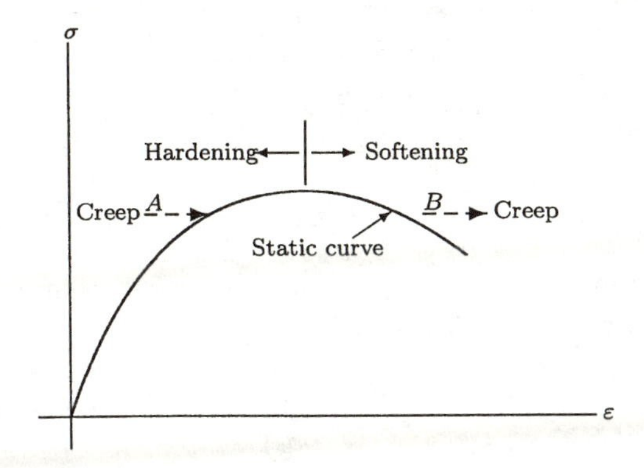

Figure 3.1.1. Hardening and softening in viscoplasticity: relation to creep and static curve.

If the stress is held constant at a value above the static curve, creep occurs, resulting in increasing strain as shown by the dashed horizontal lines. If the initial point is, like A, above the rising portion of the static stress-strain curve, then the creep tends toward the static curve and is bounded, while if it is, like B, above the falling portion, then the creep tends away from the static curve and is unbounded. Since the points on the static stress-strain curve are in effect those on the yield surface, we may generalize from the uniaxial case as follows: creep *toward* the yield surface, characterizing hardening, means that at constant stress and temperature, the yield function f decreases from a positive value toward zero, that is, $\dot{f} < 0$. Similarly, softening is characterized by $\dot{f} > 0$. But

$$\dot{f}\Big|_{\boldsymbol{\sigma}=\text{const},T=\text{const}} = \sum_\alpha \frac{\partial f}{\partial \xi_\alpha}\dot{\xi}_\alpha = \phi \sum_\alpha \frac{\partial f}{\partial \xi_\alpha} h_\alpha$$

$$= -\phi H,$$

where, by definition,

$$H = -\sum_\alpha \frac{\partial f}{\partial \xi_\alpha} h_\alpha. \tag{3.1.1}$$

Thus $H > 0$ and $H < 0$ for hardening and softening materials (or hardening and softening ranges of the same material), respectively. The limiting case $H = 0$, which in particular occurs when f is independent of the ξ_α, describes a *perfectly plastic* material.

Viscoplastic Potential

If a viscoplastic material has a flow potential in the sense of 1.5.3 (not necessarily in the stricter sense of Rice or Moreau), then it may also be shown to have a viscoplastic potential in the following sense. Let h_{ij} be defined by

$$h_{ij} = \sum_{\alpha} \frac{\partial \varepsilon^i_{ij}}{\partial \xi_\alpha} h_\alpha.$$

The flow equations then are

$$\dot{\varepsilon}^i_{ij} = \phi h_{ij}. \tag{3.1.2}$$

If there exists a function $g(\boldsymbol{\sigma}, T, \boldsymbol{\xi})$, continuously differentiable with respect to $\boldsymbol{\sigma}$ wherever $f(\boldsymbol{\sigma}, T, \boldsymbol{\xi}) > 0$, such that

$$h_{ij} = \frac{\partial g}{\partial \sigma_{ij}},$$

then g is called a *viscoplastic potential*. (The relation $h_{ij} = \lambda \partial g/\partial \sigma_{ij}$ is not more general, since the factor λ can be absorbed in ϕ.) Perzyna [1963] and many others have assumed the existence of a viscoplastic potential identical with the yield function f, or at least such that $\partial g/\sigma_{ij} \propto \partial f/\sigma_{ij}$; this identity is of no great significance in viscoplasticity, but becomes highly important after the transition to rate-independent plasticity.

Specific Models Based on J_2 Flow Potential

In 1.5.3 a flow potential was discussed that depends on the stress only through J_2, leading to the flow equation $\dot{\varepsilon}^i_{ij} = \phi s_{ij}$. A yield criterion having the same stress dependence, that is, one that can be represented by the equation

$$\sqrt{J_2} - k = 0$$

(where k depends on T and $\boldsymbol{\xi}$ and equals the yield stress in shear) is known as the *Mises* (sometimes *Huber–Mises*) yield criterion. A model of viscoplasticity incorporating this yield criterion and a J_2 flow potential was first proposed by Hohenemser and Prager [1932] as a generalization to three-dimensional behavior of the Bingham model described in 2.1.3. The flow equation is

$$\dot{\varepsilon}^i_{ij} = \frac{1}{2\eta} < 1 - \frac{k}{\sqrt{J_2}} > s_{ij}, \tag{3.1.3}$$

where η is a temperature-dependent viscosity, and the *Macauley bracket* $<\cdot>$ is defined by $<x> = xH(x)$, where $H(\cdot)$ is the Heaviside step function:

$$H(x) = \begin{cases} 0, & x \le 0, \\ 1, & x > 0. \end{cases}$$

In other words,

$$<x> = \begin{cases} 0, & x \le 0, \\ x, & x > 0. \end{cases}$$

The previously discussed model of Perzyna [1963] is a generalization of the Hohenemser–Prager model in which $<f>$ is replaced by $H(f)\Phi(f)$, or $<\Phi(f)>$ in Perzyna's notation. It will be noted that as $k \to 0$, the Hohenemser–Prager and Perzyna models reduce to the Maxwell model of linear viscoelasticity discussed in 1.5.1.

A generalized potential Ω, as discussed in 1.5.4, may be associated with the Hohenemser–Prager model if it takes the form $\Omega(\boldsymbol{\sigma}) = <f>^2/(2\eta)$, where $f = \sqrt{J_2} - k$, and with the Perzyna model if it is $\Omega(\boldsymbol{\sigma}) = H(f)\Omega_0(f)$. Hardening can be included in a simple manner by letting k be a variable. If the generalized potential is viewed as a function $\Omega(\boldsymbol{\sigma}, k)$, then the effective inelastic strain $\bar{\varepsilon}^i$ defined by (1.5.7) can easily be shown to obey the rate equation

$$\dot{\bar{\varepsilon}}^i = -\frac{1}{\sqrt{3}}\frac{\partial \Omega}{\partial k}.$$

It is convenient to let $k = k_0 + R/\sqrt{3}$, where k_0 is the initial value of k, and to treat Ω as a function of $(\boldsymbol{\sigma}, R)$. Then $\dot{\bar{\varepsilon}}^i = -\partial\Omega/\partial R$, and $-R$ may be regarded as the thermodynamic force conjugate to the internal variable $\bar{\varepsilon}^i$.

A more sophisticated model developed by Chaboche [1977] uses as internal variables $\bar{\varepsilon}^i$ and a strain-like symmetric second-rank tensor $\boldsymbol{\alpha}$. The thermodynamic forces conjugate to these variables are the stress-like variables $-R$ and $-\boldsymbol{\rho}$, respectively, and the yield surface is assumed to be given by

$$f(\boldsymbol{\sigma}, \boldsymbol{\rho}, R) = \sqrt{\bar{J}_2} - \frac{R}{\sqrt{3}} - k_0 = 0,$$

where

$$\bar{J}_2 = \frac{1}{2}(s_{ij} - \rho'_{ij})(s_{ij} - \rho'_{ij}),$$

$\boldsymbol{\rho}'$ being the deviator of $\boldsymbol{\rho}$. The yield surface is thus again of the Mises type, but capable not only of expansion (as measured by R) but also of translation (as shown by $\boldsymbol{\rho}'$, which locates the center of the elastic region). The hardening described by the expansion of the yield surface is called *isotropic*, while that described by the translation is called *kinematic*. The significance of the terms is discussed in Section 3.2.

If a generalized potential is again assumed in the Perzyna form, $\Omega(\boldsymbol{\sigma}, R, \boldsymbol{\rho}) = H(f)\Omega_0(f)$, then $\dot{\varepsilon}^i_{ij} = \partial\Omega/\partial\sigma_{ij}$ and $\dot{\bar{\varepsilon}}^i = -\partial\Omega/\partial R$ as before, the flow equations being

$$\dot{\varepsilon}^i_{ij} = \frac{\partial \Omega}{\partial \sigma_{ij}} = H(f)\Omega'_0(f)\frac{s_{ij} - \rho'_{ij}}{2\sqrt{\bar{J}_2}}.$$

In addition, $\dot{\boldsymbol{\alpha}} = \dot{\boldsymbol{\varepsilon}}^i$, so that the kinematic-hardening variable $\boldsymbol{\alpha}$, though it must be treated as a distinct variable, coincides with the inelastic strain.

Chaboche, however, assumes the generalized potential in the form $\Omega(\boldsymbol{\sigma}, R, \boldsymbol{\rho}) = H(f)\Omega_0(f) + \Omega_r(\boldsymbol{\rho})$, where the second term represents recovery (see 2.1.2). Moreover, Chaboche abandons the generalized normality hypothesis for $\boldsymbol{\alpha}$ by introducing an additional term representing a concept called *fading strain memory*, due to Il'iushin [1954], the better to describe the Bauschinger effect. The rate equation for $\boldsymbol{\alpha}$ is therefore taken as

$$\dot{\alpha}_{ij} = -\frac{\partial \Omega}{\partial \rho_{ij}} - F(\bar{\varepsilon}^i)\dot{\bar{\varepsilon}}^i \rho_{ij} = \dot{\varepsilon}^i_{ij} - F(\bar{\varepsilon}^i)\dot{\bar{\varepsilon}}^i \rho_{ij} - \frac{\partial \Omega_r}{\partial \rho_{ij}},$$

where $F(\bar{\varepsilon}^i)$ is a function to be specified, along with $\Omega_0(f)$, $\Omega_r(\boldsymbol{\rho})$, and the free-energy density $\psi(T, \boldsymbol{\varepsilon}, \bar{\varepsilon}^i, \boldsymbol{\alpha})$, from which R and $\boldsymbol{\rho}$ can be derived in accordance with Equation (1.5.4): $R = \rho\partial\psi/\partial\bar{\varepsilon}^i$, $\rho_{ij} = \rho\partial\psi/\partial\alpha_{ij}$.

3.1.2. Transition to Rate-Independent Plasticity

Aside from the previously discussed limit of the Hohenemser–Prager model as the yield stress goes to zero, another limiting case is of great interest, namely, as the viscosity η goes to zero. Obviously, if $\mathbf{s} \neq 0$ then the inelastic strain rate would become infinite, unless $\sqrt{J_2}$ simultaneously tends to k, in which case the quantity $(1/\eta)\langle 1 - k/\sqrt{J_2}\rangle$ becomes indeterminate but may remain finite and positive.

Supposing for simplicity that k in Equation (3.1.3) is constant, for a given input of stress we can solve this equation for $\boldsymbol{\varepsilon}^i$ as a function of time, and the dependence on time is through the variable t/η. In other words, decreasing the viscosity is equivalent to slowing down the process of inelastic deformation, and the limit of zero viscosity is equivalent to the limit of "infinitely slow" processes. Thus a slow process can take place if J_2 is slightly larger than k^2. We can also see this result by forming the scalar product $\dot{\varepsilon}^i_{ij}\dot{\varepsilon}^i_{ij}$ from Equation (3.1.3), from which we obtain

$$\sqrt{J_2} = k + \eta\sqrt{2\dot{\varepsilon}^i_{ij}\dot{\varepsilon}^i_{ij}}, \qquad \dot{\boldsymbol{\varepsilon}}^i \neq \mathbf{0},$$

an equation that is sometimes interpreted as a *rate-dependent yield criterion*.

Let us return to the more general model of viscoplasticity considered above, and particularly one in which ϕ increases with f. The rate equations (3.1.2) indicate that, in the same sense as in the Hohenemser–Prager model, the rate of a process in which inelastic deformation takes place increases with distance from the yield surface. If such a process is *very slow*, then it takes place *very near but just outside* the yield surface, so that ϕ is very small. In the limit as $f \to 0+$ we can eliminate ϕ (and thus no longer need to concern ourselves with the actual rate at which the process takes place) as follows: if f remains equal to zero (or a very small positive constant), then

$$\dot{f} = \frac{\partial f}{\partial \sigma_{ij}}\dot{\sigma}_{ij} + \sum_{\alpha} \frac{\partial f}{\partial \xi_\alpha}\phi h_\alpha = 0.$$

We define

$$\overset{\circ}{f} = \frac{\partial f}{\partial \sigma_{ij}} \dot{\sigma}_{ij} \tag{3.1.4}$$

and assume $H > 0$ (i.e., hardening), with H as defined by Equation (3.1.1); then the condition $\dot{f} = \overset{\circ}{f} - \phi H = 0$ is possible together with $\phi > 0$ only if $\overset{\circ}{f} > 0$; this last condition is called *loading*. Thus we have the result

$$\phi = \frac{1}{H} <\overset{\circ}{f}>,$$

and therefore

$$\dot{\xi}_\alpha = \frac{1}{H} <\overset{\circ}{f}> h_\alpha. \tag{3.1.5}$$

Note that both sides of Equation (3.1.5) are derivatives with respect to time, so that a change in the time scale does not affect the equation. Such an equation is called *rate-independent*. If it is assumed that this equation describes material behavior over a sufficiently wide range of loading rates, then the behavior is called *rate-independent plasticity*, also called *inviscid plasticity* (since it corresponds to the zero-viscosity limit of the Hohenemser–Prager model), or just plain *plasticity*. Rate-independent plasticity constitutes the principal topic of the remainder of this book. The inelastic strain occurring in rate-independent plasticity is usually denoted ε^p rather than ε^i, and is called the plastic strain. The flow equation for the plastic strain may be written as

$$\dot{\varepsilon}^p_{ij} = \frac{1}{H} <\overset{\circ}{f}> h_{ij}. \tag{3.1.6}$$

For purposes of computation, however, it is sometimes advantageous to remain within the framework of viscoplasticity without making the full transition, even when the problem to be solved is regarded as rate-independent. In other words, a fictitious viscoplastic material of very low viscosity is "associated" with a given rate-independent plastic material, with rate equations given, for example (Nguyen and Bui [1974]), by

$$\dot{\xi}_\alpha = \frac{<f>}{\eta} h_\alpha, \tag{3.1.7}$$

with the viscosity η taken as constant. Computations are then performed under time-independent loads and boundary conditions until all strain rates vanish. It was shown by Zienkiewicz and Cormeau [1974], among others, that the results are equivalent to those of rate-independent plasticity.

Combined Viscoplasticity and Rate-Independent Plasticity

At extremely high rates of deformation or loading, the internal variables do not have enough time to change and consequently the deformation can be only elastic. However, the various rate processes responsible for plastic deformation, corresponding to the generation of dislocations and the many different kinds of obstacles that dislocations must overcome, may have very different characteristic times. This means that not only do metals differ greatly among one another in their rate-sensitivity, but different mechanisms in the same metal may respond with very different speeds. Thus, those mechanisms whose characteristic times are very short compared with a typical loading time produce what appears to be instantaneous inelastic deformation, while the others produce rate-dependent deformation as discussed so far in this section. If both phenomena occur in a metal over a certain range of loading times, then the total inelastic strain ε^i may be decomposed as

$$\varepsilon^i = \varepsilon^{vp} + \varepsilon^p, \tag{3.1.8}$$

where ε^{vp} is the viscoplastic strain equivalent to that governed by Equation (3.1.2), and ε^p is the apparently rate-independent plastic strain, governed by Equation (3.1.6). It is important to note that the yield functions f and flow tensors h_{ij} are, in general, different for the two inelastic strain tensors. In particular, the viscoplastic yield surface is always assumed to be inside the rate-independent plastic (or "dynamic") yield surface.

3.1.3. Viscoplasticity Without a Yield Surface

As we have seen, in both classical viscoplasticity and rate-independent plasticity the yield surface is a central ingredient; in the latter it is indispensable. The significance of the yield surface has, however, repeatedly been questioned. Consider the following remarks by Bell [1973]: "Among the many matters pertaining to the plastic deformation of crystalline solids, yield surfaces and failure criteria early became subjects of overemphasis... Indeed most of the outstanding 19th century experimentists doubted that such a phenomenon as an elastic limit, let alone a yield surface, existed... well over a half-century of experiment, and the study of restricted plasticity theories for the 'ideal solid,' have not disposed of most of the original questions."

"Unified" Viscoplasticity Models

According to Bodner [1968], "yielding is not a separate and independent criterion but is a consequence of a general constitutive law of the material behavior." Since the 1970s several constitutive models for the rate-dependent inelastic behavior of metals have been formulated without a formal hypothesis of a yield surface, but with the feature that at sufficiently low rates the resulting stress-strain curves may resemble those of materials with fairly well defined yield stresses. In fact, with yield based on the offset definition (see 2.1.1), these models can *predict* yield surfaces in accordance with Bodner's dictum, particularly if offset strains of the order of 10^{-6} to 10^{-5} are used, in contrast to the conventional 10^{-3} to 10^{-2}.

In addition to describing the behavior traditionally called plasticity, in both monotonic and cyclic loading, these models also aim to describe creep, especially at higher temperatures, without a decomposition such as (3.1.8). They have consequently come to be known as "unified" viscoplasticity models, and are particularly useful for the description of bodies undergoing significant temperature changes — for example, spacecraft. Perhaps the simplest such model is due to Bodner and Partom [1972, 1975], in which the flow equations are given by Equation (3.1.2) with $h_{ij} = s_{ij}$ and ϕ a function of J_2 and (in order to describe hardening) of the inelastic work W_i defined by Equation (1.5.6) as the only internal variable. The rate equation is obviously

$$\dot{W}_i = 2J_2\phi(W_i,\, J_2).$$

The hardening in this case is purely isotropic, since $\sqrt{3J_2}$ is the value of the effective stress necessary to maintain a given inelastic work rate $\dot{W}_i$.[1]

More sophisticated "unified" viscoplasticity models, that describe many features of the behavior of metals at elevated temperatures, have been developed since 1975 by, among others, Miller [1976], Hart [1976], Krieg, Swearengen, and Jones [1978], Walker [1981], and Krieg, Swearengen, and Rohde [1987] (see reviews by Chan, Bodner, Walker, and Lindholm [1984], Krempl [1987], and Bammann and Krieg [1987]). The essential internal variables in these models are the *equilibrium stress* tensor $\boldsymbol{\rho}$, and the scalar *drag stress* or *friction stress* σ_D;[2] the terminology is loosely related to that of dislocation theory, and is an example of "physical" nomenclature for phenomenological internal variables.

In the "unified" models the stress-like variables σ_D and $\boldsymbol{\rho}$ are used directly as internal variables, rather than as conjugate thermodynamic forces. The equilibrium stress $\boldsymbol{\rho}$, like its counterpart in the Chaboche model, describes kinematic hardening. Some writers, following Kochendörfer [1938], relate it to the back stress due to stuck dislocations (see 2.2.3), and consequently the equilibrium stress is also termed *back stress*; see Krempl [1987] for a discussion (the relationship between constitutive theory and crystal behavior has also been discussed by Kocks [1987]). For isotropic behavior $\boldsymbol{\rho}$ is assumed as purely deviatoric, and the rate equation for inelastic strain takes the form

$$\dot{\varepsilon}^i_{ij} = \dot{e}^i_{ij} = \frac{3}{2}\frac{\phi(\Gamma/\sigma_D)}{\Gamma}(s_{ij} - \rho_{ij}), \tag{3.1.9}$$

[1] Or, equivalently, a given effective inelastic strain rate, since in this model $\dot{W}_i = \sqrt{3J_2}\dot{\bar{\varepsilon}}^i$ is an increasing function of W_i (or of $\bar{\varepsilon}^i$).

[2] A model developed by Krempl and coworkers, known as **viscoplasticity based on overstress**, dispenses with drag stress as a variable (see Yao and Krempl [1985] and Krempl, McMahon, and Yao [1986]). In another model called **viscoplasticity based on total strain** (Cernocky and Krempl [1979]), the equilibrium stress is not an internal variable but a function of total strain; this model is therefore a nonlinear version of the "standard solid" model of linear viscoelasticity (see 1.5.1).

where $\Gamma = \sqrt{3\bar{J}_2}$, with $\bar{J}_2$ as defined in 3.1.1, and ϕ is a function (whose values have the dimensions of inverse time) that increases rapidly with its argument. The evolution of the equivalent inelastic strain is given by

$$\dot{\bar{\varepsilon}}^i = \phi(\Gamma/\sigma_D), \tag{3.1.10}$$

and, in uniaxial stress,

$$\dot{\varepsilon}^i = \phi(|\sigma - \rho|/\sigma_D).$$

Typical forms of $\phi(x)$ are Ax^n, $A(e^x - 1)$, and $A[\sinh(x^m)]^n$, where A, m, and n are constants, n in particular being a large exponent. For an extension to initially anisotropic behavior, see, for example, Helling and Miller [1987].

A variety of forms has been proposed for the rate equations for ρ and σ_D; a typical set is due to Walker [1981]:

$$\dot{\rho}_{ij} = a_1\dot{\varepsilon}^i_{ij} - [a_2\dot{\bar{\varepsilon}}^i + a_3(2\rho_{kl}\rho_{kl}/3)^{(m-1)/2}]\rho_{ij},$$

$$\dot{\sigma}_D = [a_4 - a_5(\sigma_D - \sigma_{D0})]\dot{\bar{\varepsilon}}^i - a_6(\sigma_D - \sigma_{D0})^p,$$

where $\dot{\varepsilon}^i_{ij}$ and $\dot{\bar{\varepsilon}}^i$ are substituted from (3.1.9)–(3.1.10), and a_1, ..., a_6, m, p and σ_{D0} are constants.

Endochronic Theory

A different "theory of viscoplasticity without a yield surface" is the **endochronic theory** of Valanis [1971], originally formulated by him (Valanis [1971]) for "application to the mechanical behavior of metals," though its range of application has recently been extended to other materials, such as concrete (Bažant [1978]). The basic concept in the theory is that of an *intrinsic time* (hence the name) that is related to the deformation history of the material point, the relation itself being a material property. An *intrinsic time measure* ζ is defined, for example, by

$$d\zeta^2 = A_{ijkl}\, d\varepsilon^i_{ij}\, d\varepsilon^i_{kl} + B^2\, dt^2,$$

where the tensor $\mathbf{A}$ and scalar B may depend on temperature. (In the original theory of Valanis [1971], the total strain ε rather than the inelastic strain ε^i appeared in the definition.) A model in which $B = 0$ describes rate-independent behavior and thus defines the **endochronic theory of plasticity**.

An *intrinsic time scale* is next defined as $z(\zeta)$, a monotonically increasing function, and the behavior of the material is assumed to be governed by constitutive relations having the same structure as those of linear viscoelasticity, as described in 1.5.2, but with z replacing the real time t. As in linear viscoelasticity, the internal variables can be eliminated, and the stress can be related to the strain history by means of a pseudo-relaxation function. The uniaxial relation is

$$\sigma = \int_0^z R(z - z') \frac{d\varepsilon}{dz'} dz', \tag{3.1.11}$$

while the multiaxial relation describing isotropic behavior is

$$\sigma_{ij} = \int_0^z \left[R_1(z - z')\delta_{ij} \frac{d\varepsilon_{kk}}{dz'} + 2R_2(z - z') \frac{d\varepsilon_{ij}}{dz'} \right] dz'.$$

With a pseudo-relaxation function analogous to that of the "standard solid," that is, $R(z) = E_1 + E_2 e^{-\alpha z}$, and with $z(\zeta)$ given by

$$z = \frac{1}{\beta} \ln(1 + \beta\zeta),$$

where α and β are positive constants, Valanis [1971] was able to fit many experimental data on repetitive uniaxial loading-unloading cycles and on coupling between tension and shear.

More recently, Valanis [1980] showed that Equation (3.1.11) can be replaced by

$$\sigma = \sigma_0 \frac{d\varepsilon^i}{dz} + \int_0^z \rho(z - z') \frac{d\varepsilon^i}{dz'} dz'. \tag{3.1.12}$$

For rate-independent uniaxial behavior, $d\zeta = |d\varepsilon^i|$ with no loss in generality. If the last integral is called α, then the stress must satisfy

$$|\sigma - \alpha| = \sigma_0 h(z), \tag{3.1.13}$$

where $h(z) = d\zeta/dz$. Equation (3.1.12) can be used to construct stress-strain curves showing hardening depending both on the effective inelastic strain [through $h(z)$] and on the strain path (through α). The equation has a natural extension to multiaxial stress states, which for isotropic materials is

$$(s_{ij} - \alpha_{ij})(s_{ij} - \alpha_{ij}) = [s_0 h(z)]^2.$$

This equation represents a yield surface capable of both expansion and translation in stress space, thus exhibiting both isotropic and kinematic hardening. An endochronic model unifying viscoplasticity and plasticity was presented by Watanabe and Atluri [1986].

Exercises: Section 3.1

1. Suppose that the yield function has the form $f(\boldsymbol{\sigma}, T, \boldsymbol{\xi}) = F(\boldsymbol{\sigma}) - k(T, \kappa)$, where κ is the hardening variable defined by either (1.5.6) or (1.5.7), and the flow equations are assumed as in the form (3.1.2). What is the "hardening modulus" H, defined by Equation (3.1.1)?

2. If the only stress components are $\sigma_{12} = \sigma_{21} = \tau$, with $\tau > 0$, write the equation for the shear rate $\dot{\gamma} = 2\dot{\varepsilon}_{12}$ given by the Hohenemser–Prager model (3.1.3). Discuss the special case $k = 0$.

3. Generalize the Hohenemser–Prager model to include isotropic and kinematic hardening. Compare with the Chaboche model.

4. Find the flow equation for a viscoplastic solid with a rate-dependent yield criterion given by

$$\sqrt{J_2} = k + \eta(2\dot{\varepsilon}_{ij}\dot{\varepsilon}_{ij})^{\frac{1}{2m}}.$$

5. Construct a simple model for combined viscoplasticity and plasticity, with a perfectly plastic Mises yield criterion and associated flow rule in both.

6. Derive (3.1.12) from (3.1.11).

Section 3.2 Rate-Independent Plasticity

3.2.1. Flow Rule and Work-Hardening

Flow Rule

In keeping with the formulation of rate-independent plasticity as the limit of classical viscoplasticity for infinitely slow processes, we henceforth consider all processes to be "infinitely" slow (compared with the material relaxation time τ), and correspondingly regard the material as "inviscid plastic," "rate-independent plastic," or simply *plastic*. The inelastic strain ε^i will from now on be called the *plastic strain* and denoted ε^p instead of ε^i. The flow equations (3.1.6) may be written as

$$\dot{\varepsilon}_{ij}^p = \dot{\lambda} h_{ij}, \tag{3.2.1}$$

where

$$\dot{\lambda} = \begin{cases} \dfrac{1}{H} < \overset{\circ}{f} >, & f = 0, \\ 0, & f < 0, \end{cases} \tag{3.2.2}$$

with H as defined by Equation (3.1.1). The rate equations (3.1.5) analogously become

$$\dot{\xi}_\alpha = \dot{\lambda} h_\alpha.$$

If $\partial f / \partial \xi_\alpha \equiv 0$, then, as mentioned before, the material is called *perfectly plastic*. In this case $H = 0$, but $\overset{\circ}{f} = \dot{f}$, and therefore the condition $\overset{\circ}{f} > 0$

is impossible. Plastic deformation then occurs only if $(\partial f/\partial \sigma_{ij})\dot{\sigma}_{ij} = 0$ (*neutral loading*), and the definition (3.2.2) of $\dot{\lambda}$ cannot be used. Instead, $\dot{\lambda}$ is an indeterminate positive quantity when $f = 0$ and $(\partial f/\partial \sigma_{ij})\dot{\sigma}_{ij} = 0$, and is zero otherwise.

In either case, $\dot{\lambda}$ and f can easily be seen to obey the **Kuhn–Tucker conditions** of optimization theory:

$$\dot{\lambda} f = 0, \quad \dot{\lambda} \geq 0, \quad f \leq 0.$$

The specification of the tensor function $\mathbf{h}$ in Equation (3.2.1), at least to within a multiplicative scalar, is known as the **flow rule**, and if there exists a function g (analogous to a viscoplastic potential) such that $h_{ij} = \partial g/\partial \sigma_{ij}$, then such a function is called a *plastic potential*.

Deformation Theory

The plasticity theory in which the plastic strain is governed by rate equations such as (3.2.1) is known as the *incremental* or *flow* theory of plasticity. A *deformation* or *total-strain* theory was proposed by Hencky [1924]. In this theory the plastic strain tensor itself is assumed to be determined by the stress tensor, provided that the yield criterion is met. Elastic unloading from and reloading to the yield surface are in principle provided for, although a contradiction is seen as soon as one considers reloading to a stress other than the one from which unloading took place, but located on the same yield surface; clearly, no plastic deformation could have occurred during the unloading-reloading process, yet the theory requires different values of the plastic strain at the two stress states. The deformation theory, which is mathematically much simpler than the flow theory, gives results that coincide with those of the latter only under highly restricted circumstances. An obvious example is the uniaxial case, provided that no reverse plastic deformation occurs; the equivalence is implicit in the use of relations such as (2.1.2).

A more general case is that of a material element subject to *proportional* or *radial loading*, that is, loading in which the ratios among the stress components remain constant, provided the yield criterion and flow rule are sufficiently simple (for example, the Mises yield criterion and the flow rule with $h_{ij} = s_{ij}$). A rough definition of "nearly proportional" loading, for which the deformation theory gives satisfactory results, is discussed by Rabotnov [1969]. It was shown by Kachanov [1954] (see also Kachanov [1971]) that the stress states derived from the two theories converge if the deformation develops in a definite direction.

Another example of a range of validity of the deformation theory, discussed in separate developments by Budiansky [1959] and by Kliushnikov [1959], concerns a material whose yield surface has a singular point or corner, with the stress point remaining at the corner in the course of loading.

For a simplified discussion, see Chakrabarty [1987], pp. 91–94.

The deformation theory has recently been the subject of far-reaching mathematical developments (Temam [1985]). It has also been found to give better results than the incremental theory in the study of the plastic buckling of elements under multiaxial stress, as is shown in Section 5.3.

Work-Hardening

The hardening criterion $H > 0$, and the corresponding criteria $H = 0$ for perfect plasticity and $H < 0$ for softening, were formulated in 3.1.1 for viscoplastic materials on the basis of rate-dependent behavior at states outside the yield surface. An alternative derivation can be given entirely in the context of rate-independent plasticity.

For given $\boldsymbol{\xi}$, $f(\boldsymbol{\sigma}, \boldsymbol{\xi}) = 0$ is the equation describing the yield surface in stress space. If $f(\boldsymbol{\sigma}, \boldsymbol{\xi}) = 0$ and $\dot{f}|_{\sigma=\text{const}} < 0$ (i.e. $H > 0$) at a given time t, then at a slightly later time $t + \Delta t$ we have $f(\boldsymbol{\sigma}, \boldsymbol{\xi} + \dot{\boldsymbol{\xi}}\Delta t) < 0$; the yield surface is seen to have moved so that $\boldsymbol{\sigma}$ is now inside it. In other words, $H > 0$ implies that, at least locally, *the yield surface is expanding* in stress space. The expansion of the yield surface is equivalent, in uniaxial stress, to a rising stress-strain curve (see Figure 3.2.1).

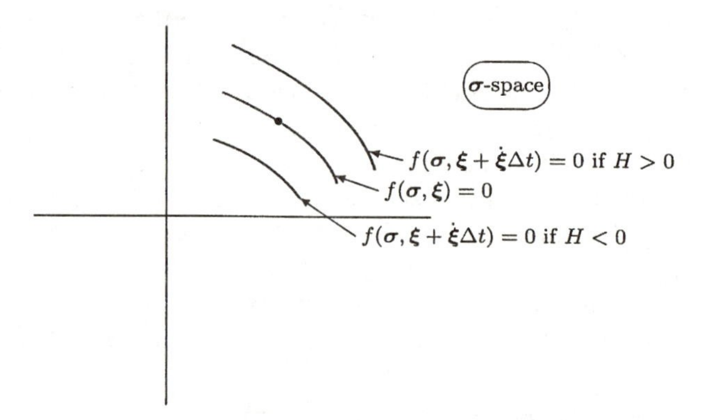

Figure 3.2.1. Hardening and softening in rate-independent plasticity: motion of yield surface in stress space

Conversely, a contracting yield surface denotes work-softening, and a stationary yield surface perfect plasticity. The description of work-softening materials is best achieved in strain space rather than stress space, and is discussed later. For now, we treat work-hardening materials only, with perfectly plastic materials as a limiting case.

In the simplest models of plasticity the internal variables are taken as (1) the plastic strain components ε_{ij}^p themselves, and (2) the hardening variable κ, defined by either Equation (1.5.6) or (1.5.7) (in rate-independent plasticity, $\bar{\varepsilon}^p$ is written in place of $\bar{\varepsilon}^i$). When the yield function is taken to have the form

$$f(\boldsymbol{\sigma}, \boldsymbol{\varepsilon}^p, \kappa) = F(\boldsymbol{\sigma} - \boldsymbol{\rho}(\boldsymbol{\varepsilon}^p)) - k(\kappa),$$

both isotropic and kinematic hardening, as discussed in Section 3.1, can be described; the hardening is isotropic if $\boldsymbol{\rho} \equiv 0$ and $dk/d\kappa > 0$, and purely kinematic if $dk/d\kappa \equiv 0$ and $\boldsymbol{\rho} \neq 0$. The condition $dk/d\kappa \equiv 0$ *and* $\boldsymbol{\rho} \equiv 0$ represents perfect plasticity. The simplest model of kinematic hardening — that of Melan [1938] — has $\boldsymbol{\rho}(\boldsymbol{\varepsilon}^p) = c\boldsymbol{\varepsilon}^p$, with c a constant. More sophisticated hardening models are discussed in Section 3.3.

Drucker's Postulate

A more restricted definition of work-hardening was formulated by Drucker [1950, 1951] by generalizing the characteristics of uniaxial stress-strain curves. With a single stress component σ, the conjugate plastic strain rate $\dot{\varepsilon}^p$ clearly satisfies [see Figure 3.2.2(a)]

$$\dot{\sigma}\dot{\varepsilon}^p \begin{cases} \geq 0, & \text{hardening material,} \\ = 0, & \text{perfectly plastic material,} \\ \leq 0, & \text{softening material.} \end{cases}$$

The inequalities are unchanged if the stress and plastic-strain rates are multiplied by the infinitesimal time increment dt, so that they hold equally well for $d\sigma\, d\varepsilon^p$. This product has the dimensions of work per unit volume, and was given by Drucker the following interpretation: *if a unit volume of an elastic-plastic specimen under uniaxial stress is initially at stress* σ *and plastic strain* ε^p, *and if an "external agency"* (one that is independent of whatever has produced the current loads) *slowly applies an incremental load resulting in a stress increment* $d\sigma$ (which causes the elastic and plastic strain increments $d\varepsilon^e$ and $d\varepsilon^p$, respectively) *and subsequently slowly removes it, then* $d\sigma\, d\varepsilon = d\sigma\, (d\varepsilon^e + d\varepsilon^p)$ *is the work*[1] *performed by the external agency in the course of incremental loading, and* $d\sigma\, d\varepsilon^p$ *is the work performed in the course of the cycle consisting of the application and removal of the incremental stress.* (Note that for $d\varepsilon^p \neq 0$, σ must be the current yield stress.)

Since $d\sigma\, d\varepsilon^e$ is always positive, and for a work-hardening material $d\sigma\, d\varepsilon^p \geq 0$, it follows that for such a material $d\sigma d\varepsilon > 0$. Drucker accordingly *defines* a work-hardening (or "stable") plastic material as one in which the work done during incremental loading is positive, and the work done in the loading-unloading cycle is nonnegative; this definition is generally known in the literature as **Drucker's postulate** (see also Drucker [1959]).

Having defined hardening in terms of work, Drucker naturally extends the definition to general three-dimensional states of stress and strain, such that

$$d\sigma_{ij}\, d\varepsilon_{ij} > 0 \quad \textit{and} \quad d\sigma_{ij}\, d\varepsilon_{ij}^p \geq 0,$$

[1] Actually it is *twice* the work.

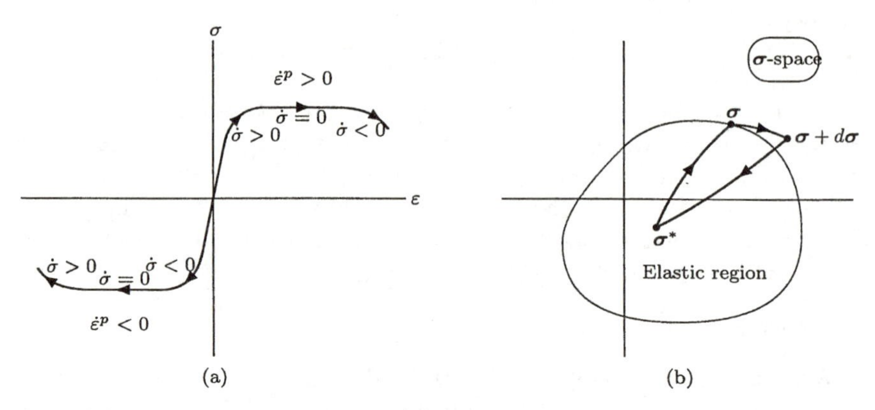

Figure 3.2.2. Drucker's postulate: (a) illustration in the uniaxial stress-strain plane; (b) illustration in stress space.

the equality holding only if $d\varepsilon^p = 0$. For perfectly plastic ("neutrally stable") materials Drucker's inequalities are $d\sigma_{ij}\, d\varepsilon_{ij} \geq 0$ and $d\sigma_{ij}\, d\varepsilon_{ij}^p = 0$. It can be seen that the inequality

$$\dot{\sigma}_{ij}\dot{\varepsilon}_{ij}^p \geq 0, \tag{3.2.3}$$

sometimes known simply as **Drucker's inequality**, is valid for both work-hardening and perfectly plastic materials.

Because it uses the concept of work, Drucker's postulate is often referred to as a quasi-thermodynamic postulate, although it is quite independent of the basic laws of thermodynamics. Drucker's inequality (3.2.3) may also be given an interpretation that is free of any considerations of incremental work: the left-hand side represents the scalar product $\dot{\boldsymbol{\sigma}} \cdot \dot{\boldsymbol{\varepsilon}}^p$, and the inequality therefore expresses the hypothesis that *the plastic strain rate cannot oppose the stress rate.*

We should note, lastly, that Drucker's definition of work-hardening is in a sense circular. The definition *assumes* an external agency that is capable of applying arbitrary stress increments. But as can readily be seen from stress-strain diagrams, this assumption is not valid for softening or perfectly plastic materials; for example, in a tension test no increase in stress is possible. In other words, such materials are *unstable under stress control.* On the other hand, they are stable under strain control (or displacement control[1]), since arbitrary strain increments that do not violate internal constraints may, in principle, be applied. This fact points to the applicability of strain-space plasticity, to be discussed later, to a wider class of materials.

Drucker's statement of his work-hardening postulate is broader than summarized above, in that the additional stress produced by the external

[1]Stability under strain control and displacement control are equivalent when deformations are infinitesimal, but not when they are finite.

agency need not be a small increment. In particular, the initial stress, say $\boldsymbol{\sigma}^*$, may be inside the elastic region, or at a point on the yield surface far away from $\boldsymbol{\sigma}$, and the process followed by the external agency may consist of elastic loading to a stress $\boldsymbol{\sigma}$ on the current yield surface, a small stress increment $d\boldsymbol{\sigma}$ producing an incremental plastic strain $d\boldsymbol{\varepsilon}$, and finally, elastic unloading back to $\boldsymbol{\sigma}^*$; the path is illustrated in Figure 3.2.2(b). With $d\boldsymbol{\sigma}$ neglected alongside $\boldsymbol{\sigma} - \boldsymbol{\sigma}^*$, the work per unit volume done by the external agency is $(\sigma_{ij} - \sigma_{ij}^*)\, d\varepsilon_{ij}^p$. Drucker's postulate, consequently, implies

$$(\sigma_{ij} - \sigma_{ij}^*)\, \dot{\varepsilon}_{ij}^p \geq 0. \tag{3.2.4}$$

3.2.2. Maximum-Plastic-Dissipation Postulate and Normality

Maximum-Plastic-Dissipation Postulate

Inequality (3.2.4) is, as we have just seen, a necessary condition for Drucker's postulate, but it is not a sufficient one. In other words, its validity is not limited to materials that are work-hardening in Drucker's sense. Its significance may best be understood when we consider its uniaxial counterpart,

$$(\sigma - \sigma^*)\dot{\varepsilon}^p \geq 0.$$

As is seen in Figure 3.2.3, the inequality expresses the property that the

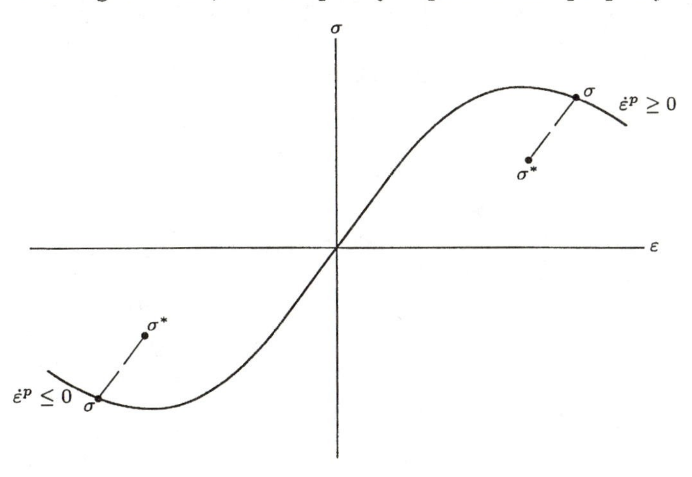

Figure 3.2.3. Maximum-plastic-dissipation postulate: illustration in the uniaxial stress-strain plane.

plastic strain rate is positive (negative) only if the current stress σ is not less than (not greater than) any stress σ^* in the current elastic range — in other words, if σ equals the current tensile (compressive) yield stress.

Clearly, work-softening and perfectly plastic materials have this property as well. Inequality (3.2.4) thus constitutes a postulate in its own right, called the **postulate of maximum plastic dissipation**. It was proposed independently by Mises [1928], Taylor [1947] and Hill [1948a]; it was derived from considerations of crystal plasticity by Bishop and Hill [1951], and is shown later to follow also from Il'iushin's postulate of plasticity in strain space.

Consequences of Maximum-Plastic-Dissipation Postulate

Inequality (3.2.4) has consequences of the highest importance in plasticity theory. To examine them, we represent symmetric second-rank tensors as vectors in a six-dimensional space, as in 1.3.5, but using boldface rather than underline notation, and using the dot-product notation for the scalar product. Our inequality may thus be written as

$$(\boldsymbol{\sigma} - \boldsymbol{\sigma}^*) \cdot \dot{\boldsymbol{\varepsilon}}^p \geq 0.$$

We suppose at first that the yield surface is everywhere smooth, so that a well-defined tangent hyperplane and normal direction exist at every point. It is clear from the two-dimensional representation in Figure 3.2.4(a) that if (3.2.4) is to be valid for all $\boldsymbol{\sigma}^*$ to the inward side of the tangent to the yield surface at $\boldsymbol{\sigma}$, then $\dot{\boldsymbol{\varepsilon}}^p$ *must be directed along the outward normal there*; this consequence is known as the **normality rule**. But as can be seen in Figure 3.2.4(b), if there are any $\boldsymbol{\sigma}^*$ lying to the outward side of the tangent, the inequality is violated. In other words, the entire elastic region must lie to one side of the tangent. As a result, **the yield surface is convex**.

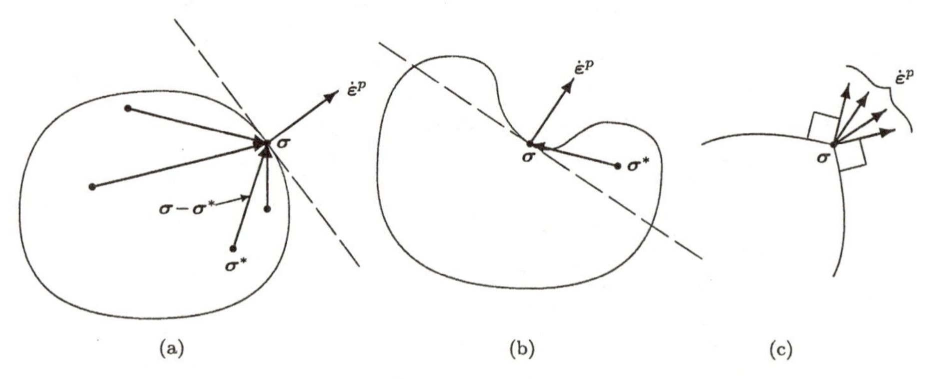

Figure 3.2.4. Properties of yield surface with associated flow rule: (a) normality; (b) convexity; (c) corner.

Let us define $D_p(\dot{\boldsymbol{\varepsilon}}^p;\, \boldsymbol{\xi})$ by

$$D_p(\dot{\boldsymbol{\varepsilon}}^p;\, \boldsymbol{\xi}) = \max_{\boldsymbol{\sigma}^*} \sigma^*_{ij}\dot{\varepsilon}^p_{ij},$$

the maximum being taken over all $\boldsymbol{\sigma}^*$ such that $f(\boldsymbol{\sigma}^*, \boldsymbol{\xi}) \le 0$. It follows from (3.2.4) that

$$\sigma_{ij}\dot{\varepsilon}^p_{ij} = D_p(\dot{\boldsymbol{\varepsilon}}^p; \boldsymbol{\xi}). \tag{3.2.5}$$

To make it clear that $D_p(\dot{\boldsymbol{\varepsilon}}^p; \boldsymbol{\xi})$ depends only on $\dot{\boldsymbol{\varepsilon}}^p$ and $\boldsymbol{\xi}$ and not on $\boldsymbol{\sigma}$, we note that, if the yield surface is strictly convex at $\boldsymbol{\sigma}$ (whether this point is regular or singular), then this is the only stress that corresponds to a given normal direction in stress space and hence to a given $\dot{\boldsymbol{\varepsilon}}^p$. If the yield surface has a flat portion, then all points on this portion have the same normal, that is, different stresses correspond to the same $\dot{\boldsymbol{\varepsilon}}^p$, but the scalar product $\boldsymbol{\sigma} \cdot \dot{\boldsymbol{\varepsilon}}^p = \sigma_{ij}\dot{\varepsilon}^p_{ij}$ is the same for all of them. $D_p(\dot{\boldsymbol{\varepsilon}}^p; \boldsymbol{\xi})$ will be called simply the plastic dissipation. Inequality (3.2.4) may now be rewritten as

$$D_p(\dot{\boldsymbol{\varepsilon}}^p; \boldsymbol{\xi}) \ge \sigma^*_{ij}\dot{\varepsilon}^p_{ij}, \tag{3.2.6}$$

giving explicit meaning to the name "principle of maximum plastic dissipation."

Normality

The normality rule is now discussed in more detail. At any point of the yield surface $f(\boldsymbol{\sigma}, \boldsymbol{\xi}) = 0$ where the surface is smooth, the outward normal vector is proportional to the gradient of f (in stress space), and therefore, reverting to indicial notation, we may express the normality rule as

$$h_{ij} = \frac{\partial f}{\partial \sigma_{ij}}, \tag{3.2.7}$$

where h_{ij} is the tensor function appearing in the flow equation (3.2.1). Equation (3.2.7) expresses the result that the function f defining the yield surface is itself a plastic potential, and therefore the normality rule is also called a flow rule that is *associated with the yield criterion*, or, briefly, an **associated** (sometimes **associative**) **flow rule**. A flow rule derivable from a plastic potential g that is distinct from f (more precisely, such that $\partial g/\partial \sigma_{ij}$ is not proportional to $\partial f/\partial \sigma_{ij}$) is accordingly called a **nonassociated** flow rule. In the French literature, materials obeying an associated flow rule are usually called *standard* materials, and this term will often be used here.

We are now in a position to say that Drucker's postulate applies to *standard work-hardening* (or, in the limit, perfectly plastic) materials. The frequently expressed notion that Drucker's postulate is required for the convexity of the yield surface and for the normality rule is clearly erroneous, as is the idea that work-hardening materials are necessarily standard.

If the yield surface is not everywhere smooth but has one or more singular points (corners) at which the normal direction is not unique, then at such a point $\dot{\boldsymbol{\varepsilon}}^p$ must lie in the cone formed by the normal vectors meeting there [see Figure 3.2.4(c)]. The argument leading to the convexity of the yield surface is not affected by this generalization. As will be seen, Equation (3.2.7) can still be formally used in this case, provided that the partial derivatives are properly interpreted. In a rigorous treatment, the concept of gradient must be replaced by that of *subgradient*, due to Moreau [1963]; its application in plasticity theory was formulated by Moreau [1976].

Another treatment of singular yield surfaces was proposed by Koiter [1953a], who supposed the yield surface to be made of a number — say n — of smooth surfaces, each defined by an equation $f_k(\boldsymbol{\sigma}, \boldsymbol{\xi}) = 0$ $(k = 1, \ldots, n)$; the elastic region is the intersection of the regions defined by $f_k(\boldsymbol{\sigma}, \boldsymbol{\xi}) < 0$, and $\boldsymbol{\sigma}$ is on the yield surface if at least one of the f_k vanishes there, it being a singular point only if two or more of the f_k vanish. Equation (3.2.7) is replaced by

$$h_{ij} = \sum_k \alpha_k \frac{\partial f_k}{\partial \sigma_{ij}},$$

the summation being over those k for which $f_k(\boldsymbol{\sigma}, \boldsymbol{\xi}) = 0$, and the α_k are nonnegative numbers that may, with no loss of generality, be constrained so that $\sum_k \alpha_k = 1$.

3.2.3. Strain-Space Plasticity

As we noted above, it is only in work-hardening materials, which are stable under stress control, that we may consider processes with arbitrary stress increments, and therefore it is only for such materials — with perfect plasticity as a limiting case — that a theory in which stress is an independent variable may be expected to work. No such limitation applies to theories using strain as an independent variable. Surprisingly, such theories were not proposed until the 1960s, beginning with the pioneering work of Il'iushin [1961], followed by papers by Pipkin and Rivlin [1965], Owen [1968], Lubliner [1974], Nguyen and Bui [1974], Naghdi and Trapp [1975], and others.

To see that strain-space yield surfaces have the same character whether the material is work-hardening or work-softening, let us consider one whose stress-strain diagram in tension/compression or in shear is as shown in Figure 3.2.5(a); such a material exemplifies Melan's linear kinematic hardening if $E' > 0$, and is perfectly plastic if $E' = 0$. The stress σ is on the yield surface if $\sigma = E'\varepsilon \pm (1 - E'/E)\sigma_E$, and therefore the condition on the strain ε is $E(\varepsilon - \varepsilon^p) = E'\varepsilon \pm (1 - E'/E)\sigma_E$, or, equivalently, $\varepsilon = [E/(E - E')]\varepsilon^p \pm \varepsilon_E$, where $\varepsilon_E = \sigma_E/E$ [see Figure 3.2.5(b)]. A yield surface in ε-space is thus given by the pair of points corresponding to a given value of ε^p, and the ε-ε^p diagram has a "stable" form (i.e., a positive slope) for all $E' < E$, even if negative.

Yield Criterion and Flow Rule in Strain Space

To formulate the three-dimensional yield criterion in strain space, let C

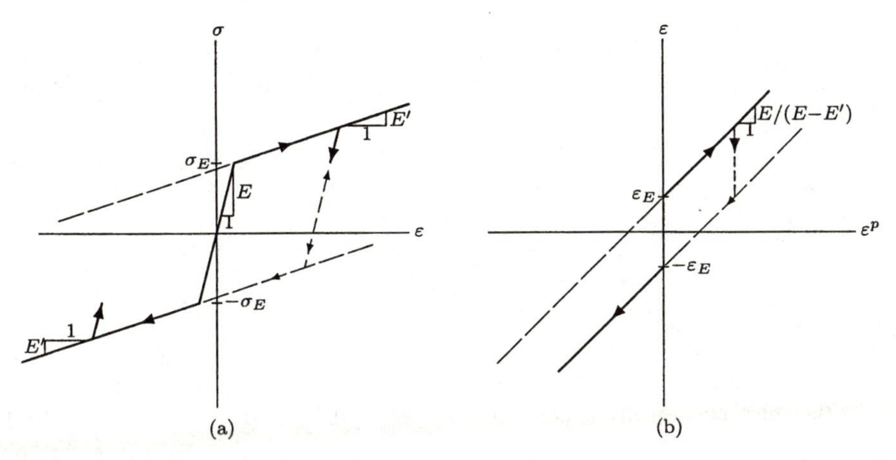

Figure 3.2.5. Material with linear hardening: (a) stress-strain diagram; (b) ε-ε^p diagram.

denote the elastic modulus tensor, so that the $\boldsymbol{\sigma}$-$\boldsymbol{\varepsilon}$-$\boldsymbol{\varepsilon}^p$ relation may be written $\boldsymbol{\sigma} = \mathsf{C}\cdot(\boldsymbol{\varepsilon} - \boldsymbol{\varepsilon}^{\mathrm{p}})$ [i.e., $\sigma_{ij} = C_{ijkl}(\varepsilon_{kl} - \varepsilon^p_{kl})$]. If $\hat{f}(\boldsymbol{\varepsilon}, \boldsymbol{\xi}) \stackrel{\text{def}}{=} f(\mathsf{C}\cdot(\boldsymbol{\varepsilon} - \boldsymbol{\varepsilon}^{\mathrm{p}}), \boldsymbol{\xi})$, then the strain-space yield criterion is just

$$\hat{f}(\boldsymbol{\varepsilon}, \boldsymbol{\xi}) = 0.$$

Since

$$\frac{\partial \hat{f}}{\partial \varepsilon_{ij}} = C_{ijkl}\left.\frac{\partial f}{\partial \sigma_{kl}}\right|_{\boldsymbol{\sigma}=\mathsf{C}\cdot(\boldsymbol{\varepsilon}-\boldsymbol{\varepsilon}^{\mathrm{p}})},$$

the same logic that led to (3.2.1)–(3.2.2) produces the flow equations

$$\dot{\varepsilon}^p_{ij} = \begin{cases} \dfrac{1}{L} h_{ij} < C_{ijkl}\dfrac{\partial f}{\partial \sigma_{ij}}\dot{\varepsilon}_{kl} >, & f = 0, \\ 0, & f < 0, \end{cases} \tag{3.2.8}$$

where

$$L = -\sum_\alpha \frac{\partial \hat{f}}{\partial \xi_\alpha} h_\alpha = H + C_{ijkl}\frac{\partial f}{\partial \sigma_{ij}} h_{kl}. \tag{3.2.9}$$

The normality rule (3.2.7), when translated into the strain-space formulation, takes the form

$$h_{ij} = C^{-1}_{ijkl}\frac{\partial \hat{f}}{\partial \varepsilon_{kl}}.$$

Note that L may very well be, and normally may be assumed to be, positive even when H is zero or negative, that is, for perfectly plastic or work-softening materials. It is thus not necessary to distinguish between these material types, the only restriction being $L > 0$. This condition describes stability under strain control in the same sense that the work-hardening criterion $H > 0$ describes stability under stress control; it will here be called *kinematic stability.*

The flow equation given by (3.2.8)–(3.2.9), when combined with the relation $\dot{\sigma}_{ij} = C_{ijkl}(\dot{\varepsilon}_{kl} - \dot{\varepsilon}^p_{kl})$, yields

$$\dot{\sigma}_{ij} = C_{ijkl}\dot{\varepsilon}_{kl} - \begin{cases} \dfrac{1}{L} C_{ijmn} h_{mn} < C_{pqkl} \dfrac{\partial f}{\partial \sigma_{pq}} \dot{\varepsilon}_{kl} >, & f = 0, \\ 0, & f < 0. \end{cases} \tag{3.2.10}$$

This is an explicit expression for $\dot{\boldsymbol{\sigma}}$ in terms of $\dot{\boldsymbol{\varepsilon}}$, which may be regarded as an inversion of $\dot{\varepsilon}_{ij} = C^{-1}_{ijkl}\dot{\sigma}_{kl} + \dot{\varepsilon}^p_{ij}$ with $\dot{\boldsymbol{\varepsilon}}^p$ given by Equation (3.2.1). In this sense the result, which was first derived by Hill [1958] [for standard materials, i.e. with $\mathbf{h}$ given by (3.2.7)], is not necessarily based on strain-space plasticity.

Work-hardening (Plastic) Modulus

It is easy to show that when $\dot{\boldsymbol{\varepsilon}}^p \neq 0$,

$$H = (C_{ijkl} f_{ij} h_{kl}) \frac{f_{ij}\dot{\sigma}_{ij}}{C_{ijkl} f_{ij} \dot{\varepsilon}_{kl}},$$

where $f_{ij} = \partial f / \sigma_{ij}$. Thus H may be related to the so-called *work-hardening modulus* or *plastic modulus* $d\sigma / d\varepsilon^p$ obtained in a simple tension test. If the material has (a) elastic isotropy, (b) plastic incompressibility, and (c) sufficient plastic symmetry so that $\sigma_{ij} = \sigma\delta_{i1}\delta_{j1}$ implies that $\dot{\varepsilon}^p_{22} = \dot{\varepsilon}^p_{33} = -\frac{1}{2}\dot{\varepsilon}^p_{11}$ and $\dot{\varepsilon}^p_{ij} = 0$ for $i \neq j$, then (with $\varepsilon^p = \varepsilon^p_{11}$)

$$H = h_{11} f_{11} \frac{d\sigma}{d\varepsilon^p}.$$

Il'iushin's Postulate

It can also be shown that the normality rule follows from a "postulate of plasticity" in strain space first proposed by Il'iushin [1961], namely, that in any cycle that is closed in strain space,

$$\oint \sigma_{ij}\, d\varepsilon_{ij} \geq 0, \tag{3.2.11}$$

where the equality holds only if the process is elastic; we show this by proving that (3.2.11) implies the maximum-plastic-dissipation postulate (3.2.4).

Consider a state $(\boldsymbol{\varepsilon}^1, \boldsymbol{\xi}^1)$ with $\boldsymbol{\varepsilon}^1$ on the yield surface, and any strain $\boldsymbol{\varepsilon}^*$ on or inside both the current yield surface and the subsequent yield surface obtained after a brief plastic process of duration Δt from $(\boldsymbol{\varepsilon}^1, \boldsymbol{\xi}^1)$ to $(\boldsymbol{\varepsilon}^1 + \dot{\boldsymbol{\varepsilon}}\Delta t, \boldsymbol{\xi}^1 + \dot{\boldsymbol{\xi}}\Delta t)$, that is,

$$\hat{f}(\boldsymbol{\varepsilon}^1, \boldsymbol{\xi}^1) = 0, \quad \hat{f}(\boldsymbol{\varepsilon}^1 + \dot{\boldsymbol{\varepsilon}}\Delta t, \boldsymbol{\xi}^1 + \dot{\boldsymbol{\xi}}\Delta t) = 0,$$

and

$$\hat{f}(\boldsymbol{\varepsilon}^*, \boldsymbol{\xi}^1) \leq 0, \quad \hat{f}(\boldsymbol{\varepsilon}^*, \boldsymbol{\xi}^1 + \dot{\boldsymbol{\xi}}\Delta t) \leq 0.$$

In the cycle

$$(\boldsymbol{\varepsilon}^*, \boldsymbol{\xi}^1) \ \stackrel{1}{\rightarrow}\ (\boldsymbol{\varepsilon}^1, \boldsymbol{\xi}^1) \ \stackrel{2}{\rightarrow}\ (\boldsymbol{\varepsilon}^1 + \dot{\boldsymbol{\varepsilon}}\Delta t, \boldsymbol{\xi}^1 + \dot{\boldsymbol{\xi}}\Delta t) \ \stackrel{3}{\rightarrow}\ (\boldsymbol{\varepsilon}^*, \boldsymbol{\xi}^1 + \dot{\boldsymbol{\xi}}\Delta t),$$

segments 1 and 3 are elastic, so that, if the process is isothermal,

$$\sigma_{ij}\dot{\varepsilon}_{ij} = \begin{cases} \rho\dot{\psi} & \text{in 1 and 3} \\ \rho\dot{\psi} + D & \text{in 2,} \end{cases}$$

where $\psi(\boldsymbol{\varepsilon}, \boldsymbol{\xi})$ is the free energy per unit mass at the given temperature and $D(\boldsymbol{\varepsilon}, \boldsymbol{\xi}, \dot{\boldsymbol{\xi}}) = -\rho\sum_{\alpha}(\partial\psi/\partial\xi_\alpha)\,\dot{\xi}_\alpha$ is the dissipation per unit volume. Now

$$\oint \sigma_{ij}\, d\varepsilon_{ij} = \oint \rho\, d\psi + \int_2 D\, dt;$$

but

$$\oint \rho\, d\psi = \rho\psi(\boldsymbol{\varepsilon}^*, \boldsymbol{\xi}^1 + \dot{\boldsymbol{\xi}}\Delta t) - \rho\psi(\boldsymbol{\varepsilon}^*, \boldsymbol{\xi}^1) \doteq -D(\boldsymbol{\varepsilon}^*, \boldsymbol{\xi}^1, \dot{\boldsymbol{\xi}})\Delta t,$$

and $size - 1int_2 D\, dt \doteq D(\boldsymbol{\varepsilon}^1, \boldsymbol{\xi}^1, \dot{\boldsymbol{\xi}})\Delta t$, the approximations being to within $o(\Delta t)$. It follows that

$$D(\boldsymbol{\varepsilon}, \boldsymbol{\xi}, \dot{\boldsymbol{\xi}}) \geq D(\boldsymbol{\varepsilon}^*, \boldsymbol{\xi}, \dot{\boldsymbol{\xi}}) \tag{3.2.12}$$

(the superscripts 1 can now be dropped) if $(\boldsymbol{\varepsilon}, \boldsymbol{\xi})$ is a state with $\boldsymbol{\varepsilon}$ on the current yield surface with corresponding $\dot{\boldsymbol{\xi}}$, and $\boldsymbol{\varepsilon}^*$ is any strain on or inside the yield surface.

With the free-energy density decomposed as in Equation (1.5.5) (and $\boldsymbol{\varepsilon}^p$ written in place of $\boldsymbol{\varepsilon}^i$), $D = \sigma_{ij}\dot{\varepsilon}_{ij}^p - \rho\dot{\psi}^i$ and therefore, if $\boldsymbol{\sigma}^* = \mathsf{C}\cdot(\boldsymbol{\varepsilon}^* - \boldsymbol{\varepsilon}^\mathrm{p})$ is any stress on or inside the yield surface, inequality (3.2.12) is equivalent to (3.2.4), which is thus seen to be a consequence of Il'iushin's postulate.

It remains to be investigated whether the converse holds. Consider an arbitrary process that is closed in strain space, going from $(\boldsymbol{\varepsilon}^*, \boldsymbol{\xi}^1)$ to $(\boldsymbol{\varepsilon}^*, \boldsymbol{\xi}^2)$, with $\boldsymbol{\xi}^2$ not necessarily infinitesimally close to $\boldsymbol{\xi}^1$. At any state of the process,

$$\sigma_{ij}\dot{\varepsilon}_{ij} = \rho\dot{\psi} + D(\boldsymbol{\varepsilon}, \boldsymbol{\xi}, \dot{\boldsymbol{\xi}}),$$

with $D = 0$ whenever the process is instantaneously elastic, and therefore

$$\oint \sigma_{ij}\, d\varepsilon_{ij} = \oint [D(\boldsymbol{\varepsilon}, \boldsymbol{\xi}, \dot{\boldsymbol{\xi}}) - D(\boldsymbol{\varepsilon}^*, \boldsymbol{\xi}, \dot{\boldsymbol{\xi}})]\, dt.$$

According to the principle of maximum plastic dissipation, the integrand is nonnegative whenever $\boldsymbol{\varepsilon}^*$ is on or inside the strain-space yield surface at the current value of $\boldsymbol{\xi}$, and therefore Il'iushin's postulate is satisfied for processes in which the original yield surface is inside all subsequent yield surfaces. The last condition is satisfied in materials with isotropic hardening, but not in general. Consequently Il'iushin's postulate is a stronger (i.e. less general) hypothesis than the principle of maximum plastic dissipation.

Nguyen–Bui Inequality

On the other hand, an inequality first explicitly stated by Nguyen and Bui [1974] may be shown to be weaker than the maximum-plastic-dissipation principle. It is readily seen that this principle, as expressed in the form (3.2.12), is equivalent to

$$C_{ijkl}(\varepsilon_{kl} - \varepsilon^*_{kl})\dot{\varepsilon}^p_{ij} \geq 0 \tag{3.2.13}$$

for any strain $\boldsymbol{\varepsilon}^*$ that is on or inside the current yield surface in strain space. Suppose, in particular, that $\boldsymbol{\varepsilon}^*$ is close to $\boldsymbol{\varepsilon}$ and is given by $\boldsymbol{\varepsilon}^* = \boldsymbol{\varepsilon} \pm \dot{\boldsymbol{\varepsilon}}\,dt$, with $dt > 0$ and $\dot{\boldsymbol{\varepsilon}}$ the strain-rate tensor in a possible process. With the plus sign chosen, the process goes from $\boldsymbol{\varepsilon}$ to $\boldsymbol{\varepsilon}^*$ and is necessarily elastic, so that $\dot{\boldsymbol{\varepsilon}}^p = 0$ and therefore Inequality (3.2.13) is satisfied as an equality. Thus $\dot{\boldsymbol{\varepsilon}} \neq 0$ only if the minus sign is taken, and therefore (3.2.13) takes the local form

$$C_{ijkl}\dot{\varepsilon}^p_{ij}\dot{\varepsilon}_{kl} \geq 0, \tag{3.2.14}$$

or the equivalent form given by Nguyen and Bui,

$$\dot{\sigma}_{ij}\dot{\varepsilon}^p_{ij} \geq C_{ijkl}\dot{\varepsilon}^p_{ij}\dot{\varepsilon}^p_{kl}.$$

Inequality (3.2.14), like Drucker's inequality (3.2.3), may be interpreted as a stability postulate, this time in strain space: if we take $C_{ijkl}a_{ij}b_{kl}$ as defining a scalar product between two tensors **a** and **b** in strain-increment space, then (3.2.14) expresses the notion that *the plastic strain rate cannot oppose the total strain rate*. Inequality (3.2.14) is by itself sufficient for the associated flow rule to follow, and consequently describes *standard kinematically stable* materials.

Exercises: Section 3.2

1. A work-hardening plastic solid is assumed to obey the Mises yield criterion with isotropic hardening, that is, $f(\boldsymbol{\sigma}, \boldsymbol{\xi}) = \sqrt{J_2} - k(\bar{\varepsilon}^p)$, and the flow rule $h_{ij} = s_{ij}$. Show that

$$\dot{\varepsilon}^p_{ij} = \frac{\sqrt{3}s_{ij}\langle s_{kl}\dot{s}_{kl}\rangle}{4k^2k'(\bar{\varepsilon}^p)}.$$

2. Show that the solid described in Exercise 1 obeys Drucker's inequality (3.2.3) if and only if $k'(\bar{\varepsilon}^p) > 0$.

3. If $\boldsymbol{\sigma}$ and $\boldsymbol{\sigma}^*$ are stresses such that $J_2 = k^2$ and $J_2^* \le k^2$, show that $(s_{ij} - s_{ij}^*)s_{ij} \ge 0$, and hence that the solid of Exercise 1 obeys the maximum-plastic-dissipation postulate (3.2.4) independently of $k'(\bar{\varepsilon}^p)$, that is, whether the solid hardens or softens.

4. For a work-hardening solid with the yield criterion of Exercise 1, but with the nonassociated flow rule $h_{ij} = s_{ij} + t_{ij}$, where $s_{ij}t_{ij} = 0$, show that for some $\dot{\boldsymbol{\sigma}}$ Drucker's inequality (3.2.3) is violated.

5. For the standard isotropically hardening Mises solid of Exercise 1, show that
$$\dot{\varepsilon}_{ij}^p = \frac{s_{ij}<s_{kl}\dot{\varepsilon}_{kl}>}{2k^2\left(1 + \dfrac{k'(\bar{\varepsilon}^p)}{\sqrt{3}G}\right)}.$$

6. Show that the standard Mises solid obeys the Nguyen–Bui inequality (3.2.14) whether it hardens or softens.

Section 3.3 Yield Criteria, Flow Rules and Hardening Rules

3.3.1. Introduction

The yield function f in stress space may be written with no loss of generality in terms of the stress deviator and the first invariant of stress, that is, $f(\boldsymbol{\sigma}, \boldsymbol{\xi}) = \bar{f}(\mathbf{s}, I_1, \boldsymbol{\xi})$, where $I_1 = \sigma_{kk} = \delta_{ij}\sigma_{ij}$, so that $\partial I_1/\partial\sigma_{ij} = \delta_{ij}$. Since
$$s_{kl} = \sigma_{kl} - \frac{1}{3}I_1\delta_{kl} = \left(\delta_{ik}\delta_{jl} - \frac{1}{3}\delta_{ij}\delta_{kl}\right)\sigma_{ij},$$
it follows that $\partial s_{kl}/\partial\sigma_{ij} = \delta_{ik}\delta_{jl} - \frac{1}{3}\delta_{ij}\delta_{kl}$. Consequently,
$$\frac{\partial f}{\partial \sigma_{ij}} = \frac{\partial \bar{f}}{\partial s_{kl}}\frac{\partial s_{kl}}{\partial \sigma_{ij}} + \frac{\partial \bar{f}}{\partial I_1}\frac{\partial I_1}{\partial \sigma_{ij}} = (\bar{f}_{ij} - \tfrac{1}{3}\delta_{ij}\bar{f}_{kk}) + \frac{\partial f}{\partial I_1}\delta_{ij},$$
where $\bar{f}_{ij} = \partial\bar{f}/\partial s_{ij}$. Accordingly, in a standard material plastic volume change ("dilatancy") occurs if and only if the yield criterion depends on I_1, i.e. on the mean stress, and, conversely, plastic incompressibility obtains if and only if the yield criterion depends on $\mathbf{s}$ but not on I_1. If the yield criterion of a plastically incompressible material is significantly affected by mean stress, then the material is necessarily nonstandard.

Isotropic Yield Criteria

If the yield criterion is initially isotropic, then the dependence of f on $\boldsymbol{\sigma}$ must be through the stress invariants I_1, I_2, and I_3, or, equivalently, on the principal stresses σ_I ($I = 1, 2, 3$), provided this dependence is symmetric, that is, invariant under any change of the index I. Similarly, the dependence of $\overline{f}$ on s must be through the stress-deviator invariants J_2 and J_3; the equivalent dependence on the principal stresses may be exhibited in the so-called principal stress-deviator plane or π-plane, namely, the plane in $\sigma_1\sigma_2\sigma_3$-space given by $\sigma_1 + \sigma_2 + \sigma_3 = 0$, shown in Figure 3.3.1.

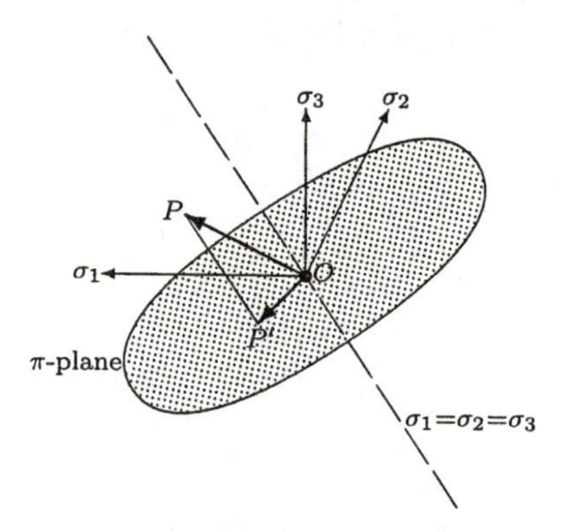

Figure 3.3.1. π-plane.

Indeed, if a point $P(\sigma_1, \sigma_2, \sigma_3)$ in $\sigma_1\sigma_2\sigma_3$-space (the *Haigh–Westergaard space*) is represented by the vector $\overrightarrow{OP}$, then its projection $\overrightarrow{OP'}$ onto the π-plane is the vector whose components are the principal stress deviators s_1, s_2, s_3. The magnitude of this projection — that is, the distance from P to the axis $\sigma_1 = \sigma_2 = \sigma_3$ — is just $\sqrt{2J_2}$. A yield surface that is independent of I_1 has, in this space, the form of a cylinder perpendicular to the π-plane, and therefore may be specified by a single curve in this plane. A yield surface that depends on I_1 may be described by a family of curves in the π-plane, each corresponding to a different value of I_1 and forming the intersection of the yield surface in $\sigma_1\sigma_2\sigma_3$-space with a plane I_1 = constant, that is, a plane parallel to the π-plane.

A curve in the π-plane can also be described in terms of the polar coordinates $(\sqrt{2J_2}, \theta)$, where the polar angle θ may defined as that measured from the projection of the σ_1-axis toward the projection of the σ_2-axis, and can be shown to be given by

$$\tan\theta = \frac{\sqrt{3}(\sigma_2 - \sigma_3)}{2\sigma_1 - \sigma_2 - \sigma_3} = \frac{s_2 - s_3}{\sqrt{3}s_1}.$$

Using some trigonometric identities and the fact that $s_1 + s_2 + s_3 = 0$, it is also possible to define θ in terms of the deviatoric stress invariants J_2 and

J_3:

$$\cos 3\theta = \frac{3\sqrt{3}J_3}{2J_2^{3/2}}.$$

A point with $\theta = 0$ corresponds to $\sigma_1 > \sigma_2 = \sigma_3$; the locus of such points on the yield surface is said to represent one of the three *tensile meridians* of the surface. A point with $\theta = \pi/3$ corresponds to $\sigma_1 = \sigma_2 > \sigma_3$, and lies on a *compressive meridian*.

3.3.2. Yield Criteria Independent of the Mean Stress

Since the concept of plasticity was first applied to metals, in which the influence of mean stress on yielding is generally negligible (Bridgman [1923, 1950]), the oldest and most commonly used yield criteria are those that are independent of I_1. Such criteria have an alternative two-dimensional representation: since their dependence on the principal stresses must be through the differences $\sigma_1 - \sigma_2$, $\sigma_1 - \sigma_3$ and $\sigma_2 - \sigma_3$, and since $\sigma_1 - \sigma_2 = (\sigma_1 - \sigma_3) - (\sigma_2 - \sigma_3)$, the yield criterion can be plotted in a plane with $\sigma_1 - \sigma_3$ and $\sigma_2 - \sigma_3$ as coordinate axes.

Tresca Criterion

The **Tresca** yield criterion is historically the oldest; it embodies the assumption that plastic deformation occurs when the maximum shear stress over all planes attains a critical value, namely, the value of the current yield stress in shear, denoted $k(\boldsymbol{\xi})$. Because of Equation (1.3.11), this criterion may be represented by the yield function

$$f(\boldsymbol{\sigma}, \boldsymbol{\xi}) = \frac{1}{2}\max\left(|\sigma_1 - \sigma_2|, |\sigma_2 - \sigma_3|, |\sigma_3 - \sigma_1|\right) - k(\boldsymbol{\xi}), \tag{3.3.1}$$

or, equivalently,

$$f(\boldsymbol{\sigma}, \boldsymbol{\xi}) = \frac{1}{4}(|\sigma_1 - \sigma_2| + |\sigma_2 - \sigma_3| + |\sigma_3 - \sigma_1|) - k(\boldsymbol{\xi}). \tag{3.3.2}$$

The projection of the Tresca yield surface in the π-plane is a regular hexagon, shown in Figure 3.3.2(a), whose vertices lie on the projections of the positive and negative σ_1, σ_2 and σ_3-axes, while in the $(\sigma_1 - \sigma_3)(\sigma_2 - \sigma_3)$-plane it takes the form of the irregular hexagon shown in Figure 3.3.2(b).

Of course, the forms (3.3.1) and (3.3.2) for the Tresca yield function are not unique. The form

$$f(\boldsymbol{\sigma}) = [(\sigma_1 - \sigma_2)^2 - 4k^2][(\sigma_2 - \sigma_3)^2 - 4k^2][(\sigma_1 - \sigma_3)^2 - 4k^2]$$

(with the dependence on $\boldsymbol{\xi}$ not indicated) has the advantage of being analytic and, moreover, expressible in terms of the principal stress-deviator invariants J_2 and J_3:

$$f(\boldsymbol{\sigma}) = 4J_2{}^3 - 27J_3^2 - 36k^2J_2{}^2 + 96k^4J_2 - 64k^6.$$

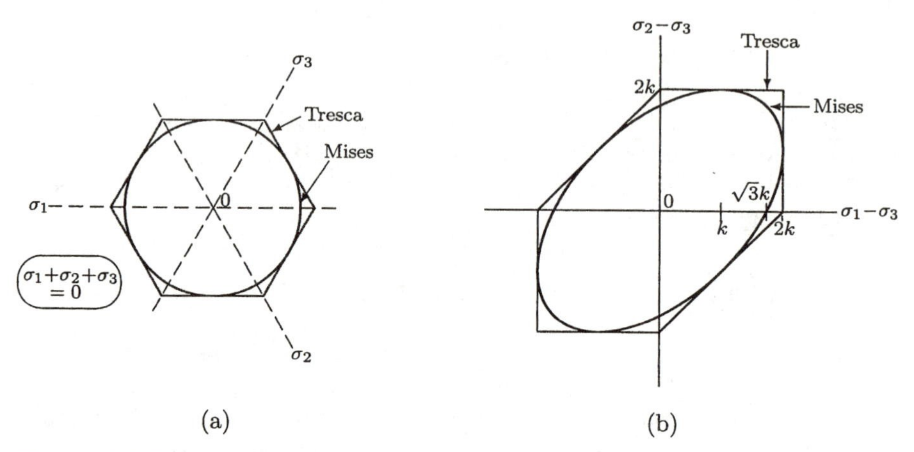

Figure 3.3.2. Projections of Tresca and Mises yield surfaces: (a) π-plane; (b) $\sigma_1 - \sigma_3$-$\sigma_2 - \sigma_3$ plane.

Tresca Criterion: Associated Flow Rule

Although the Tresca yield surface is singular, we can nonetheless derive its associated flow rule by means of a formal application of Equation (3.2.7) to the second form, Equation (3.3.2). We write

$$\frac{d}{dx}|x| = \operatorname{sgn} x, \tag{3.3.3}$$

where

$$\operatorname{sgn} x = 2H(x) - 1 = \begin{cases} +1, & x > 0 \\ -1, & x < 0 \end{cases}$$

defines the *signum* function for $x \neq 0$. It is conventional to define $\operatorname{sgn} 0 = 0$, but in the present context it is more convenient if $\operatorname{sgn} x$ does not have a unique value at 0 but can have any value between -1 and 1. Strictly speaking, then, it is not a function in the usual sense but a *set-valued function* or *multifunction*, a concept with considerable use in convex analysis.[1] In this way we obtain

$$\dot{\varepsilon}_1^p = \frac{1}{4}\dot{\lambda}[\operatorname{sgn}(\sigma_1 - \sigma_2) + \operatorname{sgn}(\sigma_1 - \sigma_3)],$$

where, in accordance with Equations (3.2.1)–(3.2.2), for work-hardening materials $\dot{\lambda} = <\overset{\circ}{f}>/H$, with

$$H = \sum_\alpha \frac{\partial k}{\partial \xi_\alpha} h_\alpha,$$

[1] In fact, our use of Equation (3.3.3) comes rather close to the subdifferential calculus of Moreau.

while for the perfectly plastic material $\dot{\lambda}$ is indeterminate. Similar expressions are obtained for $\dot{\varepsilon}_2^p$ and $\dot{\varepsilon}_3^p$. Thus, if the principal stresses are all distinct, and ordered such that $\sigma_1 > \sigma_2 > \sigma_3$, then $\dot{\varepsilon}_1^p = \frac{1}{2}\dot{\lambda}$, $\dot{\varepsilon}_2^p = 0$, $\dot{\varepsilon}_3^p = -\frac{1}{2}\dot{\lambda}$. If, on the other hand, $\sigma_1 = \sigma_2 > \sigma_3$, then $\dot{\varepsilon}_1^p = \frac{1}{4}\dot{\lambda}(1+\beta)$, $\dot{\varepsilon}_2^p = \frac{1}{4}\dot{\lambda}(1-\beta)$, $\dot{\varepsilon}_3^p = -\frac{1}{2}\dot{\lambda}$, where β is any real number between -1 and 1. Analogous expressions can be obtained for all other combinations of principal stresses.

The Tresca flow rule can also be obtained by the method due to Koiter, discussed in 3.2.2.

It can be seen that for every combination of principal stresses, $|\dot{\varepsilon}_1^p| + |\dot{\varepsilon}_2^p| + |\dot{\varepsilon}_3^p| = \dot{\lambda}$, and therefore the plastic dissipation is given by

$$D_p(\dot{\varepsilon}^p; \boldsymbol{\xi}) = \dot{\lambda}k(\boldsymbol{\xi}) = k(\boldsymbol{\xi})(|\dot{\varepsilon}_1^p| + |\dot{\varepsilon}_2^p| + |\dot{\varepsilon}_3^p|).$$

If it is desired to use an effective plastic strain $\bar{\varepsilon}^p$ as an internal variable in conjunction with the Tresca criterion and its associated flow rule, then the definition

$$\bar{\varepsilon}^p = \frac{1}{2}\int (|\dot{\varepsilon}_1^p| + |\dot{\varepsilon}_2^p| + |\dot{\varepsilon}_3^p|)\, dt$$

is more appropriate than (1.5.7). In fact, if it is assumed that k is a function of $\bar{\varepsilon}^p$ as thus defined, then $W_p = 2\int k\, d\bar{\varepsilon}^p$, so that a one-to-one correspondence can be established between W_p and $\bar{\varepsilon}^p$.

Lévy Flow Rule and Mises Yield Criterion

In the nineteenth century Saint-Venant and others used the Tresca yield criterion together with the (nonassociated) flow rule derived from the J_2 potential (see 1.5.4) whose general form was first proposed (for total rather than plastic strain) by Lévy, namely,

$$\dot{\varepsilon}_{ij}^p = \dot{\lambda}s_{ij},$$

with $\dot{\lambda}$ defined as above. As seen in Section 3.1 in connection with viscoplasticity, and as first pointed out by Mises [1913], the yield criterion with which this flow rule is associated is the **Mises criterion**, represented by the yield function

$$f(\boldsymbol{\sigma}, \boldsymbol{\xi}) = \sqrt{J_2} - k(\boldsymbol{\xi}),$$

where $k(\boldsymbol{\xi})$ is again the yield stress in shear at the current values of $\boldsymbol{\xi}$. In view of the relation (1.3.5) between J_2 and the octahedral shear stress, the Mises criterion is also known as the *maximum-octahedral-shear-stress* criterion, and as a result of Equation (1.4.17), which shows the complementary energy of an isotropic, linearly elastic material to be uncoupled into volumetric and distortional parts, it is also called the *maximum-distortional-energy* criterion.

An alternative — and analytic — form of the Mises yield function (with the dependence on $\boldsymbol{\xi}$ not shown explicitly) is

$$f(\boldsymbol{\sigma}) = J_2 - k^2.$$

Expressing J_2 in terms of the principal stresses (see Section 1.3), we may formulate the Mises yield criterion in the form

$$(\sigma_1 - \sigma_2)^2 + (\sigma_2 - \sigma_3)^2 + (\sigma_1 - \sigma_3)^2 = 6k^2$$

or

$$\sigma_1^2 + \sigma_2^2 + \sigma_3^2 - \sigma_2\sigma_3 - \sigma_3\sigma_1 - \sigma_1\sigma_2 = 3k^2.$$

The form taken by the Mises yield surface in the π-plane is that of a circle of radius $\sqrt{2}k$, and in the $(\sigma_1 - \sigma_3)(\sigma_2 - \sigma_3)$-plane that of an ellipse. Both forms are shown, along with those for the Tresca criterion, in Figure 3.3.2 (page 138).

The plastic dissipation for the Mises criterion and associated flow rule is given by

$$\begin{aligned} D_p(\dot{\varepsilon}^p; \boldsymbol{\xi}) &= \sigma_{ij}\dot{\varepsilon}_{ij}^p = \dot{\lambda}s_{ij}s_{ij} = \sqrt{2J_2}\sqrt{\dot{\varepsilon}_{ij}^p\dot{\varepsilon}_{ij}^p} \\ &= k(\boldsymbol{\xi})\sqrt{2\dot{\varepsilon}_{ij}^p\dot{\varepsilon}_{ij}^p}. \end{aligned}$$

If k is a function of $\overline{\varepsilon}^p$ as defined by (1.5.7), then W_p is in one-to-one correspondence with $\overline{\varepsilon}_p$.

As mentioned above, Lévy and Mises formulated the flow rule bearing their name for the total, rather than merely the plastic, strain rate; in this form it is valid as an approximation for problems in which elastic strains are vanishingly small, or, equivalently, for materials whose elastic moduli are infinite — the so-called *rigid-plastic* materials (see Section 3.4). The generalization allowing for nonvanishing elastic strains is due to Prandtl [1924] and Reuss [1930]; expressed in terms of total strain rate, with isotropic linear elasticity, the result is known as the **Prandtl–Reuss equations**:

$$\begin{aligned} \dot{\varepsilon}_{kk} &= \frac{1}{3K}\dot{\sigma}_{kk}, \\ \dot{e}_{ij} &= \frac{1}{2G}\dot{s}_{ij} + \dot{\lambda}s_{ij}. \end{aligned} \tag{3.3.4}$$

Some generalizations of the Mises yield function have been proposed so that dependence on J_3 is included. A typical form is

$$f(\boldsymbol{\sigma}) = \left(1 - c\frac{J_3^2}{J_2^3}\right)^{\alpha} J_2 - k^2.$$

The exponent α has variously been taken as $\frac{1}{3}$ and 1, k is as usual the yield stress in simple shear, and c is a parameter that is to be determined so as to optimize the fit with experimental data.

Anisotropic Yield Criteria

Anisotropy in yielding may be of two types: *initial* anisotropy and *induced* anisotropy. The former exists in materials that are structurally anisotropic, even before any plastic deformation has taken place; the latter appears, even in initially isotropic materials, as a result of work-hardening (Section 2.1). An example of an initially anisotropic yield criterion is Schmid's law (Section 2.2), according to which yielding in single crystals occurs when the shear stress on certain preferred planes (the slip planes) reaches a critical value; in the special case when every plane is a slip plane, Schmid's law reduces to the Tresca criterion.

An anisotropic generalization of the Mises criterion is due to Hill [1950]; it replaces J_2 with a general quadratic function of $\boldsymbol{\sigma}$ and therefore has the form

$$\tfrac{1}{2}A_{ijkl}\sigma_{ij}\sigma_{kl} = k^2,$$

where $\mathbf{A}$ is a fourth-rank tensor which has the same symmetries as the elasticity tensors ($A_{ijkl} = A_{jikl} = A_{klij}$). If the yield criterion is independent of mean stress, then $\mathbf{A}$ also obeys $A_{ijkk} = 0$, so that it has at most fifteen independent components (like a symmetric 5×5 matrix); the isotropic (Mises) case corresponds to $A_{ijkl} = \delta_{ik}\delta_{jl} - \frac{1}{3}\delta_{ij}\delta_{kl}$. A special case considered by Hill [1948b] refers to a material with three mutually perpendicular planes of symmetry; if a Cartesian basis is chosen so that the coordinate planes are parallel to the planes of symmetry, then in this basis the components of $\mathbf{A}$ coupling normal stresses with shear stresses (e.g. A_{1112}, A_{1123}, etc. — nine independent components altogether) are zero, and $\mathbf{A}$ is given by

$$\begin{aligned} A_{ijkl}\sigma_{ij}\sigma_{kl} &= A(\sigma_{22}-\sigma_{33})^2 + B(\sigma_{11}-\sigma_{33})^2 + C(\sigma_{11}-\sigma_{22})^2 \\ &\quad + 4D\sigma_{23}^2 + 4E\sigma_{13}^2 + 4F\sigma_{12}^2, \end{aligned}$$

where $A, \ldots, F$ are constants; clearly $A_{1111} = B + C$, $A_{1122} = -C$, $A_{1133} = -B$, $A_{1212} = F$, and so on.

3.3.3. Yield Criteria Dependent on the Mean Stress

A yield criterion depending on the mean stress becomes necessary when it is desired to apply plasticity theory to soils, rocks, and concrete, as discussed in Section 2.3. One such criterion has its origin in the **Mohr theory of rupture**, according to which failure (rupture) occurs on a plane in a body if the shear stress and normal stress on that plane achieve a critical combination. Since the strength properties of an isotropic material are unchanged when the direction of the shear stress is reversed, the critical combination may be expressed by the functional equation $\tau = \pm g(\sigma)$. This equation represents a pair of curves (each being the other's reflection through the σ-axis) in the Mohr plane, and a state of stress, as determined by the three Mohr's circles, is safe if all three circles lie between the curves, while it is a critical state if one of the three is tangent to the curves. These curves are thus the envelopes of the Mohr's circles representing failure and are therefore called the *Mohr failure (rupture)* envelopes. The point (σ, τ) is a point of tangency — say the upper one — if it obeys (1) the equation $\tau = g(\sigma)$, (2) the equation of the Mohr's circle [centered at $(\sigma_m, 0)$ and of radius τ_m], and (3) the tangency condition. If σ and τ are eliminated between these three equations, there remains an equation in terms of σ_m and τ_m, which constitutes the failure criterion.[1] Concretely, if a point in principal-stress space is located in the sextant $\sigma_1 > \sigma_2 > \sigma_3$, then $\sigma_m = \frac{1}{2}(\sigma_1 + \sigma_3)$ and $\tau_m = \frac{1}{2}|\sigma_1 - \sigma_3|$; the equation consequently represents a cylindrical surface parallel to the σ_2-axis, and the failure surface is formed by six such surfaces.

Mohr–Coulomb Criterion

The equations can be reduced explicitly if the Mohr envelopes are straight lines, that is, if

$$g(\sigma) = c - \mu\sigma.$$

This is just Equation (2.3.3), with the sign of σ changed to the usual convention whereby it is positive in tension; c is the cohesion, and $\mu = \tan\phi$ is the coefficient of internal friction in the sense of the Coulomb model of friction. The resulting criterion is consequently known as the **Mohr–Coulomb criterion**. It is convenient to represent the Mohr's circle parametrically:

$$\sigma = \sigma_m + \tau_m \cos 2\alpha, \quad \tau = \tau_m \sin 2\alpha,$$

where α is the angle between the failure plane and the axis of the least tensile (greatest compressive) stress. The tangency condition is then $\mu = \cot 2\alpha$, so that $\alpha = \frac{1}{4}\pi - \frac{1}{2}\phi$, $\sin 2\alpha = \cos\phi$, and $\cos 2\alpha = \sin\phi$. The equation in terms of σ_m and τ_m becomes

$$\tau_m + \sigma_m \sin\phi = c\cos\phi,$$

from which it is seen that the failure stress in simple shear is $c\cos\phi$. (Needless to say, when $\phi = 0$ the Mohr–Coulomb criterion reduces to that of Tresca.) In terms of the principal stresses the criterion takes the form

$$\max_{i \neq j}[|\sigma_i - \sigma_j| + (\sigma_i + \sigma_j)\sin\phi] = 2c\cos\phi,$$

so that the yield stresses in tension and compression are respectively $2c\cos\phi/(1+\sin\phi) = 2c\tan\alpha$ and $2c\cos\phi/(1-\sin\phi) = 2c\cot\alpha$. The associated plastic dissipation was shown by Drucker [1953] to be

$$D_p(\dot{\varepsilon}^p; \boldsymbol{\xi}) = c\cot\phi(\dot{\varepsilon}_1^p + \dot{\varepsilon}_2^p + \dot{\varepsilon}_3^p).$$

[1] It is pointed out by Hill [1950] that tangency between the Mohr envelope and Mohr's circle does not necessarily occur at real (σ, τ), and that it is the failure criterion and not the envelope that is fundamental.

The failure surfaces in $\sigma_1\sigma_2\sigma_3$-space are obviously planes that intersect to form a hexagonal pyramid; the plane in the sextant $\sigma_1 > \sigma_2 > \sigma_3$, for example, is described by

$$\sigma_1 - \sigma_3 + (\sigma_1 + \sigma_3)\sin\phi = 2c\cos\phi.$$

A form valid in all six sextants is

$$\sigma_{\max} - \sigma_{\min} + (\sigma_{\max} + \sigma_{\min})\sin\phi = 2c\cos\phi,$$

where $\sigma_{\max}$ and $\sigma_{\min}$ denote respectively the (algebraically) largest and smallest principal stresses.

The last equation may be rewritten as

$$\sigma_{\max} - \sigma_{\min} + \frac{1}{3}[(\sigma_{\max} - \sigma_{\text{int}}) - (\sigma_{\text{int}} - \sigma_{\min})]\sin\phi = 2c\cos\phi - \frac{2}{3}I_1\sin\phi,$$

where σ_{int} denotes the intermediate principal stress. The left-hand side, being an isotropic function of the stress deviator, is therefore a function of J_2 and J_3. The Mohr–Coulomb criterion is therefore seen to be a special case of the family of criteria based on Coulomb friction and described by equations of the form

$$\bar{F}(J_2,\, J_3) = c - \lambda I_1,$$

where c and λ are constants.

Drucker–Prager Criterion

Another yield criterion of this family, combining Coulomb friction with the Mises yield criterion, was proposed by Drucker and Prager [1952] and has become known as the **Drucker–Prager criterion**. With the Mises criterion interpreted in terms of the octahedral shear stress, it may be postulated that yielding occurs on the octahedral planes when $\tau_{\text{oct}} = \sqrt{\frac{2}{3}}k - \frac{1}{3}\mu I_1$, so that, in view of Equation (1.3.5), the criterion may be represented by the yield function $\overline{f}(\mathbf{s},\, I_1) = \sqrt{J_2} + \mu I_1/\sqrt{6} - k$. The yield surface in Haigh–Westergaard space is a right circular cone about the mean-stress axis, subtending the angle $\tan^{-1}(\sqrt{3}\mu)$. The yield stresses in simple shear, tension, and compression are respectively k, $\sqrt{3}k/(1 + \mu/\sqrt{2})$ and $\sqrt{3}k/(1 - \mu/\sqrt{2})$; note that for this criterion to be physically meaningful, μ must be less than $\sqrt{2}$. The associated plastic dissipation is

$$D_p(\dot{\boldsymbol{\varepsilon}}^p; \boldsymbol{\xi}) = \frac{k\sqrt{2\dot{\varepsilon}^p_{ij}\dot{\varepsilon}^p_{ij}}}{\sqrt{1+\mu^2}}.$$

Projections of the yield surfaces corresponding to the Mohr–Coulomb and Drucker–Prager criteria onto a plane parallel to the π-plane (i.e. one with $\sigma_1 + \sigma_2 + \sigma_3 = \text{constant}$) are shown in Figure 3.3.3(a).

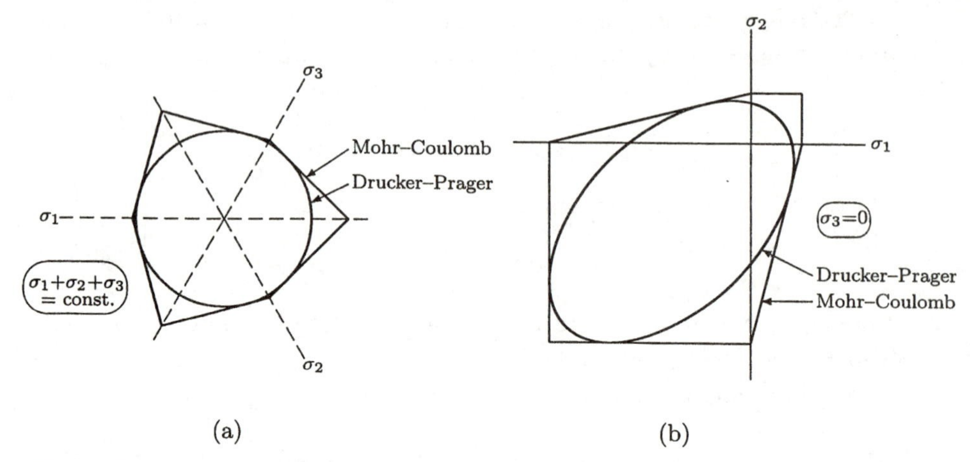

Figure 3.3.3. Mohr–Coulomb and Drucker–Prager criteria: (a) plane parallel to π-plane; (b) plane stress.

Mises–Schleicher Criterion

A yield criterion that takes into account the difference between the yield strengths in tension and compression was discussed by Mises [1926] and Schleicher [1926]. If σ_T and σ_C denote, respectively, the tensile and compressive yield stresses, then the criterion may be expressed in the form

$$3J_2 + (\sigma_C - \sigma_T)I_1 - \sigma_T\sigma_C = 0.$$

The associated plastic dissipation is

$$D_p(\dot{\varepsilon}^p; \boldsymbol{\xi}) = \frac{\sigma_C - \sigma_T}{2}\frac{\dot{e}^p_{ij}\dot{e}^p_{ij}}{\dot{\varepsilon}^p_{kk}} + \frac{\sigma_C\sigma_T}{3(\sigma_C - \sigma_T)}\dot{\varepsilon}^p_{kk}.$$

3.3.4. Yield Criteria Under Special States of Stress or Deformation

Equibiaxial Stress

A state of stress is called *equibiaxial* if two of the principal stresses are equal, as, for example, in the triaxial soil test described in Section 2.3. If $\sigma_2 = \sigma_3$, then

$$\sqrt{J_2} = \sqrt{\tfrac{1}{3}}|\sigma_1 - \sigma_3| = 2\sqrt{\tfrac{1}{3}}\tau_m,$$

so that the Mises and Tresca yield criteria are formally equivalent, as are the **Mohr–Coulomb criterion**Mohr–Coulomb and Drucker–Prager criteria.

Plane Stress

The criteria for *plane stress* are obtained simply by setting $\sigma_3 = 0$. Thus the Tresca and Mises criteria are just as they appear in Figure 3.3.2(b) (page 138). In a state of plane stress in the x_1x_2-plane with $\sigma_{22} = 0$ (i.e., a state of stress that may be represented as a superposition of simple tension or compression and shear), it can further be shown that both the Mises and the Tresca yield criteria can be expressed in the form

$$\left(\frac{\sigma}{\sigma_Y}\right)^2 + \left(\frac{\tau}{\tau_Y}\right)^2 = 1, \tag{3.3.5}$$

where $\sigma = \sigma_{11}$, $\tau = \sigma_{12}$, and σ_Y and τ_Y are respectively the yield stresses in simple tension or compression and in shear, that is, $\tau_Y = k$, and $\sigma_Y = \sqrt{3}k$ or $2k$, depending on the criterion. The **Mohr–Coulomb criterion**Mohr–Coulomb and Drucker–Prager criteria in plane stress are shown in Figure 3.3.3(b).

In general, an isotropic yield criterion with $\sigma_3 = 0$ may be written (with dependence on $\boldsymbol{\xi}$ not indicated explicitly) as

$$f_0(\sigma_1, \sigma_2) = 0,$$

or equivalently, upon transforming the independent variables, as

$$f_1[n, \tfrac{1}{2}(\sigma_1 - \sigma_2)] = 0,$$

where $n \stackrel{\text{def}}{=} \frac{1}{2}(\sigma_1 + \sigma_2)$. Because of isotropy, the dependence of f_1 on its second argument must be through the absolute value $r = \frac{1}{2}|\sigma_1 - \sigma_2|$. The preceding equation can then be solved for r as a function of n:

$$r = h(n). \tag{3.3.6}$$

In particular, $h(n)$ takes the form

$$h(n) = \sqrt{k^2 - \tfrac{1}{3}n^2}$$

for the Mises criterion and

$$h(n) = \begin{cases} k, & |n| < k \\ 2k - |n|, & |n| > k \end{cases}$$

for the Tresca criterion.

Plane Strain

In plane strain, as defined, for example, by $\dot{\varepsilon}_3 = 0$, the situation is more complicated, since the plane-strain condition requires $\dot{\varepsilon}_3^p = -\dot{\varepsilon}_3^e$, in turn involving the stress rates. If, however, the elastic strain rates may be equated to zero (the condition for this is discussed later), then we have

$\dot{\varepsilon}_3 = \dot{\varepsilon}_3^p = 0$. Assuming a plastic potential $g(\sigma_1, \sigma_2, \sigma_3)$, we obtain the equation

$$\frac{\partial}{\partial \sigma_3} g(\sigma_1, \sigma_2, \sigma_3) = 0,$$

which when combined with the yield criterion in terms of the principal stresses, permits the elimination of σ_3 and hence the formulation of a yield criterion in terms of σ_1 and σ_2, leading once more to Equation (3.3.6).

Consider, for example, the Mises criterion with its associated flow rule $\dot{\varepsilon}^p = \dot{\lambda}\mathbf{s}$, which requires $s_3 = \frac{2}{3}[\sigma_3 - \frac{1}{2}(\sigma_1 + \sigma_2)] = 0$, that is, $\sigma_3 = \frac{1}{2}(\sigma_1 + \sigma_2)$. Substituting this into the Mises yield criterion yields $\frac{3}{4}(\sigma_1 - \sigma_2)^2 = 3k^2$, or $|\sigma_1 - \sigma_2| = 2k$. According to the Tresca flow rule, on the other hand, for $\dot{\varepsilon}_3^p$ to be zero, σ_3 must be the intermediate principal stress, that is, either $\sigma_1 > \sigma_3 > \sigma_2$ or $\sigma_1 < \sigma_3 < \sigma_2$, so that

$$\max(|\sigma_1 - \sigma_3|, |\sigma_2 - \sigma_3|, |\sigma_1 - \sigma_2|) = |\sigma_1 - \sigma_2| = 2k.$$

Consequently the two criteria coincide, and may be expressed by Equation (3.3.6) with $h(n) = k$.

Consider next the **Mohr–Coulomb criterion**Mohr–Coulomb criterion, with a nonassociated flow rule that is governed by a plastic potential having the same form as the yield function, that is,

$$g(\boldsymbol{\sigma}) = \sigma_{\max} - \sigma_{\min} + (\sigma_{\max} + \sigma_{\min}) \sin \psi,$$

where ψ is known as the angle of dilatancy, since $\psi = 0$ describes a plastically incompressible solid; the special case $\psi = \phi$ represents the associated flow rule. The plane-strain condition $\dot{\varepsilon}_3^p = 0$ again requires that σ_3 be the intermediate principal stress. The criterion therefore may be described by Equation (3.3.6) with

$$h(n) = c \cos \phi - n \sin \phi.$$

3.3.5. Hardening Rules

A specification of the dependence of the yield criterion on the internal variables, along with the rate equations for these variables, is called a *hardening rule*. In this subsection we first review in more detail the significance of the two models of hardening — isotropic and kinematic — previously discussed for viscoplasticity in Section 3.1. Afterwards we look at some more general hardening rules.

Isotropic Hardening

The yield functions that we have studied so far in this section are all reducible to the form

$$f(\boldsymbol{\sigma}, \boldsymbol{\xi}) = F(\boldsymbol{\sigma}) - k(\boldsymbol{\xi}).$$

Since it is only the yield stress that is affected by the internal variables, no generality is lost if it is assumed to depend on only one internal variable, say ξ_1, and this is invariably identified with the hardening variable κ, defined as either the plastic work W_p by Equation (1.5.6) or as the effective plastic strain $\bar{\varepsilon}^p$ by Equation (1.5.7). The function h_1 corresponding to ξ_1 [see Equation (3.1.5)] is given by $\sigma_{ij}h_{ij}$ or $\sqrt{\frac{2}{3}h_{ij}h_{ij}}$, respectively, for each of the two definitions of κ, so that the work-hardening modulus H is

$$H = \begin{cases} k'(W_p)\sigma_{ij}h_{ij}, \\ k'(\bar{\varepsilon}^p)\sqrt{\frac{2}{3}h_{ij}h_{ij}}. \end{cases}$$

As was pointed out in Section 3.2, work-hardening in rate-independent plasticity corresponds to a local expansion of the yield surface. The present behavior model (which, as we said in Section 3.1, is called isotropic hardening) represents a *global* expansion, with no change in shape. Thus for a given yield criterion and flow rule, hardening behavior in any process can be predicted from the knowledge of the function $k(\kappa)$, and this function may, in principle, be determined from a single test (such as a tension test).

The most attractive feature of the isotropic hardening model, which was introduced by Odqvist [1933], is its simplicity. However, its usefulness in approximating real behavior is limited. In uniaxial stressing it predicts that when a certain yield stress σ has been attained as a result of work-hardening, the yield stress encountered on stress reversal is just $-\sigma$, a result clearly at odds with the Bauschinger effect (Section 2.1). Furthermore, if $F(\boldsymbol{\sigma})$ is an isotropic function, then the yield criterion remains isotropic even after plastic deformation has taken place, so that the model cannot describe induced anisotropy.

Kinematic Hardening

In Sections 3.1 and 3.2 we saw, however, that if f can be written in the form

$$f(\boldsymbol{\sigma}, \boldsymbol{\xi}) = F(\boldsymbol{\sigma} - \boldsymbol{\rho}) - k(\boldsymbol{\xi}), \tag{3.3.7}$$

then more general hardening behavior can be described. Isotropic hardening is a special case of (3.3.7) if $\boldsymbol{\rho} \equiv 0$ and if k depends only on κ, while purely kinematic hardening corresponds to constant k but nonvanishing variable $\boldsymbol{\rho}$. Kinematic hardening represents a translation of the yield surface in stress space by shifting its reference point from the origin to $\boldsymbol{\rho}$, and with uniaxial stressing this means that the the length of the stress interval representing the elastic region (i.e., the difference between the current yield stress and the one found on reversal) remains constant. This is in fairly good agreement with the Bauschinger effect for those materials whose stress-strain curve in the work-hardening range can be approximated by a straight line ("linear hardening"), and it is for such materials that Melan [1938] proposed the model in which $\boldsymbol{\rho} = c\boldsymbol{\varepsilon}^p$, with c a constant. A similar idea was also proposed by Ishlinskii [1954], and a generalization of it is due to Prager [1955a, 1956a], who coined the term "kinematic hardening" on the basis of his use of a mechanical model in explaining the hardening rule (Figure 3.3.4). A kinematic

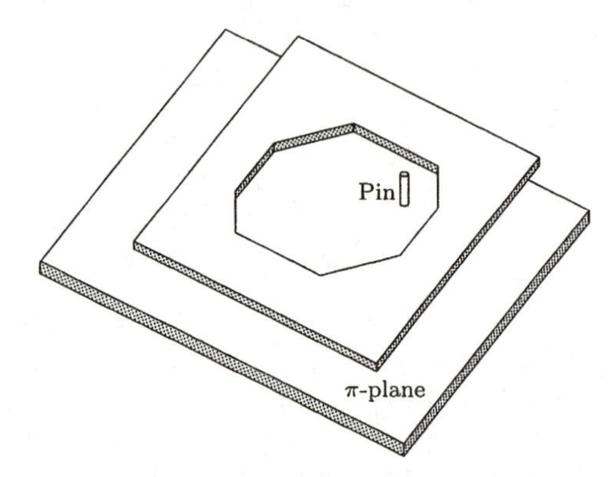

Figure 3.3.4. Prager's mechanical model of kinematic hardening.

hardening model is also capable of representing induced anisotropy, since a function $F(\boldsymbol{\sigma} - \boldsymbol{\rho})$ that depends only on the invariants of its argument stops being an isotropic function of the stress tensor as soon as $\boldsymbol{\rho}$ differs from zero.

It should be pointed out that, since $\boldsymbol{\rho}$ is a tensor in stress space (sometimes called the *back stress*, as discussed in 3.1.3), the equation $\rho_{ij} = c\varepsilon_{ij}^p$ does not imply proportionality between the vectors representing $\boldsymbol{\rho}$ and $\boldsymbol{\varepsilon}^p$ in any space other than the nine-dimensional space of second-rank tensors, and particularly not in the six-dimensional space in which symmetric tensors are represented, since the mappings of stress and strain into this space must be different [see Equations (1.4.9)] in order to preserve the scalar product $\boldsymbol{\sigma} \cdot \boldsymbol{\varepsilon} = \sigma_{ij}\varepsilon_{ij}$; consequently, as was pointed out by Hodge [1957a], the translation of the yield surface for a material with an associated flow rule is not necessarily in the direction of the normal to the yield surface, as was assumed by Prager in constructing his model.

In more sophisticated kinematic hardening models, internal variables other than $\boldsymbol{\varepsilon}^p$ and κ are included; in particular, the back stress $\boldsymbol{\rho}$ may be treated as a tensorial internal variable with its own rate equation. Indeed, the Melan–Prager model falls into this category when its equation is rewritten as

$$\dot{\rho}_{ij} = c\dot{\varepsilon}_{ij}^p; \tag{3.3.8}$$

here c need not be a constant but may itself depend on other internal variables. In the model described by Backhaus [1968], for example, c depends on the effective plastic strain $\bar{\varepsilon}^p$. Lehmann [1972] replaces the isotropic relation

(3.3.8) between $\dot{\rho}$ and $\dot{\varepsilon}^p$ by a more general one,

$$\dot{\rho}_{ij} = c_{ijkl}(\boldsymbol{\sigma}, \boldsymbol{\rho})\dot{\varepsilon}^p_{kl}.$$

Another example of a kinematic hardening model is that due to Ziegler [1959],

$$\dot{\rho}_{ij} = \dot{\mu}(\sigma_{ij} - \rho_{ij}),$$

where

$$\dot{\mu} = \frac{< \frac{\partial f}{\partial \sigma_{kl}} \dot{\sigma}_{kl} >}{\partial f/\partial \sigma_{mn}(\sigma_{mn} - \rho_{mn})}$$

in order to satisfy the consistency condition $\dot{f} = 0$.

A modification of Equation (3.3.8) that better reproduces the real Bauschinger effect consists of including in its right-hand side a term representing "fading strain memory," so that the rate equation takes the form

$$\dot{\rho}_{ij} = c\dot{\varepsilon}^p_{ij} - a\rho_{ij}\dot{\bar{\varepsilon}}^p.$$

The more general kinematic hardening models can be similarly modified.

Generalized Hardening Rules

The hardening represented by Equation (3.3.7) with both $\boldsymbol{\rho}$ and k variable seems to have been first studied by Kadashevich and Novozhilov [1952]; it is called *combined hardening* by Hodge [1957a]. The combined hardening model proposed for viscoplasticity by Chaboche [1977], presented in Section 3.1, has been applied by Chaboche and his collaborators to rate-independent plasticity as well.

A model with a family of back stresses $\boldsymbol{\rho}_{(l)}$ ($l = 1, 2, ..., n$) is due to Mróz [1967]; a similar model is due to Iwan [1967]. Both models describe materials whose stress-strain curves are piecewise linear. For materials whose stress-strain curves in the work-hardening range are smooth with straight-line asymptotes, a class of models known as *two-surface* models have been proposed by Dafalias [1975] (see also Dafalias and Popov [1975]), Krieg [1975], and others. In these models the yield surface in stress space is constrained to move inside an outer surface, known variously as *bounding surface, loading surface*, or *memory surface*, given by, say, $\bar{f}(\boldsymbol{\sigma}, \boldsymbol{\xi}) = 0$. The work-hardening modulus H at a given state is assumed to be an increasing function of a suitably defined distance, in stress space, between the current stress $\boldsymbol{\sigma}$ and a stress $\bar{\boldsymbol{\sigma}}$ on the outer surface, called the *image stress* of $\boldsymbol{\sigma}$. When this distance vanishes, the work-hardening modulus attains its minimum value, and further hardening proceeds linearly, with the two surfaces remaining in contact at $\boldsymbol{\sigma} = \bar{\boldsymbol{\sigma}}$.

The various two-surface models differ from one another in the definition of the bounding surface, in the way the image stress depends on the current state, and in the variation of work-hardening modulus. In the model of Dafalias and Popov, both surfaces are given similar combined-hardening structures, with a "back stress" $\boldsymbol{\beta}$ playing the same role for the outer surface that $\boldsymbol{\rho}$ plays for the yield surface, and $\overline{\boldsymbol{\sigma}} = c(\boldsymbol{\sigma} - \boldsymbol{\rho}) + \boldsymbol{\beta}$, where c is a constant. H is assumed to depend on $\delta = \sqrt{(\overline{\boldsymbol{\sigma}} - \boldsymbol{\sigma}) : (\overline{\boldsymbol{\sigma}} - \boldsymbol{\sigma})}$ in such a way that $H = \infty$ at initial yield, producing a smooth hardening curve.

Experiments by Phillips and Moon [1977] showed that when yield surfaces are defined on the basis of a very small offset strain, they undergo considerable distortion, in addition to the expansion and translation considered thus far. In order to describe such distortion in initially isotropic materials, Equation (3.3.7) must be modified to

$$f(\boldsymbol{\sigma}, \boldsymbol{\xi}) = F(\boldsymbol{\sigma} - \boldsymbol{\rho}, \boldsymbol{\xi}) - k(\boldsymbol{\xi}),$$

where F is initially an isotropic function of its first argument but becomes anisotropic as plastic deformation takes place. An example of such a function is that proposed by Baltov and Sawczuk [1965] for a Mises-type yield surface:

$$F(\boldsymbol{\sigma} - \boldsymbol{\rho}, \boldsymbol{\xi}) = \frac{1}{2} A_{ijkl}(\boldsymbol{\xi})(\sigma_{ij} - \rho_{ij})(\sigma_{kl} - \rho_{kl}),$$

where

$$A_{ijkl}(\boldsymbol{\xi}) = \delta_{ik}\delta_{ij} - \frac{1}{3}\delta_{ij}\delta_{kl} + A\varepsilon_{ij}^p\varepsilon_{kl}^p,$$

A being a constant. Other proposals are reviewed by Bergander [1980]. Extensive experimental investigation of the hardening of metals were carried out by Phillips and co-workers; their work, along with that of others, is reviewed by Phillips [1986].

Exercises: Section 3.3

1. Show that the forms (3.3.1) and (3.3.2) for the Tresca yield function are equivalent.

2. Derive the associated flow rule for the Tresca yield criterion by means of Koiter's method (see 3.2.2).

3. Show that for any combination of principal stresses, the associated flow rule for the Tresca yield criterion gives $|\dot{\varepsilon}_1^p| + |\dot{\varepsilon}_2^p| + |\dot{\varepsilon}_3^p| = \phi$.

4. An elastic–perfectly plastic solid with a uniaxial yield stress of 300 MPa is assumed to obey the Tresca yield criterion and its associated flow rule. If the rate of plastic work per unit volume is 1.2 MW/m^3, find the principal plastic strain-rate components when

   (a) $\sigma_1 = 300$ MPa, $\sigma_2 = 100$ MPa, $\sigma_3 = 0$,

(b) $\sigma_1 = 200$ MPa, $\sigma_2 = -100$ MPa, $\sigma_3 = 0$,

(c) $\sigma_1 = 200$ MPa, $\sigma_2 = -100$ MPa, $\sigma_3 = -100$ MPa.

5. Derive the associated flow rule for the general isotropic yield criterion given by $F(J_2, J_3) - k^2 = 0$, and in particular (a) for the one given by the equation following (3.3.4) and (b) for the analytic form of the Tresca criterion.

6. A work-hardening elastic-plastic solid is assumed to obey the Mises criterion with the associated flow rule and isotropic hardening. If the virgin curve in uniaxial tension can be described in the small-deformation range by $\sigma = F(\varepsilon^p)$, state the rate equations (3.2.1)–(3.2.2) explicitly when k is assumed to depend (a) on $\bar{\varepsilon}^p$ and (b) on W_p.

7. Derive the Mohr-Coulomb criterion as follows.

   (a) Using the theory of Mohr's circles in plane stress, in particular Equations (1.3.9)–(1.3.10), find the direction θ such that $\tau_\theta - \mu(-\sigma_\theta)$ is maximum.

   (b) Show that this maximum value is $\sqrt{1+\mu^2}|\sigma_1 - \sigma_2|/2 + \mu(\sigma_1 + \sigma_2)/2$, and that the Mohr-Coulomb criterion results when this value is equated to the cohesion c, with $\mu = \tan\phi$.

   (c) Show that the Mohr circles whose parameters σ_1, σ_2 are governed by this criterion are bounded by the lines $\pm\tau_\theta = \sigma_\theta \tan\phi - c$.

8. Derive the associated flow rule and plastic dissipation for the Drucker-Prager yield criterion.

9. Given the yield stresses σ_T and σ_C in uniaxial tension and compression, respectively, find the yield stress in shear resulting from the following yield criteria: (a) Mohr–Coulomb, (b) Drucker–Prager, (c) Mises–Schleicher.

10. Show that in a state of plane stress with $\sigma_{11} = \sigma$, $\sigma_{12} = \tau$ and $\sigma_{22} = 0$, both the Tresca and the Mises yield criteria can be expressed in the form (3.3.5).

11. Derive the form of Equation (3.3.6) for the Mohr-Coulomb criterion in plane stress.

12. If the function F in Equation (3.3.7) equals $\sqrt{\bar{J}_2}$, with $\bar{J}_2$ defined as in 3.1.1, while k is constant, and if the evolution of $\boldsymbol{\rho}$ is governed by (3.3.8), show that the rate equation of $\boldsymbol{\rho}$ is

$$\dot{\rho}_{ij} = (s_{ij} - \rho_{ij})\frac{(s_{kl} - \rho_{kl})\dot{s}_{kl}}{2k^2}.$$

13. Generalize the preceding result to the case where k depends on $\bar{\varepsilon}^p$, obtaining rate equations for both $\boldsymbol{\rho}$ and $\boldsymbol{\varepsilon}^p$.

Section 3.4 Uniqueness and Extremum Theorems

3.4.1. Uniqueness Theorems

Uniqueness Theorems in Elastic Bodies

Consider a body made of a linearly elastic material with no internal constraints, occupying a region R and subject to prescribed tractions $\mathbf{t}^a$ on ∂R_t, prescribed displacements $\mathbf{u}^a$ on ∂R_u, and a prescribed body-force field $\mathbf{b}$ in R. For convenience, the body force per unit volume is defined as $\mathbf{f} = \rho\mathbf{b}$. We suppose that a stress field $\boldsymbol{\sigma}$ and a displacement field $\mathbf{u}$ in R have been found such that $\sigma_{ij} = C_{ijkl}\varepsilon_{kl}$ (where $\boldsymbol{\varepsilon}$ is the strain field derived from $\mathbf{u}$) and $\sigma_{ij},_j + f_i = 0$ in R, $n_j\sigma_{ij} = t_i^a$ on ∂R_t, and $\mathbf{u} = \mathbf{u}^a$ on ∂R_u. In other words, $(\boldsymbol{\sigma}, \mathbf{u})$ constitutes a solution of the static boundary-value problem. Is this solution unique? As a result of the classical uniqueness theorem due to Kirchhoff, the answer is "yes" as regards the stress field and "almost" as regards the displacement field. For, if $(\boldsymbol{\sigma}^{(1)}, \mathbf{u}^{(1)})$ and $(\boldsymbol{\sigma}^{(2)}, \mathbf{u}^{(2)})$ are two different solutions, and if we write $\overline{\phi} = \phi^{(1)} - \phi^{(2)}$ for any ϕ, then $\overline{\sigma}_{ij} = C_{ijkl}\overline{\varepsilon}_{kl}$, $\overline{\sigma}_{ij},_j = 0$ in R, and $n_j\overline{\sigma}_{ij}\bar{u}_i = 0$ on ∂R. Consequently, by the divergence theorem,

$$
\begin{aligned}
0 &= \int_R (\overline{\sigma}_{ij}\bar{u}_i),_j \, dV \\
&= \int_R (\overline{\sigma}_{ij},_j \bar{u}_i + \overline{\sigma}_{ij}\bar{u}_i,_j) \, dV \\
&= \int_R \overline{\sigma}_{ij}\overline{\varepsilon}_{ij} \, dV \\
&= \int_R C_{ijkl}\overline{\varepsilon}_{ij}\overline{\varepsilon}_{kl} \, dV.
\end{aligned}
$$

It follows from the positive-definiteness of C (see 1.4.3) that $\overline{\boldsymbol{\varepsilon}}$ must vanish throughout R. Consequently, $\overline{\boldsymbol{\sigma}}$ must vanish as well, while $\bar{\mathbf{u}}$ may have at most the form of a rigid-body displacement; full uniqueness of the displacement field depends on having sufficient external constraints.

If the material were nonlinearly elastic, the same method could be applied, but *incrementally*. Suppose that the stress field $\boldsymbol{\sigma}$ and displacement field $\mathbf{u}$ have been found under the current $\mathbf{f}$, $\mathbf{t}^a$ and $\mathbf{u}^a$. We may then prove the uniqueness of *infinitesimal increments* $d\boldsymbol{\sigma}$ resulting from increments $d\mathbf{f}$, $d\mathbf{t}^a$ and $d\mathbf{u}^a$, provided that C is interpreted as the tangent elastic modulus tensor as defined by Equation (1.4.8), so that

$$d\sigma_{ij} = C_{ijkl}\, d\varepsilon_{kl}.$$

Incremental uniqueness implies global uniqueness, since any state of loading can be attained by the successive imposition of small incremental loads. Consequently, the stress and strain fields are uniquely determined (and the displacement field determined to within a rigid-body displacement) so long as C is positive definite.

Uniqueness of Stress Field in an Elastic–Plastic Body

The positive-definiteness of C means that, for an elastic material, $d\sigma_{ij}\, d\varepsilon_{ij} > 0$ whenever $d\boldsymbol{\sigma} \neq 0$. The last inequality is equivalent to Drucker's first inequality for a work-hardening plastic material (see Section 3.2). Indeed, we know that as long as no unloading occurs, no distinction can be made between plastic and nonlinearly elastic materials. It was shown by Melan [1938] that incremental uniqueness of stress and strain can be established for work-hardening standard materials when unloading has taken place, provided that the hypothesis of infinitesimal strains is valid. The reason for the proviso is that, with finite deformations, a distinction must be made between increments at a fixed material point and those at a fixed point in space (see Hill [1950] and Chapter 8 of the present book).

By analogy with the proof for elastic bodies, it can be shown that a sufficient condition for the uniqueness of the stress field in a plastic body is that if $d\boldsymbol{\sigma}^{(1)}$ and $d\boldsymbol{\sigma}^{(2)}$ are two possible incremental stress fields and $d\boldsymbol{\varepsilon}^{(1)}$ and $d\boldsymbol{\varepsilon}^{(2)}$ are the associated incremental strain fields, then

$$(d\sigma_{ij}^{(1)} - d\sigma_{ij}^{(2)})(d\varepsilon_{ij}^{(1)} - d\varepsilon_{ij}^{(2)}) > 0 \qquad (3.4.1)$$

unless $d\boldsymbol{\sigma}^{(1)} = d\boldsymbol{\sigma}^{(2)}$. It was shown by Valanis [1985] that condition (3.4.1) applies in dynamic as well as in quasi-static problems.

Consider next an elastic-plastic body made of standard material, occupying the region R. Let R_e and R_p denote the parts of R where $f < 0$ and $f = 0$, respectively. With linear elasticity assumed, the inequality is clearly satisfied in R_e, while in R_p we may use the general flow equation (3.2.1) together with (3.2.2) and the normality rule (3.2.7). Converting rates into increments by multiplying them by the infinitesimal time increment dt, we obtain, at any point in R_p,

$$d\varepsilon_{ij}^{(\alpha)} = C_{ijkl}^{-1} d\sigma_{kl}^{(\alpha)} + \frac{<df^{(\alpha)}>}{H}\frac{\partial f}{\partial \sigma_{ij}}, \quad \alpha = 1,\, 2, \qquad (3.4.2)$$

where

$$df^{(\alpha)} = \frac{\partial f}{\partial \sigma_{kl}} d\sigma_{kl}^{(\alpha)}.$$

Hence

$$(d\sigma_{ij}^{(1)} - d\sigma_{ij}^{(2)})(d\varepsilon_{ij}^{(1)} - d\varepsilon_{ij}^{(2)}) = C_{ijkl}^{-1}(d\sigma_{ij}^{(1)} - d\sigma_{ij}^{(2)})(d\sigma_{kl}^{(1)} - d\sigma_{kl}^{(2)})$$
$$+\frac{1}{H}(df^{(1)} - df^{(2)})(<df^{(1)}> - <df^{(2)}>).$$

It is easy to see that any two real numbers a, b satisfy $<a> - <b> = \beta(a-b)$ for some β, $0 \le \beta \le 1$, and that therefore the second term on the right-hand side is never negative. Since the first term is positive unless $d\boldsymbol{\sigma}^{(1)} = d\boldsymbol{\sigma}^{(2)}$, the uniqueness of $d\boldsymbol{\sigma}$ (and therefore of $d\boldsymbol{\varepsilon}$) is proved, and hence the uniqueness of the stress and strain fields under a given *history* of $\mathbf{f}$, $\mathbf{t}^a$, and $\mathbf{u}^a$.

If the material is perfectly plastic, then Equation (3.4.2) for the strain increments at points where $f = 0$ must be replaced by

$$d\varepsilon_{ij}^{(\alpha)} = C_{ijkl}^{-1}\, d\sigma_{kl}^{(\alpha)} + d\lambda^{(\alpha)}\frac{\partial f}{\partial \sigma_{ij}}, \quad \alpha = 1,\, 2, \tag{3.4.3}$$

where $d\lambda^{(\alpha)} = 0$ if $df^{(\alpha)} < 0$, and $d\lambda^{(\alpha)} > 0$ only if $df^{(\alpha)} = 0$ but is otherwise undetermined. Thus

$$(d\sigma_{ij}^{(1)} - d\sigma_{ij}^{(2)})(d\varepsilon_{ij}^{(1)} - d\varepsilon_{ij}^{(2)}) = C_{ijkl}^{-1}(d\sigma_{ij}^{(1)} - d\sigma_{ij}^{(2)})(d\sigma_{kl}^{(1)} - d\sigma_{kl}^{(2)})$$
$$+(d\lambda^{(1)} - d\lambda^{(2)})(df^{(1)} - df^{(2)}).$$

The second term evidently vanishes either if both $df^{(1)}$ and $df^{(2)}$ vanish, or if both are negative (leading to $d\lambda^{(1)} = d\lambda^{(2)} = 0$). If $df^{(1)} = 0$ and $df^{(2)} < 0$, then the term equals $-d\lambda^{(1)}\, df^{(2)}$ and is positive, as it is when (1) and (2) are interchanged. The uniqueness of $d\boldsymbol{\sigma}$ — and hence that of the stress field — follows, but not that of $d\boldsymbol{\varepsilon}$.

If normality is not obeyed, then work-hardening (i.e. the positiveness of H) is not sufficient for uniqueness. The corresponding sufficient condition is, instead,

$$H > H_{cr},$$

where H_{cr} is a critical value of the hardening modulus given by Raniecki [1979] as

$$H_{cr} = \frac{1}{2}\left(\sqrt{C_{ijkl}\frac{\partial f}{\partial \sigma_{ij}}\frac{\partial f}{\partial \sigma_{kl}}}\sqrt{C_{ijkl}h_{ij}h_{kl}} - C_{ijkl}h_{ij}\frac{\partial f}{\partial \sigma_{kl}}\right).$$

Clearly, $H_{cr} = 0$ when the normality rule (3.2.7) is obeyed.

3.4.2. Extremum and Variational Principles

In Section 1.4 we derived the two fundamental variational principles of elastostatics, which form a dual pair. The principle of minimum potential energy teaches that in a body in stable equilibrium the correct displacement field — that is, the one that, with its associated stress field, forms the solution of the boundary-value problem — is the one which, among all the kinematically admissible displacement fields that are close to it, minimizes the total potential energy Π. Similarly, the principle of minimum complementary energy asserts that the correct stress field is the one which, among all the neighboring stress fields that are statically admissible, minimizes the total complementary energy Π^c.

If the material is linear, then unless large displacements come into play, the respective energies depend at most quadratically on their variables, and it is easy to see that the restriction to admissible fields that are close to the correct one may be removed. Indeed, it can be shown directly that, if Π^* is the total potential energy evaluated at the arbitrary kinematically admissible displacement field $\mathbf{u}^*$, and if $\Pi = \Pi^*|_{\mathbf{u}^*=\mathbf{u}}$, where $\mathbf{u}$ is the correct displacement field, then

$$\Pi^* > \Pi \quad \text{unless} \quad \mathbf{u}^* = \mathbf{u}.$$

For

$$\Pi^* - \Pi = \frac{1}{2}\int_R C_{ijkl}(\varepsilon_{ij}^*\varepsilon_{kl}^* - \varepsilon_{ij}\varepsilon_{kl})\, dV - \int_R f_i(u_i^* - u_i)\, dV - \int_{\partial R_t} t_i^a(u_i^* - u_i)\, dS;$$

but the surface integral may be changed into one over all of ∂R, since $\mathbf{u}^* = \mathbf{u}$ on ∂R_u, and consequently, by the divergence theorem, into

$$\int_R [\sigma_{ij},_j (u_i^* - u_i) + \sigma_{ij}(\varepsilon_{ij}^* - \varepsilon_{ij})]\, dV,$$

where $\sigma_{ij} = C_{ijkl}\varepsilon_{kl}$ is the correct stress field. If we define $\sigma_{ij}^* = C_{ijkl}\varepsilon_{kl}^*$ then with the help of the equilibrium equations, we obtain

$$\Pi^* - \Pi = \frac{1}{2}\int_R (\sigma_{ij}^*\varepsilon_{ij}^* + \sigma_{ij}\varepsilon_{ij} - 2\sigma_{ij}\varepsilon_{ij}^*)\, dV.$$

The integrand, however, is

$$C_{ijkl}(\varepsilon_{ij}\varepsilon_{kl} + \varepsilon_{ij}^*\varepsilon_{kl}^* - 2\varepsilon_{ij}^*\varepsilon_{kl}) = C_{ijkl}(\varepsilon_{ij}^* - \varepsilon_{ij})(\varepsilon_{kl}^* - \varepsilon_{kl}),$$

and is positive except when $\boldsymbol{\varepsilon}^* = \boldsymbol{\varepsilon}$.

Similarly,

$$\Pi^{c*} - \Pi^c = \frac{1}{2}\int_R C_{ijkl}^{-1}(\sigma_{ij}^*\sigma_{kl}^* - \sigma_{ij}\sigma_{kl}) - \int_{\partial R_u} n_j(\sigma_{ij}^* - \sigma_{ij})u_i^a\, dS,$$

and with the help of analogous transformations, we obtain

$$\Pi^{c*} - \Pi^c = \frac{1}{2}\int_R C_{ijkl}^{-1}[\sigma_{ij}^*\sigma_{kl}^* - \sigma_{ij}\sigma_{kl} - 2(\sigma_{ij}^* - \sigma_{ij})\varepsilon_{ij}]\, dV,$$

where $\varepsilon_{ij} = C^{-1}_{ijkl}\sigma_{kl}$, so that the integrand is

$$C^{-1}_{ijkl}(\sigma^*_{ij} - \sigma_{ij})(\sigma^*_{kl} - \sigma_{kl}),$$

also positive except when $\boldsymbol{\sigma}^* = \boldsymbol{\sigma}$.

In view of the resemblance between the uniqueness proof and the proofs of the extremum principles, it appears reasonable that such principles may be derived for displacement and stress increments (or, equivalently, velocities and stress rates) in elastic-plastic bodies that are either work-hardening or perfectly plastic. Such is indeed the case. The theorems to be presented have been derived by Handelman [1944], Markov [1947], Greenberg [1949], and Hill [1950];[1] the proofs are Hill's.

Extremum Principle for Displacement

Let $d\mathbf{u}^*$ denote a kinematically admissible displacement increment, that is, one which obeys the internal constraints, if any, and which satisfies $d\mathbf{u}^* = d\mathbf{u}^a$ on ∂R_u. The corresponding incremental strain and stress field are $d\boldsymbol{\varepsilon}^*$ and $d\boldsymbol{\sigma}^*$, where $d\boldsymbol{\sigma}^*$ is not in general statically admissible, but is related to $d\boldsymbol{\varepsilon}^*$ through the associated flow rule; we thus use Equation (3.2.10) with $h_{ij} = \partial f/\partial\sigma_{ij}$, to obtain Hill's [1958] result

$$\dot{\sigma}_{ij} = \begin{cases} C_{ijkl}\dot{\varepsilon}_{kl}, & f < 0, \\ C_{ijkl}\left(\dot{\varepsilon}_{kl} - \dfrac{1}{L} < C_{pqmn}\dfrac{\partial f}{\partial \sigma_{pq}}\dot{\varepsilon}_{mn} > \dfrac{\partial f}{\partial \sigma_{kl}}\right), & f = 0. \end{cases}$$

Here L is given by Equation (3.2.9) for a standard material, namely,

$$L = H + C_{ijkl}\frac{\partial f}{\partial \sigma_{ij}}\frac{\partial f}{\partial \sigma_{kl}},$$

and the perfectly plastic case corresponds to $H = 0$.

Let the functional Λ^* be defined by

$$\Lambda^* = \frac{1}{2}\int_R d\sigma^*_{ij} d\varepsilon^*_{ij}\, dV - \int_R df_i\, du^*_i\, dV - \int_{\partial R_t} dt^a_i\, du^*_i\, dS,$$

with

$$\Lambda = \Lambda^*|_{d\mathbf{u}^*=d\mathbf{u}}\,.$$

Note that the form of Λ is essentially that of the potential energy Π of the linear elastic body, with incremental rather than total displacements and with the total (elastic-plastic) tangent modulus tensor in place of C. By

[1] Some extensions have been proposed by, among others, Ceradini [1966], Maier [1969, 1970] and Martin [1975].

means of the transformations used in the elastic case, we can therefore show that

$$\Lambda^* - \Lambda = \frac{1}{2}\int_R (d\sigma^*_{ij}\, d\varepsilon^*_{ij} + d\sigma_{ij}\, d\varepsilon_{ij} - 2d\sigma_{ij}\, d\varepsilon^*_{ij})\, dV,$$

and it remains to be shown that the integrand is positive unless $d\boldsymbol{\sigma}^* = d\boldsymbol{\sigma}$. With the decomposition $d\boldsymbol{\varepsilon} = \mathsf{C}^{-1}d\boldsymbol{\sigma} + d\boldsymbol{\varepsilon}^p$, the integrand becomes

$$C^{-1}_{ijkl}(d\sigma^*_{ij} - d\sigma_{ij})(d\sigma^*_{kl} - d\sigma_{kl}) + d\sigma^*_{ij}d\varepsilon^{p*}_{ij} + d\sigma_{ij}d\varepsilon_{ij} - 2d\sigma_{ij}d\varepsilon^{p*}_{ij}.$$

The term in C^{-1} is clearly nonnegative, and vanishes only if $d\boldsymbol{\sigma}^* = d\boldsymbol{\sigma}$.

As for the remaining terms, in the case of a work-hardening material their sum becomes

$$\phi = \frac{1}{H}(df^* <df^*> + df <df> - 2df <df^*>),$$

where $df = (\partial f/\partial\sigma_{ij})\, d\sigma_{ij}$ and $df^* = (\partial f/\partial\sigma_{ij})\, d\sigma^*_{ij}$. Clearly, ϕ vanishes if df^* and df are both nonpositive, and equals $H^{-1}(df^* - df)^2$ if both are positive. If $df^* > 0$ and $df \le 0$, $\phi = H^{-1}df^*(df^* - 2df) > 0$, while in the opposite case $\phi = H^{-1}(df)^2$. For the perfectly plastic material, we replace $H^{-1}<df>$ by $d\lambda$, with $d\lambda$ related to df as in the uniqueness proof, and similarly $H^{-1}<df^*>$ by $d\lambda^*$, so that

$$\phi = df^*\, d\lambda^* + df\, d\lambda - 2df\, d\lambda^*.$$

The first two terms always vanish, as does the last term except when $df < 0$ and $df^* = 0$, in which case $d\lambda^* > 0$, so that $\phi = -2df\, d\lambda^* > 0$. Consequently ϕ is never negative, so that

$$\Lambda^* > \Lambda \tag{3.4.4}$$

except when $d\boldsymbol{\sigma}^* = d\boldsymbol{\sigma}$. For the work-hardening (but not the perfectly plastic) material this also means that $d\boldsymbol{\varepsilon}^* = d\boldsymbol{\varepsilon}$ and therefore $d\mathbf{u}^* = d\mathbf{u}$ if sufficient constraints exist.

Extremum Principle for Stress

The complementary extremum principle concerns a statically admissible incremental stress field $d\boldsymbol{\sigma}^*$. In the case of a work-hardening material this determines an incremental strain field $d\boldsymbol{\varepsilon}^*$, related to $d\boldsymbol{\sigma}^*$ through the associated flow rule — Equation (3.4.2) with the superscript (α) replaced by $*$ — but not, in general, derivable from a continuous displacement field. Clearly,

$$d\sigma^*_{ij}\, d\varepsilon^*_{ij} = C^{-1}_{ijkl}\, d\sigma^*_{ij}\, d\sigma^*_{kl} + \frac{1}{H}<df^*>^2.$$

If the material is perfectly plastic, on the other hand, then $d\varepsilon^*_{ij} = C^{-1}_{ijkl}\, d\sigma^*_{kl} + d\varepsilon^{p*}_{ij}$, with $d\boldsymbol{\varepsilon}^{p*}$ not determined by $d\boldsymbol{\sigma}^*$. However, in such a material $d\sigma^*_{ij}d\varepsilon^{p*}_{ij} = 0$, and therefore

$$d\sigma^*_{ij}\, d\varepsilon^*_{ij} = C^{-1}_{ijkl}\, d\sigma^*_{ij}\, d\sigma^*_{kl}.$$

We now define the functional

$$\Omega^* = \frac{1}{2}\int_R d\sigma^*_{ij}\, d\varepsilon^*_{ij}\, dV - \int_{\partial R_u} n_j\, d\sigma^*_{ij}\, du^a_i\, dS,$$

with $d\sigma^*_{ij}\, d\varepsilon^*_{ij}$ defined by the appropriate formula above. We also define, as before,

$$\Omega = \Omega^*|_{d\boldsymbol{\sigma}^*=d\boldsymbol{\sigma}}\,.$$

Note that the form of Ω is, *mutatis mutandis*, that of Π^c, and also that

$$\Omega = -\Lambda.$$

The surface integral in Ω^* is transformed into

$$\int_R d\sigma^*_{ij}\, d\varepsilon_{ij}\, dV,$$

and therefore

$$\Omega^* - \Omega = \frac{1}{2}\int_R (d\sigma^*_{ij}\, d\varepsilon^*_{ij} + d\sigma_{ij}\, d\varepsilon_{ij} - 2d\sigma^*_{ij}\, d\varepsilon_{ij})\, dV.$$

The integrand differs from the one in $\Lambda^* - \Lambda$ only in the interchange of starred and unstarred quantities, and may by the same method be shown to be nonnegative and to vanish only if $d\boldsymbol{\sigma}^* = d\boldsymbol{\sigma}$, so that

$$\Omega^* > \Omega \quad \text{unless} \quad d\boldsymbol{\sigma}^* = d\boldsymbol{\sigma}. \tag{3.4.5}$$

We furthermore have the double inclusion

$$-\Lambda^* \leq -\Lambda = \Omega \leq \Omega^*.$$

Variational Principles

As mentioned earlier, an extremum principle is stronger than a variational principle because it asserts an extremum over all admissible functions of a certain class, not only over those that are infinitesimally close to the extremal. Going further, we see that a variational principle need not assert an extremum at all, even a local one, but only the condition that the functional obeying it is stationary. For example, the function $f(x) = x^3$ is stationary at $x = 0$, since $f'(0) = 0$, but has neither a minimum nor a maximum there. Likewise, $f(x, y) = x^2 - y^2$ is stationary at (0, 0); the point is a *saddle point* — a maximum when viewed along the x-axis and a minimum along the y-axis.

Extremum principles are useful for many reasons, one of them being that they allow us to evaluate approximate solutions when the exact solution is unknown: between two incremental displacement fields $d\mathbf{u}^*$, in the absence of other information we choose the one that produces the smaller value of Λ^*. But variational principles are useful even when they are not extremum principles: they permit compact statements of boundary-value problems, and they are useful in formulating the "weak form" of such problems, which is necessary for consistent discretization, as shown in 1.3.5 and as further shown in Section 4.5 which deals with numerical methods.

The extremum principles derived above clearly imply the corresponding variational principles:

$$\delta\Lambda = 0, \quad \delta\Omega = 0.$$

We should remember, however, that the extremum principles rely on the positive-definiteness of certain local quantities, whose proof requires that the material be nonsoftening. Consider, on the other hand, the following functional:

$$\Theta = \frac{1}{2}\int_R C_{ijkl}\dot{\varepsilon}_{ij}\dot{\varepsilon}_{kl}\, dV - \frac{1}{2}\int_{R_p} \frac{1}{L}<C_{ijkl}\frac{\partial f}{\partial \sigma_{kl}}\dot{\varepsilon}_{ij}>^2 dV$$

$$- \int_R \dot{b}_i \dot{u}_i\, dV - \int_{\partial R_t} \dot{t}_i^a \dot{u}_i\, dV.$$

It is easy to see that Θ is just Λ with rates in place of increments; the change is made to avoid mixing the differential operator d with the variation operator δ. Clearly, with $\dot{\boldsymbol{\sigma}}$ given in terms of $\dot{\boldsymbol{\varepsilon}}$ by the strain-space flow rule,

$$\delta\Theta = \int_R \dot{\sigma}_{ij}\, \delta\dot{\varepsilon}_{ij}\, dV - \int_R \dot{b}_i\, \delta\dot{u}_i\, dV - \int_{\partial R_t} \dot{t}_i^a\, \delta\dot{u}_i\, dV,$$

and $\delta\Theta = 0$ for an arbitrary kinematically admissible $\delta\dot{\mathbf{u}}$ only if $\dot{\boldsymbol{\sigma}}$ is statically admissible.

3.4.3. Rigid–Plastic Materials

In discussing the Saint-Venant–Lévy–Mises flow rule in the preceding section, we mentioned in passing that these authors equated $\dot{\boldsymbol{\varepsilon}}$ with $\dot{\boldsymbol{\varepsilon}}^p$, in effect neglecting the elastic strain rate — a treatment tantamount to treating the nonvanishing elastic moduli as infinite. Any solutions obtained on this basis are, theoretically, valid for idealized materials called *rigid-plastic*.[1] Practically, however, they are useful approximations for two classes, not mutually exclusive, of problems: (a) those in which the elastic strain rates may be neglected, and (b) those in which the elastic strains are significantly smaller than the plastic ones. Problems of class (a) include those of *impending collapse* or *incipient plastic flow* of elastic–perfectly plastic bodies, for which it

[1]In many references, the theory of rigid-plastic materials obeying the Mises yield criterion and the associated flow rule is called the **Mises theory**, while the corresponding theory of elastic-plastic materials is called the **Prandtl–Reuss theory**.

is shown later in this section that, when deformation proceeds at constant loads, the elastic strain rates vanish identically. In order to accommodate problems of class (b), the hypothesis of infinitesimal deformations should be abandoned in formulating any general theorems for this theory. Since the problems are frequently those of flow, it appears natural to use an Eulerian formulation, in which R is the region currently occupied by the body and $\boldsymbol{\sigma}$ is the true (Cauchy) stress tensor. No reference configuration is used in this approach, and therefore no displacement field $\mathbf{u}$ appears; in its place we have the velocity field $\mathbf{v}$. The boundary ∂R is accordingly partitioned into ∂R_t and ∂R_v, with $\mathbf{v} = \mathbf{v}^a$ on ∂R_v. No strain tensor is introduced, but rather the Eulerian deformation-rate tensor (also called *stretching tensor*) $\mathbf{d}$, defined by

$$d_{ij} = \tfrac{1}{2}(v_{i,j} + v_{j,i}).$$

With infinitesimal deformations, of course, $\mathbf{d}$ approximates $\dot{\boldsymbol{\varepsilon}}$.

Uniqueness of Stress Field

In general plasticity theory, the decomposition of the deformation rate into elastic and plastic parts is far from unequivocal, as is shown in Chapter 8. In a rigid-plastic material, however, $\mathbf{d}$ may be identified with the plastic strain rate, and thus the plastic dissipation per unit volume, D_p, defined in Section 3.2, satisfies

$$D_p(\mathbf{d}) = \sigma_{ij} d_{ij}. \tag{3.4.6}$$

The maximum-plastic-dissipation principle may therefore be written

$$(\sigma_{ij} - \sigma_{ij}^*) d_{ij} \geq 0. \tag{3.4.7}$$

If $\mathbf{d} \neq 0$, then the equality holds only if $\boldsymbol{\sigma}$ and $\boldsymbol{\sigma}^*$ are *plastically equivalent*, that is, if $\mathbf{d}$ is related to both of them through the associated flow rule. In a Mises material, plastically equivalent stresses differ at most by a hydrostatic pressure, but in a Tresca material two stresses are plastically equivalent if they lie on the same facet of the hexagonal cylinder in principal-stress space.

If, now, $(\boldsymbol{\sigma}^{(\alpha)}, \mathbf{v}^{(\alpha)})$, $\alpha = 1, 2$, represent two admissible states of a rigid-plastic body, corresponding to the same body force and boundary conditions, then

$$n_j(\sigma_{ij}^{(1)} - \sigma_{ij}^{(2)})(v_i^{(1)} - v_i^{(2)}) = 0 \quad \text{on } \partial R,$$

and therefore

$$\int_{\partial R} n_j(\sigma_{ij}^{(1)} - \sigma_{ij}^{(2)})(v_i^{(1)} - v_i^{(2)})\, dS = \int_R (\sigma_{ij}^{(1)} - \sigma_{ij}^{(2)})(d_{ij}^{(1)} - d_{ij}^{(2)})\, dV = 0.$$

But the last integrand may be written as

$$(\sigma_{ij}^{(1)} - \sigma_{ij}^{(2)}) d_{ij}^{(1)} + (\sigma_{ij}^{(2)} - \sigma_{ij}^{(1)}) d_{ij}^{(2)},$$

and this is positive unless either $\mathbf{d}^{(1)}$ and $\mathbf{d}^{(2)}$ both vanish, or $\boldsymbol{\sigma}^{(1)}$ and $\boldsymbol{\sigma}^{(2)}$ are plastically equivalent. The general conclusion is thus that two admissible stress fields $\boldsymbol{\sigma}^{(1)}$ and $\boldsymbol{\sigma}^{(2)}$ must be plastically equivalent everywhere except in their common rigid region, that is, at points where $f(\boldsymbol{\sigma}^{(\alpha)}) < 0$, $\alpha = 1, 2$. If the body is made of a Mises material (or any other material whose yield surface in stress-deviator space is strictly convex) and deforms plastically in its entirety, then the two stress fields can differ at most by a hydrostatic pressure field, which must be uniform in order to satisfy equilibrium, and must vanish if a surface traction is prescribed anywhere on ∂R. Thus there is not more than one admissible stress field for which the whole body is plastic (Hill [1948a]), unless $\partial R = \partial R_v$, in which case the stress field is determined only to within a uniform hydrostatic pressure. On the other hand, uniqueness of the velocity field is not established.

Extremum Principle for Velocity

The degree to which the velocity field is determined may be learned from the kinematic extremum principle to be shown next, first proposed by Markov [1947] for a Mises material. Given a kinematically admissible velocity field $\mathbf{v}^*$, we define a functional Γ^* by

$$\Gamma^* = \int_R D_p(\mathbf{d}^*)\, dV - \int_R f_i v_i^*\, dV - \int_{\partial R_t} t_i^a v_i^*\, dS,$$

and, as usual,

$$\Gamma = \Gamma^*|_{\mathbf{v}^*=\mathbf{v}}.$$

Using the standard transformations we can show that

$$\Gamma = \int_{\partial R_v} n_j \sigma_{ij} v_i^a\, dS \tag{3.4.8}$$

and that

$$\Gamma^* - \Gamma = \int_R [D_p(\mathbf{d}^*) - D_p(\mathbf{d}) - \sigma_{ij}(d_{ij}^* - d_{ij})]\, dV.$$

Because of (3.4.6), the integrand is just $D_p(\mathbf{d}^*) - \sigma_{ij} d_{ij}^*$, and this is non-negative as a result of the maximum-plastic-dissipation principle, Equation (3.4.7), since the actual stress $\boldsymbol{\sigma}$ necessarily obeys the yield criterion. Consequently,

$$\Gamma^* \geq \Gamma. \tag{3.4.9}$$

It is not possible, in general, to strengthen the inequality by asserting that the equality holds only when $\mathbf{v}^* = \mathbf{v}$. The most we can say is that if $\Gamma^* = \Gamma$, then $\mathbf{v}^*$ is kinematically admissible, and $\mathbf{d}^*$ is associated with a stress field $\boldsymbol{\sigma}^*$ that is statically admissible and obeys the yield criterion everywhere. More particularly, however, if the body is one for which the stress field is unique (see above) and if the stress determines the deformation rate to within a scale factor (as is true of the Mises material, or any other material with a smooth yield surface), then the entire deformation-rate field is determined to within a scale factor. The indeterminacy may be eliminated if a nonzero velocity is prescribed anywhere on ∂R_v.

Principle of Maximum Plastic Work

Given a statically admissible stress field $\boldsymbol{\sigma}^*$ which nowhere violates the yield criterion, then, if $\boldsymbol{\sigma}$ and $\mathbf{v}$ are the actual stress and velocity fields, respectively, we clearly have

$$\int_R (\sigma_{ij} - \sigma^*_{ij}) d_{ij}\, dV = \int_{\partial R_v} n_j(\sigma_{ij} - \sigma^*_{ij}) v^a_i\, dS.$$

From the principle of maximum plastic dissipation there immediately follows the result

$$\int_{\partial R_v} n_j \sigma_{ij} v^a_i\, dS \geq \int_{\partial R_v} n_j \sigma^*_{ij} v^a_i\, dS,$$

due to Hill [1948a] and dubbed by him the **principle of maximum plastic work.** Here, again, the equality holds only if $\boldsymbol{\sigma}^*$ and $\boldsymbol{\sigma}$ are plastically equivalent, and therefore only if $\boldsymbol{\sigma}^* = \boldsymbol{\sigma}$ whenever $\boldsymbol{\sigma}$ is unique. Note that the left-hand side of the inequality is equal to Γ, Equation (3.4.8), and is therefore bounded both above and below.

Many results relating to variational principles in both elastic-plastic and rigid-plastic solids are contained in Washizu [1975].

Exercises: Section 3.4

1. Show that for any two real numbers a, b,

$$<a> - <b> = \beta(a - b)$$

   for some β, $0 \leq \beta \leq 1$.

2. Show that the Hu–Washizu principle (Exercise 14 of Section 1.4) may be extended to elastic-plastic materials if $W(\boldsymbol{\varepsilon})$ is replaced by $W(\boldsymbol{\varepsilon} - \boldsymbol{\varepsilon}^p)$.

3. Derive Equation (3.4.8) and the one following it.

Section 3.5 Limit-Analysis and Shakedown Theorems

3.5.1. Standard Limit-Analysis Theorems

The extremum principles for standard rigid-plastic materials that were discussed in the preceding section can be reformulated as the **theorems of**

limit analysis, which give upper and lower bounds on the loads under which a body that may be approximately modeled as elastic–perfectly plastic reaches a critical state. By a critical state we mean one in which large increases in plastic deformation — considerably greater than the elastic deformation — become possible with little if any increase in load. In the case of perfectly plastic bodies this state is called *unrestricted plastic flow*,[1] and the loading state at which it becomes possible is called *ultimate* or *limit loading*. It will be shown that, in a state of unrestricted plastic flow, elasticity may be ignored, and therefore a theory based on rigid-plastic behavior is valid for elastic-plastic bodies.

The proof of the limit-analysis theorems is based on the principle of maximum plastic dissipation, and consequently they are valid only for standard materials; a limited extension to nonstandard materials is discussed in 3.5.2.

It should be noted that the "loads" in the present context include not only the prescribed surface tractions $\mathbf{t}^a$ but all the surface tractions $\mathbf{t}$ operating at points at which the displacement (or velocity) is not constrained to be zero. In other words, the loads include *reactions that do work*; the definition of ∂R_t is accordingly extended. The reason for the extension is that the velocity fields used in the limit-analysis theorems are kinematically admissible velocity fields, not virtual velocity fields. The latter is, as we recall, the difference between two kinematically admissible velocity fields, and must therefore vanish wherever the velocity is prescribed, whereas a kinematically admissible velocity field takes on the values of the prescribed velocity.

We begin by defining a state of impending plastic collapse or incipient plastic flow as one in which a nonvanishing strain rate ($\dot{\varepsilon} \neq 0$) occurs under constant loads ($\dot{\mathbf{f}} = 0$, $\dot{\mathbf{t}} = 0$). The qualification "impending" or "incipient" is important: we are looking at the very beginning of such a state, which means that (1) all prior deformation has been of the same order of magnitude as elastic deformation, so that changes of geometry can be neglected, and (2) acceleration can be neglected and the problem can be treated as quasi-static.

Vanishing of Elastic Strain Rates

In addition to the preceding assumptions, it is assumed that the equations of equilibrium and the traction boundary conditions can be differentiated with respect to time with no change in form; consequently, the principle of virtual work is valid with $\dot{\boldsymbol{\sigma}}$, $\dot{\mathbf{f}}$, and $\dot{\mathbf{t}}$ replacing $\boldsymbol{\sigma}$, $\mathbf{f}$, and $\mathbf{t}$. For a virtual

[1]A rigorous formulation of the theorems in the context of convex analysis is due to Frémond and Friaâ [1982], who show that the concept of unrestricted plastic flow is too general for materials whose elastic range is unbounded in the appropriate stress space, and must be replaced by the weaker concept of "almost unrestricted" plastic flow (*écoulement presque libre*).

displacement field we take $\mathbf{v}\,\delta t$, where $\mathbf{v}$ is the actual velocity field and δt is a small time increment. The virtual strain field is, accordingly, $\dot{\varepsilon}_{ij}\,\delta t$, where $\dot{\varepsilon}_{ij} = \frac{1}{2}(v_{i,j} + v_{j,i})$. At impending collapse, then,

$$0 = \int_R \dot{\mathbf{f}} \cdot \mathbf{v}\, dV + \int_{\partial R_t} \dot{\mathbf{t}} \cdot \mathbf{v}\, dS = \int_R \dot{\sigma}_{ij}\dot{\varepsilon}_{ij}\, dV = \int_R \dot{\sigma}_{ij}(\dot{\varepsilon}^p_{ij} + C^{-1}_{ijkl}\dot{\sigma}_{kl})\, dV.$$

The positive definiteness of the elastic complementary energy implies $C^{-1}_{ijkl}\dot{\sigma}_{ij}\dot{\sigma}_{kl} \geq 0$ unless $\dot{\boldsymbol{\sigma}} = 0$. This fact, combined with Drucker's inequality (3.2.3), implies that at impending collapse or incipient plastic flow the stress rates vanish, so that $\dot{\boldsymbol{\varepsilon}}^e = 0$ and $\dot{\boldsymbol{\varepsilon}} = \dot{\boldsymbol{\varepsilon}}^p$. In other words, a body experiencing plastic collapse or flow behaves as though it were rigid-plastic rather than elastic-plastic. This result, first noted by Drucker, Greenberg, and Prager [1951], makes possible the rigorous application to elastic-plastic bodies of the theorems of limit analysis that had previously been formulated for rigid-plastic bodies. The following presentation of the theorems follows Drucker, Greenberg and Prager.[1]

Lower-Bound Theorem

Suppose that at collapse the actual loads are $\mathbf{f}$, $\mathbf{t}$ and the actual stress, velocity and strain-rate fields (in general unknown) are $\boldsymbol{\sigma}$, $\mathbf{v}$ and $\dot{\boldsymbol{\varepsilon}}$. Suppose further that we have somehow determined a stress field $\boldsymbol{\sigma}^*$ which does not violate the yield criterion anywhere and which is in equilibrium with the loads $\mathbf{f}^* = (1/s)\mathbf{f}$, $\mathbf{t}^* = (1/s)\mathbf{t}$, where s is a numerical factor. By virtual work, we have

$$\begin{aligned}\int_R \sigma^*_{ij}\dot{\varepsilon}_{ij}\, dV &= \frac{1}{s}\left(\int_{\partial R} \mathbf{t}\cdot\mathbf{v}\, dS + \int_R \mathbf{f}\cdot\mathbf{v}\, dV\right)\\ &= \frac{1}{s}\int_R \sigma_{ij}\dot{\varepsilon}_{ij}\, dV = \frac{1}{s}\int_R D_p(\dot{\boldsymbol{\varepsilon}})\, dV.\end{aligned}$$

But, by the principle of maximum plastic dissipation, $D_p(\dot{\boldsymbol{\varepsilon}}) \geq \sigma^*_{ij}\dot{\varepsilon}_{ij}$, so that $s \geq 1$. In other words, the factor s (the so-called "static multiplier") is in fact a **safety factor**.

Upper-Bound Theorem

Let us suppose next that instead of $\boldsymbol{\sigma}^*$, we somehow determine a velocity field $\mathbf{v}^*$ (a *collapse mechanism*), with the corresponding strain-rate field $\dot{\boldsymbol{\varepsilon}}^*$, and loads $\mathbf{f}^* = c\mathbf{f}$, $\mathbf{t}^* = c\mathbf{t}$ that satisfy

$$\int_{\partial R_t} \mathbf{t}^* \cdot \mathbf{v}^*\, dS + \int_R \mathbf{f}^* \cdot \mathbf{v}^*\, dV = \int_R D_p(\dot{\boldsymbol{\varepsilon}}^*)\, dV,$$

[1] See also Drucker, Prager and Greenberg [1952], Hill [1951, 1952], and Lee [1952].

provided the right-hand side (the total plastic dissipation) is positive;[2] then, again by virtual work,

$$\int_R D_p(\dot{\varepsilon}^*)\,dV = c\int_R \sigma_{ij}\dot{\varepsilon}_{ij}^*\,dV,$$

where $\boldsymbol{\sigma}$ is, as before, the actual stress field at collapse. The principle of maximum plastic dissipation, however, also implies that $D_p(\dot{\varepsilon}^*) \geq \sigma_{ij}\dot{\varepsilon}_{ij}^*$. Consequently $c \geq 1$, that is, c (the "kinematic multiplier") is an **overload factor**.

Alternative Formulation for One-Parameter Loading

Rather than using multipliers, the theorems can also be expressed in terms of a single reference load to which all the loads on the body are proportional. Let this reference load be denoted P, and let the loading (consisting of applied loads and working reactions) be expressed as

$$\mathbf{f} = P\tilde{\mathbf{f}}, \quad \mathbf{t} = P\tilde{\mathbf{t}},$$

where $\tilde{\mathbf{f}}$ and $\tilde{\mathbf{t}}$ are known functions of position in R and on ∂R_t, respectively. Let, further, P_U denote the value of P at collapse. If a plastically admissible stress field $\boldsymbol{\sigma}^*$ is in equilibrium with $P\tilde{\mathbf{f}}$ and $P\tilde{\mathbf{t}}$ for some value of P, say ${P_{LB}}^*$, then this value is a lower bound, that is, ${P_{LB}}^* \leq P_U$. An upper bound ${P_{UB}}^*$ can be found explicitly for a kinematically admissible velocity field $\mathbf{v}^*$:

$$P_U \leq {P_{UB}}^* = \frac{\int_R D_p(\dot{\varepsilon}^*)dV}{\int_R \tilde{\mathbf{f}}\cdot\mathbf{v}^*\,dV + \int_{\partial R_t}\tilde{\mathbf{t}}\cdot\mathbf{v}^*\,dS}. \tag{3.5.1}$$

For loading governed by a single parameter, therefore, the two theorems may also be expressed as follows: *The loads that are in equilibrium with a stress field that nowhere violates the yield criterion do not exceed the collapse loads, while the loads that do positive work on a kinematically admissible velocity field at a rate equal to the total plastic dissipation are at least equal to the collapse loads.* If the loads produced by the application of the two theorems are equal to each other, then they equal the collapse loads.

In particular, if one has succeeded in finding both (a) a statically and plastically admissible stress field, and (b) a kinematically admissible velocity field such that the strain rate produced by it is everywhere[1] associated to the stress, then a *complete solution* is said to have been found. This solution is not necessarily unique and hence cannot be called an exact solution, but, as a result of the theorems of limit analysis, it predicts the correct collapse load. Some applications of this concept are given in Chapter 5. In Chapter

[2]If the total plastic dissipation is negative, $\mathbf{v}^*$ can be replaced by $-\mathbf{v}^*$; if it is zero, $\mathbf{v}^*$ does not represent a collapse mechanism.

[1]If the velocity field involves regions that move as rigid bodies (*rigid regions*), then the strain rate there is of course zero and the question of association does not arise.

6 the theorems are used to obtain estimates of collapse loads in problems for which no complete solutions have been found.

Multiparameter Loadings

We now consider loadings that are governed by several parameters that can vary independently. These parameters will be called *generalized loads*; some of them may be applied loads, and others may be reactions that do work. Let them be denoted P_I ($I = 1, ..., N$), so that

$$\mathbf{f} = \sum_{I=1}^{N} P_I \tilde{\mathbf{f}}^{(I)}, \quad \mathbf{t} = \sum_{I=1}^{N} P_I \tilde{\mathbf{t}}^{(I)},$$

the $\tilde{\mathbf{f}}^{(I)}$ and $\tilde{\mathbf{t}}^{(I)}$ again being known functions. For a kinematically admissible velocity field $\mathbf{v}$, generalized velocities $\dot{p}_I$ can be defined by

$$\dot{p}_I = \int_R \tilde{\mathbf{f}}^{(I)} \cdot \mathbf{v} dR + \int_{\partial R_t} \tilde{\mathbf{t}}^{(I)} \cdot \mathbf{v} dS,$$

so that, by virtual work,

$$\int_R \sigma_{ij} \dot{\varepsilon}_{ij} \, dV = \sum_{I=1}^{N} P_I \dot{p}_I \stackrel{\text{def}}{=} \mathbf{P} \cdot \dot{\mathbf{p}}, \tag{3.5.2}$$

where $\mathbf{P}$ and $\dot{\mathbf{p}}$ are the N-dimensional vectors representing the P_I and $\dot{p}_I$. Any combination of generalized loads P_I thus represents a point in $\mathbf{P}$-space (a *load point*), while a fixed proportion among the P_I represents a direction in this space (a *loading direction*).

A combination of P_I at which unrestricted plastic flow occurs represents a *limit point* (or *flow point* or *yield point* — the last designation makes sense only in terms of a rigid-plastic material), and the set of all such points is the *limit locus* (or *flow locus* or *yield locus*), given by, say,

$$\Phi(\mathbf{P}) = 0.$$

It follows from Equation (3.5.2) and the principle of maximum plastic dissipation that if $\mathbf{P}$ and $\mathbf{P}^*$ are on or inside the limit locus, and if $\dot{\mathbf{p}}$ is a generalized velocity vector that is possible under a load vector $\mathbf{P}$, then

$$(\mathbf{P} - \mathbf{P}^*) \cdot \dot{\mathbf{p}} \geq 0. \tag{3.5.3}$$

By arguments identical with those of 3.2.2 for the maximum-plastic-dissipation principle in terms of stresses and strain rates, it follows from inequality (3.5.3) that the limit locus is convex, and that the generalized velocity vector is normal to the limit locus, that is,

$$\dot{p}_I = \dot{\lambda} \frac{\partial \Phi}{\partial P_I}$$

wherever $\Phi(\mathbf{P})$ is regular. The appropriate generalization, following either the Koiter or the Moreau formalism, may be formed for singular limit loci.

A fixed loading direction is described by $N-1$ parameters, and keeping these constant is equivalent to one-parameter loading. Consequently, along any direction lower-bound and upper-bound load points can be found, and since they depend on the $N-1$ parameters, they form parametric representations of upper-bound and lower-bound loci.

A simple illustration of multi-parameter loading is provided by an "ideal sandwich beam," composed of two equal, very thin flanges of cross-sectional area A, separated by a distance h that is spanned by a web of negligible area [see Figure 3.5.1(a)]. The beam is subject to an axial force P, whose line of

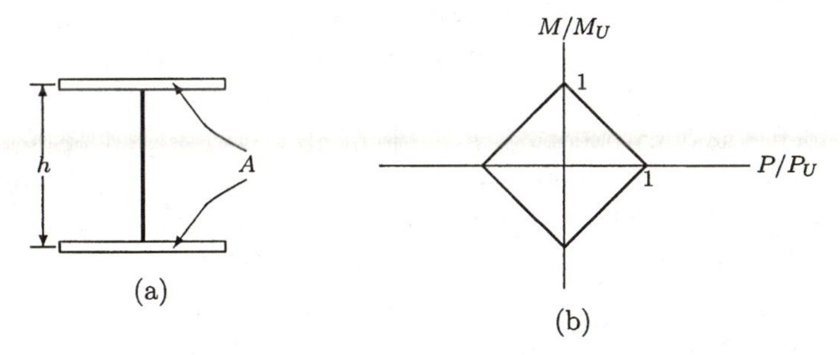

Figure 3.5.1. Ideal sandwich beam: (a) cross-section; (b) flow locus.

action is defined as midway between the flanges, and a bending moment M; P and M may be treated as generalized loads, and the conjugate generalized velocities are respectively the average elongation rate $\dot{\Delta}$ and the rotation rate $\dot{\theta}$. The stresses in the flanges may be assumed to be purely axial, and equilibrium requires that they be

$$\sigma = \frac{P}{2A} \pm \frac{M}{Ah}.$$

If the yield stress is σ_Y, then the yield inequality $|\sigma| \le \sigma_Y$ in each flange is equivalent to

$$\left|\frac{P}{P_U} \pm \frac{M}{M_U}\right| \le 1,$$

where $P_U = 2\sigma_Y A$, $M_U = \sigma_Y Ah$. The limit locus, shown in Figure 3.5.1(b), is thus given by

$$\Phi(M,\, P) = \max\left(\left|\frac{P}{P_U} + \frac{M}{M_U}\right|, \left|\frac{P}{P_U} - \frac{M}{M_U}\right|\right) - 1 = 0.$$

Since this locus was found on the basis of a stress distribution, it is strictly speaking a lower-bound locus. The stress distribution is, of course, unique, and therefore the lower bound must in fact give the true limit locus.

It can easily be shown that the same limit locus results from an application of the upper-bound theorem. Consider a velocity field in which only one of the flanges elongates while the other remains rigid. In order for the mean elongation rate of the beam to be $\dot{\Delta}$, that of the deforming flange must be $2\dot{\Delta}$, and the rotation rate, to within a sign, is $\dot{\theta} = 2\dot{\Delta}/h$. For the sake of definiteness, let us take both $\dot{\Delta}$ and $\dot{\theta}$ as positive. The strain rate in the deforming flange is $2\dot{\Delta}/L$, so that the total plastic dissipation — the numerator on the right-hand side of (3.5.1) — is $\sigma_Y(2\dot{\Delta}/L)(AL) = P_U\dot{\Delta}$. We take P as the reference load, and pick a loading direction by letting $M = \alpha Ph/2$. The denominator in (3.5.1) is thus $\dot{\Delta} + (\alpha h/2)(2\dot{\Delta}/h) = (1+\alpha)\dot{\Delta}$, and the upper bound for P is $P_U/(1+\alpha)$. Since $M_U = P_U h/2$, the upper bound for M is $\alpha M_U/(1+\alpha)$. The upper-bound values satisfy $M/M_U + P/P_U = 1$, an equation describing the first quadrant of the previously found limit locus. The remaining quadrants are found by varying the signs of $\dot{\Delta}$ and $\dot{\theta}$.

A velocity field with both flanges deforming leads to an upper-bound load point lying outside the limit locus just found, with two exceptions: one where the elongation rates of the flanges are the same, and one where they are equal and opposite. Details are left to an exercise.

3.5.2. Nonstandard Limit-Analysis Theorems

The theorems of limit analysis can be stated in a form that does not directly refer to any concepts from plasticity theory:

A body will not collapse under a given loading if a possible stress field can be found that is in equilibrium with a loading greater than the given loading.

A body will collapse under a given loading if a velocity field obeying the constraints (or a mechanism) can be found that so that the internal dissipation is less than the rate of work of the given loading.

In this form, the theorems appear intuitively obvious. In fact, the concepts underlying the theorems were used long before the development of plasticity theory. Use of what is essentially the upper-bound theorem goes back to the eighteenth century: it was used in 1741 by a group of Italian mathematicians to design a reinforcement method for the crumbling dome of Saint Peter's Church, and in 1773 by Coulomb to investigate the collapse strength of soil. The latter problem was also studied by Rankine in the mid-nineteenth century by means of a technique equivalent to the lower-bound theorem.

The simple form of the theorems given above hides the fact that the postulate of maximum plastic dissipation (and therefore the normality of the flow rule) is an essential ingredient of the proof. It was therefore necessary to find a counterexample showing that the theorems are not universally applicable to nonstandard materials. One such counterexample, in which plasticity is combined with Coulomb friction at an interface, was presented by Drucker [1954a]. Another was shown by Salençon [1973].

Radenkovic's Theorems

A theory of limit analysis for nonstandard materials, with a view toward its application to soils, was formulated by Radenkovic [1961, 1962], with modifications by Josselin de Jong [1965, 1974], Palmer [1966], Sacchi and Save [1968], Collins [1969], and Salençon [1972, 1977]. **Radenkovic's first theorem** may be stated simply as follows: *The limit loading for a body made of a nonstandard material is bounded from above by the limit loading for the standard material obeying the same yield criterion.*

The proof is straightforward. Let $\mathbf{v}^*$ denote any kinematically admissible velocity field, and $\mathbf{P}^*$ the upper-bound load point obtained for the standard material on the basis of this velocity field. If $\boldsymbol{\sigma}$ is the actual stress field at collapse in the real material, then, since this stress field is also statically and plastically admissible in the standard material,

$$D_p(\dot{\varepsilon}^*) \geq \sigma_{ij}\dot{\varepsilon}_{ij}{}^*,$$

and therefore, by virtual work,

$$\mathbf{P}^* \cdot \dot{\mathbf{p}}^* \geq \mathbf{P} \cdot \dot{\mathbf{p}}^*.$$

Since $\mathbf{v}^*$ may, as a special case, coincide with the correct collapse velocity field in the fictitious material, $\mathbf{P}^*$ may be the correct collapse loading in this material, and the theorem follows.

Radenkovic's second theorem, as modified by Josselin de Jong [1965], is based on the existence of a function $g(\boldsymbol{\sigma})$ with the following properties:

1. $g(\boldsymbol{\sigma})$ is a convex function (so that any surface $g(\boldsymbol{\sigma}) = constant$ is convex);
2. $g(\boldsymbol{\sigma}) = 0$ implies $f(\boldsymbol{\sigma}) \leq 0$ (so that the surface $g(\boldsymbol{\sigma}) = 0$ lies entirely within the yield surface $f(\boldsymbol{\sigma}) = 0$);
3. to any $\boldsymbol{\sigma}$ with $f(\boldsymbol{\sigma}) = 0$ there corresponds a $\boldsymbol{\sigma}'$ such that (a) $\dot{\boldsymbol{\varepsilon}}^p$ is normal to the surface $g(\boldsymbol{\sigma}) = 0$ at $\boldsymbol{\sigma}'$, and (b)

$$(\sigma_{ij} - \sigma'_{ij})\dot{\varepsilon}_{ij} \geq 0. \tag{3.5.4}$$

The theorem may then be stated thus: *The limit loading for a body made of a nonstandard material is bounded from below by the limit loading for the standard material obeying the yield criterion* $g(\boldsymbol{\sigma}) = 0$.

The proof is as follows. Let $\boldsymbol{\sigma}$ denote the actual stress field at collapse, $\mathbf{P}$ the limit loading, $\mathbf{v}$ the actual velocity field at collapse, $\dot{\boldsymbol{\varepsilon}}$ the strain-rate field, and $\dot{\mathbf{p}}$ the generalized velocity vector conjugate to $\mathbf{P}$. Thus, by virtual work,

$$\mathbf{P} \cdot \dot{\mathbf{p}} = \int_R \sigma_{ij} \dot{\varepsilon}_{ij} \, dV.$$

Now, the velocity field $\mathbf{v}$ is kinematically admissible in the fictitious standard material. If $\boldsymbol{\sigma}'$ is the stress field corresponding to $\boldsymbol{\sigma}$ in accordance with the definition of $g(\boldsymbol{\sigma})$, then it is the stress field in the fictitious material that is plastically associated with $\dot{\boldsymbol{\varepsilon}}$, and, if $\mathbf{P}'$ is the loading that is in equilibrium with $\boldsymbol{\sigma}'$, then

$$\mathbf{P}' \cdot \dot{\mathbf{p}} = \int_R \sigma'_{ij} \dot{\varepsilon}_{ij} \, dV.$$

It follows from inequality (3.5.4) that

$$\mathbf{P}' \cdot \dot{\mathbf{p}} \leq \mathbf{P} \cdot \dot{\mathbf{p}}.$$

Again, $\boldsymbol{\sigma}'$ may, as a special case, coincide with the correct stress field at collapse in the standard material, and therefore $\mathbf{P}'$ may be the correct limit loading in this material. The theorem is thus proved.

In the case of a Mohr–Coulomb material, the function $g(\boldsymbol{\sigma})$ may be identified with the plastic potential if this is of the same form as the yield function, but with an angle of dilatation that is less than the angle of internal friction (in fact, the original statement of the theorem by Radenkovic [1962] referred to the plastic potential only). The same is true of the Drucker–Prager material.

It should be noted that neither the function g, nor the assignment of $\boldsymbol{\sigma}'$ to $\boldsymbol{\sigma}$, is unique. In order to achieve the best possible lower bound, g should be chosen so that the surface $g(\boldsymbol{\sigma}) = 0$ is as close as possible to the yield surface $f(\boldsymbol{\sigma}) = 0$, at least in the range of stresses that are expected to be encountered in the problem studied. Since the two surfaces do not coincide, however, it follows that the lower and upper bounds on the limit loading, being based on two different standard materials, cannot be made to coincide. The correct limit loading in the nonstandard material cannot, therefore, be determined in general. This result is consistent with the absence of a uniqueness proof for the stress field in a body made of a nonstandard perfectly plastic material (see 3.4.1).

3.5.3. Shakedown Theorems

The collapse discussed thus far in the present section is known as *static collapse*, since it represents unlimited plastic deformation while the loads remain constant in time. If the loads are applied in a cyclic manner, without ever reaching the static collapse condition, other forms of collapse may occur. If the strain increments change sign in every cycle, with yielding on both sides of the cycle, then *alternating plasticity* is said to occur; the *net* plastic deformation may remain small, but weakening of the material may occur nevertheless — a phenomenon called *low-cycle fatigue* — leading to breaking of the most highly stressed points after a certain number of cycles.

It may also happen that plastic deformation in each cycle accumulates so that after enough cycles, the displacements are large enough to be equivalent to collapse; this is called *incremental collapse*. On the other hand, it may happen that no further plastic deformation occurs after one or a few cycles — that is, all subsequent unloading-reloading cycles are elastic. In that case the body is said to have experienced *shakedown* or *adaptation*. It is obvious that for bodies subject to repeated loading, shakedown is more relevant than static collapse, and that criteria for shakedown are of great importance.

Residual Stresses: Example

If an initially stress-free body has been loaded into the plastic range, but short of collapse, and the loads are then reduced to zero, then the stress field in the unloaded body does not in general vanish. As an example, consider the four-flange sandwich beam shown in Figure 3.5.2, subject to a bending moment M only. If the flanges behave similarly in tension and compression,

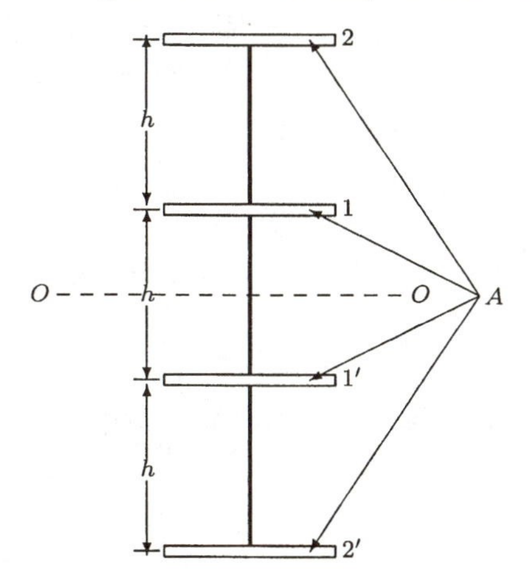

Figure 3.5.2. Ideal four-flange beam: geometry.

then the stresses in the flanges satisfy $\sigma_{1'} = -\sigma_1$ and $\sigma_{2'} = -\sigma_2$, and the moment is

$$M = -\sigma_1 Ah - 3\sigma_2 Ah.$$

In accordance with elementary beam theory (which is discussed further in Section 4.4), the longitudinal strain in the flanges varies linearly with distance from the so-called *neutral plane*, which in the present case may be shown to be the plane OO. Thus $\varepsilon_2 = 3\varepsilon_1$, with $\varepsilon_{1'} = -\varepsilon_1$ and $\varepsilon_{2'} = -\varepsilon_2$. As long as all flanges are elastic, stress is proportional to strain, and therefore

$\sigma_2 = 3\sigma_1$. It follows that

$$\sigma_1 = -\frac{M}{10Ah} \stackrel{\text{def}}{=} \sigma_1^e, \quad \sigma_2 = -\frac{3M}{10Ah} \stackrel{\text{def}}{=} \sigma_2^e.$$

By definition, the notation $\boldsymbol{\sigma}^e$ will be used for stresses calculated on the basis of assumed elastic behavior of a body, or *elastic stresses*.

The outer flanges will yield when $|M| = (10/3)\sigma_Y Ah \stackrel{\text{def}}{=} M_E$. Provided that the web holds, however, the deformation will remain contained so long as the inner flanges remain elastic. Collapse occurs when all four flanges yield, that is, when $\sigma_1 = \sigma_2 = -\sigma_Y$, so that $|M| = 4\sigma_Y Ah = M_U$. The range $M_E \le |M| \le M_U$ is called the range of *contained plastic deformation*.

Suppose, for example, that $M = (11/3)\sigma_Y Ah = M^*$. The actual stresses are $\sigma_1 = -(2/3)\sigma_Y$, $\sigma_2 = -\sigma_Y$, while the elastic stresses are $\sigma_1^e = -(11/30)\sigma_Y$, $\sigma_2^e = -(11/10)\sigma_Y$. When the moment is removed, the flanges will unload elastically, a process equivalent to subtracting the elastic stresses from the actual stresses. The resulting *residual stresses* are $\sigma_1^r = -(3/10)\sigma_Y$ and $\sigma_2^r = (1/10)\sigma_Y$. Their resultant moment is, of course, zero. If the beam is now repeatedly reloaded with a bending moment of the same sign as before, the response will be elastic provided that $M \le M^*$. Consequently, for the given load amplitude, shakedown takes place in the first cycle.

Residual Stress and Displacement

A stress field $\boldsymbol{\rho}$ that is in equilibrium with zero body force and zero prescribed surface tractions is called *self-equilibrated*, or, more simply, a field of *self-stress*. Such a field must clearly satisfy

$$\rho_{ij},_j = 0 \text{ in } R, \quad \rho_{ij} n_j = 0 \text{ on } \partial R_t.$$

In an elastic-plastic body under given loads $\mathbf{f}$, $\mathbf{t}^a$, the stress field can always be written as

$$\boldsymbol{\sigma} = \boldsymbol{\sigma}^e + \boldsymbol{\rho}, \tag{3.5.5}$$

where $\boldsymbol{\sigma}^e$ is the elastic stress field corresponding to the given loads, and $\boldsymbol{\rho}$ is the self-equilibrated field of residual stresses. The strain field can accordingly be written as

$$\varepsilon_{ij} = C_{ijkl}^{-1}\sigma_{kl}^e + C_{ijkl}^{-1}\rho_{kl} + \varepsilon_{ij}^p. \tag{3.5.6}$$

The first term, which will be written ε'_{ij}, represents the strain field in the hypothetical elastic body under the prescribed loads. This strain field is compatible with a displacement field that will be denoted $\mathbf{u}^e$, that is,

$$\varepsilon'_{ij} = \frac{1}{2}(u_i^e,_j + u_j^e,_i).$$

Since the total strain field $\boldsymbol{\varepsilon}$ is also compatible, the remaining terms of (3.5.6) are also derivable from a displacement field $\mathbf{u}^r$ (the *residual displacement* field):

$$C_{ijkl}^{-1}\rho_{kl} + \varepsilon_{ij}^p = \frac{1}{2}(u_i^r,_j + u_j^r,_i), \tag{3.5.7}$$

and $\mathbf{u} = \mathbf{u}^e + \mathbf{u}^r$.

It may be shown that the plastic strain field $\boldsymbol{\varepsilon}^p$ uniquely determines the residual stress field $\boldsymbol{\rho}$ and, with sufficient constraints to prevent rigid-body displacement, also the residual displacement field $\mathbf{u}^r$. The method of proof is analogous to that of the elastic uniqueness theorem of 3.4.1. If $\boldsymbol{\rho}$ and $\boldsymbol{\rho}+\overline{\boldsymbol{\rho}}$ are two different residual stress fields, then, from the principle of virtual work,

$$\int_R \rho_{ij}(C_{ijkl}^{-1}\rho_{kl} + \varepsilon_{ij}^p)\, dV = \int_R (\rho_{ij} + \overline{\rho}_{ij})(C_{ijkl}^{-1}\rho_{kl} + \varepsilon_{ij}^p)\, dV$$

$$= \int_R (\rho_{ij} + \overline{\rho}_{ij})[C_{ijkl}^{-1}(\rho_{kl} + \overline{\rho}_{kl}) + \varepsilon_{ij}^p]\, dV = 0.$$

Rearrangement of terms leads to

$$\int_R \overline{\rho}_{ij} C_{ijkl}^{-1} \overline{\rho}_{kl}\, dV = 0,$$

and hence to $\overline{\boldsymbol{\rho}} = 0$ in view of the positive-definiteness of C^{-1}.

With $\boldsymbol{\rho}$ uniquely determined by $\boldsymbol{\varepsilon}^p$, it follows from (3.5.7) that $\mathbf{u}^r$ is determined to within a rigid-body displacement.

Quasi-Static Shakedown Theorem

Suppose that an elastic-plastic body has already shaken down under a loading that is varying in time (but sufficiently slowly so that inertia may be neglected) within a certain range of the generalized loads. It follows from the definition of shakedown that the plastic strain field $\boldsymbol{\varepsilon}^p$ remains constant in time and defines a time-independent residual stress field $\boldsymbol{\rho}$ such that the total stress field $\boldsymbol{\sigma}$, given by (3.5.5), does not violate the yield criterion anywhere:

$$f(\boldsymbol{\sigma}^e + \boldsymbol{\rho}) \le 0.$$

Clearly, the existence of such a residual stress field is a necessary condition for shakedown.

It was shown by Melan [1938], however, that this is also a sufficient condition: shakedown will occur in the given load range if a time-independent self-stress field $\boldsymbol{\rho}^*$, not necessary equal to the actual residual stress field $\boldsymbol{\rho}$, can be found such that

$$f(\boldsymbol{\sigma}^e + \boldsymbol{\rho}^*) < 0$$

for all elastic stress fields $\boldsymbol{\sigma}^e$ corresponding to loadings within the given range.

To prove the theorem, we consider the nonnegative quantity

$$Y = \frac{1}{2}\int_R C^{-1}_{ijkl}(\rho_{ij} - \rho_{ij}{}^*)(\rho_{kl} - \rho_{kl}{}^*)\,dV.$$

Since the body may not yet have shaken down, the actual residual stress field $\boldsymbol{\rho}$, and hence Y, may be time-dependent, with

$$\dot{Y} = \int_R C^{-1}_{ijkl}(\rho_{ij} - \rho_{ij}{}^*)\dot{\rho}_{kl}\,dV.$$

Since $\boldsymbol{\rho}$ and $\boldsymbol{\rho}^*$ are both self-equilibrated, and since the left-hand side of (3.5.7) forms a compatible strain field, it follows from the principle of virtual work that

$$\int_R (\rho_{ij} - \rho_{ij}{}^*)(C^{-1}_{ijkl}\dot{\rho}_{kl} + \dot{\varepsilon}^p_{ij})dV = 0.$$

Consequently

$$\dot{Y} = -\int_R (\rho_{ij} - \rho_{ij}{}^*)\dot{\varepsilon}^p_{ij}dV = -\int_R (\sigma_{ij} - \sigma_{ij}{}^*)\dot{\varepsilon}^p_{ij}dV,$$

where $\boldsymbol{\sigma}^* = \boldsymbol{\sigma}^e + \boldsymbol{\rho}^*$. By hypothesis, $\boldsymbol{\sigma}^*$ does not violate the yield criterion, and therefore, as a result of the maximum-plastic-dissipation postulate (3.2.4), $\dot{Y} \leq 0$, where the equality holds only in the absence of plastic flow. Since $Y \geq 0$, the condition $\dot{Y} = 0$ must eventually be reached, and this condition corresponds to shakedown.

An extension of the theorem to work-hardening materials is due to Mandel [1976] (see also Mandel, Zarka, and Halphen [1977]). The yield surface is assumed to be of the form

$$f(\boldsymbol{\sigma}, \boldsymbol{\xi}) = F(\boldsymbol{\sigma} - c\boldsymbol{\varepsilon}^p) - k^2(\kappa),$$

where F is a homogeneous quadratic function of its argument, and k is a nondecreasing function of κ. A body made of such a material will shake down if, in addition to a time-independent residual stress field $\boldsymbol{\rho}^*$, there exist time-independent internal-variable fields $\boldsymbol{\xi}^* = (\boldsymbol{\varepsilon}^{p*}, \kappa^*)$ such that

$$f(\boldsymbol{\sigma}^e + \boldsymbol{\rho}^*, \boldsymbol{\xi}^*) < 0$$

everywhere.

Kinematic Shakedown Theorem

A kinematic criterion for shakedown was derived by Koiter [1956, 1960]. A strong version of Koiter's theorem states that shakedown has not taken place if a kinematically admissible velocity field $\mathbf{v}^*$, satisfying $\mathbf{v}^* = 0$ on ∂R_v, can be found so that

$$\int_R \mathbf{f}\cdot\mathbf{v}^*\,dV + \int_{\partial R_t} \mathbf{t}^a\cdot\mathbf{v}^*\,dS > \int_R D_p(\dot{\boldsymbol{\varepsilon}}^*)\,dV.$$

This inequality, with the principle of virtual work applied to its left-hand side, can be transformed into

$$\int_R \sigma_{ij}^e \dot{\varepsilon}_{ij}{}^* \, dV > \int_R D_p(\dot{\varepsilon}^*) \, dV. \tag{3.5.8}$$

Suppose, now, that shakedown has taken place, with a time-independent residual stress field $\boldsymbol{\rho}$. From the maximum-plastic-dissipation postulate,

$$D(\dot{\varepsilon}^*) \geq (\sigma_{ij}^e + \rho_{ij}) \dot{\varepsilon}_{ij}{}^*,$$

and therefore

$$\int_R D(\dot{\varepsilon}^*) dV \geq \int_R \sigma_{ij}^e \dot{\varepsilon}_{ij}{}^* \, dV + \int_R \rho_{ij} \dot{\varepsilon}_{ij}{}^* \, dV. \tag{3.5.9}$$

An application of the virtual-work principle to the last integral shows that it vanishes, since $\boldsymbol{\rho}$ is self-equilibrated and $\mathbf{v}^* = 0$ on ∂R_v. Inequalities (3.5.8) and (3.5.9) are therefore in contradiction, that is, shakedown cannot have taken place.

A weaker version of the theorem requires only that a strain rate $\dot{\boldsymbol{\varepsilon}}^*$ be found during a time interval $(0, T)$ such that the strain field

$$\boldsymbol{\varepsilon}^* = \int_0^T \dot{\boldsymbol{\varepsilon}}^* \, dt$$

is compatible with a displacement field $\mathbf{u}^*$ that satisfies $\mathbf{u}^* = 0$ on ∂R_u, and

$$\int_0^T \left(\int_R \mathbf{f} \cdot \dot{\mathbf{u}}^* \, dV + \int_{\partial R_t} \mathbf{t}^a \cdot \dot{\mathbf{u}}^* dS \right) dt > \int_0^T \int_R D_p(\dot{\varepsilon}^*) \, dV \, dt.$$

With these conditions met, the body will not shake down during the interval.

Recent developments in shakedown theory have included taking into account the effects of temperature changes, creep, inertia, and geometric nonlinearities. For a review, see the book by König [1987]

Exercises: Section 3.5

1. Find the ultimate load F_U for the structure shown in Figure 4.1.2(a) (page 185), assuming that all the bars have the same cross-sectional area, are made of the same elastic–perfectly plastic material with uniaxial yield stress σ_Y, and act in simple tension.

2. Using both the lower-bound and the upper-bound theorems, find the limit locus for the beam having the idealized section shown in Figure 3.5.2 (page 171) subject to combined axial force P and bending moment M.

3. In a body made of a standard Mohr–Coulomb material with cohesion c and internal-friction angle ϕ under a load P, lower and upper bounds to the ultimate load P_U have been found as $P_U^- = ch^-(\phi)$ and $P_U^+ = ch^+(\phi)$. Show how the results can be used to find the best bounds on P_U if the material is nonstandard but has a plastic potential of the same form as the yield function, with a dilatation angle $\psi \neq \phi$.

4. For the beam of Figure 3.5.2 subject to a bending moment M only, find the range of M within which shakedown occurs on the basis of the following assumed self-stress distributions.

   (a) $\rho_1{}^* = \frac{1}{2}\sigma_Y = -\rho_1'{}^*$, $\rho_2{}^* = -\frac{1}{6}\sigma_Y = -\rho_2'{}^*$
   (b) $\rho_1{}^* = -\frac{1}{2}\sigma_Y = -\rho_1'{}^*$, $\rho_2{}^* = \frac{1}{6}\sigma_Y = -\rho_2'{}^*$
   (c) $\rho_1{}^* = \frac{3}{4}\sigma_Y = -\rho_1'{}^*$, $\rho_2{}^* = -\frac{1}{4}\sigma_Y = -\rho_2'{}^*$

5. For the beam of Figure 3.5.2 under combined axial force P and bending moment M, (a) find the elastic stresses σ^e in each flange; (b) given the self-stress distribution $\rho_1{}^* = -\frac{3}{5}\sigma_Y$, $\rho_2{}^* = \frac{4}{5}\sigma_Y$, $\rho_1'{}^* = \frac{1}{5}\sigma_Y$, $\rho_2'{}^* = -\frac{2}{5}\sigma_Y$ find the range of P and M moments under which shakedown occurs by ensuring that $|\rho^* + \sigma^e| \leq \sigma_Y$ in each flange.

6. Using the result of Exercise 2 for the beam of Figure 3.5.2 under combined axial force and bending moment, find, if possible, loading cycles between pairs of points on the limit locus such that there occurs (a) incremental plastic deformation, (b) alternating plastic deformation, and (c) shakedown.

Chapter 4

Problems in Contained Plastic Deformation

Section 4.1 Elementary Problems

4.1.1. Introduction: Statically Determinate Problems

There are a few static or quasi-static boundary-value problems in solid mechanics that are approximately statically determinate. That is, much of the stress field may be determined from the applied loading (whether prescribed tractions or initially unknown reactions), independently of the material properties. These problems include the following:

(a) Uniaxial tension or compression of a straight rod or tube with an axial force P.

(b) Torsion of a thin-walled tube with a torque T.

(c) Simultaneous tension or compression and torsion of a thin-walled tube.

(d) Axially symmetric loading of a thin-walled shell of revolution.

The solution of these problems, as regards stress, is the stuff of elementary solid mechanics.

In problem (a), away from the ends the only significant stress component is the axial stress $\sigma = P/A$, where A is the cross-sectional area, and this stress is uniform if A is constant, that is, if the bar is prismatic. This solution constitutes the basis of the simple tension or compression test discussed in Section 2.1; the load-elongation diagram essentially reproduces the conventional stress-strain diagram.

Similarly, in problem (b) the only significant stress component is the shear stress $\tau = T/2Ah$, where A is the area enclosed by the curve tracing the midpoints of the wall, and h is the wall thickness. This stress is constant if the wall thickness is uniform. If the tube, moreover, is circular, with radius

c, then $\tau = T/2\pi c^2 h$, while the shear strain is given by $\gamma = c\phi/L$, where ϕ is the relative angle of twist between the ends of the tube, and L is its length. The T-ϕ diagram of the tube thus reproduces the τ-γ diagram of the material. The torsion of a thin-walled circular tube is consequently a simple means of determining the pure shearing behavior of solids that are sufficiently strong in tension, such as metals and hard plastics. As discussed in Section 2.3, geomaterials require different tests of shear behavior. A complication of the torsion test on thin-walled tubes is that if the thickness is too small in relation to the radius, the induced compressive stress may provoke buckling.

It is characteristic of these two problems that if the material is taken as perfectly plastic, then the applied load (force or torque) cannot exceed the value at which yielding first occurs. When this value is reached, the body continues to deform under constant load — a process that, depending on the field of application, is called *plastic flow* or *plastic collapse* — until significant changes in geometry take place, or until inertial effects become important.

The solution of problem (c) *for stresses* is the superposition of those of (a) and (b). The resulting stress field forms, locally, a state of plane stress determined by σ and τ, so that if the material yields according to either the Mises or the Tresca criterion, the yielding is described by Equation (3.3.5). Measurements of changes in elongation and twist angle provide a test of the flow rule, and the movement of the yield curve in the σ-τ plane may elucidate the hardening rule. Tests based on this problem were first performed by Taylor and Quinney [1931].

The solution of problem (d) is particularly simple when the shell is a sphere (or a portion of a sphere), of radius c and wall thickness h, under uniform internal and external pressure. This loading produces a state of uniform, essentially plane stress, with both principal stresses equal to $\sigma = pc/2h$, p being the pressure difference (internal less external). Another important case is that of a cylindrical tube under a combination of uniform pressure and an axial force P; in this case the principal stresses are $\sigma_1 = pc/h$ (circumferential) and $\sigma_2 = pc/2h + P/2\pi ch$ (axial). Since arbitrary combinations of principal stresses can be produced by varying the axial force and the pressure, this case has formed the basis of numerous experiments, beginning with those of Lode [1925], to study yield criteria, flow rules, and hardening.

4.1.2. Thin-Walled Circular Tube in Torsion and Extension

We consider, as a special case of problem (c), a circular tube made of a rate-independent plastic material governed by either the Mises or the Tresca yield criterion, given in terms of σ and τ by Equation (3.3.5). We furthermore assume isotropic hardening based on a hardening variable κ. The criterion may therefore be written in the form

$$f(\sigma, \tau, \kappa) = \bar{\sigma} - \sigma_Y(\kappa) = 0, \tag{4.1.1}$$

where $\bar{\sigma} = \sqrt{\sigma^2 + \alpha\tau^2}$, and $\alpha = (\sigma_Y/\tau_Y)^2$ is a constant equal to 3 for Mises criterion and 4 for the Tresca criterion. Note that $finte = \dot{\bar{\sigma}}$.

The flow rule will be assumed to be associated and hence derivable from a plastic potential which may be identified with $\bar{\sigma}$. The associated flow rule is equivalent to the postulate of maximum plastic dissipation, Equation (3.2.4), which in the present state of stress reduces to

$$(\sigma - \sigma^*)\dot{\varepsilon}^p + (\tau - \tau^*)\dot{\gamma}^p >= 0$$

for any σ^* and τ^* such that $f(\sigma^*, \tau^*, \kappa) \leq 0$. Thus the plastic elongation rate $\dot{\varepsilon}^p$ and shearing rate $\dot{\gamma}^p$ are the strain rates conjugate to the stresses σ and τ, respectively, and are given by

$$\dot{\varepsilon}^p = \dot{\lambda}\frac{\sigma}{\bar{\sigma}}, \qquad \dot{\gamma}^p = \dot{\lambda}\frac{\alpha\tau}{\bar{\sigma}}.$$

If the hardening variable κ is defined by

$$\dot{\kappa} = \sqrt{(\dot{\varepsilon}^p)^2 + \frac{1}{\alpha}(\dot{\gamma}^p)^2},$$

then $\dot{\kappa} = \dot{\lambda}$, and therefore the hardening modulus H, as defined by Equation (3.1.1), is just $\sigma'_Y(\kappa)$. In view of Equation (4.1.1), H is a function of $\bar{\sigma}$ in any plastic loading process. In particular, $H(\bar{\sigma})$ is the slope of the σ-ε^p curve obtained from a simple tension or compression test at a point where $|\sigma| = \bar{\sigma}$; it is also $\frac{1}{\alpha}$ times the slope of the τ-γ^p curve obtained from the simple torsion test, problem (b), at a point where $|\tau| = \bar{\sigma}/\sqrt{\alpha}$.

Equations (3.2.1)–(3.2.2) for this problem thus reduce to

$$\begin{aligned} \dot{\varepsilon} &= \frac{\dot{\sigma}}{E} + \frac{\sigma}{H\bar{\sigma}} < \dot{\bar{\sigma}} >, \\ \dot{\gamma} &= \frac{\dot{\tau}}{G} + \frac{\alpha\tau}{H\bar{\sigma}} < \dot{\bar{\sigma}} > . \end{aligned} \tag{4.1.2}$$

When a stress path $\sigma(t)$, $\tau(t)$ is prescribed, Equations (4.1.2) can be used to find the corresponding strain path $\varepsilon(t)$, $\gamma(t)$ by integration, in general numerical. An explicit form may be obtained when the stress path is radial, that is, when $\sigma = a\bar{\sigma}$ and $\tau = b\bar{\sigma}$, a and b being constants constrained by the requirement $a^2 + \alpha b^2 = 1$. With $\bar{\varepsilon}(\bar{\sigma})$ defined by

$$\bar{\varepsilon}(\bar{\sigma}) = \int \frac{d\bar{\sigma}}{H(\bar{\sigma})}$$

(note that this describes the σ-ε^p curve in simple tension and compression), the strains are given by

$$\varepsilon = \frac{a}{E}\bar{\sigma} + a\bar{\varepsilon}(\bar{\sigma}),$$
$$\gamma = \frac{b}{G}\bar{\sigma} + \alpha b\bar{\varepsilon}(\bar{\sigma}).$$

When the strain path is given, Equations (4.1.2) form a pair of coupled nonlinear differential equations for the stresses; they must in general be solved numerically. If, however, the work-hardening is very slight (this includes the limit of perfect plasticity), then σ_Y may be regarded, at least over a certain range of strain, as a constant, and σ and τ are no longer independent but are coupled through the yield criterion. The quantity $\dot{\bar{\sigma}}/H\bar{\sigma}$ may be eliminated between the two equations, resulting in the single equation

$$\frac{\alpha\tau\dot{\sigma}}{E} - \frac{\sigma\dot{\tau}}{G} = \alpha\tau\dot{\varepsilon} - \sigma\dot{\gamma}.$$

Suppose, for example, that the tube is stretched until it just yields ($\sigma = \sigma_Y$, $\tau = 0$); thereafter, it is twisted, with further axial deformation prevented, that is, $\dot{\varepsilon} = 0$. After σ is eliminated, the equation becomes

$$\frac{d\gamma}{d\tau} = \frac{1}{G} + \frac{\alpha\tau^2}{E(\tau_Y^2 - \tau^2)},$$

and may be integrated to yield

$$\gamma = \frac{1}{G}\left\{\tau + \frac{\alpha}{2(1+\nu)}\left[\frac{\tau_Y}{2}\ln\frac{\tau_Y + \tau}{\tau_Y - \tau} - \tau\right]\right\}.$$

This result can also be in terms of the ratio of plastic to elastic shear strain:

$$\frac{\gamma^p}{\gamma^e} = \frac{\alpha}{2(1+\nu)}\left[\frac{\tau_Y}{2\tau}\ln\frac{1+\tau/\tau_Y}{1-\tau/\tau_Y} - 1\right].$$

It is clear that the shear strain, and therefore the angle of twist, grows indefinitely as the torque approaches its ultimate value $2\pi c^2 h\tau_Y$. Unlike the simple problems (a) and (b), however, in this case the growth is asymptotic. Indeed, so long as the shear stress is less than, say, 99its yield value, the plastic strain remains of the same order of magnitude as the elastic strain. The tube can then be said to be in a state of *contained plastic deformation.* In the present chapter we concentrate on problems of this type; problems in plastic flow and collapse are considered in Chapter 5.

As the shear stress approaches the yield value, the axial stress approaches zero, and a state of virtually pure torsion is attained. The constant axial strain σ_Y/E changes, in the process, from purely elastic to plastic. Since these changes take place over a rather narrow range of strain, the validity of the assumption of negligible work-hardening, which was necessary to obtain the solution, may be limited to this range. It therefore follows that the τ-γ curve for this problem will asymptotically attain the curve for simple shear.

If desired, the remaining strain-rate components $\dot{\varepsilon}_r$ and $\dot{\varepsilon}_\theta$ may be found through the flow rule. Since $\sigma_r = \sigma_\theta = 0$, and therefore $s_r = s_\theta = -\frac{1}{2}s_z$, the Mises flow rule ($\alpha = 3$) implies

$$\dot{\varepsilon}_r = \dot{\varepsilon}_\theta = -\tfrac{1}{2}\dot{\varepsilon}_z.$$

Since, however, $\sigma_r = 0$ is the intermediate principal stress, the Tresca flow rule ($\alpha = 4$) implies

$$\dot{\varepsilon}_r = 0, \quad \dot{\varepsilon}_\theta = -\dot{\varepsilon}_z.$$

A completely analogous result is produced if the twist is held constant while extension proceeds. For this and related problems, see Hill [1950], pp. 71–75, or Chakrabarty [1987], Section 3.3(i). Hill's results relating to large deformation must be taken with caution, because, as was pointed out in 2.1.1, the use of logarithmic strains is inappropriate when the principal strain axes rotate. Moreover, the results are based on the use of nonobjective stress rates; this matter is discussed further in Chapter 8.

4.1.3. Thin-Walled Cylinder Under Pressure and Axial Force

In a cylindrical shell under a uniform internal pressure p and a tensile axial force P, the principal stresses are $\sigma_1 = pc/h$ (circumferential) and $\sigma_2 = \frac{1}{2}\sigma_1 + \sigma_a$ (axial), where $\sigma_a = P/2\pi ch$. If the material is governed by the Mises criterion with isotropic hardening, this criterion can once more be written in the form (4.1.1), but this time with $\bar{\sigma} = \sqrt{\sigma_a^2 + \frac{3}{4}\sigma_1^2}$.

It is convenient to use σ_a and σ_1 as the variables defining the stress state, since the yield function is defined in terms of these stresses. The associated flow rule will be assumed, and its form may again be derived from the postulate of maximum plastic dissipation. Let ε_z denote the axial and ε_θ the circumferential strain. Since the plastic dissipation takes the form

$$D_p = \sigma_1\dot{\varepsilon}_1^p + \sigma_2\dot{\varepsilon}_2^p = \sigma_1(\dot{\varepsilon}_\theta^p + \tfrac{1}{2}\dot{\varepsilon}_z^p) + \sigma_a\dot{\varepsilon}_z^p,$$

it follows that

$$\dot{\varepsilon}_z^p = \frac{\sigma_a}{H\bar{\sigma}} < \dot{\bar{\sigma}} >,$$

where H is again the uniaxial hardening modulus, and

$$\dot{\varepsilon}_\theta^p + \frac{1}{2}\dot{\varepsilon}_z^p = \frac{3}{4}\frac{\sigma_1}{H\bar{\sigma}} < \dot{\bar{\sigma}} >,$$

from which

$$\dot{\varepsilon}_\theta^p = \frac{\frac{3}{4}\sigma_1 - \frac{1}{2}\sigma_a}{H\bar{\sigma}} < \dot{\bar{\sigma}} > .$$

The equations for the total strain rates are therefore

$$\dot{\varepsilon}_z = \frac{\dot{\sigma}_a + (\frac{1}{2} - \nu)\dot{\sigma}_1}{E} + \frac{\sigma_a}{H\bar{\sigma}} < \dot{\bar{\sigma}} >,$$
$$\dot{\varepsilon}_\theta = \frac{(1 - \frac{1}{2}\nu)\dot{\sigma}_1 - \nu\dot{\sigma}_a}{E} + \frac{\frac{3}{4}\sigma_1 - \frac{1}{2}\sigma_a}{H\bar{\sigma}} < \dot{\bar{\sigma}} > . \tag{4.1.3}$$

The calculations may now proceed as in the previous example. However, since in the present case the principal axes of strain do not rotate, there is no difficulty in extending the results into the large-deformation range by interpreting the strains as logarithmic. The strain rates are given by

$$\dot{\varepsilon}_z = \frac{\dot{l}}{l}, \quad \dot{\varepsilon}_\theta = \frac{\dot{c}}{c}.$$

The volume enclosed by the shell is $\pi c^2 l$ (if h is neglected next to c), and its change is given by

$$\frac{\dot{V}}{V} = \frac{\dot{l}}{l} + \frac{2\dot{c}}{c} = \dot{\varepsilon}_z + 2\dot{\varepsilon}_\theta = \frac{1}{E}\left[(1 - 2\nu)\dot{\sigma}_a + \left(\frac{5}{2} - 2\nu\right)\dot{\sigma}_1\right] + \frac{3\sigma_1\dot{\bar{\sigma}}}{2H\bar{\sigma}}.$$

When the deformations are sufficiently great, the elastic strains may be neglected. The shell material may then be regarded as incompressible, so that the solid volume $2\pi chl$ remains constant, and the thickness h may be determined from this condition. Consider, for example, a closed cylindrical tank containing gas under pressure and undergoing an axial extension at a constant absolute temperature T. It follows from the ideal-gas law,

$$pV = nRT,$$

where n is the number of moles of gas and R is the universal gas constant (Avogadro's number × Boltzmann's constant), that plc^2 is constant. But

$$plc^2 = \frac{pc}{h} lch = \sigma_1 lch,$$

and therefore σ_1 is also a constant, given by the initial value of pc/h. We may now write

$$\frac{\dot{V}}{V} = \frac{1 - 2\nu}{E}\dot{\sigma}_a + \frac{3\sigma_1\dot{\bar{\sigma}}}{2H\bar{\sigma}}.$$

We assume, first, linear work-hardening (H = constant) after an initial yield stress σ_E. We further suppose that the gas pressure alone is not sufficient to cause yielding, that is, $\sigma_1 < 2\sigma_E/\sqrt{3}$. We may then integrate the preceding equation to obtain

$$\ln\frac{V}{V_0} = \frac{1 - 2\nu}{E}\sigma_a + \frac{3\sigma_1}{2H} < \ln\sqrt{\frac{\sigma_a^2 + \frac{3}{4}\sigma_1^2}{\sigma_E}} >.$$

If, on the other hand, the work-hardening is described by the Ramberg–Osgood equation (2.1.2), then

$$\ln\frac{V}{V_0} = \frac{1-2\nu}{E}\sigma_a + \frac{3m\sigma_1}{2(m-1)E}\left[\left(\frac{\sigma_a^2+\frac{3}{4}\sigma_1^{2\,2}}{\sigma \quad R}\right)^{(m-1)/2} - \left(\frac{\sqrt{3}\sigma_1}{2\sigma_R}\right)^m - 1\right].$$

Note that the elastic part of the volumetric strain has the same sign as σ_a, but the plastic part is never negative, whether the tank undergoes extension or compression. Note also that the expression for the volume becomes infinite in the limit as $H- > 0$ for the linear work-hardening case and as $m- > inf$ for the Ramberg–Osgood case; both of these limits represent perfect plasticity. In this example, then, contained plastic deformation cannot occur if the material is perfectly plastic. The difference between the present problem and the preceding one does not lie in the fact that one involves torsion and extension, and the other extension and gas pressure. Rather, the crucial difference is that the twisted tube was subject to the *kinematic constraint* of no axial extension; it is this constraint that prevented unrestricted plastic deformation immediately upon yielding. Had the axial force, rather than the length, been held constant, contained plastic deformation would not have taken place.

In the case of linear work-hardening an explicit expression can also be found for the axial extension and the radial expansion. The former is governed by

$$\frac{\dot{l}}{l} = \frac{\dot{\sigma}_a}{E} + \frac{\sigma_a^2\dot{\sigma}_a}{(\sigma_a^2+\frac{3}{4}\sigma_1^2)H}, \tag{4.1.4}$$

which may be integrated to give

$$\ln\frac{l}{l_0} = \frac{\sigma_a}{E} + \frac{1}{H}<\sigma_a - \sqrt{\sigma_Y^2 - \tfrac{3}{4}\sigma_1^2} - \sqrt{\frac{3}{2}}\sigma_1\left[\tan^{-1}\frac{2\sigma_a}{\sqrt{3}\sigma_1} - \cos^{-1}\frac{\sqrt{3}\sigma_1}{2\sigma_Y}\right]>.$$

The axial force at a given value of l may then be found from

$$P = 2\pi c_0 h_0 \frac{l_0}{l}\sigma_a. \tag{4.1.5}$$

A similar integration for the radius change leads to

$$\begin{aligned}\ln\frac{c}{c_0} = &-\frac{\nu\sigma_a}{E} + \frac{3\sigma_1}{4H}<\ln\sqrt{\frac{\sigma_a^2+\frac{3}{4}\sigma_1^2}{\sigma_E}}> \\ &-\frac{1}{2H}<\sigma_a - \sqrt{\sigma_Y^2 - \tfrac{3}{4}\sigma_1^2} - \sqrt{\frac{3}{2}}\sigma_1\left[\tan^{-1}\frac{2\sigma_a}{\sqrt{3}\,\sigma_1} - \cos^{-1}\frac{\sqrt{3}\,\sigma_1}{2\sigma_Y}\right]>.\end{aligned}$$

The force P given by Equation (4.1.5) becomes stationary ($\dot{P} = 0$), indicating an *instability*, when

$$\frac{\dot{l}}{l} = \frac{\dot{\sigma}_a}{\sigma_a}.$$

Equating the right-hand side of this equation to that of (4.1.4), and assuming that $H \ll E$ so that the elastic strain rate can be neglected, we find a critical value of the hardening modulus above which instability will not occur, namely

$$H_c r = \frac{\sigma_a^3}{\bar{\sigma}^2}.$$

Note that this is positive only when $\sigma_a > 0$, that is, when the axial force is tensile; when this force is compressive, any nonnegative value of H is sufficient for stability. Note further that, in the absence of pressure, $H_c r = \sigma_a$; this is just the necking condition discussed in 2.1.2. The pressure tends to lower the critical value of the hardening modulus, so that it is a stabilizing factor.

Different values of $H_c r$ are found for problems in which the ratio σ_a/σ_1 or the ratio P/p is held fixed (see, e.g., Chakrabarty [1987], Section 3.3(iii)].

4.1.4. Statically Indeterminate Problems

In contrast to the statically determinate problems just discussed, in statically *indeterminate* problems contained plastic deformation occurs even without kinematic constraints. Consider the simple pin-connected three-bar assemblage shown in Figure 4.1.1(a). Since a pin joint cannot transfer moments, the bars carry axial forces only and each bar can be assumed to be in a state of uniaxial tension or compression, the stress being $\sigma = P/A$, where P is the axial force in the bar. If all three bars have the same cross-sectional area A and are made of the same elastic–perfectly plastic material, with Young's modulus E and yield stress σ_Y, the elastic solution for the bar forces P_0 and P_1 ($= P_{1'}$) is

$$P_0 = \frac{F}{1 + 2\cos^3\theta}, \quad P_1 = \frac{F\cos^2\theta}{1 + 2\cos^3\theta},$$

and is valid while $P_0 \le P_U$ (where $P_U = \sigma_Y A$), or $F \le (1+2\cos^3\theta)P_U \stackrel{\text{def}}{=} F_E$; here F_E denotes, as usual, the elastic-limit value of the load F. When $F = F_E$, the downward displacement Δ of the joint equals $\sigma_Y L/E \stackrel{\text{def}}{=} \Delta_E$. The range $0 < F < F_E$, or $0 < \Delta < \Delta_E$, represents the *elastic range*.

When $F > F_E$, P_0 remains constant at P_U, and P_1 may be determined by statics alone to be $\frac{1}{2}(F - P_U)\sec\theta$, provided that this value does not reach P_U. The latter condition occurs when $F = (1 + 2\cos\theta)P_U \stackrel{\text{def}}{=} F_U$. In the range $F_E < F < F_U$, the displacement Δ is determined by the deformation of bars 1 and 1′ alone, a deformation that remains purely elastic. It is this elastic deformation that also determines the plastic deformation of bar 0. The range in question is thus the range of contained plastic deformation.

Contained plastic deformation ends when $\Delta = \Delta_E \sec^2\theta$. Beyond this value, the displacement grows indefinitely while F remains at the value F_U,

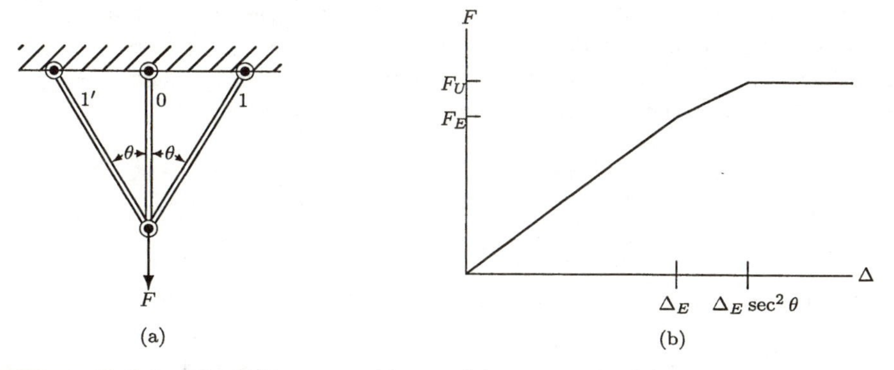

Figure 4.1.1. Three-bar assemblage: (a) structure; (b) load-deflection diagram.

precisely the definition of plastic collapse given in Section 3.4, and the load F_U is the ultimate load of the structure. The complete F-Δ diagram is shown in Figure 4.1.1(b).

It is not difficult to see that if the number of bars in the assemblage is multiplied, as in Figure 4.1.2(a), the load-deflection diagram has the form shown in Figure 4.1.2(b). The range of contained plastic deformation is now represented by numerous segments, whose initial and final points must be calculated step by step.

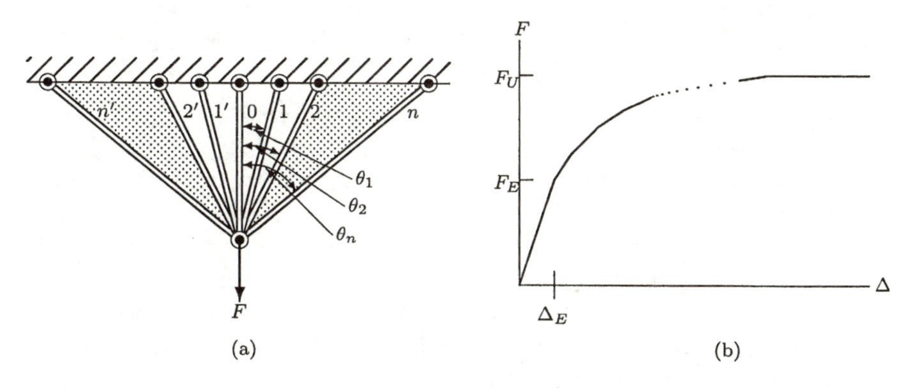

Figure 4.1.2. Multi-bar assemblage: (a) structure; (b) load-deflection diagram.

It can be shown that in the elastic range, the bar forces (if all bar areas are equal) are given by

$$P_0 = \frac{F}{1 + 2\sum_{k=1}^{n} \cos^3 \theta_k}, \quad P_k = \frac{F \cos^2 \theta_k}{1 + 2\sum_{k=1}^{n} \cos^3 \theta_k}. \tag{4.1.6}$$

Note that the bar forces decrease with distance from the central bar. It can thus be assumed that the order of yielding of the bars is outward from bar 0. If bar 0 and $m-1$ bars on either side of it have yielded (so that $P_k = P_U$, $k = 0, ..., m-1$), then it can be shown that for $k = m, ..., n$,

$$P_k = \frac{\cos^2 \theta_k}{2 \sum_{l=m}^{n} \cos^3 \theta_l} \left[F - P_U \left(1 + 2 \sum_{l=1}^{m-1} \cos \theta_l \right) \right], \tag{4.1.7}$$

provided that $P_m \le P_U$, that is,

$$F \le P_U \left(1 + 2 \sum_{k=1}^{m-1} \cos \theta_k + \frac{2}{\cos^2 \theta_m} \sum_{k=m}^{n} \cos^3 \theta_k \right).$$

When all but the last pair of bars have yielded, that is, when $m = n$, the upper bound on F is just the ultimate load F_U, given by

$$F_U = P_U \left(1 + 2 \sum_{k=1}^{n} \cos \theta_k \right).$$

Of course, as we learned in 3.4.1, this value could easily have been determined without going through the step-by-step process.

If, on the other hand, the structure were reduced to bar 0 only, then it would be statically determinate and naturally there would be no contained plastic deformation; the elastic range would be followed immediately by unrestricted plastic flow. Contained plastic deformation, therefore, is due either to constraints or to the introduction of *redundant* elements that make the structure statically indeterminate.

A thin-walled tube under torsion or pressure forms, as we have seen, a statically determinate problem. A thick-walled tube may be thought of as the limit of an assemblage of a large number of concentric thin-walled tubes, all but one of which are redundant. The range of contained plastic deformation is therefore characterized by a torque-twist or pressure-expansion diagram which is the limit of that of Figure 4.1.2(b) as the segments become infinitesimal — that is, a smooth curve in which the ultimate load may be attained asymptotically, as in the problem of 4.1.2 [see Figure 4.1.3(a)] or at a finite displacement, as in Figure 4.1.3(b).

Elastic-Plastic Boundary

At any stage of contained plastic deformation in the problem of Figure 4.1.2, the spaces between the plastic and elastic groups of bars may be said to form the *elastic-plastic boundary*. As the load is increased and an additional pair of bars yields, the boundary moves to the next pair of spaces.

In a relatively simple problem such as this one, the elastic-plastic boundary is defined by a single parameter, for example the integer m. Problems

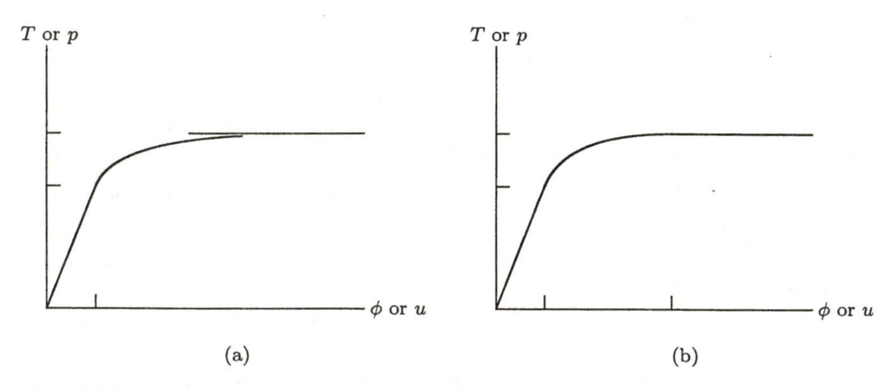

Figure 4.1.3. General load-displacement diagrams: (a) ultimate load reached asymptotically; (b) ultimate load reached at a finite displacement.

with radial symmetry, as shown in the next two sections, are similarly simple: the elastic and plastic zones can be separated only by a spherical or a cylindrical surface, and the boundary is determined by its radius. It is the simplicity of the elastic-plastic boundary that makes problems of this nature tractable analytically.

In general, the elastic-plastic boundary is one of the unknown quantities of a problem, and when it is not defined by a small number of parameters then it can only be determined by trial and error, step by step, in the course of an incremental solution. Typically, the given loading is scaled down until it is small enough so that the whole body is elastic. Since the elastic problem is linear, the stresses are directly proportional to the load (in the case of one-parameter loading). With the help of the elastic solution it is therefore possible to find the value of the load at which the body first yields, such as F_E in the preceding example, as well as the point or points where yielding first occurs. The elastic-plastic boundary originates at these points.

Following the initiation of yielding, the loading must be increased by small increments, and a trial solution is found for a statically admissible stress-increment field $d\boldsymbol{\sigma}$. The resulting stress $fat\sigma$ must then be tested to determine whether it is plastically admissible for the current yield surface. If it is not, then the trial solution must be corrected so as to bring all stress values inside or on the yield surface; the procedure is usually an iterative one. The points at which $\boldsymbol{\sigma}$ is inside the yield surface constitute the current elastic domain, while those where $\boldsymbol{\sigma}$ is on the yield surface form the plastic domain. The elastic-plastic boundary is thus a part of the solution at each step. The solution must usually be carried out numerically, and methods are discussed in Section 4.5. Sections 4.2, 4.3, and 4.4 deal with problems that can be treated analytically.

Another feature of the problem of Figure 4.1.2, as well as of thick-walled tubes under pressure or torsion, is that when the ultimate load is reached, the entire body is plastic; in other words, the elastic-plastic boundary passes out of the body. In the case of our bar assemblage, this is a result of the symmetry of the problem, with respect to both geometry and loading. In the absence of such symmetry it is not necessary for all the bars of an assemblage of this type to be plastic at collapse: one bar may remain elastic, since it can rotate rigidly as the other bars deform. There are other problems as well in which an elastic-plastic boundary is present at collapse. An example is the cantilever beam studied in Section 4.4, and other examples are given in Section 5.1.

Exercises: Section 4.1

1. Solve Equations (4.1.2) for the stress-strain relation in tension when the tube is twisted until it just yields, and thereafter it is stretched with further twisting prevented. Assume negligible work-hardening.

2. If the load on the assemblage of Figure 4.1.1(a) is not vertical but has both a vertical component V directed downward and a horizontal load H directed to the right, the problem is no longer symmetric and the three bar forces must be assumed to be distinct. Discuss the behavior of the assemblage when (a) $V = cH$, with c constant and H increasing, and (b) V is held constant at a value at which the assemblage is elastic, and then H is gradually increased from zero.

3. Suppose the assemblage of Figure 4.1.1(a) to carry no load but to be subject to a monotonically increasing temperature change, the temperature being the same in all bars. If the Young's modulus E, the thermal expansion coefficient α, and the yield stress σ_Y (equal in tension and compression) are independent of temperature, find the temperature change ΔT at which yielding begins.

4. Do Exercise 3 when σ_Y is a linearly decreasing function of temperature; that is, $\sigma_Y = \sigma_Y 0 - a(T - T_0)$, where a is a constant and T_0 is the initial temperature.

5. Show that Equations (4.1.6) describe the elastic state of the assemblage of Figure 4.1.2(a).

6. Show that Equations (4.1.7) give the forces in the bars of the assemblage of Figure 4.1.2(a) that have remained elastic when the central bar and $m - 1$ bars on either side of it have yielded. Find the value attained by the load F so that bars m are just about to yield.

7. When the load F has attained the value found in Exercise 6 and is then removed, find the residual bar forces. Which of the bars go into compression?

Section 4.2 Elastic-Plastic Torsion

4.2.1. The Torsion Problem

The twisting of prismatic bars under the action of equal and opposite torques applied at the ends is one of the classical topics of solid mechanics. For elastic bars of circular cross-section, the relation between torque and twist was first derived by Coulomb in 1784. A general theory for bars of arbitrary cross-section was published by Saint-Venant in 1855; this theory makes it possible to predict where the stress is maximum, and therefore where, and at what values of the torque and twist, yielding will first occur.

Saint-Venant's theory is based on the **semi-inverse** method originated by him. An **inverse** method is one in which a solution (a stress field or a displacement field) is assumed, while the boundary conditions that it satisfies, and in particular the resulting surface tractions, are determined afterwards. In a semi-inverse method, a part of the solution is assumed, and the rest is left to be determined by solving appropriate differential equations. In the torsion problem, assumptions can be made on both the stress and displacement fields. It is important to note that the value of the torque is not a part of the boundary data; instead, it is the angle of twist that is prescribed, and the torque results as a *reaction* to the imposed twist, not a load.

The Boundary-Value Problem

We consider a prismatic bar whose longitudinal axis (e.g., the centroidal axis) is the x_3-axis and whose cross-section (assumed simply connected) is defined by a closed curve C in the x_1x_2-plane (see Figure 4.2.1). The bar is defined as being in a state of torsion or twist if one end section, viewed in its plane (i.e., with any warping ignored), rotates rigidly with respect to the other. The end sections are assumed to be free to warp, so that there is no normal traction on them. The lateral surface is traction-free. Consequently, if the rotation of the section at $x_3 = 0$ is defined as zero and if the relative angle of rotation between the two end sections is θL, then the boundary conditions are

$$\begin{aligned}
&\text{at } x_3 = 0: && u_1 = u_2 = 0,\ \sigma_{33} = 0;\\
&\text{at } x_3 = L: && u_1 = -\theta L x_2,\ u_2 = \theta L x_1,\ \sigma_{33} = 0;\\
&\text{on } C,\ 0 < x_3 < L: && T_i = \sigma_{ij}n_j = \sigma_{i\alpha}n_\alpha = 0, \quad \text{where } n_1 = \frac{dx_2}{ds},\ n_2 = -\frac{dx_1}{ds}.
\end{aligned}$$

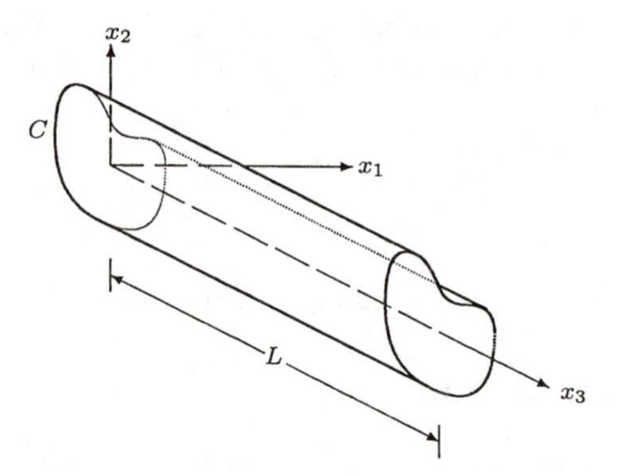

Figure 4.2.1. Prismatic shaft: geometry.

Stress Assumptions

If we assume that $\sigma_{11} = \sigma_{22} = \sigma_{33} = \sigma_{12} = 0$, then the equilibrium equations (with zero body force) reduce to

$$\tau_{\alpha,3} = 0 \quad (\alpha = 1,\, 2), \quad \tau_{\alpha,\alpha} = 0,$$

where $\tau_\alpha = \sigma_{\alpha 3}$. The first two equations imply that τ_1 and τ_2 are functions of x_1 and x_2 only, so that the problem is mathematically (though not physically) two-dimensional. The third equation implies that there exists a function $\phi(x_1,\, x_2)$ such that $\tau_1 = \phi_{,2}$ and $\tau_2 = -\phi_{,1}$. To satisfy the traction boundary conditions, we need

$$\tau_\alpha n_\alpha = \phi_{,\alpha} \left.\frac{dx_\alpha}{ds}\right|_C = \left.\frac{d\phi}{ds}\right|_C = 0,$$

so that ϕ must be constant on C. With no loss of generality we may let the constant be 0. If the section is multiply connected, with interior contours C_1, C_2,..., then we can set $\phi = 0$ on the exterior boundary curve C, but on each of the interior curves we can only set it equal to some unknown constant.

The torque T is given by

$$\begin{aligned} T &= \int_A (x_1\tau_2 - x_2\tau_1)\, dA = -\int_A x_\alpha \phi_{,\alpha}\, dA \\ &= -\int_A [(x_\alpha\phi)_{,\alpha} - 2\phi]\, dA = -\oint_C n_\alpha x_\alpha \phi\, dA + 2\int_A \phi\, dA \\ &= 2\int_A \phi\, dA, \end{aligned}$$

since ϕ vanishes on C. We may think of the surface formed by the stress function $\phi(x_1,\, x_2)$ as a "tent" pitched over the base formed by the boundary curve C; then the torque is twice the volume enclosed by the tent.

The contour lines of the tent (curves along which ϕ has a constant value) are also significant: if at a given point of the cross-section, a local Cartesian coordinate system is established so that the x_1-axis is parallel to the contour, then locally $\tau_1 = \phi_{,2}$ and $\tau_2 = 0$. Thus every point of the shaft is stresswise in a state of simple shear, and the shear stress is everywhere directed along the contour, while its magnitude is just the slope of the tent surface. If the shaft is made of an elastic-plastic material, then the yield criterion may be expressed in terms of the stress function ϕ as $|\nabla\phi| = k$, where $k = \tau_Y$ is the yield stress in shear.

Displacement Assumptions

If the material is isotropic, then regardless of whether it is elastic, plastic, or whatever, the only nonvanishing strain components (under the assumption of infinitesimal deformation) consistent with the stress assumptions can be the shear strains $\gamma_1 \stackrel{\text{def}}{=} 2\varepsilon_{13}$ and $\gamma_2 \stackrel{\text{def}}{=} 2\varepsilon_{23}$. A displacement field consistent with this condition and with the displacement boundary conditions is

$$u_1 = -\theta x_2 x_3, \quad u_2 = \theta x_1 x_3, \quad u_3 = \theta\psi(x_1, x_2),$$

where ψ is known as the *warping function*; the factor θ is placed in front of it for convenience. Note that θ, which is simply called the *twist*, is the angle of twist per unit length, assumed uniform. The dependence of u_3 on θ is not linear in all situations, but it is so in many of the ones that we are going to study. The strain components are

$$\gamma_1 = \theta(\psi_{,1} - x_2), \quad \gamma_2 = \theta(\psi_{,2} + x_1).$$

4.2.2. Elastic Torsion

The isotropic linear elastic stress-strain relations $\tau_\alpha = G\gamma_\alpha$ translate to

$$\begin{aligned} \tau_1 &= \phi_{,2} = G\theta(\psi_{,1} - x_2), \\ \tau_2 &= -\phi_{,1} = G\theta(\psi_{,2} + x_1). \end{aligned} \tag{4.2.1}$$

These are two equations in the two unknown functions ϕ and ψ. We may eliminate either one and be left with one differential equation in one unknown function.

The Stress Method

If we eliminate ψ, so that ϕ is the unknown of the problem, we obtain

$$\nabla^2\phi = -2G\theta, \tag{4.2.2}$$

which, together with the boundary condition $\phi = 0$ on C, constitutes a standard mathematical problem. The conventional method of solution consists of, first, finding a particular solution ϕ_1 which satisfies the differential equation but not necessarily the boundary conditions, and next, assuming that

$\phi = \phi_1 + \phi_2$, so that ϕ_2 is the solution of the differential equation $\nabla^2\phi_2 = 0$ (the **Laplace equation**) with the boundary condition $\phi_2 = -\phi_1$ on C — a classical problem known as the **Dirichlet problem**.

Equation (4.2.2) together with the boundary condition is mathematically identical with the problem of the deflection of a membrane, tightly stretched in a frame occupying the contour C by an isotropic tension H and subject to a transverse pressure p; if this deflection is denoted $w(x_1, x_2)$, then the equation governing it is

$$\nabla^2 w = -\frac{p}{H},$$

and the correspondence, first discussed by Prandtl [1903], is known as the *membrane* or *soap-film analogy*. With appropriate numerical scaling, it is possible to use the results of experiments on soap films in order to calculate the torsional properties of shafts (see, e.g., Timoshenko and Goodier [1970], pp. 324–325]. Even without actual experiments, the analogy is useful in helping to visualize the nature of the solution. In particular, it is often intuitively obvious where the slope of the membrane deflection, and hence the stress, is maximum. Typically, this maximum occurs at the points on the exterior boundary nearest the axis.

For some cross-sections, such as an ellipse and an equilateral triangle, a complete solution ϕ can be found in closed form. Using the conventional notation x, y in place of x_1, x_2, we may write the solution for an ellipse described by $(x/a)^2 + (y/b)^2 = 1$ as

$$\phi = G\theta\frac{a^2b^2}{a^2+b^2}\left(1 - \frac{x^2}{a^2} - \frac{y^2}{b^2}\right).$$

The torque on a shaft made of an isotropic, linearly elastic material with shear modulus G always has the form $T = GJ\theta$, where J has dimensions of length^4 and depends on the geometry of the cross-section alone; the product GJ is known as the *torsional stiffness*.

For the ellipse, we have $J = \pi a^3b^3/(a^2+b^2)$. The maximum stress is

$$\tau_{\max} = \frac{2T}{\pi a^2 b^2}\max(a, b).$$

The circle is, of course, a special case of the ellipse with $a = b$.

For the equilateral triangle, a stress function having the form

$$\phi(x, y) = Cf_1(x, y)f_2(x, y)f_3(x, y),$$

where

$$f_1(x, y) = 0, \quad f_2(x, y) = 0, \quad f_3(x, y) = 0$$

are the equations of the three lines forming the triangle, obviously satisfies the boundary condition $\phi = 0$, and can easily be shown to satisfy the differential equation (4.2.2) with a suitable value of the constant C. If c denotes the side of the triangle, then the torsional stiffness is $\sqrt{3}Gc^4/80$, and the maximum shear stress is $\tau_{\max} = 20T/c^3$.

Rectangular Shaft

A closed-form solution does not exist for any other polygonal shaft. A Fourier series solution can be obtained for a rectangular shaft. Consider the rectangle bounded by $x = +-a/2$, $y = +-b/2$, with $b >= a$. If we assume that

$$\phi_1 = G\theta\left(\frac{a^2}{4} - x^2\right),$$

then this satisfies the boundary condition on the longer sides but not on the shorter sides; it is reasonable to expect that the larger b/a, the closer ϕ_1 will be to ϕ. The "correction" ϕ_2 may be assumed in the form of a Fourier series in x that is even in x and vanishes on $x = +-a/2$, and whose coefficients depend on y:

$$\phi_2 = G\theta \sum_{m=0}^{\infty} f_m(y) \cos\frac{(2m+1)\pi x}{a}.$$

Now

$$\nabla^2\phi_2 = G\theta \sum_{m=0}^{\infty} \left[f_m''(y) - \frac{(2m+1)^2\pi^2}{a^2} f_m(y)\right] \cos\frac{(2m+1)\pi x}{a},$$

and if this is to vanish for all (x, y), then $f_m'' = [(2m+1)\pi/a]^2 f_m$; the solution that is even in y is $f_m(y) = A_m \cosh(2m+1)\pi y/a$, where the coefficients A_m must be such that $\phi_2 = -\phi_1$ on $y = +-b/2$. To satisfy this condition we need the appropriate Fourier series expansion for ϕ_1, that is, we need

$$\frac{a^2}{4} - x^2 = \sum_{m=0}^{\infty} B_m \cos\frac{(2m+1)\pi x}{a};$$

then the solution is complete if $A_m \cosh\left[(2m+1)\pi b/2a\right] = -B_m$. In accordance with the theory of Fourier series,

$$B_m = \frac{4}{a}\int_0^{a/2} \left(\frac{a^2}{4} - x^2\right) \cos\frac{(2m+1)\pi x}{a} dx = \frac{8(-1)^m a^2}{(2m+1)^3\pi^3}.$$

Hence, finally,

$$\phi_2 = -G\theta\frac{8a^2}{\pi^3} \sum_{m=0}^{\infty} \frac{(-1)^m}{(2m+1)^3} \frac{\cosh(2m+1)\pi y/a}{\cosh(2m+1)\pi b/2a} \cos\frac{(2m+1)\pi x}{a}.$$

Integration leads to

$$J = a^3 b\left[\frac{1}{3} - \frac{64a}{\pi^5 b}\sum_{m=0}^{\infty}\frac{1}{(2m+1)^5}\tanh\frac{(2m+1)\pi b}{2a}\right].$$

Since the hyperbolic tangent function approaches unity as its argument increases, the series converges very rapidly.

The maximum shear stress occurs at $(\pm\frac{1}{2}a,\ 0)$ and is given by

$$\tau_{\max} = Ga\theta\left[1 - \frac{8}{\pi^2}\sum_{m=0}^{\infty}\frac{1}{(2m+1)^2}\operatorname{sech}\frac{(2m+1)\pi b}{2a}\right].$$

It is seen from the formula for J that as $b/a \to \infty$, $J \to a^3b/3$, which is the result based on ϕ_1 alone. For such infinitely narrow rectangles we also obtain $\tau_{\max} = G\theta a = Ta/J$. These results are often applied to the torsion of shafts of open thin-walled section made up of several rectangles (possibly even curved) of dimensions a_i, b_i $(i = 1, \ldots, n)$ with $b_i \gg a_i$. For such sections we have the approximations

$$J \doteq \frac{1}{3}\sum_{i=1}^{n} a_i^3 b_i, \quad \tau_{\max} \doteq \frac{Ta_{\max}}{J}.$$

The Displacement Method

If we choose the warping function ψ rather than the stress function ϕ as the unknown function of the problem, the governing differential equation is the Laplace equation

$$\nabla^2\psi = 0.$$

The traction boundary conditions in terms of ψ are

$$n_1(\psi,_1 - y) + n_2(\psi,_2 + x) = \frac{\partial\psi}{\partial\nu} - y\frac{dy}{ds} - x\frac{dx}{ds} = 0,$$

or

$$\frac{\partial\psi}{\partial\nu} = \frac{1}{2}\frac{d}{ds}r^2 \quad (r^2 = x^2 + y^2).$$

This result makes it clear that the warping vanishes if and only if the cross-section is circular. For an ellipse, we find that $\psi = [(b^2 - a^2)/(b^2 + a^2)]xy$ is an exact solution of the problem, but in general this problem is more difficult to work with than the one with ϕ as the unknown. Moreover, as is shown next, it is only with ϕ that a boundary-value problem for elastic-plastic shafts can be formulated.

4.2.3. Plastic Torsion

It is clear that in the elastic range the stress function is of the form $\phi(x_1,\ x_2) = G\theta \times f(x_1,\ x_2)$, where $f(x_1,\ x_2)$ is determined by the cross-section geometry alone. As the twist θ is increased, the stresses grow proportionately, until at one or more points of the cross-section the yield criterion is met; these are the points — necessarily located on C — where $|\nabla f|$ attains its maximum value, and the value of θ when this occurs is $\theta_E = \tau_E/(G|\nabla f|_{\max})$, where τ_E is the elastic-limit shear stress. As the twist is further increased beyond θ_E, one or more plastic regions form and expand, while the elastic regions shrink.

If the shaft material is elastic–perfectly plastic, with $\tau_E = \tau_Y = k$, then the equation satisfied by the stress function in the plastic regions is $|\nabla\phi| = k$. Eventually this equation becomes valid everywhere, a condition known as *fully plastic torsion.* In a solid (simply connected) shaft, this state is attained in the limit as $\theta \to \infty$, while in a hollow (multiply connected) shaft it may occur at a finite value of θ.

The torsion of elastic-plastic shafts with work-hardening has generally been studied by means of deformation theory, it being argued that under monotonic loading such a treatment would provide a good approximation to the solutions of the incremental theory,[1] although the two coincide exactly only when the shaft is circular. A formulation according to Prandtl–Reuss theory is due to Prager [1947]. In either case, the governing equations must be solved numerically.

Fully Plastic Torsion

The mathematical problem given by

$$|\nabla\phi| = k \text{ in } A, \quad \phi = 0 \text{ on } C, \tag{4.2.3}$$

has a unique solution (Ting [1966a]), which will be denoted ϕ_p. The solution describes a roof of constant slope, and $\phi_p(x_1, x_2)$ is simply k times the distance from (x_1, x_2) to the nearest point on C. A point (x_1, x_2) is a *ridge point* whenever there is more than one such nearest point. The ridge points represent the remnants of the elastic zone as it shrinks to zero area, so that the plastic strain rates vanish there. At a ridge point, therefore, $\dot\gamma_1 = \dot\gamma_2 = 0$ even in the limit as $\theta \to \infty$.

A line consisting of ridge points is called a *ridge line.* Since a ridge point is the meeting place of contours having different directions, a ridge line is the locus of a discontinuity in $\nabla\phi$, and therefore the stress is discontinuous across the plane formed by the x_3-axis and the ridge line. A stress discontinuity does not, by itself, violate equilibrium: consider an element ΔS of a surface perpendicular to the page and a very thin volume element containing it, as shown in Figure 4.2.2. In the limit as the thickness goes to zero, the element is in equilibrium if the traction is continuous, that is, if the stress components σ_n (the normal stress) and τ (the shear stress) are continuous; the lateral stress σ_t, on the other hand, may be discontinuous. Now, if $\mathbf{n}$ is a unit vector normal to a ridge line, then the traction on the plane normal

[1] See, e.g., Mendelson [1968], Section 11-6; Kachanov [1971], Section 30; Chakrabarty [1987], Section 3.6(vi).

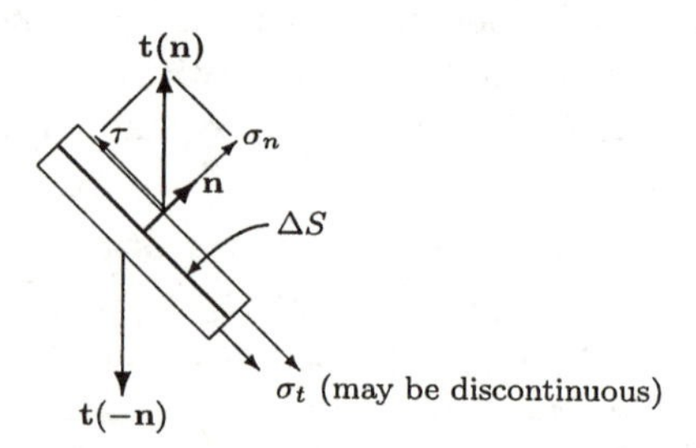

Figure 4.2.2. Stress discontinuity

to $\mathbf{n}$ is $\mathbf{e}_3(n_1\tau_1 + n_2\tau_2)$, but $n_1\tau_1 + n_2\tau_2$ is just the component of $\nabla\phi$ parallel to the ridge line, and this is continuous if the ridge line bisects the angle between the two ϕ contours that meet there.

An experimental method for finding the solution is furnished by the *sand-heap analogy*: if sand is piled onto a horizontal table having the shape of the cross-section, the slope of the resulting heap cannot exceed the angle of internal friction of the sand, and the maximal heap is formed when the critical slope obtains everywhere except at a vertex or on ridge lines; any additional sand poured will just slide off.

If C is a regular polygon, then the ϕ_p surface takes the form of a pyramid. The limit of a regular polygon as the number of sides becomes infinite is a circle; the corresponding surface is a cone. For any other convex polygon, it is formed by ridge lines which bisect the angles between nearest sides, as seen in Figure 4.2.3.

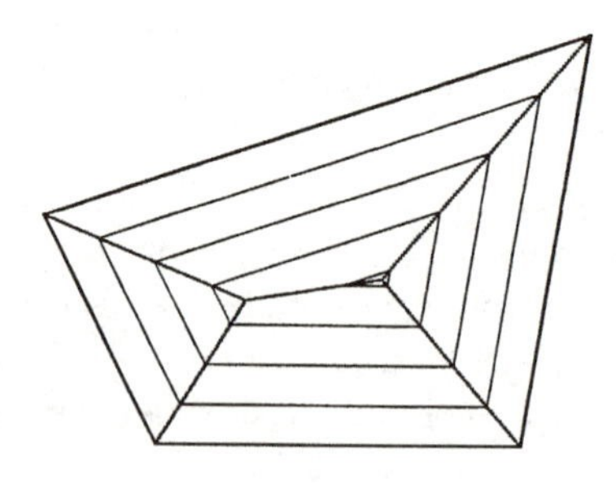

Figure 4.2.3. Ridge lines and contours in a convex polygon.

The torque given by $2\int_A \phi_p\, dA$ is the *ultimate torque* T_U. Thus, for a circle of radius a, $T_U = \frac{2}{3}\pi k a^3$. For a rectangle of sides a and b, with $b > a$, $T_U = \frac{1}{6}ka^2(3b - a)$.

In a nonconvex polygon, that is, one with re-entrant corners, the ridge lines and contours are not all straight. Consider the cross-section shown in Figure 4.2.4. For a point (x, y) with $x > y > 0$, the nearest point on the

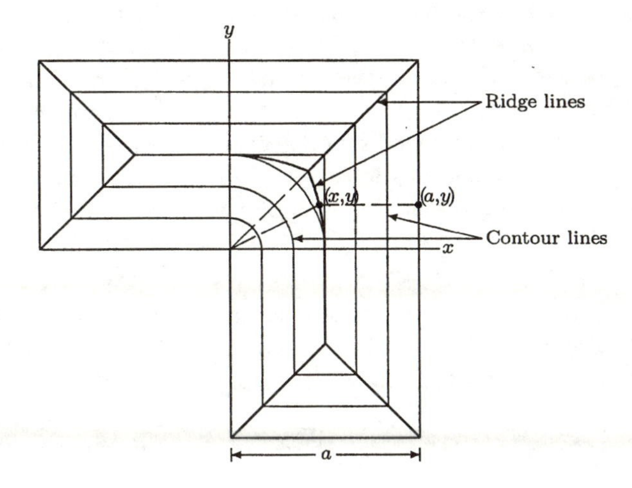

Figure 4.2.4. Ridge lines and contours in a nonconvex polygon.

side with the re-entrant corner is just the corner, while the nearest point on the other side is (a, y); (x, y) is therefore a ridge point if $\sqrt{x^2 + y^2} = a - x$, that is, if it lies on the parabola described by

$$y^2 = a^2 - 2ax.$$

To the left of this parabola, the ϕ_p contours are circular arcs centered about the corner, and to the right they are straight lines parallel to the y-axis. The structure is, of course, symmetric about the line $y = x$.

Contained Plastic Torsion

For values of the twist θ such that $\theta_E < \theta < \infty$, the cross-sectional area A consists of elastic and plastic regions, with Equation (4.2.2) governing the stress function ϕ in the former and Equation (4.2.3) in the latter. The boundary of each region consists of those parts of C (if any) that belong to it, and of the elastic-plastic boundary Γ, which is in general unknown and whose determination, as discussed in Section 4.1, constitutes a part of the problem. Moreover, values of ϕ on Γ are unknown. However, Equation (4.2.2) is an elliptic partial differential equation, and such an equation requires the specification of boundary data on a known, closed boundary. Consequently, the solution in the elastic region cannot be constructed. Equation (4.2.3), on the other hand, is parabolic, and its solution can be obtained by working inward from the plastic part of C. This solution is therefore the same as the solution for the fully plastic cross-section, provided that the plastic region lies entirely between the normals to C at the points where Γ meets C (see Prager and Hodge [1951], Section 3.2); this result was proved by T. W. Ting [1966a]. We are thus led to the *membrane-roof analogy* due to Nadai [1923]: we imagine a rigid roof having the shape of the surface of the ultimate sand heap, and a membrane stretched underneath it and subjected to a uniform upward pressure. Under sufficiently low pressures the membrane does not touch the roof, a state corresponding to elastic torsion. As the pressure is increased, parts of the membrane become tangent to the roof and remain there, representing the plastic regions, while those parts of the membrane that are still free represent the elastic region. The membrane-roof analogy was used by Nadai [1950] for the experimental solution of elastic-plastic torsion problems.

It is clear from the analogy that the elastic region consists of those points where $\phi < \phi_p$ *and* $|\nabla\phi| < k$. It was shown by Ting [1966b] that the correct solution is the one that minimizes the functional

$$K[\phi^*] \stackrel{\text{def}}{=} \int_A (|\nabla\phi^*|^2 - 4G\theta\phi^*)\, dA,$$

the minimum being taken either over all those functions ϕ^* that satisfy $\phi^* \le \phi_p$ or those that satisfy $|\nabla\phi^*| \le k$, in addition to satisfying $\phi^* = 0$ on C. The properties of the solution for various cross-sections were further investigated in a series of papers by T. W. Ting [1967, 1969a,b, 1971]. Ting's minimum principle also permits the application of numerical methods based on discretization, similar to that discussed for elastic bodies in 1.4.3, except for the presence of inequality constraints.

Elastic–Plastic Warping

An important feature of the elastic-plastic solution is that once a point of the cross-section has become plastic, the stress state there remains constant. Any strain increment in the plastic region is therefore purely plastic. The flow rule $\dot{\gamma}_\alpha^p = \dot{\lambda}\tau_\alpha$, which follows from isotropy, implies that in the plastic zone,

$$\frac{\dot{\gamma}_1}{\dot{\gamma}_2} = \frac{\psi_{,1} - x_2}{\psi_{,2} + x_1} = -\frac{\phi_{,2}}{\phi_{,1}}.$$

Let the normal to the boundary subtend an angle χ with the x_1-axis; then, in the plastic zone,

$$\phi_{,1} = -k\cos\chi, \quad \phi_{,2} = -k\sin\chi.$$

Consequently,

$$\psi_{,1}\cos\chi + \psi_{,2}\sin\chi = x_2\cos\chi - x_1\sin\chi,$$

or

$$\frac{\partial\psi}{\partial n} = d,$$

where n denotes the directed distance, measured outward, along a line normal to the boundary, and d denotes the distance from this line to the origin (see Figure 4.2.5). If n is measured from the intersection point R of the

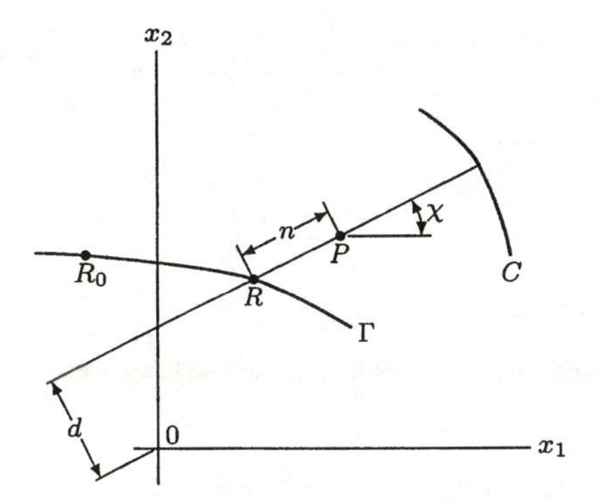

Figure 4.2.5. Elastic-plastic warping: illustration of the notation used in the theory.

given normal line with the elastic-plastic boundary, then at a point P on the line such that n is the distance from R to P,

$$\psi(P) = \psi(R) + nd.$$

To obtain $\psi(R)$, we solve Equations (4.2.1) for the $\psi_{,\alpha}$. Along Γ,

$$d\psi = \psi_{,\alpha}\, dx_\alpha = x_2\, dx_1 - x_1\, dx_2 + \frac{1}{G\theta}\tau_\alpha\, dx_\alpha,$$

so that

$$\psi(R) = \int_{R_0}^{R} (x_2\, dx_1 - x_1\, dx_2) + \int_{R_0}^{R} \tau_\alpha\, dx_\alpha,$$

where R_0 is a point on Γ such that $\psi(R_0) = 0$ — for example, the intersection of Γ with a line of symmetry, if such a line exists, or else an arbitrary point.

Warping of a Fully Plastic Rectangular Shaft

The warping of a fully plastic rectangular shaft can be obtained more simply. Because of symmetry, it is sufficient to consider only the quadrant $0 < x < \frac{1}{2}a$, $0 < y < \frac{1}{2}b$, shown in Figure 4.2.6, which we divide into the regions I and II, respectively above and below the ridge line whose equation is $y = [(b - a)/2] + x$. In region I, $\tau_2 = 0$ and therefore

$$\dot{\gamma}_2 = \dot{\theta}(\psi_{,2} + x) = 0.$$

The general solution of the partial differential equation $\psi_{,2} + x = 0$ is

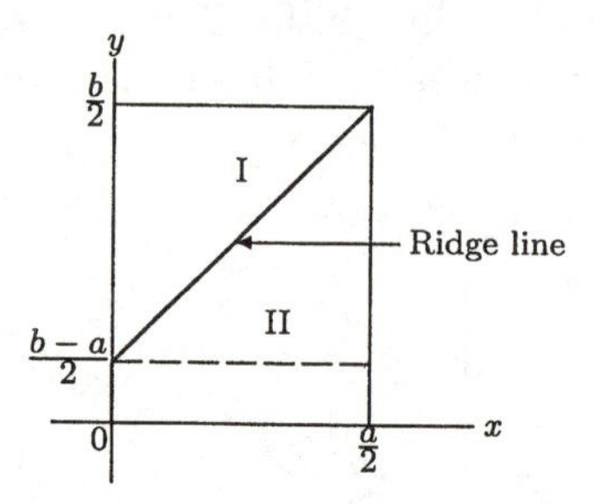

Figure 4.2.6. Fully plastic rectangular shaft: first quadrant, regions separated by ridge line.

$\psi(x, y) = -xy + f(x)$, f being an arbitrary continuously differentiable function. Symmetry requires zero warping on the y-axis, so that $f(0) = 0$. The nonvanishing shearing-rate component is $\dot{\gamma}_1 = \dot{\theta}[-2y + f'(x)]$, but since the shearing rate must vanish on the ridge line, it follows that $f'(x) = b - a + 2x$ (so that $\dot{\gamma}_1 = \dot{\theta}[b - a + 2(x - y)]$, and, consequently, $f(x) = (b - a)x + x^2$.

In region II the warping function is governed by $\psi_{,1} - y = 0$, so that $\psi = xy + g(y)$ with $\dot{\gamma}_2 = \dot{\theta}[2x + g'(y)]$. For this to vanish on the ridge line, we must have $g'(y) = b - a - 2y$ for $y > \frac{1}{2}(b - a)$, and if we integrate g' in the form $g(y) = -[y - \frac{1}{2}(b - a)]^2$, then the warping function is also continuous on the ridge line. For $y < \frac{1}{2}(b - a)$, however, the vanishing of ψ on the y-axis requires $g(y) = 0$, and therefore the complete solution for g is $g(y) = -\langle y - \frac{1}{2}(b - a)\rangle^2$, $\langle \cdot \rangle$ being the Macauley bracket, and the shearing rate is $\dot{\gamma}_2 = 2\dot{\theta}[(x - \langle y - \frac{1}{2}(b - a)\rangle]$.

The results may be used to obtain the value of the ultimate torque by means of the upper-bound theorem (Section 3.4). The plastic dissipation is

$$D_p(\dot{\varepsilon}) = \begin{cases} 2k|\dot{\theta}|x, & 0 < y < b - \dfrac{a}{2}, \\ 2k|\dot{\theta}|\left(x - y + \dfrac{b-a}{2}\right), & b - \dfrac{a}{2} < y < b - \dfrac{a}{2} + x, \\ 2k|\dot{\theta}|\left(y - x - \dfrac{b-a}{2}\right), & b - \dfrac{a}{2} + x < y < \dfrac{b}{2}. \end{cases}$$

The relation

$$T_U\dot{\theta} = \int_A D_p(\dot{\varepsilon})\, dA$$

yields the previously obtained ultimate torque $\frac{1}{6}ka^2(3b - a)$.

Circular Shaft

For a circular shaft, the derivation of the elastic-plastic solution is simple. By symmetry, the elastic-plastic boundary must be a circle of radius, say,

r^*. The elastic region is thus $r < r^*$ and the plastic region $r > r^*$. In the latter region we have the fully plastic solution, $\phi = k(c - r)$ (where c is the shaft radius), while in the former,

$$\nabla^2\phi = \frac{1}{r}\frac{d}{dr}\left(r\frac{d\phi}{dr}\right) = -2G\theta.$$

Integrating once, we obtain

$$-\frac{d\phi}{dr} = \tau_{z\theta} = \tau = G\theta r$$

in the elastic region; continuity of stress at the elastic-plastic boundary, where $\tau = k$, requires $r^* = k/G\theta$; thus $\theta_E = k/Gc$. Integrating again, we find that for $r < r^*$,

$$\phi(r) = k\left(c - \frac{r^*}{2} - \frac{r^2}{2r^*}\right),$$

and the torque is

$$T = \frac{2\pi}{3}k(c^3 - r^{*3}/4) = T_U\left[1 - \frac{1}{4}\left(\frac{\theta_E}{\theta}\right)^3\right], \tag{4.2.4}$$

where $T_U = 2\pi kc^3/3$ as before. The torque-twist diagram is therefore one in which the ultimate torque is approached asymptotically as $\theta \to \infty$ [Figure 4.2.7(a)]. As usual in problems of contained plastic deformation, the approach is quite rapid: $T/T_U = 0.99$ when $\theta/\theta_E = 3$.

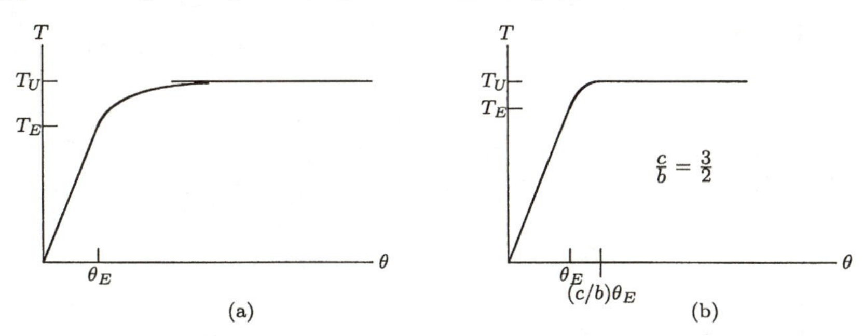

Figure 4.2.7. Torque-twist diagrams for circular shaft: (a) solid; (b) hollow. (Compare with Figure 4.1.3.)

If the shaft is hollow, with inner radius b, then the expressions derived above for ϕ is the elastic and plastic regions are valid as long as $b < r^* < c$. The onset of plasticity occurs when $T = \pi(c^4 - b^4)k/2c = T_E$, while the ultimate torque is $T_U = 2\pi(c^3 - b^3)k/3$. The torque in contained plastic deformation is given by

$$T = 2\pi k\left(\frac{c^3}{3} - \frac{b^4}{4r^*} - \frac{r^{*3}}{12}\right).$$

As can be seen in Figure 4.2.7(b), the ultimate torque is attained when $r^* = b$, that is, when $\theta = k/Gb = (c/b)\theta_E$. In this case the approach is not asymptotic.

The elastic-plastic torsion of a circular shaft of variable diameter was studied by Sokolovskii [1945], extending the elastic theory due to Michell [1900]. In such a shaft the shear strain $\gamma_{r\theta}$ and shear stress $\tau_{r\theta}$ are present in addition to $\gamma_{z\theta}$ and $\tau_{z\theta}$. Equilibrium requires that the stresses be derived from a stress function $\bar{\phi}(r, z)$ by means of

$$\tau_{z\theta} = \frac{1}{r^2}\frac{\partial\bar{\phi}}{\partial r}, \quad \tau_{r\theta} = -\frac{1}{r^2}\frac{\partial\bar{\phi}}{\partial z},$$

where $\bar{\phi}$ obeys

$$\frac{\partial^2\bar{\phi}}{\partial r^2} - \frac{3}{r}\frac{\partial\bar{\phi}}{\partial r} + \frac{\partial^2\bar{\phi}}{\partial z^2} = 0$$

in the elastic region and

$$|\nabla\bar{\phi}| = kr^2$$

in the plastic region. The equations must in general be solved numerically.

A problem mathematically similar to the preceding is that of the torsion of a segment of a torus by means of equal and opposite forces acting over the end sections and perpendicular to the plane of the torus. Both the elastic and the elastic-plastic problems were treated by Freiberger [1949, 1956a].

Sokolovskii's Oval Shaft

While no known solution exists for the contained plastic torsion of an elliptic shaft, an inverse method has been applied by Sokolovskii [1942] to the torsion of a shaft of an oval cross-section that differs only slightly from an elliptic one, and whose form is determined from the solution. A form of ϕ is assumed in the elastic region, namely,

$$\phi(x, y) = \frac{k}{2}\left(c - \frac{x^2}{\bar{a}} - \frac{y^2}{\bar{b}}\right). \tag{4.2.5}$$

The angle of twist is given in accord with Equation (4.2.2) as

$$\theta = \frac{k(\bar{a} + \bar{b})}{2G\bar{a}\bar{b}}. \tag{4.2.6}$$

Furthermore,

$$|\nabla\phi| = k\sqrt{\left(\frac{x}{\bar{a}}\right)^2 + \left(\frac{y}{\bar{b}}\right)^2},$$

so that the elastic-plastic boundary Γ is given by the ellipse

$$\left(\frac{x}{\bar{a}}\right)^2 + \left(\frac{y}{\bar{b}}\right)^2 = 1,$$

or, parametrically, by

$$x = \bar{a}\cos\chi, \qquad y = \bar{b}\sin\chi.$$

In terms of this representation,

$$\nabla\phi = -k(\mathbf{e}_1\cos\chi + \mathbf{e}_2\sin\chi).$$

The angle χ thus defines the direction of *steepest descent* of the ϕ surface at the elastic-plastic boundary; note that this direction is not, in general, normal to this boundary, except where it coincides with the principal axes of the ellipse.

The form of ϕ in the plastic zone may now be obtained by means of the previously discussed general solution for fully plastic torsion. Thus, if straight lines are drawn outward from Γ in the direction of steepest descent, then along these lines ϕ has the constant slope $-k$, and these lines are normal to the external boundary curve C. Consider one such line, drawn from the point (ξ, η) on Γ to a point (x, y) on the external boundary C. Since $\phi(x, y) = 0$, the distance between the two points is $\phi(\xi, \eta)/k$. If $\xi = \bar{a}\cos\chi$ and $\eta = \bar{b}\sin\chi$, then

$$x = \bar{a}\cos\chi + \frac{1}{2}(c - \bar{a}\cos^2\chi - \bar{b}\sin^2\chi)\cos\chi,$$

$$y = \bar{b}\sin\chi + \frac{1}{2}(c - \bar{a}\cos^2\chi - \bar{b}\sin^2\chi)\sin\chi.$$

From these expressions we can determine the semiaxes of C, namely,

$$a = x|_{\chi=0} = \frac{1}{2}(\bar{a} + c), \qquad b = y|_{\chi=\pi/2} = \frac{1}{2}(\bar{b} + c);$$

note that

$$\bar{a} - \bar{b} = 2(a - b). \tag{4.2.7}$$

The boundary curve may now be represented in terms of a and b as

$$x = (2a - b)\cos\chi - (a - b)\cos^3\chi,$$

$$y = (2b - a)\sin\chi + (a - b)\sin^3\chi.$$

These equations describe a closed oval shape if $\frac{1}{2}a < b < 2a$. For most ratios a/b within this range, the oval is almost indistinguishable from the ellipse [see Figure 4.2.8(a)].

Equations (4.2.6) and (4.2.7) can now be solved simultaneously to give the the semiaxes $\bar{a}$, $\bar{b}$ of the elastic-plastic boundary as functions of the twist θ:

$$\bar{a} = a - b + \frac{k}{2G\theta} + \sqrt{(a-b)^2 + \left(\frac{k}{2G\theta}\right)^2},$$

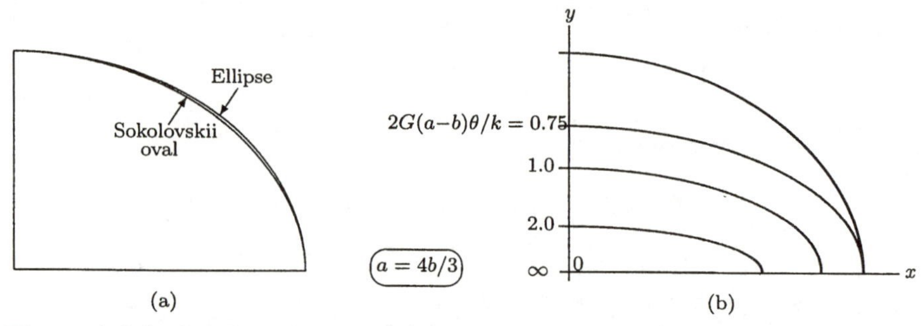

Figure 4.2.8. Sokolovskii oval: (a) comparison with ellipse; (b) elastic-plastic boundary at various values of the twist.

$$\bar{b} = b - a + \frac{k}{2G\theta} + \sqrt{(a-b)^2 + \left(\frac{k}{2G\theta}\right)^2}.$$

For the elastic-plastic boundary to lie entirely within C, it is necessary that $\bar{a} \le a$ and $\bar{b} \le b$, or

$$\frac{k}{2G\theta} + \sqrt{(a-b)^2 + \left(\frac{k}{2G\theta}\right)^2} \le \min(a,\, b).$$

Supposing that $a > b$, we find the minimum value of θ for which the solution is valid to be

$$\theta_{\min} = \frac{kb}{Ga(2b-a)}.$$

As θ increases above this value, the ellipse shrinks; in the limit as $\theta \to \infty$, $\bar{b}$ goes to zero, and $\bar{a}$ goes to $2(a-b)$, so that in the fully plastic state a ridge line extending over $-2(a-b) \le x \le 2(a-b)$ is formed. Elastic-plastic boundaries corresponding to different values of θ are shown in Figure 4.2.8(b) for $a = 4b/3$.

By means of a different inverse method, Galin [1949] was able to solve a series of elastic-plastic torsion problems for shafts of nearly polygonal cross-section.

Exercises: Section 4.2

1. Show that for the torsion problem, the compatibility conditions (1.2.4) are equivalent to the single equation

$$\gamma_{2,1} - \gamma_{1,2} = \text{constant},$$

and that the constant is 2θ.

2. If the stress field in the plastic region is expressed as

$$\sigma_{13} = k\cos\alpha, \quad \sigma_{23} = k\sin\alpha,$$

find the forms taken by the equilibrium equation and the boundary conditions on C in terms of α. Interpret α in relation to the plastic stress function ϕ_p.

3. Formulate the torsion problem in terms of the deformation theory of plasticity (see 3.2.1). Show that for a circular shaft the solution coincides with that given by the incremental theory.

4. Find the ultimate torque and the fully plastic warping for a shaft whose cross-section is an equilateral triangle.

5. If $\bar{\phi}(r, z)$ is the stress function for the torsion of a circular shaft of variable diameter,

   (a) (a) find the boundary conditions that must be satisfied by $\bar{\phi}$ on the outer surface (say $r = \bar{r}(z)$) and on the axis $r = 0$, and

   (b) (b) show that the torque at a section z is

$$T = 2\pi[\bar{\phi}(\bar{r}(z), z) - \bar{\phi}(0, z)].$$

6. Discuss the contour lines of the fully plastic stress function in the Sokolovskii oval shaft. Plot for $a/b = 4/3$.

7. Use the formulation of Exercise 2 to show that the fully plastic stress distribution for a shaft of elliptic cross-section, bounded by $(x/a)^2 + (y/b)^2 = 1$, is given implicitly by

$$x \cos\alpha + y \sin\alpha = \frac{(a^2 - b^2)\sin 2\alpha}{2\sqrt{a^2 \sin^2\alpha + b^2 \cos^2\alpha}}.$$

   Plot the contours of the stress function for $a/b = 4/3$, and compare the result with that of Exercise 6.

Section 4.3 The Thick-Walled Hollow Sphere and Cylinder

Like the problem of a tube under torsion, that of an axisymmetrically loaded shell of revolution is statically determinate when the shell is thin-walled but ceases to be so when the shell is thick-walled. In the present section we present problems of pressurized thick-walled elastic-plastic shells that are spherical or cylindrical. Problems of the hollow sphere and cylinder are sufficiently different to warrant presentation of both, but sufficiently similar so that not all topics need be treated for both. Treatment of the sphere includes the solution for finite strain and thermal stresses, while treatment of the cylinder includes the effects of unloading and reloading and of work-hardening. More details can be found in books such as Johnson and Mellor [1973] and Chakrabarty [1987].

4.3.1. Elastic Hollow Sphere Under Internal and External Pressure

Basic Equations

In a hollow sphere of inner radius a and outer radius b, subject to normal pressures on its inner and outer surfaces, and made of an isotropic material, the displacement and stress fields must be spherically symmetric. The only nonvanishing displacement component is the radial displacement u, a function of the radial coordinate r only. The only nonvanishing strains are, by Equations (1.2.2), the radial strain $\varepsilon_r = du/dr$ and the circumferential strains $\varepsilon_\theta = \varepsilon_\phi = u/r$. The strains obviously satisfy the compatibility condition

$$\varepsilon_r = \frac{d}{dr}(r\varepsilon_\theta). \tag{4.3.1}$$

The only nonvanishing stress components are the radial stress σ_r and the circumferential stresses $\sigma_\theta = \sigma_\phi$, which satisfy the equilibrium equation

$$\frac{d\sigma_r}{dr} + 2\frac{\sigma_r - \sigma_\theta}{r} = 0. \tag{4.3.2}$$

Elastic Solution

The strain-stress relations for an isotropic linearly elastic solid, Equations (1.4.11), reduce in the present case to

$$\varepsilon_r = \frac{1}{E}(\sigma_r - 2\nu\sigma_\theta),$$

$$\varepsilon_\theta = \frac{1}{E}[(1-\nu)\sigma_\theta - \nu\sigma_r].$$

The compatibility equation (4.3.1) may now be rewritten in terms of the stresses to read

$$\frac{d}{dr}[(1-\nu)\sigma_\theta - \nu\sigma_r] + \frac{1+\nu}{r}(\sigma_\theta - \sigma_r) = 0,$$

and with the help of (4.3.2) reduces to

$$\frac{d}{dr}(\sigma_r + 2\sigma_\theta) = 0.$$

The quantity $\sigma_r + 2\sigma_\theta$ is accordingly equal to a constant, say $3A$. Furthermore, $d\sigma_\theta/dr = -\frac{1}{2}d\sigma_r/dr$, so that $\frac{2}{3}d(\sigma_\theta - \sigma_r)/dr = d\sigma_r/dr$, and Equation (4.3.2) may be rewritten as

$$\frac{d}{dr}(\sigma_\theta - \sigma_r) + \frac{3}{r}(\sigma_\theta - \sigma_r) = 0,$$

leading to the solution

$$\sigma_\theta - \sigma_r = \frac{3B}{r^3},$$

where B is another constant. The stress field is therefore given by

$$\sigma_r = A - \frac{2B}{r^3}, \quad \sigma_\theta = A + \frac{B}{r^3}.$$

With the boundary conditions

$$\sigma_r|_{r=a} = -p_i, \quad \sigma_r|_{r=b} = -p_e,$$

where p_i and p_e are the interior and exterior pressures, respectively, the constants A and B can be solved for, and the stress components σ_r, σ_θ can be expressed as

$$\sigma_r = -\frac{1}{2}(p_i + p_e) + \frac{p_i - p_e}{2[1 - (a/b)^3]}\left[1 + \left(\frac{a}{b}\right)^3 - 2\left(\frac{a}{r}\right)^3\right],$$

$$\sigma_\theta = -\frac{1}{2}(p_i + p_e) + \frac{p_i - p_e}{2[1 - (a/b)^3]}\left[1 + \left(\frac{a}{b}\right)^3 + \left(\frac{a}{r}\right)^3\right],$$

that is, the stress field is the superposition of (1) a uniform stress field equal to the negative of the average of the external and internal pressures, and (2) a variable stress field proportional to the pressure difference.

Sphere Under Internal Pressure Only

The preceding solution is due to Lamé. It becomes somewhat simpler if the sphere is subject to an internal pressure only, with $p_i = p$ and $p_e = 0$. The stresses are then

$$\sigma_r = -\frac{p}{(b/a)^3 - 1}\left(\frac{b^3}{r^3} - 1\right),$$

$$\sigma_\theta = \frac{p}{(b/a)^3 - 1}\left(\frac{b^3}{2r^3} + 1\right).$$

If the sphere material is elastic-plastic, then the largest pressure for which the preceding solution is valid is that at which the stresses at some r first satisfy the yield criterion; this limiting pressure will be denoted p_E. Since two of the principal stresses are equal, the stress state is equibiaxial (see 3.3.4), and both the Tresca and Mises criteria reduce to

$$\sigma_\theta - \sigma_r = \sigma_Y, \tag{4.3.3}$$

where σ_Y is the tensile yield stress, since $\sigma_\theta > \sigma_r$ everywhere. The value of $\sigma_\theta - \sigma_r$ is maximum at $r = a$, where it attains $3p/2[1 - (a/b)^3]$. The largest pressure at which the sphere is wholly elastic is therefore

$$p_E = \frac{2}{3}\sigma_Y\left(1 - \frac{a^3}{b^3}\right). \tag{4.3.4}$$

4.3.2. Elastic-Plastic Hollow Sphere Under Internal Pressure

Stress Field

When the pressure in the hollow sphere exceeds p_E, a spherical domain of inner radius a and outer radius, say, c becomes plastic. The elastic domain $c < r < b$ behaves like an elastic shell of inner radius c that is just yielding at $r = c$, so that σ_r and σ_θ are given by

$$\sigma_r = -\frac{p_c}{(b/c)^3 - 1}\left(\frac{b^3}{r^3} - 1\right),$$

$$\sigma_\theta = \frac{p_c}{(b/c)^3 - 1}\left(\frac{b^3}{2r^3} + 1\right).$$

where $p_c = -\sigma_r(c)$ is such that the yield criterion is met at $r = c$, that is, it is given by the right-hand side of (4.3.4) with a replaced by c:

$$p_c = \frac{2}{3}\sigma_Y\left(1 - \frac{c^3}{b^3}\right).$$

Therefore,

$$\left.\begin{aligned} \sigma_r &= -\frac{2}{3}\sigma_Y\left(\frac{c^3}{r^3} - \frac{c^3}{b^3}\right), \\ \sigma_\theta &= \frac{2}{3}\sigma_Y\left(\frac{c^3}{2r^3} + \frac{c^3}{b^3}\right). \end{aligned}\right\} \quad c < r < b \tag{4.3.5}$$

In particular, $\sigma_\theta(b) = \sigma_Y(c/b)^3$.

In the plastic domain, the yield criterion (4.3.3) holds everywhere, so that the equilibrium equation (4.3.2) may be integrated for σ_r, subject to continuity with the elastic solution at $r = c$, to yield

$$\sigma_r = -\frac{2}{3}\sigma_Y\left(1 - \frac{c^3}{b^3} + \ln\frac{c^3}{r^3}\right).$$

We immediately obtain $\sigma_\theta = \sigma_r + \sigma_Y$ as well, namely,

$$\sigma_\theta = \frac{1}{3}\sigma_Y\left(1 + 2\frac{c^3}{b^3} - 2\ln\frac{c^3}{r^3}\right).$$

The radius c marking the extent of the plastic domain is obtained, at a given pressure p, from the condition that $\sigma_r(a) = -p$, or

$$p = \frac{2}{3}\sigma_Y\left(1 - \frac{c^3}{b^3} - \ln\frac{c^3}{a^3}\right). \tag{4.3.6}$$

When $c = b$, the shell is completely plastic. The corresponding pressure is the *ultimate pressure*, given by

$$p_U = 2\sigma_Y \ln\frac{b}{a}.$$

Displacement

The displacement u in the elastic domain is given by $u(r) = (r/E)[(1-\nu)\sigma_\theta - \nu\sigma_r]$. In particular,

$$u(c) = \frac{\sigma_Y c}{3E}\left[(1+\nu) + 2(1-2\nu)\frac{c^3}{b^3}\right].$$

In the plastic domain the displacement is determined from the fact that the volume strain is purely elastic, that is,

$$\frac{du}{dr} + 2\frac{u}{r} = \frac{1-2\nu}{E}(\sigma_r + 2\sigma_\theta) = \frac{2(1-2\nu)\sigma_Y}{3E}\left(2 + \frac{c^3}{b^3} - \ln\frac{c^3}{r^3}\right).$$

The left-hand member is equal to $r^{-2}d(r^2u)/dr$, and therefore the equation can be integrated subject to continuity at $r = c$. In particular

$$u(a) = \frac{\sigma_Y a}{E}\left[(1-\nu)\frac{c^3}{a^3} - \frac{2}{3}(1-2\nu)\left(1 - \frac{c^3}{b^3} + \ln\frac{c^3}{a^3}\right)\right].$$

When $p = p_U$, $c = b$, and the expression simplifies to

$$u(a) = \frac{\sigma_Y a}{E}\left[(1-\nu)\frac{b^3}{a^3} - 2(1-2\nu)\ln\frac{b}{a}\right].$$

Solution for Finite Strains

If b/a is sufficiently large, the strains become of order $(b/a)^3\sigma_Y/E$ and may be too large to be regarded as infinitesimal. A solution for finite strains may be obtained by regarding the strains as logarithmic, a procedure that is permissible in the present case, since the principal strain directions do not change (see Section 2.1).

If r is the current radius of a spherical surface whose initial radius is r_0, then the current volume of the infinitesimal shell contained between r and $r + dr$ is $4\pi r^2\, dr$, while its initial volume is $4\pi r_0^2\, dr_0$. The volume strain is consequently $\ln(r^2\, dr/r_0^2\, dr_0)$. Assuming plastic incompressibility, we can equate this strain to $(1-2\nu)(\sigma_r + 2\sigma_\theta)/E$, or

$$\frac{r_0^2}{r^2}\frac{dr_0}{dr} = \exp\left[-\frac{1-2\nu}{E}(\sigma_r + 2\sigma_\theta)\right] \doteq 1 - \frac{1-2\nu}{E}(\sigma_r + 2\sigma_\theta), \qquad (4.3.7)$$

since the stresses are small compared to E. The equilibrium equation (4.3.2) is exact if it is interpreted as Eulerian, that is, if r denotes the current radius and σ_r, σ_θ are the true stresses. Eliminating σ_θ between (4.3.2) and (4.3.7), we obtain

$$r_0^2\frac{dr_0}{dr} = r^2 - \frac{1-2\nu}{E}\frac{d}{dr}(r^3\sigma_r).$$

Integration leads to

$$\frac{r_0^3}{r^3} = 1 - 3(1-2\nu)\frac{\sigma_r}{E} - \frac{3A}{r^3},$$

where A is a constant of integration. Note that this relation holds in both the elastic and plastic domains. At $r = b$ we have $r_0 = b_0$, and since $\sigma_r(b) = 0$,

$$\frac{b_0}{b} = \left[1 - \frac{3A}{b^3}\right]^{1/3} \doteq 1 - \frac{A}{b^3}.$$

But $b_0 = b - u(b)$, where $u(b)$ may be obtained from the previous solution for the elastic domain, namely, $u(b) = (1-\nu)b\sigma_\theta(b)/E = (1-\nu)b(c/b)^3\sigma_Y/E$. Consequently, $A = (1-\nu)c^3\sigma_Y/E$. The change in the internal radius can be expressed as

$$\left(\frac{a_0}{a}\right)^3 = 1 - \frac{3}{E}\left[(1-\nu)\sigma_Y\frac{c^3}{a^3} - (1-2\nu)p\right]. \qquad (4.3.8)$$

If the material is perfectly plastic, then the previously derived stress field in the plastic zone is still valid, and the relation between c and p is still given by (4.3.6). However, the preceding solution for the radius change is not limited to a perfectly plastic material. The yield stress σ_Y enters only because it is contained in the elastic solution, and reflects the fact that yielding first occurs at $r = c$. If the material work-hardens, then σ_Y need only be replaced by the initial yield stress σ_E.

At at any point of the shell, the state of stress is a superposition of the hydrostatic stress σ_θ and of the uniaxial compressive stress $\sigma_r - \sigma_\theta$ acting in the radial direction. Since it is only the latter stress that produces plastic deformation, in the plastic domain this stress is related to ε_r^p in the same way that σ is related to ε^p in a uniaxial compression test. Let this relation be denoted $\sigma = -\sigma_Y(-\varepsilon^p)$; then (4.3.3) may be written as

$$\sigma_\theta - \sigma_r = \sigma_Y(-\varepsilon_r^p).$$

Since $\varepsilon_\theta = \ln(r/r_0)$,

$$\varepsilon_r + 2\varepsilon_\theta = \frac{1-2\nu}{E}(\sigma_r + 2\sigma_\theta),$$

and

$$\varepsilon_r = \varepsilon_r^p + \frac{1}{E}(\sigma_r - 2\nu\sigma_\theta),$$

it follows that

$$\varepsilon_r^p = \frac{2}{E}[(1-\nu)\sigma_\theta - \nu\sigma_r] - 2\ln\frac{r}{r_0},$$

so that

$$\sigma_\theta - \sigma_r = \sigma_Y\left(2\ln\frac{r}{r_0} - 2\frac{(1-\nu)\sigma_\theta - \nu\sigma_r}{E}\right). \tag{4.3.9}$$

This equation, along with the equilibrium equation (4.3.2) and the elastic compressibility condition (4.3.7), constitutes the system of three simultaneous equations that govern the three variables σ_r, σ_θ, and r_0 as functions of r in the plastic domain. The equations must, in general, be solved numerically.

A simplification is achieved if the material is elastically as well as plastically incompressible, that is, if $\nu = \frac{1}{2}$. In this case (4.3.9) becomes

$$\sigma_\theta - \sigma_r = \sigma_Y\left(2\ln\frac{r}{r_0} - \frac{\sigma_\theta - \sigma_r}{E}\right),$$

and therefore may be solved for $\sigma_\theta - \sigma_r$ as

$$\sigma_\theta - \sigma_r = f\left(2\ln\frac{r}{r_0}\right).$$

Here f is the same function that gives the relation between the true stress and the *total* logarithmic strain in uniaxial compression, $\sigma = f(\varepsilon)$. Inserting the last equation into (4.3.2) and integrating leads to

$$\sigma_r = 2\int_c^r \frac{f(2\ln(r/r_0))}{r}dr - \frac{2}{3}\left(1 - \frac{c^3}{b^3}\right)\sigma_E,$$

with continuity at $r = c$ taken into account. The internal pressure is therefore

$$p = \frac{2}{3}\left(1 - \frac{c^3}{b^3}\right)\sigma_E + 2\int_a^c \frac{f(2\ln(r/r_0))}{r}dr, \tag{4.3.10}$$

a result that agrees with (4.3.6) when $f(\varepsilon) \equiv \sigma_Y$ (perfect plasticity). Since the material is assumed incompressible,

$$r^3 - r_0^3 = a^3 - a_0^3.$$

In the elastic region, however, $r^3 - r_0^3 \doteq 3r^2 u(r)$, while $u(r) = \sigma_E c^3/2Er^2$. Hence

$$c^3 = \alpha(a^3 - a_0^3), \tag{4.3.11}$$

where

$$\alpha = \frac{2E}{3\sigma_E}.$$

Equation (4.3.10) may now be rewritten with the help of the variable $s = (a^3 - a_0^3)/r^3$ to read

$$p = \frac{2}{3}\left(1 - \frac{c^3}{b^3}\right)\sigma_E + \frac{2}{3}\int_\alpha^{1-(a_0/a)^3} \frac{f(-\frac{2}{3}\ln(1-s))}{s} ds. \tag{4.3.12}$$

Equations (4.3.11)–(4.3.12) may be used to calculate the pressure as a function of the expansion ratio $\rho = a/a_0$.

When large plastic strains are taken into account, it is found that the pressure attains a maximum at a finite value of the expansion, representing *plastic instability* analogous to necking of tensile specimens. Differentiating (4.3.12) with respect to a we obtain

$$\frac{dp}{da} = -2\sigma_E \frac{c^2}{b^3}\frac{dc}{da} + \frac{2}{a} f(2\ln\rho)\frac{1}{\rho^3 - 1}.$$

From (4.3.11),

$$c^2 \frac{dc}{da} = \frac{2E}{3\sigma_E} a^2,$$

so that $dp/da = 0$ when ρ is given by

$$\frac{2E\rho^3(\rho^3 - 1)}{3f(2\ln\rho)} = \left(\frac{b}{a_0}\right)^3,$$

an equation that can be easily solved graphically for a given stress-strain curve. In particular, for the perfectly plastic material the equation is quadratic in ρ^3 and can be solved explicitly.

If the material is work-hardening and the shell is sufficiently thin, however, the instability does not occur before the shell has become completely plastic. The limiting wall ratio, for which the pressure is maximum just when the shell becomes fully plastic, is obtained by setting $c = b$, so that $\rho^3 = 1 + \alpha(b/a_0)^3$. Defining

$$\varepsilon_0 = \frac{2}{3}\ln\left(1 + \alpha\frac{b^3}{a_0^3}\right),$$

we find that the limiting value of ε_0 (and hence of b/a_0) is given by

$$f(\varepsilon_0) = \sigma_E e^{3\varepsilon_0/2}.$$

Spherical Cavity in an Infinite Solid

In all the foregoing, the limit as $b \to inf$ describes the expansion of a spherical cavity in an infinite solid. When a/a_0 is very large, the left-hand side of (4.3.8) is negligible next to unity, as is the last term in the brackets on the right-hand side (since $p \ll E$). Consequently c/a becomes approximately constant, with the value

$$\frac{c}{a} = \left[\frac{E}{3(1-\nu)\sigma_Y}\right]^{1/3},$$

which holds in a work-hardening solid as well if σ_Y is replaced by σ_E. If the material is incompressible then the pressure is given by (4.3.12), with $c^3/b^3 = 0$ and the upper limit of the integral equal to unity. At this upper limit, corresponding to $r = a$, the argument of f becomes infinite. Nonetheless, the integral converges if the derivative of f approaches a constant value for large arguments. In particular, if $f'(\varepsilon) = H_1$ (constant), then the pressure is given by

$$p = \frac{2}{3}\left(1 + \ln\frac{2E}{3\sigma_E}\right)\sigma_E + \frac{2}{27}\pi^3 H_1.$$

Bishop, Hill and Mott [1945], who derived this result, applied it to obtain a theoretical estimate of the maximum steady-state pressure in the deep penetration of a smooth punch into a quasi-infinite medium, and found close agreement with an experiment using cold-worked copper.

4.3.3. Thermal Stresses in an Elastic–Perfectly Plastic Hollow Sphere

Thermoelastic Stresses

If the inner and outer surfaces of the sphere are at different temperatures, say T_a and T_b, respectively, then the temperature inside the shell varies with r. The thermoelastic stress-strain-temperature relations (1.4.13) are, in the present case,

$$\begin{aligned} \varepsilon_r &= \frac{1}{E}(\sigma_r - 2\nu\sigma_\theta) + \alpha T, \\ \varepsilon_\theta &= \frac{1}{E}[(1-\nu)\sigma_\theta - \nu\sigma_r] + \alpha T. \end{aligned} \tag{4.3.13}$$

The compatibility equation (4.3.1) now reads

$$\frac{d}{dr}[(1-\nu)\sigma_\theta - \nu\sigma_r] + \frac{1+\nu}{r}(\sigma_\theta - \sigma_r) + E\alpha\frac{dT}{dr} = 0,$$

and with the help of (4.3.2), reduces to

$$\frac{d}{dr}(\sigma_r + 2\sigma_\theta) - 2\frac{E\alpha}{1-\nu}\frac{dT}{dr} = 0,$$

or, upon integrating,

$$\sigma_r + 2\sigma_\theta = 3A - 2\frac{E\alpha}{1-\nu}T.$$

Combining this again with (4.3.2) leads to

$$\frac{d}{dr}(r^3\sigma_r) = 3Ar^2 - 2\frac{E\alpha}{1-\nu}r^2 T,$$

which can be integrated to yield $\sigma_r(r)$ when the temperature distribution is known. Equation (4.3.2) can then be used to obtain $\sigma_\theta(r)$, and the displacement field $u(r)$ can be obtained from Equation $(4.3.13)_2$ through the relation $u = r\varepsilon_\theta$.

If the temperature difference $T_a - T_b$ is not too great, then the steady-state temperature field $T(r)$ obeys the spherically symmetric Laplace equation,

$$\frac{d^2T}{dr^2} + \frac{2}{r}\frac{dT}{dr} = 0.$$

The solution is

$$T(r) = \frac{(b/a)T_b + T_a}{(b/a) - 1} + \frac{T_a - T_b}{(b/a) - 1}\frac{b}{r}. \tag{4.3.14}$$

If the constant A is redefined so as to include the contribution of the constant term in (4.3.14), then the solution for the stresses is

$$\begin{aligned} \sigma_r &= A + \frac{2B}{r^3} - \lambda\frac{b}{r}, \\ \sigma_\theta &= A - \frac{B}{r^3} - \frac{\lambda}{2}\frac{b}{r}, \end{aligned} \tag{4.3.15}$$

where

$$\lambda = \frac{E\alpha(T_a - T_b)}{(1-\nu)[(b/a) - 1]}.$$

If the stresses are due to the temperature difference only, then the constants A, B may be determined from the conditions $\sigma_r(a) = \sigma_r(b) = 0$. The results are

$$\sigma_r = -\lambda\left[\frac{b}{r} - \frac{1}{(b/a)^2 + (b/a) + 1}\left(\frac{b^3}{r^3} + \frac{b^2}{a^2} + \frac{b}{a}\right)\right],$$

$$\sigma_\theta = -\lambda\left[\frac{b}{2r} + \frac{1}{(b/a)^2 + (b/a) + 1}\left(\frac{b^3}{2r^3} - 2\frac{b^2}{a^2} - 2\frac{b}{a}\right)\right],$$

so that

$$\sigma_\theta - \sigma_r = \frac{\lambda}{2}\left[\frac{b}{r} - \frac{3}{(b/a)^2 + (b/a) + 1}\frac{b^3}{r^3}\right].$$

The greatest numerical value of the right-hand side occurs at $r = a$, where it equals

$$|\sigma_\theta - \sigma_r|_{\max} = \frac{|\lambda|}{2}\frac{(b/a)(b/a - 1)(2b/a + 1)}{(b/a)^2 + (b/a) + 1}.$$

Yielding therefore occurs when this quantity first equals σ_Y, so that the temperature difference required for initial yielding is

$$|T_a - T_b|_E = \frac{(1-\nu)\sigma_Y}{E\alpha}\frac{1 + (a/b) + (a/b)^2}{1 + (a/2b)}. \tag{4.3.16}$$

The combined effect of internal pressure and temperature gradient was analyzed by Derrington and Johnson [1958]. The stress field is given by the superposition of the previously obtained stress fields in the shell under pressure alone and temperature gradient alone. However, in the shell under combined pressure and temperature gradient the maximum of $|\sigma_\theta - \sigma_r|$ does not always occur at $r = a$. It does so when p/λ (with λ positive) is sufficiently large *or* sufficiently small, the upper and lower limiting values depending on the wall ratio b/a. At intermediate values of p/λ, yielding may begin in the interior of the shell or (if $b/a \le 2$ and $\lambda \ge 3\sigma_Y$) at $r = b$. At the lower limiting value of p/λ, initial yielding occurs simultaneously at $r = a$ and at $r = \min(b, 2a)$.

Thermoplastic Stresses

We return to the shell under temperature gradient alone. When $|T_a - T_b|$ exceeds $|T_a - T_b|_E$ as given by (4.3.16), a plastic region $a < r < c$ is formed, in which the equilibrium equation (4.3.2), the yield criterion $\sigma_\theta - \sigma_r = \sigma_Y$ and the boundary condition $\sigma_r(a) = 0$ together produce the stress field

$$\left.\begin{aligned} \sigma_r(r) &= -2\sigma_Y \ln\frac{r}{a}, \\ \sigma_\theta &= -\sigma_Y\left(1 + \ln r\frac{r}{a}\right), \end{aligned}\right\} \quad a \le r \le c. \tag{4.3.17}$$

In the elastic region $c < r < b$, the stresses are given by Equations (4.3.15). The boundary condition $\sigma_r(b) = 0$ leads to

$$A + \frac{2B}{b^3} = \lambda,$$

and the continuity of the stresses at $r = c$ yields two additional equations for B and λ in terms of c, which may be solved to give

$$B = c^3\sigma_Y\frac{1 - c/b + \ln(c/a)}{(2 + c/b)(1 - c/b)^2},$$

$$\lambda = \sigma_Y \frac{c}{b} \frac{2(1 - c^3/b^3) + \ln(c/a)^3}{(2 + c/b)(1 - c/b)^2}.$$

Note that c/b attains the value 1 only as λ becomes infinite.

It was observed by Cowper [1960] that the elastic domain $c < r < b$ is equivalent to a shell that is just yielding at the inner radius c under the combined action of a temperature difference and an internal pressure $2\sigma_Y \ln(c/a)$. It follows from the previous discussion that if λ, and hence c, is sufficiently large, a second plastic region is formed, starting either at $r = b$ (if $c \geq b/2$) or at $r = 2c$ (if $c \leq b/2$). The former case occurs if $b/a \leq 2e^{1/3}$ and the latter if $b/a \geq 2e^{1/3}$. Even when the second plastic domain is taken into account, the sphere can become completely plastic only at an infinite temperature difference. For more discussion and further details on thermal stresses in elastic-plastic hollow spheres, see Johnson and Mellor [1973], pp. 204–214, and Chakrabarty [1987], pp. 343–350.

4.3.4. Hollow Cylinder: Elastic Solution and Initial Yield Pressure

Basic Equations

The problem of a thick-walled cylindrical tube of inner radius a and outer radius b, subject to normal pressures on its inner and outer surfaces, is not as simple as the corresponding problem of the hollow sphere, because the mechanical state in general varies not only with the radial coordinate r but also the axial coordinate z. If, however, the cylinder is of sufficiently great length compared to b, then some simplifying assumptions may be made. First, the stresses and strains on sections far enough away from the ends may be regarded as independent of z. In addition, plane sections perpendicular to the tube axis may be assumed to remain plane, so that the axial strain ε_z is constant.

As a result of symmetry about the z-axis, the shear-stress components (in cylindrical coordinates) $\tau_{r\theta}$ and $\tau_{z\theta}$ vanish, and all remaining stress components are independent of θ as well, and hence functions of r only. The equilibrium equations are therefore

$$\frac{d\sigma_r}{dr} + \frac{\sigma_r - \sigma_\theta}{r} = 0 \tag{4.3.18}$$

and

$$\frac{d\tau_{rz}}{dr} + \frac{\tau_{rz}}{r} = 0.$$

The latter equation has the solution $r\tau_{rz} =$ constant, and since the traction on the tube boundaries consists of pressures only, τ_{rz} must vanish. Equation (4.3.18) is therefore the only equilibrium equation of the problem.

If $u(r)$ is the radial displacement, then the radial and circumferential strain components are, from Equations (1.2.1), $\varepsilon_r = du/dr$ and $\varepsilon_\theta = u/r$, so that the compatibility condition is once more Equation (4.3.1).

Elastic Solution

The elastic stress-strain relations reduce to

$$\varepsilon_r = \frac{1}{E}[\sigma_r - \nu(\sigma_\theta + \sigma_z)],$$

$$\varepsilon_\theta = \frac{1}{E}[\sigma_\theta - \nu(\sigma_r + \sigma_z)],$$

$$\varepsilon_z = \frac{1}{E}[\sigma_z - \nu(\sigma_r + \sigma_\theta)].$$

If the third equation is solved for σ_z and the result is substituted in the first two, these equations become

$$\varepsilon_r = \frac{1+\nu}{E}[(1-\nu)\sigma_r - \nu\sigma_\theta] - \nu\varepsilon_z,$$

$$\varepsilon_\theta = \frac{1+\nu}{E}[(1-\nu)\sigma_\theta - \nu\sigma_r] - \nu\varepsilon_z.$$

Substitution in the compatibility relation (4.3.1) results in

$$\frac{d}{dr}[(1-\nu)\sigma_\theta - \nu\sigma_r] = \frac{\sigma_r - \sigma_\theta}{r},$$

which, when combined with (4.3.18), becomes

$$\frac{d}{dr}(\sigma_\theta + \sigma_r) = 0.$$

Consequently $\sigma_\theta + \sigma_r$ is constant, as is σ_z. Equation (4.3.18) can now be rewritten as

$$\frac{d}{dr}(\sigma_\theta - \sigma_r) + 2\frac{\sigma_\theta - \sigma_r}{r} = 0.$$

The solution of this equation is $\sigma_\theta - \sigma_r = 2B/r^2$, where B is a constant. If the constant value of $\sigma_\theta + \sigma_r$ is $2A$, then

$$\sigma_r = A - \frac{B}{r^2}, \quad \sigma_\theta = A + \frac{B}{r^2}.$$

With the boundary conditions

$$\sigma_r|_{r=a} = -p_i, \quad \sigma_r|_{r=b} = -p_e,$$

where p_i and p_e have the same meaning as for the sphere, the constants A and B can be solved for, and the stress components σ_r, σ_θ can be expressed as

$$\sigma_r = -\frac{1}{2}(p_i + p_e) + \frac{p_i - p_e}{2[1-(a/b)^2]}\left[1 + \left(\frac{a}{b}\right)^2 - 2\left(\frac{a}{r}\right)^2\right],$$

$$\sigma_\theta = -\frac{1}{2}(p_i + p_e) + \frac{p_i - p_e}{2[1-(a/b)^2]}\left[1 + \left(\frac{a}{b}\right)^2 + 2\left(\frac{a}{r}\right)^2\right].$$

Under internal pressure only, these equations simplify to

$$\begin{aligned} \sigma_r &= -\frac{p}{(b/a)^2 - 1}\left(\frac{b^2}{r^2} - 1\right), \\ \sigma_\theta &= \frac{p}{(b/a)^2 - 1}\left(\frac{b^2}{r^2} + 1\right). \end{aligned} \tag{4.3.19}$$

The axial stress σ_z is given by

$$\sigma_z = \frac{2\nu p}{(b/a)^2 - 1} + E\varepsilon_z,$$

and the radial displacement u is

$$u(r) = \frac{(1+\nu)p}{E[(b/a)^2 - 1]}\left[(1-2\nu)r + \frac{b^2}{r}\right] - \nu\varepsilon_z r.$$

Since the axial stress σ_z is constant, it is equal to $P/\pi(b^2 - a^2)$, where P is the resultant axial force. If P is prescribed, then ε_z may be determined accordingly. Alternatively, the value of ε_z may be prescribed; for example, a condition of *plane strain* ($\varepsilon_z = 0$) may be assumed.

If the tube is **open-ended**, then $P = 0$ and therefore $\sigma_z = 0$, so that this condition is equivalent to **plane stress** (as is shown later, the equivalence no longer holds when the tube is partially plastic). If the tube is **closed-ended**, then P must balance the resultant of the interior pressure over the interior cross-sectional area πa^2, so that $\sigma_z = p/[(b/a)^2 - 1] = \frac{1}{2}(\sigma_r + \sigma_\theta)$. The tube may also be assumed to be in a state of **plane strain**, that is, $\varepsilon_z = 0$, and thus $\sigma_z = \nu(\sigma_r + \sigma_\theta) = 2\nu p/[(b/a)^2 - 1]$. Note that this result coincides with the preceding one if and only if $\nu = \frac{1}{2}$, that is, if the material is elastically incompressible. The axial strain is given, for the three cases, by

$$E\varepsilon_z = \frac{\alpha p}{b^2/a^2 - 1} \tag{4.3.20}$$

where

$$\alpha = \begin{cases} 1 - 2\nu, & \text{closed end}, \\ 0, & \text{plane strain}, \\ -2\nu, & \text{open end}. \end{cases} \tag{4.3.21}$$

The axial stress is

$$\sigma_z = \frac{(2\nu + \alpha)p}{b^2/a^2 - 1}. \tag{4.3.22}$$

Initial Yield Pressure

The largest pressure for which the preceding solution is valid is that at which the stresses at some r first satisfy the yield criterion; this limiting pressure will again denoted p_E. It can be readily seen that in all three conditions — open end, closed end, and plane strain — the axial stress σ_z is the intermediate principal stress. Consequently, if the **Tresca criterion** is assumed, then it takes the simple form $\sigma_\theta - \sigma_r = 2k$, since $\sigma_\theta > 0$ and $\sigma_r < 0$. Since $\sigma_\theta - \sigma_r = pa^2/[1-(a/b)^2]r^2$, the maximum is reached at $r = a$, and therefore

$$p_E = k\left(1 - \frac{a^2}{b^2}\right). \tag{4.3.23}$$

The **Mises criterion** in terms of σ_r, σ_θ, σ_z is

$$\sigma_r^2 + \sigma_\theta^2 + \sigma_z^2 - \sigma_r\sigma_\theta - \sigma_r\sigma_z - \sigma_\theta\sigma_z = 3k^2,$$

and may be rewritten as

$$\left(\frac{\sigma_\theta - \sigma_r}{2}\right)^2 = k^2 - \frac{1}{3}\left(\sigma_z - \frac{\sigma_r + \sigma_\theta}{2}\right)^2.$$

Since σ_z is explicitly involved, the form taken by the criterion in the present case depends on the end conditions. In each case, however, the right-hand side is constant, so that as in the Tresca case, yielding first occurs at $r = a$, where the left-hand side takes the value $p_E^2/[1-(a/b)^2]^2$.

In the closed-end condition, substitution of $\sigma_z = \frac{1}{2}(\sigma_r + \sigma_\theta)$ leads to p_E given by (4.3.23). In the open-end condition, the criterion is the one that applies in plane stress, namely, $\sigma_r^2 - \sigma_r\sigma_\theta + \sigma_\theta^2 = 3k^2$. Hence

$$p_E = k\frac{1-(a/b)^2}{\sqrt{1+\frac{1}{3}(a/b)^4}}. \tag{4.3.24}$$

In plane strain, as we have found, $\sigma_z = \nu(\sigma_\theta + \sigma_r)$, and the criterion takes the form

$$(1 - \nu + \nu^2)(\sigma_r^2 + \sigma_\theta^2) - (1 + 2\nu - 2\nu^2)\sigma_r\sigma_\theta = 3k^2,$$

producing the critical pressure

$$p_E = k\frac{1-(a/b)^2}{\sqrt{1+\frac{1}{3}(1-2\nu)^2(a/b)^4}}. \tag{4.3.25}$$

The three results may be combined in the form

$$p_E = k\frac{1-(a/b)^2}{\sqrt{1+\frac{1}{3}(1-2\nu-\alpha)^2(a/b)^4}},$$

where α is defined by (4.3.21).

4.3.5. Elastic-Plastic Hollow Cylinder

When the pressure exceeds p_E, a plastic zone extends from $r = a$ to, say, $r = c$. The elastic zone $c < r < b$ behaves like an elastic tube of inner radius c that is just yielding at $r = c$, so that σ_r and σ_θ are given by

$$\left.\begin{aligned} \sigma_r &= -\frac{p_c}{(b/c)^2 - 1}\left(\frac{b^2}{r^2} - 1\right), \\ \sigma_\theta &= \frac{p_c}{(b/c)^2 - 1}\left(\frac{b^2}{r^2} + 1\right), \end{aligned}\right\} \quad c < r < b \tag{4.3.26}$$

where $p_c = -\sigma_r(c)$ is such that the yield criterion is met at $r = c$.

Tresca Criterion: Stress Field

If it is again assumed that $\sigma_\theta \geq \sigma_z \geq \sigma_r$ (an assumption that must be verified *a posteriori*) then the Tresca criterion yields

$$p_c = k\left(1 - \frac{c^2}{b^2}\right).$$

The associated flow rule for the Tresca criterion implies that with σ_z as the intermediate principal stress, ε_z is purely elastic and given by $[\sigma_z - \nu(\sigma_r + \sigma_\theta)]/E$, so that

$$\sigma_z = \frac{2\nu p_c}{(b/c)^2 - 1} + E\varepsilon_z.$$

In the plane-strain condition, $\varepsilon_z = 0$, the assumption that σ_z is the intermediate principal stress at $r = c$ can be immediately verified, since

$$\frac{b^2}{c^2} + 1 \geq 2\nu \geq -\frac{b^2}{c^2} + 1.$$

For the other two conditions, σ_z cannot be determined without studying the plastic zone.

The assumption $\sigma_\theta \geq \sigma_z \geq \sigma_r$ will be retained as valid in the plastic zone as well. The yield criterion $\sigma_\theta - \sigma_r = 2k$ must then be satisfied for $a \leq r \leq c$, and therefore the equilibrium equation (4.3.18) may be integrated immediately for σ_r. The solution

$$\sigma_r = -p + k \ln \frac{r^2}{a^2}$$

satisfies the boundary condition at $r = a$. Continuity of σ_r at $r = c$ leads to the result

$$p = k\left(1 - \frac{c^2}{b^2} + \ln \frac{c^2}{a^2}\right), \tag{4.3.27}$$

first derived by Turner [1909]. Equation (4.3.27) can be used to determine c for a given pressure p.

The *ultimate pressure* p_U is attained when the whole tube is plastic, that is, when $c = b$; it is given by

$$p_U = k \ln \frac{b^2}{a^2}. \tag{4.3.28}$$

If ε_z, as assumed, is purely elastic, then $\sigma_z = E\varepsilon_z + \nu(\sigma_r + \sigma_\theta)$ in the plastic zone as well. The resultant axial force is

$$P = 2\pi \int_a^b r\sigma_z \, dr = \pi E\varepsilon_z(b^2 - a^2) + 2\pi\nu \int_a^b (\sigma_r + \sigma_\theta) r \, dr.$$

But $r(\sigma_\theta + \sigma_r) = r(\sigma_\theta - \sigma_r + 2\sigma_r) = r^2 d\sigma_r/dr + 2r\sigma_r = d(r^2\sigma_r)/dr$ by the equilibrium equation (4.3.18), so that the integral on the right-hand side is $2\pi\nu r^2\sigma_r|_a^b = 2\pi\nu pa^2$. Hence

$$P = \pi[E\varepsilon_z(b^2 - a^2) + 2\nu pa^2],$$

independently of c. It follows that ε_z has the same form (4.3.20) as in the purely elastic tube, with α given by (4.3.21), and p by (4.3.27).

The stress field in the elastic zone can now be written as

$$\left.\begin{aligned} \sigma_r &= -k\left(\frac{c^2}{r^2} - \frac{c^2}{b^2}\right), \\ \sigma_\theta &= k\left(\frac{c^2}{r^2} + \frac{c^2}{b^2}\right), \\ \sigma_z &= k\left[2\nu\frac{c^2}{b^2} + \frac{\alpha}{(b/a)^2 - 1}\left(1 - \frac{c^2}{b^2} + \ln\frac{c^2}{a^2}\right)\right]. \end{aligned}\right\} \quad c < r < b \tag{4.3.29}$$

In the plastic zone, the stress field is

$$\left.\begin{aligned} \sigma_r &= -k\left(1 - \frac{c^2}{b^2} + \ln\frac{c^2}{r^2}\right), \\ \sigma_\theta &= k\left(1 + \frac{c^2}{b^2} - \ln\frac{c^2}{r^2}\right), \\ \sigma_z &= k\left[2\nu\left(\frac{c^2}{b^2} - \ln\frac{c^2}{r^2}\right) + \frac{\alpha}{(b/a)^2 - 1}\left(1 - \frac{c^2}{b^2} + \ln\frac{c^2}{a^2}\right)\right]. \end{aligned}\right\} \quad a < r < c \tag{4.3.30}$$

Limit of Validity

It is clear from the last equation that $\sigma_z \geq 0$ everywhere if $\alpha \geq 0$, that is, in the closed-end and plane-strain conditions. If $\alpha = -2\nu$ (open-end condition), then σ_z must be negative in some parts and positive in others in order that $P = 0$; note that in the elastic-plastic tube this condition is *not* equivalent to plane stress, as was remarked before. Nonetheless, the inequality $\sigma_z \geq \sigma_r$ can be verified for this condition as well. The last equation can in this case be rewritten as

$$\sigma_z = 2\nu\sigma_r + \frac{2\nu k}{(b/a)^2 - 1}\left(\frac{b^2}{a^2} + \frac{c^2}{b^2} - \ln\frac{c^2}{a^2} - 2\right).$$

The quantity in parentheses is easily seen to be nonnegative whenever $a \leq c \leq b$, and therefore, since $\sigma_r < 0$ throughout the plastic region, the inequality $\sigma_z \geq \sigma_r$ is always satisfied.

To ensure that $\sigma_\theta - \sigma_z \geq 0$ is satisfied everywhere, it is sufficient that it be satisfied where this quantity is smallest, namely, at $r = a$, where it equals

$$k\left[(1-2\nu)\left(1 + \frac{c^2}{b^2} - \ln\frac{c^2}{a^2}\right) + 2\nu - \frac{\alpha}{(b/a)^2 - 1}\left(1 - \frac{c^2}{b^2} + \ln\frac{c^2}{a^2}\right)\right].$$

For the results to be valid for all pressures up to ultimate, this quantity must be nonnegative for all c up to and including b, and therefore the limit of validity is given by the equation

$$2(1-\nu) - (1-2\nu)\ln\frac{b^2}{a^2} - \frac{\alpha}{(b/a)^2 - 1}\ln\frac{b^2}{a^2} = 0,$$

a relation between b/a and ν that depends on the end condition. For $\nu = 0.3$, the limiting values of b/a are respectively 5.43, 6.19 and 5.75 for the closed-end, open-end and plane-strain conditions.

The preceding result is due to Koiter [1953b], who pointed out that while an analysis may also be carried out for higher values of b/a, the hypothesis of infinitesimal deformations is no longer tenable.

The Plane-Stress Problem

A condition of **plane stress** ($\sigma_z = 0$) is descriptive of a thin circular disk with a concentric hole around whose edge a radial pressure p is applied. In this case the preceding solution for σ_r and σ_θ, given by Equations $(4.3.29)_{1,2}$ and Equations $(4.3.30)_{1,2}$, is valid as long as $\sigma_\theta \geq 0$, or

$$1 + \frac{c^2}{b^2} - \ln\frac{c^2}{a^2} \geq 0.$$

The entire disk can become plastic ($c = b$) only if $b/a \leq e$. For larger values of b/a, the plastic zone cannot expand beyond a limiting radius c whose value is furnished by the preceding inequality, turned into an equation. The maximum c/a turns out to be a *decreasing* function of b/a, with the limit $e^{1/2} = 1.649$ as $b/a \to inf$ (a circular hole in an infinite plate). In other words, the larger the disk in relation to the hole, the less the plastic zone can expand. Note that the pressure can never exceed $2k$.

While it may be thought that the preceding limitation is due to the discontinuous nature of the Tresca criterion, this is not the case. Qualitatively similar results are given by the Mises criterion. The equation representing the Mises criterion in plane stress,

$$\sigma_r^2 - \sigma_r\sigma_\theta + \sigma_\theta^2 = 3k^2,$$

can be solved for σ_θ:

$$\sigma_\theta = \frac{1}{2}\sigma_r + \sqrt{\frac{3}{2}}\sqrt{4k^2 - \sigma_r^2}.$$

When this expression is substituted in (4.3.18) then the differential equation is separable. Defining $s = -\sigma_r/2k$, we obtain

$$\frac{dr}{r} = -\frac{ds}{\sqrt{\frac{3}{2}}\sqrt{1-s^2} + \frac{s}{2}}.$$

The integral may be evaluated by means of the substitution $s = \sin(\theta + \pi/6)$, which turns the denominator of the right-hand side into $\cos\theta$ and the numerator into $(\sqrt{\frac{3}{2}}\cos\theta - \frac{1}{2}\sin\theta)\,d\theta$. The solution is

$$\ln r = A - \frac{\sqrt{3}}{2}\theta - \frac{1}{2}\ln\cos\theta,$$

where A is a constant of integration. Consider, for example, the question of the largest value of b/a for which a fully plastic plane-stress solution is possible. In this case $p = 2k$, so that $r = a$ corresponds to $s = 1$ or $\theta = \pi/3$, while $r = c = b$ corresponds to $s = 0$, or $\theta = -\pi/6$. It follows that b/a is given by

$$\ln\frac{b}{a} = \frac{1}{4}(\sqrt{3}\pi - \ln 3) = 1.086.$$

The fully plastic solution is therefore possible only if $b/a \le e^{1.086}$, a result not too different from the one furnished by the Tresca criterion.

To determine the greatest extent of the plastic domain in the limit as $b/a \to inf$, we again set $\theta = \pi/3$ at $r = a$, and $\theta = 0$ at $r = c$, since $\sigma_r = -k$ there. We obtain

$$\ln\frac{c}{a} = \frac{1}{2}\left(\frac{\pi}{\sqrt{3}} - \ln 2\right) = 0.560.$$

Determination of Displacement

The displacement field in a partly plastic tube governed by the Tresca yield criterion and its associated flow rule was also determined by Koiter

[1953b]. Since the volume strain $\varepsilon_r + \varepsilon_\theta + \varepsilon_z$ is elastic throughout the tube, it follows that

$$\frac{du}{r} + \frac{u}{r} = \frac{1-2\nu}{E}(\sigma_r + \sigma_\theta + \sigma_z) - \varepsilon_z = \frac{(1-2\nu)(1+\nu)}{E}(\sigma_r + \sigma_\theta) - 2\nu\varepsilon_z.$$

Substituting σ_θ from Equation (4.3.18), we obtain

$$\frac{d}{dr}(ru) = \frac{(1-2\nu)(1+\nu)}{E}\frac{d}{dr}(r^2\sigma_r) - 2\nu\varepsilon_z r,$$

or, upon integration,

$$u(r) = \frac{(1-2\nu)(1+\nu)}{E}r\sigma_r - \nu\varepsilon_z r + \frac{A}{r},$$

where A is a constant of integration. This solution is valid throughout the tube. In the elastic zone, it must coincide with that given by

$$u = r\varepsilon_\theta = \frac{r}{E}[\sigma_\theta - \nu(\sigma_r + \sigma_z)] = \frac{1+\nu}{E}[(1-\nu)\sigma_\theta - \nu\sigma_r]r - \nu\varepsilon_z r.$$

Substituting for σ_r, σ_θ from $(4.3.29)_{1,2}$ leads to

$$A = \frac{2kc^2(1-\nu^2)}{E}.$$

In the case of an *incompressible* solid ($\nu = \frac{1}{2}$) in *plane strain* ($\varepsilon_z = 0$), the displacement field reduces to

$$u(r) = \frac{kc^2}{2Gr}. \tag{4.3.31}$$

An analogous procedure for the annular disk (plane stress) leads to

$$u(r) = \frac{1-\nu}{E}r\sigma_r + \frac{2kc^2}{Er}.$$

Mises Criterion (Plane Strain)

If the tube material obeys the Mises yield criterion, then in the plane-strain condition a plastic zone $a \le r \le c$ forms when the pressure exceeds p_E as given by Equation (4.3.25). In the elastic zone the stresses σ_r and σ_θ are given by (4.3.26), and

$$\sigma_z = \nu(\sigma_r + \sigma_\theta) = 2\nu\frac{p_c}{(b/c)^2 - 1},$$

where p_c is given by the right-hand side of (4.3.25) with c replacing a. Defining

$$\bar{p} = \frac{k}{\sqrt{(b/c)^4 + \frac{1}{3}(1-2\nu)^2}},$$

we may write the stress components as

$$\sigma_r = \bar{p}\left(1 - \frac{b^2}{r^2}\right), \quad \sigma_\theta = \bar{p}\left(1 + \frac{b^2}{r^2}\right), \quad \sigma_z = 2\nu\bar{p}.$$

In the plastic region, the stress components cannot be determined without considering the displacement. Since the Mises flow rule implies that $\dot{\varepsilon}^p_{kk} = 0$, the volume strain $\varepsilon_{kk} = \varepsilon_r + \varepsilon_\theta$ is purely elastic, and hence the stresses can be written as

$$\sigma_r = s_r + K(\varepsilon_r + \varepsilon_\theta), \quad \sigma_\theta = s_\theta + K(\varepsilon_r + \varepsilon_\theta), \quad \sigma_z = -(s_r + s_\theta) + K(\varepsilon_r + \varepsilon_\theta),$$

K being the bulk modulus. The quantities s_r, s_θ, ε_r, and ε_θ form the basic un-knowns of the problem. It is convenient to regard them as functions of r and of c, with the latter as the time-like variable. s_θ, however, may be eliminated through the yield criterion, which takes the form

$$s_r^2 + s_r s_\theta + s_\theta^2 = k^2,$$

yielding

$$s_\theta = \frac{1}{2}\left(\sqrt{4k^2 - 3s_r^2} - s_r\right);$$

the positive root is chosen because $2s_\theta + s_r$ is positive in the elastic region. The equilibrium equation (4.3.18) now becomes

$$\frac{\partial s_r}{\partial r} + K\frac{\partial}{\partial r}(\varepsilon_r + \varepsilon_\theta) = \frac{1}{2r}\left(\sqrt{4k^2 - 3s_r^2} - s_r\right). \tag{4.3.32}$$

The strains ε_r and ε_θ furthermore satisfy the compatibility condition

$$\varepsilon_r = \frac{\partial}{\partial r}(r\varepsilon_\theta). \tag{4.3.33}$$

Finally, the flow rule must be invoked, in the form of the Prandtl–Reuss equations (3.3.4), which in the present case become

$$\frac{2G}{3}\frac{\partial}{\partial c}(2\varepsilon_r - \varepsilon_\theta) = \frac{\partial s_r}{\partial c} + 2G\dot{\lambda}s_r,$$

$$\frac{2G}{3}\frac{\partial}{\partial c}(2\varepsilon_\theta - \varepsilon_r) = \frac{\partial s_\theta}{\partial c} + 2G\dot{\lambda}s_\theta;$$

recall that c is used to denote time. Eliminating $\dot{\lambda}$ between the two equations and substituting for s_θ produces

$$\frac{\partial s_r}{\partial c} = \frac{G}{4k^2}\left[\left(4k^2 - 3s_r^2\right)\frac{\partial \varepsilon_r}{\partial c} - \left(2k^2 + 3s_r\sqrt{4k^2 - 3s_r^2}\right)\frac{\partial \varepsilon_r}{\partial c}\right]. \tag{4.3.34}$$

Equations (4.3.32)–(4.3.34) are three nonlinear partial differential equations in the unknown variables s_r, ε_r and ε_θ in $a \le r \le c$ for $a \le c \le b$.

Their solution can be effected numerically, as was done by Hodge and White [1950].[1] Hodge and White's results are reviewed in Section 4.5, where they are compared with a numerical solution based on the finite-element method.

A considerable simplification is achieved if the material is assumed to be elastically as well as plastically incompressible, that is, if $\nu = \frac{1}{2}$. In this case the relation $s_z = 0$ implies that both $\varepsilon_z^e = 0$ and $\varepsilon_z^p = 0$, and as we know from 3.3.4, when the latter condition holds, the Mises criterion coincides with the Tresca criterion based on k. Hence the stresses are given by Equations (4.3.29) and (4.3.30) in the elastic and plastic regions, respectively, with $\alpha = 0$ and $\nu = \frac{1}{2}$.

A comparison of the results of this simplification with those obtained numerically for $\nu < \frac{1}{2}$ shows that the values of σ_r and σ_θ are practically indistinguishable. For the axial stress and the displacement, on the other hand, the following corrections provide a good approximation:[2]

$$\sigma_z^{\text{comp}} \doteq 2\nu\sigma_z^{\text{inc}}, \quad u^{\text{comp}} \doteq 2(1-\nu)u^{\text{inc}}.$$

Unloading and Reloading

If a pressure $p > p_Y$ is applied to the interior of the tube and then reduced back to zero, the unloading process is elastic if p is not too great. The resulting field of *residual stresses* is obtained by subtracting the elastic stresses given by (4.3.19) and (4.3.22), with p given by (4.3.27), from the elastic-plastic stresses given, in the case of the Tresca criterion, by (4.3.29)–(4.3.30). Using, furthermore, Equation (4.3.23), we obtain the residual stresses in the form

$$\left.\begin{aligned}
\sigma_r &= -k\left[\frac{p}{p_E}\left(1-\frac{a^2}{r^2}\right) - \ln\frac{r^2}{a^2}\right],\\
\sigma_\theta &= -k\left[\frac{p}{p_E}\left(1+\frac{a^2}{r^2}\right) - \ln\frac{r^2}{a^2} - 2\right],\\
\sigma_z &= -2\nu k\left(\frac{p}{p_E} - 1 - \ln\frac{r^2}{a^2}\right),
\end{aligned}\right\} \quad a < r < c$$

and

$$\left.\begin{aligned}
\sigma_r &= -k\left(\frac{c^2}{a^2} - \frac{p}{p_E}\right)\left(\frac{a^2}{r^2} - \frac{a^2}{b^2}\right),\\
\sigma_\theta &= k\left(\frac{c^2}{a^2} - \frac{p}{p_E}\right)\left(\frac{a^2}{r^2} + \frac{a^2}{b^2}\right),\\
\sigma_z &= 2\nu k\left(\frac{c^2}{a^2} - \frac{p}{p_E}\right)\frac{a^2}{b^2}.
\end{aligned}\right\} \quad c < r < b$$

[1] A similar analysis of the tube under closed-end and open-end conditions was performed by Marcal [1965].

[2] Hodge and White [1950] (see also Prager and Hodge [1951], Section 16).

Note that the residual axial stress is independent of the end condition; the reason is that the axial strain ε_z, being purely elastic, is completely removed.

The largest value of $|\sigma_\theta - \sigma_r|$ can be seen to occur at $r = a$, where it equals $2k(p/p_E - 1)$. Also, at that location $\sigma_r = 0$ and $\sigma_z = \nu\sigma_\theta$, so that renewed yielding takes place if $p/p_E = 2$, provided that $2k$ remains the yield stress on reversed loading. The unloading is accordingly elastic if $p < 2p_E = 2k(1 - a^2/b^2)$, and this occurs for all pressures p up to the ultimate pressure p_U, given by (4.3.28), if the wall ratio b/a is such that $p_U < 2p_E$, that is, if it is less than that which satisfies the equation

$$\ln\frac{b^2}{a^2} = 2\left(1 - \frac{a^2}{b^2}\right).$$

The largest wall ratio for which the unloading is elastic at all pressures is found to be about 2.22; the corresponding ultimate pressure is about $1.59k$. When $b/a > 2.22$ and $p > 2p_E$, unloading produces a new plastic zone, $a < r < c'$ (with $c' < c$), in which $\sigma_\theta - \sigma_r = -2k$.

If the *shakedown pressure* is defined as $p_S = \min(p_U, 2p_E)$, then for $p \le p_S$ not only is unloading elastic, but so is any subsequent reloading with a pressure no greater than p. In other words, shakedown (as discussed in 3.5.3) takes place: the initial loading extends the elastic range of the tube, with the limiting pressure for purely elastic expansion increased from p_E to p. This strengthening can be attributed to the development of compressive hoop stresses σ_θ in the inner portion of the tube — an effect similar to that of hoops around a barrel. The process is commonly known as *autofrettage*, a French term meaning "self-hooping."

Effect of Work-Hardening

A solution for a work-hardening material will be obtained following Bland [1956]. We retain the assumption $\sigma_\theta > \sigma_z > \sigma_r$ and the Tresca yield criterion and flow rule, but assume that the yield stress k is a function of a hardening variable κ, defined, say, by (1.5.6) ("plastic work"); in the present case, this definition reduces to

$$\dot{\kappa} = k(\kappa)(|\dot{\varepsilon}_\theta^p| + |\dot{\varepsilon}_r^p|) = 2k(\kappa)|\dot{\varepsilon}_\theta^p|,$$

since $\dot{\varepsilon}_r^p = -\dot{\varepsilon}_\theta^p$. As long as $\dot{\varepsilon}_\theta^p \ge 0$ (no reverse yielding), the preceding relation can be integrated to give

$$\int_0^\kappa \frac{d\kappa'}{k(\kappa')} = 2\varepsilon_\theta^p,$$

so that κ is in a one-to-one relation with ε_θ^p, and k can also be viewed as a function of ε_θ^p.

A similar relation in uniaxial tension, with $\sigma_1 = \sigma$, $\sigma_2 = \sigma_3 = 0$, gives

$$\dot{\kappa} = k(\kappa)(|\dot{\varepsilon}_1^p| + |\dot{\varepsilon}_2^p| + |\dot{\varepsilon}_3^p|) = 2k(\kappa)(|\dot{\varepsilon}_1^p|),$$

since $\dot{\varepsilon}_2^p = \dot{\varepsilon}_3^p = -\frac{1}{2}\dot{\varepsilon}_1^p$. The relation between κ and ε_θ^p in the tube is thus the same as between κ and ε_1^p in uniaxial tension. If $\sigma_1 = \sigma_Y(\varepsilon_1^p)$ describes the uniaxial relation between stress and plastic strain, then, in the tube, $k = \frac{1}{2}\sigma_Y(\varepsilon_\theta^p)$.

In the elastic zone the stress field is given by Equations (4.3.29), with k replaced by $\frac{1}{2}\sigma_E$, where $\sigma_E = \sigma_Y(0)$ is the initial yield stress. In the plastic zone the equilibrium equation (4.3.18) becomes

$$\frac{d\sigma_r}{dr} = \frac{1}{r}\sigma_Y(\varepsilon_\theta^p), \tag{4.3.35}$$

with the boundary conditions $\sigma_r = -p$ at $r = a$ and $\sigma_r = \frac{1}{2}\sigma_E(1 - c^2/b^2)$ at $r = c$. The elastic relations for $\varepsilon_\theta^e = \varepsilon_\theta - \varepsilon_\theta^p$ and $\varepsilon_r^e = \varepsilon_r + \varepsilon_\theta^p$, together with the strain-displacement relations $\varepsilon_\theta = u/r$ and $\varepsilon_r = du/dr$, lead to

$$\frac{u}{r} = \varepsilon_\theta^p - \nu\varepsilon_z + \frac{1}{E}[(1-2\nu)(1+\nu)\sigma_r + (1-\nu^2)\sigma_Y(\varepsilon_\theta^p)]$$

and

$$\frac{du}{dr} = -\varepsilon_\theta^p - \nu\varepsilon_z + \frac{1}{E}[(1-2\nu)(1+\nu)\sigma_r - \nu(1+\nu)\sigma_Y(\varepsilon_\theta^p)].$$

Eliminating u between these equations, with the substitution from (4.3.35), leads further to

$$2\varepsilon_\theta^p + r\frac{d\varepsilon_\theta^p}{dr} = -\frac{1-\nu^2}{E}\left[2\sigma_Y(\varepsilon_\theta^p) + rH(\varepsilon_\theta^p)\frac{d\varepsilon_\theta^p}{dr}\right],$$

where $H = d\sigma_Y/d\varepsilon_\theta^p$ is the hardening modulus. The last equation can be integrated, subject to the boundary condition that $\varepsilon_\theta^p = 0$ at $r = c$, to yield

$$\varepsilon_\theta^p = \frac{1-\nu^2}{E}\left[\sigma_E\frac{c^2}{r^2} - \sigma_Y(\varepsilon_\theta^p)\right],$$

With the help of the last equation, r can be eliminated as a variable in favor of ε_θ^p, so that (4.3.32) can be integrated into a relation between σ_r and ε_θ^p. If ε_0^p denotes the value of ε_θ^p at $r = a$, then the pressure is given by

$$p = \frac{1}{2}\left(1 - \frac{c^2}{b^2}\right)\sigma_E + \frac{1}{2}\int_0^{\varepsilon_0^p} \frac{E + (1-\nu^2)H(\varepsilon_\theta^p)}{E\varepsilon_\theta^p + (1-\nu^2)\sigma_Y(\varepsilon_\theta^p)}\sigma_Y(\varepsilon_\theta^p)\,d\varepsilon_\theta^p, \tag{4.3.36}$$

with c given by

$$\frac{c^2}{a^2} = \frac{1}{\sigma_E}\left[\sigma_Y(\varepsilon_0^p) + \frac{E}{1-\nu^2}\varepsilon_0^p\right].$$

The two last equations provide the relation between p and c through the parameter ε_0^p. The integration is in general carried out numerically, but an explicit relation is easily obtained if the hardening is linear, that is, if H is constant. Details are left to an exercise.

Exercises: Section 4.3

1. Discuss the behavior of an elastic-plastic spherical shell under both external and internal pressure in comparison with its behavior under internal pressure only.

2. Evaluate Equation (4.3.10) for the case of linear work-hardening, the uniaxial relation being $\sigma_Y = H\varepsilon^p$, where the plastic modulus H is constant.

3. A spherical shell of initial inner and outer radii a_0, b_0 is made of a material whose stress-strain curve is given by $\sigma = f(\varepsilon)$ with no elastic range.

   (a) Find an equation to replace (4.3.12).

   (b) Find the condition for plastic instability.

4. A hollow cylinder of initial inner and outer radii a_0, b_0 is assumed to be in plane strain and made of an isotropically hardening Mises material that is both plastically and elastically incompressible.

   (a) Show that if the uniaxial stress-strain curve is given by $\sigma = f(\varepsilon)$, then $\sigma_\theta - \sigma_r = (2/\sqrt{3})f(2\varepsilon_\theta/\sqrt{3})$.

   (b) Assuming large strains, with $\varepsilon_\theta = \ln(r/r_0)$, find the analogues of Equations (4.3.10) and (4.3.12).

   (c) Determine the criterion for plastic instability.

5. Discuss how Equation (4.3.36) would be integrated for a cylinder made of a material with no elastic range.

6. Integrate Equation (4.3.36) and thus find an explicit p-c relation for the case of linear work-hardening [i. e., $\sigma_Y(\varepsilon^p) = \sigma_E + H\varepsilon^p$, with H constant].

Section 4.4 Elastic-Plastic Bending

4.4.1. Pure Bending of Prismatic Beams

General Concepts

A straight bar is said to be in a state of pure bending if the end sections are subject to normal tractions whose resultants are equal and opposite moments about an axis perpendicular to the longitudinal axis of the bar. It follows from equilibrium requirements that the stresses on every transverse section add up to the same moment (*constant bending moment*). In practice a condition of pure bending can be achieved by placing a bar on symmetrically located supports, with equal overhangs of length, say, a; if equal transverse forces F are applied at the ends of the bar, then the portion between the supports is in a state of constant bending moment Fa [see Figure 4.4.1(a)].

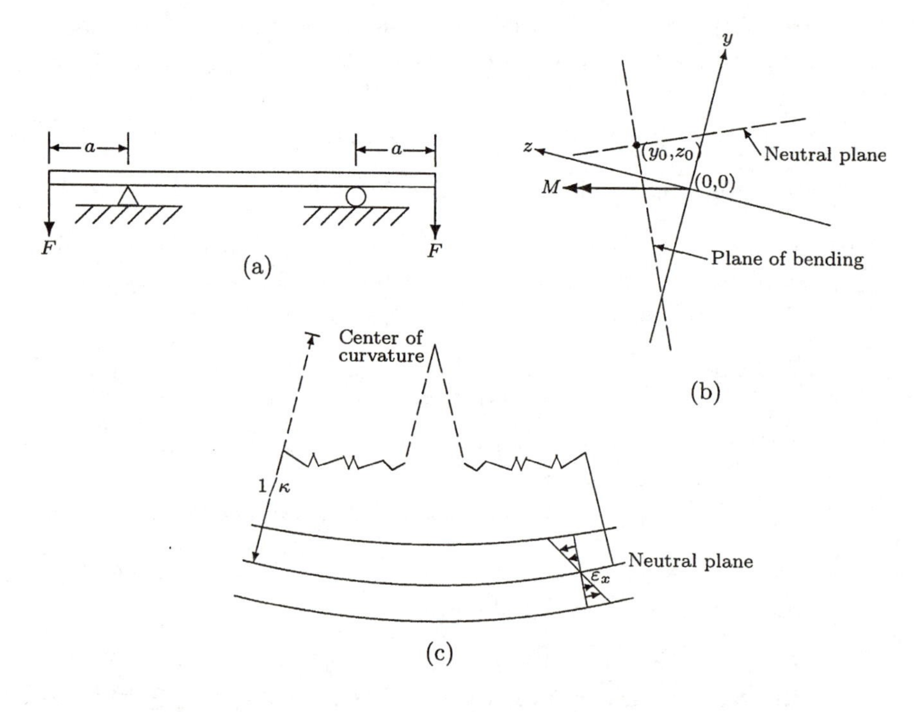

Figure 4.4.1. Pure bending: (a) possible beam geometry and loading; (b) local geometry in the cross-section plane; (c) local geometry in the plane of bending.

If the bar is prismatic (of uniform cross-section, in terms of both geometry and material properties) and sufficiently long in comparison with cross-sectional dimensions, then it can be assumed, as with the cylinder discussed in Section 4.3, that the stresses are independent of the longitudinal coordinate, except in the vicinity of the end sections. Let this coordinate be denoted x; then the preceding assumption, as well as the equilibrium equations and the boundary conditions of zero traction on the lateral surface, is satisfied if

$$\sigma_x = \sigma(y, z),$$

with all other stress components equal to zero. If the moment components about the y- and z-axes are M_y and M_z, respectively, then equilibrium is satisfied if

$$\int_A \sigma(y, z)\, dA = 0, \qquad \int_A z\sigma(y, z)\, dA = M_y, \qquad -\int_A y\sigma(y, z)\, dA = M_z. \tag{4.4.1}$$

Since every cross-section of the bar has the same stress distribution, it can be expected to have the same strain distribution as well. When the strain components are infinitesimal and independent of x, three of the compatibility conditions (1.2.4) reduce to

$$\frac{\partial^2 \varepsilon_x}{\partial y^2} = \frac{\partial^2 \varepsilon_x}{\partial z^2} = \frac{\partial^2 \varepsilon_x}{\partial y \partial z} = 0,$$

so that the longitudinal strain ε_x must be of the form $a + by + cz$, with a, b and c constant. It will be written as

$$\varepsilon_x = -\kappa[(y - y_0)\cos\alpha + (z - z_0)\sin\alpha]. \tag{4.4.2}$$

Here (y_0, z_0) are the coordinates of the intersection of the cross-section with a fiber that does not elongate (a *neutral fiber*), α is the angle between the *plane of bending* and the xy-plane, and κ is the *curvature* (the reciprocal of the radius of curvature) of the neutral fibers in the plane of bending. For an illustration of the geometry, see Figure 4.4.1(b) and (c).

The identical vanishing of the shear stresses τ_{xy} and τ_{xz} implies, if the material properties are isotropic, the vanishing of the corresponding strains γ_{xy} and γ_{xz}, so that the displacements satisfy

$$\frac{\partial v}{\partial x} + \frac{\partial u}{\partial y} = 0, \qquad \frac{\partial w}{\partial x} + \frac{\partial u}{\partial z} = 0.$$

Consequently,

$$\frac{\partial^2 v}{\partial x^2} = -\frac{\partial^2 u}{\partial x \partial y} = -\frac{\partial \varepsilon_x}{\partial y} = \kappa\cos\alpha,$$

$$\frac{\partial^2 w}{\partial x^2} = -\frac{\partial^2 u}{\partial x \partial z} = -\frac{\partial \varepsilon_x}{\partial z} = -\kappa\sin\alpha.$$

Combining, we find that

$$\kappa = \frac{\partial^2}{\partial x^2}(v\cos\alpha - w\sin\alpha).$$

The quantity in parentheses is the displacement of a fiber from its reference position in the plane of bending. Supposing this to be the xy-plane ($\alpha = 0$), we may write $\kappa = \partial^2 v/\partial x^2$. Strictly speaking, the curvature of the bent fiber is given by

$$\frac{\partial^2 v/\partial x^2}{[1 + (\partial v/\partial x)^2]^{3/2}},$$

and consequently the infinitesimal-deformation theory is valid as long as the slope of bent fibers with respect to the longitudinal axis is small, or, equivalently, as long as the radius of curvature remains large compared with the length of the bar.

Elastic Bending

If the bar is made of a single linearly elastic material with Young's modulus E, then Equation (4.4.2) can be combined with the stress-strain relation $\sigma_x = E\varepsilon_x$ to give

$$\sigma(y,\, z) = -E\kappa[(y - y_0)\cos\alpha + (z - z_0)\sin\alpha]. \qquad (4.4.3)$$

It is convenient to place the origin of the yz-plane at the centroid of the cross-section, that is,

$$\int_A y\, dA = \int_A z\, dA = 0.$$

Inserting (4.4.3) into the first of Equations (4.4.1) yields $y_0 \cos\alpha + z_0 \sin\alpha = 0$ as the equation governing the coordinates of the neutral fibers — that is, they lie in the plane (the *neutral plane*) that is perpendicular to the plane of bending and that contains the centroidal fiber. Equation (4.4.3) can accordingly be simplified to

$$\sigma(y,\, z) = -E\kappa(y\cos\alpha + z\sin\alpha). \qquad (4.4.4)$$

It is also convenient to make the y- and z-axes the *principal axes* of area, that is,

$$\int_A yz\, dA = 0.$$

Inserting (4.4.4) into the remaining Equations (4.4.1) leads to

$$M_y = -EI_y\kappa\sin\alpha, \qquad M_z = EI_z\kappa\cos\alpha, \qquad (4.4.5)$$

where

$$I_y = \int_A z^2\, dA, \quad I_z = \int_A y^2\, dA$$

are the *principal second moments of area* (often called the principal moments of inertia) of the cross-section.

The inclination of the bending plane is given, from (4.4.5), by $\tan\alpha = M_y I_z / M_z I_z$. The bending plane is perpendicular to the moment vector, whatever that may be, if $I_y = I_z$; this equality occurs whenever the cross-section has two or more *nonperpendicular* axes of symmetry, as in the case of a circle or a regular polygon (equilateral triangle, square, etc.). Otherwise, the moment vector is normal to the bending plane only if it is itself directed along a principal axis, that is, if $M_y = 0$ or $M_z = 0$.

Combining Equations (4.4.4) and (4.4.5) gives the stress distribution

$$\sigma_x = -\frac{M_z y}{I_z} + \frac{M_y z}{I_y}. \tag{4.4.6}$$

All the preceding results are valid as long as the extreme values of the stress given by (4.4.6) remain in the elastic range. More generally, we may assume a nonlinear stress-strain relation $\sigma = f(\varepsilon)$. The function f, with the right-hand side of (4.4.2) as its argument, defines $\sigma(y, z)$, which when inserted in (4.4.1) furnishes three *coupled, nonlinear* equations for the parameters κ, α, and $y_0 \cos\alpha + z_0 \sin\alpha$, the last of which is the displacement between the neutral plane and the centroidal axis. Coupled nonlinear equations are generally difficult to solve. A simplification is achieved if the bending is symmetric.

Symmetric Bending

An axis of symmetry is necessarily a principal axis of the cross-section. Let the section be symmetric about the y-axis, and let $M_y = 0$, $M_z = M$. Then $\alpha = 0$, and Equations (4.4.5)–(4.4.6) simplify to $M = EI\kappa$, $\sigma_x = -My/I$, where $I = I_z$, in the elastic range. However, symmetry alone dictates $\alpha = 0$, independently of material properties, and this condition may be assumed to hold under nonlinear material behavior as well. With the stress-strain relation $\sigma = f(\varepsilon)$ as above, the parameters κ and y_0 are governed by

$$\int_A f(-\kappa(y - y_0))\, dA = 0, \qquad \int_A f(-\kappa(y - y_0)) y\, dA = -M. \tag{4.4.7}$$

If y_0 can be eliminated between these equations, then the result is the **moment-curvature relation**

$$M = \Phi(\kappa). \tag{4.4.8}$$

Doubly Symmetric Sections

An even greater simplification is achieved if the z-axis, also, is an axis of symmetry and if the stress-strain relation is the same in tension and compression, that is, if f is an odd function. In that case $y_0 = 0$ by symmetry, and the function Φ is given explicitly by

$$\Phi(\kappa) = \int_A f(\kappa y) y\, dA.$$

Consider, for example, a bar of rectangular section of width b and depth $2c$ (so that $I = 2bc^3/3$), made of an elastic–perfectly plastic material. The elastic moment-curvature relation and stress distribution are valid as long as $|\sigma|_{\max} = 3|M|/2bc^2 \le \sigma_Y$, or $|M| \le M_E$, where $M_E = 2\sigma_Y bc^2/3$ is the initial yield moment.

When $M > M_E$, the elastic relation $\sigma = -E\kappa y$ holds at values of y at which this quantity does not exceed σ_Y in magnitude, that is, for $|y| \le y^* \stackrel{\text{def}}{=} \sigma_Y/E|\kappa|$; this range forms the *elastic core*. The *plastic zones* are $-c < y < -y^*$ and $y^* < y < c$, where $\sigma = \sigma_Y \operatorname{sgn}\kappa$ and $\sigma = -\sigma_Y \operatorname{sgn}\kappa$, respectively; the stress distribution is shown graphically in Figure 4.4.2(a). Because of symmetry, the moment is given by

$$M = -2b\int_0^c \sigma y\,dy = 2b\left[E\kappa\int_0^{y^*} y^2\,dy + \sigma_Y \operatorname{sgn}\kappa\int_{y^*}^c y\,dy\right].$$

Upon integration and substitution for y^*, this becomes

$$M = M_U \operatorname{sgn}\kappa\left[1 - \frac{1}{3}\left(\frac{\kappa_E}{\kappa}\right)^2\right], \qquad |\kappa| > \kappa_E, \tag{4.4.9}$$

where $\kappa_E = \sigma_Y/Ec$ is the absolute value of the curvature corresponding to $|M| = M_E$, and $M_U = \sigma_Y bc^2$ is the *ultimate moment* which is attained asymptotically as $|\kappa| \to \infty$. The moment-curvature relation given by (4.4.9) is shown in Figure 4.4.2(b), and is quite similar to the torque-twist relation (4.2.4) for an elastic–perfectly plastic shaft of solid circular cross-section. The difference is that while the ratio T_U/T_E between the ultimate and initial yield torques is 4/3 in the torsion of a circular shaft, in the bending of a rectangular beam the corresponding ratio M_U/M_E, known as the *shape factor*, is 3/2.

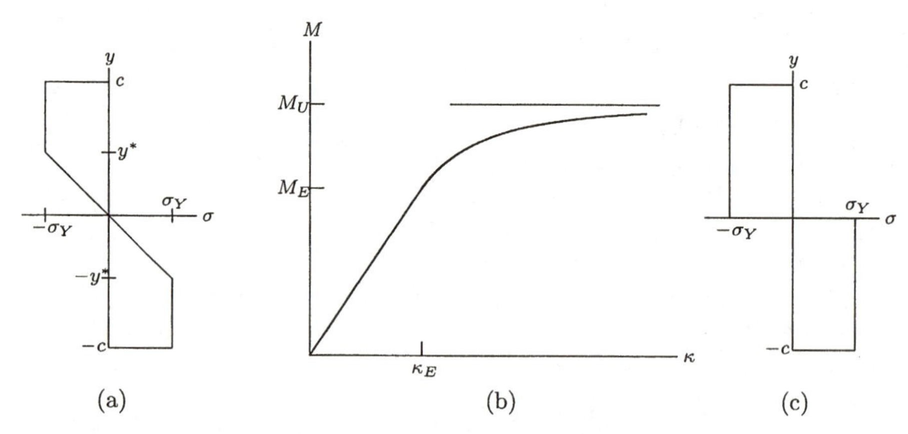

Figure 4.4.2. Elastic-plastic beam: (a) stress distribution; (b) moment-curvature relation; (c) ultimate stress distribution.

Suppose, now, that the material work-hardens according to the law (2.1.4), with $0 < n < 1$; the limiting cases $n = 0$ and $n = 1$ represent perfect plasticity and unlimited elasticity, respectively. The moment-curvature relation can easily be shown to be

$$M = \frac{M_E}{2+n} \operatorname{sgn} \kappa \left[3 \left| \frac{\kappa}{\kappa_E} \right|^n - (1-n) \left(\frac{\kappa_E}{\kappa} \right)^2 \right], \quad |\kappa| > \kappa_E, \tag{4.4.10}$$

where $\kappa_E = \sigma_E/Ec$, $M_E = 2\sigma_E bc^2/3$. Note that there is no ultimate moment when $n > 0$.

Elastic-plastic moment-curvature relations analogous to (4.4.9) can be found for other doubly symmetric sections, although the integration often gets cumbersome. It is fairly easy, however, to determine the ultimate moment and hence the shape factor. In the limit as $\kappa \to \infty$, the stress distribution becomes as shown in Figure 4.4.2(c), that is, consisting of two blocks of constant stress of value σ_Y and $-\sigma_Y$, respectively, each distributed over one-half the area, and statically equivalent to equal and opposite forces of magnitude $\sigma_Y A/2$, each acting at the centroid of the half-area. The ultimate moment is therefore

$$M_U = \frac{1}{2}\sigma_Y Ad, \tag{4.4.11}$$

where d is the *distance between the centroids of the half-areas*. Since $M_E = \sigma_Y I/c$, where c is one-half the depth, the shape factor is $Acd/2I$. The quantity $\frac{1}{2}Ad$ is often called the *plastic modulus* of the section and denoted Z.

For a bar of solid circular cross-section with radius c, we have $I = \pi c^4/4$, $A = \pi c^2$, and $d/2 = (4/3\pi)c$; hence the shape factor is $16/3\pi \approx 1.7$. For a thin-walled circular tube of wall thickness t, $I = \pi c^3 t$, $A = 2\pi ct$, and $d/2 = (2/\pi)c$, so that $M_U/M_E = 4/\pi \approx 1.27$. As a rule, the shape factor is greater the more material is concentrated near the center, and the smaller (closer to 1) the more material is concentrated near the extreme fibers. For rolled structural shapes, typical values are near 1.2 for I-beams and between 1.1 and 1.15 for wide-flange beams. The ideal sandwich beam (or ideal I-beam) discussed in 3.5.1 has a shape factor of 1. The moment-curvature relation for such a beam is

$$M = \begin{cases} EI\kappa, & |\kappa| \le M_U/EI, \\ M_U \operatorname{sgn} \kappa, & |\kappa| \ge M_U/EI, \end{cases} \tag{4.4.12}$$

analogous to the torque-twist relation for a thin-walled circular tube, discussed in 4.1.1.

Sections Without Double Symmetry

For sections without symmetry about the z-axis the determination of a moment-curvature relation is more complicated, since y_0 as a function of κ must first be found from the first of Equations (4.4.7) and then inserted in the second. For an elastic–perfectly plastic material it is easy, however, to find the asymptotic value of y_0 in the limit of infinite curvature, since the

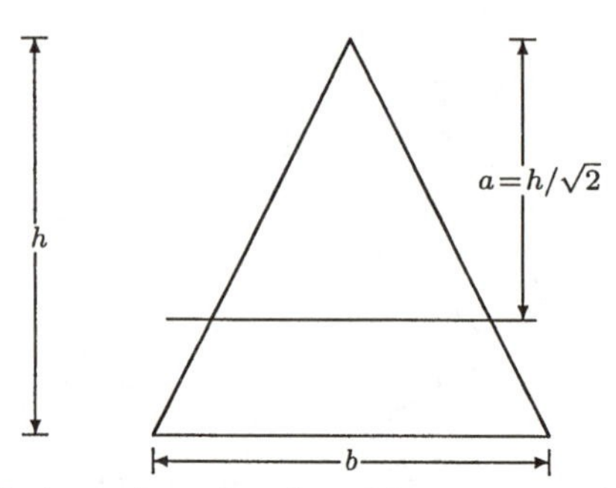

Figure 4.4.3. Isosceles triangle: ultimate neutral plane.

neutral plane must then divide the cross-section area into equal halves, as in the case of sections with double symmetry. Unlike this special case, in general this plane does not go through the centroid. The ultimate moment is still given by (4.4.11), with $d = d_1 + d_2$, where d_1 and d_2 are the distances from the centroids of each of the half-areas to the ultimate neutral plane. The relations $\kappa_E = \sigma_Y/Ec$, $M_E = \sigma_Y I/c$, and the expression $\frac{1}{2}Acd/I$ for the shape factor, are valid if c is interpreted as the distance from the centroid to the *farthest* fiber in the xy-plane. When $|\kappa|$ exceeds κ_E somewhat, a plastic zone forms on the side of the farthest fiber, but the rest of the section remains elastic as long as the magnitude of the stress at the opposite extreme fiber does not attain σ_Y. It is only when this last condition is reached that the second plastic zone forms. The calculations are often involved.

Consider, for example, an isosceles triangle of base b and height h. For a section of this shape, $A = bh/2$, $c = 2h/3$, and $I = bh^3/36$. To determine d, we must first find the location of the ultimate neutral plane. If the distance from this plane to the apex is a (see Figure 4.4.3), then the area of the isosceles triangle of height a and base ba/h must equal one-half the area of the whole triangle, that is, $\frac{1}{2}ba^2/h = \frac{1}{4}bh$; hence $a = h/\sqrt{2}$, while $d_1 = a/3 = h/3\sqrt{2}$. For a trapezoid of height $h - a = (1 - 1/\sqrt{2})h$ and major and minor bases b and $ba/h = b/\sqrt{2}$, respectively, it can be shown that the distance from the centroid to the minor base is $(8 - 5\sqrt{2})h/6$. Consequently $d = 2(2 - \sqrt{2})h/3$, and the shape factor is easily calculated to be $4(2 - \sqrt{2}) \approx 2.34$. The elastic-plastic solution for the isosceles triangle can be found, for example, in Nadai [1950], Volume I, page 358.

Unloading and Residual Stresses

Consider a symmetrically bending elastic-plastic beam whose moment-curvature relation has the general form $M = \Phi(\kappa)$, of which (4.4.9), (4.4.10) and (4.4.12) are special cases. Suppose, in particular, that initial loading has produced a curvature κ_0. If the moment $M = \Phi(\kappa_0)$ is removed, then the residual stress distribution can be calculated on the assumption that the unloading is elastic, that is, stresses equal to $\Phi(\kappa_0)y/I$ are added to those produced by the initial loading. The assumption is justified if the resulting stresses are in the elastic range.

The springback, described by the change in curvature, is likewise elastic, and given by $-\Phi(\kappa_0)/EI$. The residual curvature, which defines the permanent deformation of the beam, is

$$\kappa_{res} = \kappa_0 - \frac{\Phi(\kappa_0)}{EI}.$$

If, for example, the moment-curvature relation is (4.4.10), the ratio of residual to initial curvature is (if $\kappa_0 > 0$)

$$\frac{\kappa_{res}}{\kappa_0} = 1 - \frac{1}{2+n}\left[3\left(\frac{\kappa_E}{\kappa_0}\right)^{1-n} - \left(\frac{\kappa_E}{\kappa_0}\right)^3\right]. \tag{4.4.13}$$

In the perfectly plastic case ($n = 0$), half of the curvature is recovered when $\kappa_0 = 2.88\kappa_E$.

The calculation of residual stresses is particularly simple for doubly symmetric sections, since there is then no need to find y_0, and the residual stresses are given by

$$\sigma_{res} = \Phi(\kappa_0)\frac{y}{I} - f(\kappa_0 y).$$

A typical residual-stress distribution is shown in Figure 4.4.4. Note that the

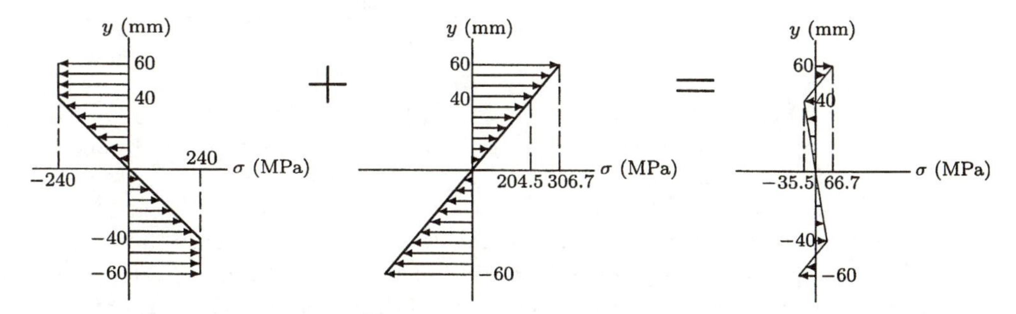

Figure 4.4.4. Residual-stress distribution in an elastic–perfectly plastic rectangular beam.

stresses near the center of the section keep their sign, while near the extreme fibers the loading is reversed. If the material is perfectly plastic, then the residual stresses at the extreme fibers are $\pm(1 - |M/M_E|)\sigma_Y$. Reverse yielding can occur only if the shape factor is greater than 2, a condition unlikely to occur in doubly symmetric sections.

Asymmetric Plastic Bending

When the moment is not perpendicular to an axis of symmetry, the elastic-plastic solution becomes exceedingly difficult even for a simple cross-section. The initial yield condition can be determined by finding, for given M_y and M_z, the maximum absolute value of the right-hand side of (4.4.6) and equating it to σ_Y, a procedure that gives a relation between M_y and M_z. For a rectangle, this relation is

$$|M_z| + \frac{2c}{b}|M_y| = M_E$$

if M_E is defined for symmetric bending in the xy-plane as before; the relation is represented by the rhombus shown in Figure 4.4.5(a). The fully plastic condition can be studied according to the following procedure, illustrated in Figure 4.4.5(b).

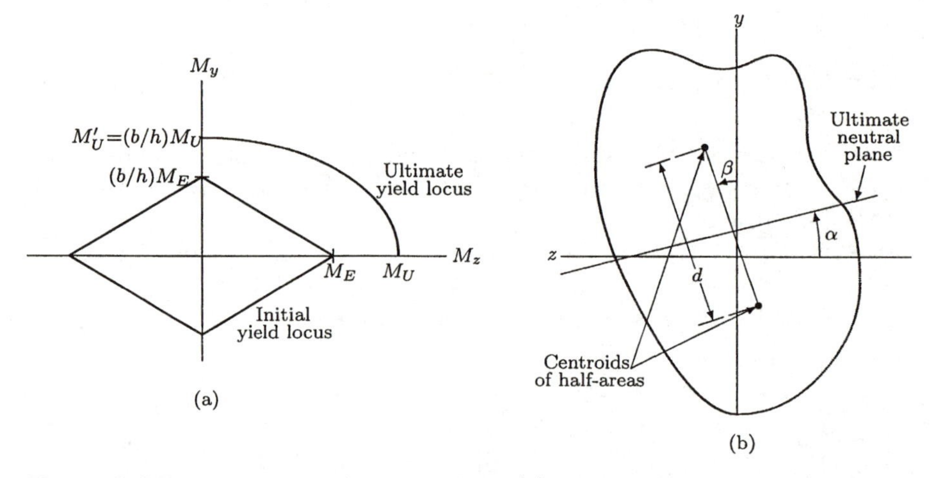

Figure 4.4.5. Asymmetric plastic bending: (a) initial and ultimate yield loci for a rectangular beam; (b) ultimate neutral place and centroids of half-areas.

The ultimate neutral plane is assumed to form an angle α with the z-axis. Since this plane divides the cross-section into two equal areas, its location can be determined by geometry, as can the locations of the centroids of the half-areas. If the distance between these centroids is once more denoted d, then the magnitude of the fully plastic moment is again $\sigma_Y Ad/2$. It must be kept in mind, however, that d depends on α. The line joining the centroids of the half-areas forms an angle β with the y-axis, which depends on α. The moment vector is perpendicular to this line, and its components are, if the half-area below the neutral plane is in tension,

$$M_y = -\frac{1}{2}\sigma_Y Ad \sin\beta, \qquad M_z = \frac{1}{2}\sigma_Y Ad \cos\beta.$$

Since both d and β depend on α, the preceding equation forms a parametric representation, in terms of α, of a closed curve in the M_yM_z-plane called the *interaction diagram.*

For a rectangle of width b and depth h, it can be shown that

$$d\cos\beta = \frac{h}{2}\left(1 - \frac{b^2}{3h^2}\tan^2\alpha\right), \quad d\sin\beta = \frac{b^2}{3h}\tan\alpha, \quad \tan\alpha \le \frac{h}{b}$$

and

$$d\cos\beta = \frac{h^2}{3b}\cot\alpha, \quad d\sin\beta = \frac{b}{2}\left(1 - \frac{h^2}{3b^2}\cot^2\alpha\right), \quad \tan\alpha \ge \frac{h}{b}.$$

It is easy to eliminate α between the expressions for M_y and M_z to obtain the interaction relation directly. If we define $M_U = \sigma_Y bh^2/4$ as before, $M'_U = \sigma_Y b^2h/4 = (b/h)M_U$, $m_y = M_y/M'_U$, and $m_z = M_z/M_U$, then the relation is given by the two equations

$$|m_z| + \frac{3}{4}m_y^2 = 1, \quad \left|\frac{m_y}{m_z}\right| \le 1, \quad \textit{and} \quad |m_y| + \frac{3}{4}m_z^2 = 1, \quad \left|\frac{m_y}{m_z}\right| \ge 1.$$

The relation is represented by the outer curve of Figure 4.4.5(a). The interaction curves for many other sections with double symmetry are found to lie fairly close to this curve.

4.4.2. Rectangular Beams Under Transverse Loads

End-Loaded Cantilever: Elastic Solution

The elastic theory of beams under transverse loads has its origin in Saint-Venant's research on the bending of a prismatic cantilever carrying a concentrated force $\mathbf{F} = \mathbf{j}F_y + \mathbf{k}F_z$ at its free end. Saint-Venant began by assuming that Equation (4.4.6) for the longitudinal stress σ_x remains valid, along with $\sigma_y = \sigma_z = \tau_{yz} = 0$. If the beam spans $0 < x < L$, with $x = 0$ denoting the built-in end, the moment acting on the cross-section at x is $(L - x)\mathbf{i} \times \mathbf{F}$, or $M_z = (L - x)F_y$, $M_y = -(L - x)F_z$. Since σ_x depends on x, equilibrium requires the presence of shear stresses τ_{xy}, τ_{xz}, independent of x, that satisfy

$$\int_A \tau_{xy}\, dA = F_y, \qquad \int_A \tau_{xz}\, dA = F_z.$$

Since the problem is linear, it may be treated as a superposition of the two problems corresponding to $F_z = 0$ and $F_y = 0$, and since the labeling of the principal axes as y or z is arbitrary, only one of these problems need be treated — for example, $F_z = 0$, $F_y = F$. It must be kept in mind that τ_{xz} *cannot* be assumed to vanish. The application of the one nontrivial equilibrium equation to the stresses, and of the compatibility conditions to the strains resulting from them, reduces the problem to one of solving the Laplace equation in the yz-plane subject to certain boundary conditions, in addition to that of finding the stress function of the elastic torsion problem (see 4.2.2); the latter is unnecessary if the y-axis is a symmetry axis of the cross-section *and* is the line of action of the force, as will be assumed. Details of the theory, along with solutions for circular, elliptic, rectangular and other cross-sections can be found in most books on elasticity.[1] In particular, for any cross-section the deflection satisfies $\partial^2 v/\partial x^2 = F(L-x)/EI_z$, so that within the limitations of small-deformation theory, the *local* moment-curvature relation is the same as in pure bending.

The results for the rectangular beam show that the z-average of τ_{xz} vanishes, while that of τ_{xy} is given by

$$\bar{\tau}_{xy} \stackrel{\text{def}}{=} \frac{1}{b}\int_{-b/2}^{b/2} \tau_{xy}\,dz = \frac{3F}{4bc^2}(c^2 - y^2),$$

where b and c are the width and half-depth of the cross-section, respectively, as before. Moreover, the shear stresses τ_{xy} become independent of z in the limit of both infinitely narrow ($b/c \to 0$) and infinitely wide ($b/c \to \infty$) rectangles, corresponding respectively to solutions for plane stress and plane strain. For such beams, then, the stress distribution is

$$\sigma = \frac{F(L-x)y}{I}, \quad \tau = \frac{F}{2I}(c^2 - y^2), \tag{4.4.14}$$

where $\sigma = \sigma_x$, $\tau = \tau_{xy}$, and $I = I_z = 2bc^3/3$.

End-Loaded Cantilever: Elastic-Plastic Solution

For the state of stress just derived, both the Mises and the Tresca yield criteria are given by Equation (3.3.5). It is not difficult to see that in the cantilever under study, the left-hand side of this equation attains its maximum wherever $|\sigma|$ is greatest, that is, at the corners $(0, \pm c)$. Consequently, plasticity begins when $F = F_E \stackrel{\text{def}}{=} 2\sigma_Y bc^2/3L$. If the load F is increased beyond this value, plastic zones develop at the top and bottom faces of the beam, spreading from the corners, as shown in Figure 4.4.6.

In the plastic zones the yield criterion requires that the stresses obey

$$\frac{\sigma}{\sigma_Y^2}\frac{\partial \sigma}{\partial x} + \frac{\tau}{\tau_Y^2}\frac{\partial \tau}{\partial x} = 0, \qquad \frac{\sigma}{\sigma_Y^2}\frac{\partial \sigma}{\partial y} + \frac{\tau}{\tau_Y^2}\frac{\partial \tau}{\partial y} = 0$$

in addition to the equilibrium equations

$$\frac{\partial \sigma}{\partial x} + \frac{\partial \tau}{\partial y} = 0, \qquad \frac{\partial \tau}{\partial x} = 0.$$

[1] For example, Love [1927], Chapter 15; Timoshenko and Goodier [1970], Chapter 11; or Boresi and Chong [1987], Chapter 7.

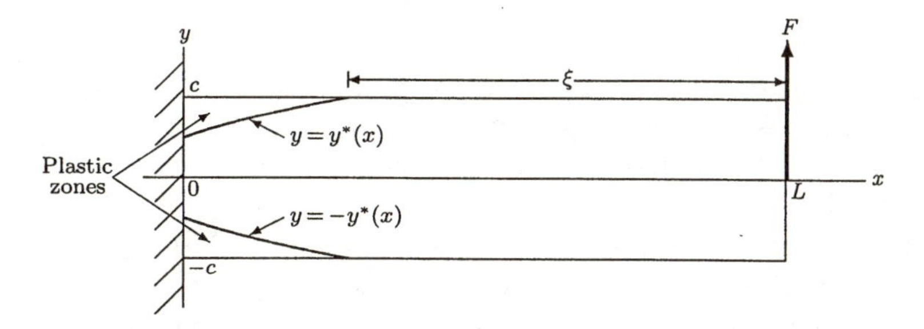

Figure 4.4.6. Elastic and plastic zones in an end-loaded cantilever.

These equations can be satisfied simultaneously only if σ and τ are both constant in each plastic zone, and since τ must vanish on the top and bottom faces, it must vanish everywhere in the plastic zones, and therefore $\sigma = \pm\sigma_Y$. In particular, if the elastic-plastic boundary at x is located at $y = \pm y^*(x)$, then $\sigma = \sigma_Y$ in $-c \le y \le -y^*(x)$ and $\sigma = -\sigma_Y$ in $y^*(x) \le y \le c$. In the elastic core $-y^*(x) \le y \le y^*(x)$ the y-distribution of σ is linear as in the elastic case, that is, $\sigma = -\sigma_Y y/y^*(x)$. The resultant moment about the z-axis at x is

$$M = -b\int_{-c}^{c} y\sigma\, dy = \sigma_Y b\left(c^2 - \frac{y^{*2}}{3}\right),$$

and since for equilibrium this must equal $F(L-x)$, it follows that the elastic-plastic boundary is given by

$$y^*(x) = \sqrt{3\left[c^2 - \frac{F(L-x)}{\sigma_Y}b\right]}. \tag{4.4.15}$$

In particular, $y^*(x) = c$ at $x = L - 2\sigma_Y bc^2/3F \stackrel{\text{def}}{=} L - \xi$; the length ξ represents the extent of the elastic zone at the given value of F. We can accordingly write, more simply,

$$y^*(x) = c\sqrt{3 - \frac{2(L-x)}{\xi}}.$$

The elastic-plastic boundary shown in Figure 4.4.6 is thus given by two symmetrically located segments of a parabola whose vertex is on the extension of the beam axis beyond the built-in section.

As in the case of the twisted shaft, the deformation in the plastic zone is contained by the elastic deformation as long as the elastic core is of non-vanishing thickness. The regime of contained plastic deformation ends when $y^*(0) = 0$, that is, when the vertex of the parabola reaches the built-in section of the beam. The corresponding value of the load is $F = \sigma_Y bc^2/L \stackrel{\text{def}}{=} F_U$, the *ultimate load*. When it is attained, unrestrained plastic flow may begin, taking the form of a rigid rotation of the remaining elastic region about the point (0, 0), with an attendant stretching and shrinking of the x-fibers in the upper and lower plastic zones, respectively. The resulting mechanism is known as a *plastic hinge*. Plastic hinges are used extensively in the plastic analysis of beams and framed structures under various loadings, as is shown later in this book.

The preceding derivation of the ultimate load neglects the growth of the shear stress in the elastic core. In fact, since $\tau = 0$ in the plastic zones, the transverse force must be balanced by the shear stress in the elastic core only. This shear stress has a parabolic distribution in $-y^*(x) \le y \le y^*(x)$, with a maximum value, at $(x, 0)$, of $3F/4by^*(x)$, so that the maximum shear stress in the beam is $\tau_{\max} = 3F/4by^*(0)$. As the elastic core shrinks, this maximum shear stress grows until it attains τ_Y, when a secondary plastic zone forms. With the help of Equation (4.4.15), the value of the force F at which this occurs is found to be given by

$$\left(\frac{F}{F_U}\right)^2 + \frac{16}{3\beta^2}\frac{F}{F_U} - \frac{16}{3\beta^2} = 0,$$

where $\beta = \sigma_Y c/\tau_Y L$. The only positive root of the equation is

$$\frac{F}{F_U} = \frac{8}{3\beta^2}\left(\sqrt{1 + \frac{3}{4}\beta^2} - 1\right).$$

For span-to-depth ratios greater than 3, we have $c/L < 1/6$, so that in a beam made of a Mises material ($\sigma_Y/\tau_Y = \sqrt{3}$), β^2 will be no greater than 1/12. The secondary plastic zone will therefore not form before $F = 0.985F_U$, and the effect of the shear stress in the elastic core on the formation of the plastic hinge may accordingly be neglected.

As a further refinement we note that since the shear stress in the elastic core varies with x, a transverse normal stress σ_y must develop there as well. However, the effect of this stress on the criterion for the initiation of the inner plastic zone, and consequently on the ultimate load for reasonably long beams, is negligible for sufficiently great span-to-depth ratios.[1]

In order to determine the deflection, it will be assumed that the moment-curvature relation derived for pure bending holds locally in the elastic-plastic portion of the beam as well. Letting $\kappa = \bar{v}''(x)$, where $\bar{v}(x) \stackrel{\text{def}}{=} v(x, 0)$, and using Equation (4.4.9) with the relevant substitutions, we obtain

$$\bar{v}''(x) = \frac{\sigma_Y}{Ec}\begin{cases} \dfrac{1}{\sqrt{3 - 2(L - x)/\xi}}, & x < L - \xi, \\ L - \dfrac{x}{\xi}, & x > L - \xi. \end{cases}$$

[1]For a "stubby" cantilever whose span is comparable to its depth, beam theory cannot be used; the problem of the collapse load for such a beam is studied as a limit-analysis problem in 6.1.2.

Note that $\bar{v}''(0) \to \infty$ as $F \to F_U$; this is consistent with the notion of a plastic hinge, since infinite curvature at a point on a curve corresponds to a slope discontinuity.

Rather than determine the whole function $\bar{v}(x)$, we calculate only the tip deflection $\Delta = \bar{v}(L) - \bar{v}(0)$ by means of the fundamental relation

$$\bar{v}(L) - \bar{v}(0) = \int_0^L \bar{v}'(x)\, dx,$$

which by integration by parts becomes

$$\Delta = \int_0^L (L - x)\bar{v}''(x)\, dx, \tag{4.4.16}$$

since $\bar{v}'(0) = 0$ and $\bar{v}'(L) = \int_0^L \bar{v}''(x)\, dx$. Hence, for $2L/3 <= \xi <= L$, a change of variable from x to $L - x$ leads to

$$\begin{aligned}\Delta &= \frac{\sigma_Y}{Ec}\left[\int_0^\xi \frac{x^2}{\xi} dx + \int_\xi^L \frac{x}{\sqrt{3 - 2x/\xi}} dx\right] \\ &= \Delta_E \left(\frac{F_E}{F}\right)^2 \left[5 - \left(3 + \frac{F}{F_E}\right)\sqrt{3 - 2\frac{F}{F_E}}\right],\end{aligned}$$

where $\Delta_E = \sigma_Y L^2/3Ec$ is the deflection when $F = F_E$. Since $F_U = 3F_E/2$, we see that the tip deflection increases by a factor of 20/9 as F goes from F_E (beginning of plastic deformation) to F_U (incipient plastic flow).

Uniformly Loaded Cantilever

A theory of elastic beams carrying a transverse load, comprising both body force and surface traction, that is uniformly distributed along the length is due to Michell. For such a beam, proportionality between moment and curvature no longer holds exactly, although it is a good approximation when the beam is long compared to its cross-sectional dimensions. For a cantilever carrying a downward load q per unit length, the resultant shear force at x (positive upward) is $Q = -q(L - x)$, and the bending moment is $M = -\frac{1}{2}q(L - x)^2$. If the beam is of narrow rectangular cross-section and carries the load on its upper surface $y = c$, then the plane-stress solution gives the following stress distribution:

$$\sigma_x = -\frac{My}{I} + \frac{q}{b}\left(\frac{3y}{10c} - \frac{y^3}{2c^3}\right), \quad \tau_{xy} = \frac{Q}{2I}(c^2 - y^2), \quad \sigma_y = \frac{q}{6I}(y^3 - 3c^2 y - 2c^3).$$

Note that σ_y equals $-q/b$ at $y = c$ and vanishes at $y = -c$, as it should. The shear stress has the same parabolic distribution as in the end-loaded cantilever, but the longitudinal stress no longer varies linearly over the cross-section. The correction, however, becomes negligible at all sections sufficiently far away from the free end, that is, where $(L - x)^2 \gg c^2$. At such sections, similarly, the magnitude of σ_y becomes negligible in comparison with that of σ_x, except near $y = 0$.

With the correction to σ_x and the effects of τ_{xy} and σ_y neglected, it is easy to determine that plasticity begins at $(0, \pm c)$ when $q = 2M_E/L^2 = 4\sigma_Y bc^2/3L^2 \stackrel{\text{def}}{=} q_E$. With ξ again denoting the length of the elastic portion of the beam and $y^*(x)$ the half-depth of the elastic core in the elastic-plastic portion, we have $\xi = \sqrt{q_E/q}L$ and

$$\left(\frac{y^*}{c}\right)^2 + 2\left(L - \frac{x}{\xi}\right)^2 = 3.$$

The elastic-plastic boundary therefore consists of two arcs of an ellipse centered at $(L, 0)$, with semimajor and semiminor axes respectively equal to $\sqrt{3}/2\xi$ and $\sqrt{3}c$. Collapse occurs when $y^*(0) = 0$, or $\xi = \sqrt{2}/3L$. Thus the ultimate load intensity is $q_U = 3q_E/2$. Indeed, the ratio between the ultimate and initial-yield values of the load is necessarily equal to the shape factor M_U/M_E in any beam that is statically determinate.

The deflection is derived as for the end-loaded cantilever. The curvature distribution is

$$\bar{v}''(x) = -\frac{\sigma_Y}{Ec}\begin{cases} \dfrac{1}{\sqrt{3 - 2(L-x)^2/\xi^2}}, & x < L - \xi, \\ \dfrac{(L-x)^2}{\xi^2}, & x > L - \xi. \end{cases}$$

The tip deflection is therefore

$$\Delta = -\frac{\sigma_Y}{Ec}\frac{\xi^2}{4}\left(3 - 2\sqrt{3 - 2\frac{L^2}{\xi^2}}\right).$$

As ξ decreases from L $(q = q_E)$ to $\sqrt{2}/3L$ $(q = q_U)$, the deflection may be seen to double.

Uniformly Loaded Simply Supported Beam

A uniformly loaded beam of span $2L$ that is simply supported at its ends is equivalent to a uniformly loaded cantilever of length L carrying, in addition, an end load that is opposed to the distributed load and equal in magnitude to the resultant of the latter (see Figure 4.4.7). The moment distribution at any state, as well as all other quantities while the beam is completely elastic, can therefore be obtained by superposing the solutions of the two previously treated problems, with $F = qL$. The moment at x is

$$M(x) = qL(L - x) - \frac{1}{2}q(L - x)^2 = \frac{1}{2}q(L^2 - x^2).$$

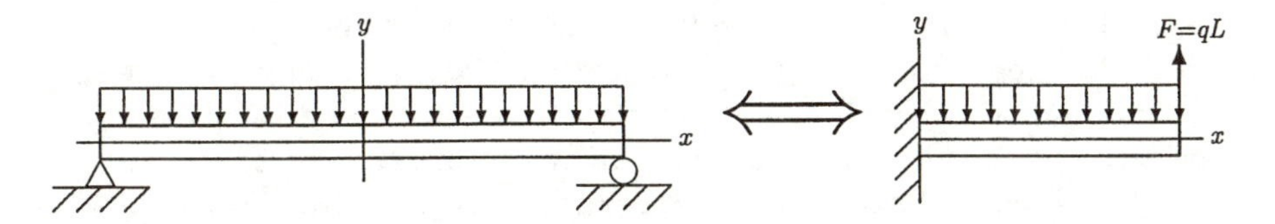

Figure 4.4.7. Uniformly loaded simply supported beam, equivalence with cantilever.

The maximum moment occurs at $x = 0$ and equals $qL^2/2$. Consequently, $q_E = 2M_E/L^2$ and, as expected, $q_U = 3q_E/2$. When $q > q_E$, the elastic-plastic portion extends over $0 < x < \eta$, where $\eta = \sqrt{1 - q/q_E}L$, and the boundary is described by

$$\left(\frac{y^*}{c}\right)^2 - 3\rho\left(\frac{x}{L}\right)^2 = 3(1-\rho),$$

where $\rho = q/q_U$. The boundary is thus given by portions of hyperbolas whose asymptotes are $y = \pm\sqrt{3}(c/L)x$. At collapse, the boundary attains the asymptotes.

The curvature distribution is

$$\bar{v}''(x) = \frac{\sigma_Y}{Ec}\begin{cases} \dfrac{1}{\sqrt{3(1-\rho+\rho x^2/L^2)}}, & x < \eta, \\ \dfrac{3}{2}\rho\left(1 - \dfrac{x^2}{L^2}\right), & x > \eta. \end{cases}$$

The deflection is once more given by Equation (4.4.16), namely,

$$\Delta = \frac{\sigma_Y}{Ec}\left[\int_0^\eta \frac{L-x}{\sqrt{3(1-\rho+\rho x^2/L^2)}}dx + \frac{3}{2}\rho\int_\eta^L (L-x)\left(1-\frac{x^2}{L^2}\right)dx\right].$$

Carrying out the integration and inserting $\eta = \sqrt{1 - 2\rho/3}L$, we obtain

$$\Delta = \frac{\sigma_Y}{Ec}L^2\left[\frac{1}{\sqrt{3\rho}}\sinh^{-1}\sqrt{\frac{3\rho-2}{3(1-\rho)}} + \sqrt{1-\frac{\rho}{\sqrt{3}}\rho+\rho-\frac{1}{2\rho}} - \frac{3\rho+1}{3}\sqrt{1-\frac{2}{3\rho}}\right].$$

In this case the deflection tends to infinity as $\rho \to 1$, but remains of the order of $\Delta_E = (5/12)\sigma_Y L^2/Ec$ until q gets quite close to q_U; for example, $\Delta \approx 2.0\Delta_E$ when $\rho = 0.95$.

4.4.3. Plane-Strain Pure Bending of Wide Beams or Plates

Initially Flat Plates

If the width b of a rectangular beam is considerably greater than its depth $2c$, then it is better to regard is as a plate, and, in the problem of bending, replace the assumption of plane stress with that of plane strain; the assumption can be justified everywhere except near the sides $z = \pm b/2$. The solutions to elastic problems in plane stress are formally identical to those in plane strain if the Young's modulus E and the Poisson's ratio ν in the plane-stress solution is replaced by $E' = E/(1-\nu^2)$ and $\nu' = \nu/(1-\nu)$, respectively; in addition, a stress $\sigma_z = \nu(\sigma_x + \sigma_y)$ must be present in order to maintain the state of plane strain. In fact, the plane-strain solutions are "exact" in the sense that all the compatibility conditions are satisfied, while in all but the simplest plane-stress solutions some of the conditions are violated, except in the limit of infinitely thin sheets (see, e.g., Timoshenko and Goodier [1970], Article 98).

The elastic moment-curvature relation in the pure bending of a wide plate is accordingly $M = E'\kappa$, while the distribution of bending stress is again given by $\sigma_x = -My/I$. If the stress $\sigma_z = \nu\sigma_x$ is taken into account, yielding occurs wherever $|\sigma_x| = \sigma_{Y'}$, where

$$\sigma_{Y'} = \begin{cases} \sigma_Y, & \text{Tresca,} \\ \dfrac{\sigma_Y}{\sqrt{1-\nu+\nu^2}}, & \text{Mises.} \end{cases}$$

A similar relation relates the elastic-limit stress in plane strain, $\sigma_{E'}$, to σ_E for a work-hardening material. With E replaced by E' and σ_Y (or σ_E) by $\sigma_{Y'}$ (or $\sigma_{E'}$), the results obtained for narrow rectangular beams in 4.4.1 apply to wide plates. For example, Equation (4.4.13) for the residual curvature upon unloading holds if $\kappa_E = \sigma_{Y'}/E'c$.

Initially Curved Plates: Elastic State

A section of a wide plate of constant curvature, subject to a constant bending moment M, is shown in Figure 4.4.8(a). The state of stress at each point may be assumed to be independent of θ and to consist of the components σ_r, σ_θ, and σ_z, with $\sigma_z = \nu(\sigma_r + \sigma_\theta)$ to maintain plane strain. The only equilibrium equation is therefore (4.3.18):

$$\frac{d\sigma_r}{dr} + \frac{\sigma_r - \sigma_\theta}{r} = 0.$$

Unlike the expansion of a cylindrical tube, however, the bending of a plate requires a tangential displacement $u_\theta = v(r, \theta)$ in addition to the radial displacement $u_r = u(r, \theta)$. In view of the strain-displacement relations (1.2.1), the strain-stress relations may be written as

$$\frac{\partial u}{\partial r} = \frac{1}{E'}(\sigma_r - \nu'\sigma_\theta), \quad \frac{u}{r} + \frac{1}{r}\frac{\partial v}{\partial \theta} = \frac{1}{E'}(\sigma_\theta - \nu'\sigma_r), \quad \frac{1}{r}\frac{\partial u}{\partial \theta} + \frac{\partial v}{\partial r} - \frac{v}{r} = 0.$$

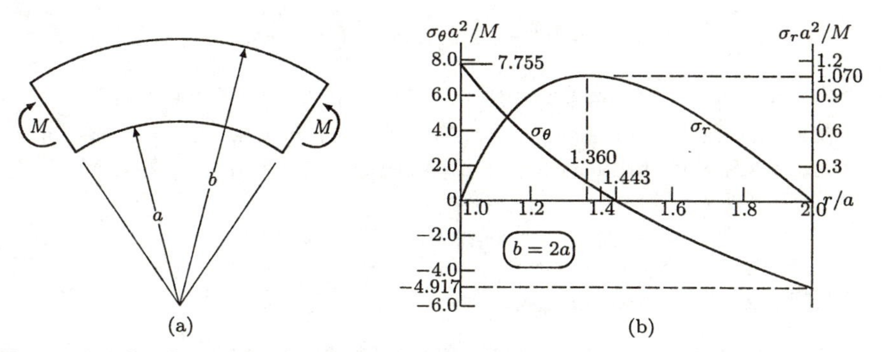

Figure 4.4.8. Curved wide plate or bar in pure bending: (a) geometry; (b) elastic stress distribution for $b = 2a$.

Eliminating u, v and σ_θ between these equations and (4.3.18) results in

$$\frac{d^3\sigma_r}{dr^3} + \frac{5}{r}\frac{d^2\sigma_r}{dr^2} + \frac{3}{r^2}\frac{d\sigma_r}{dr} = 0.$$

The general solution is

$$\sigma_r = A\ln r + B + \frac{C}{r^2},$$

and (4.3.18) immediately gives

$$\sigma_\theta = A(\ln r + 1) + B - \frac{C}{r^2}.$$

The condition that the curved boundary, $r = a$ and $r = b$, is traction-free requires that σ_r vanish there, that is,

$$Ab^2 \ln b + Bb^2 + C = 0, \quad Aa^2 \ln a + Ba^2 + C = 0.$$

The couple M is given by

$$M = -\int_a^b \sigma_\theta r\, dr = -\frac{A}{4}[b^2(1+2\ln b) - a^2(1+2\ln a)] - \frac{B}{2}(b^2 - a^2) - C\ln\frac{b}{a}.$$

Solving the three equations for A, B, and C in terms of M, we obtain the stresses

$$\begin{aligned}\sigma_r &= 4\frac{M}{\Delta}\left(b^2\ln\frac{b}{r} + a^2\ln\frac{r}{a} - \frac{a^2b^2}{r^2}\ln\frac{b}{a}\right)\\ \sigma_\theta &= 4\frac{M}{\Delta}\left(\frac{a^2b^2}{r^2}\ln\frac{b}{a} + b^2\ln\frac{b}{r} + a^2\ln\frac{r}{a} - b^2 + a^2\right),\end{aligned} \tag{4.4.17}$$

where

$$\Delta = (b^2 - a^2)^2 - 4a^2b^2\left(\ln\frac{b}{a}\right)^2.$$

The variation of the stresses with r is shown in Figure 4.4.8(b).

The maximum value of $|\sigma_r - \sigma_\theta|$ occurs at $r = a$, where it equals $\sigma_\theta(a)$. Initial yielding according to the Tresca criterion takes place when $M = M_E$, where

$$M_E = \frac{k\Delta}{2}\left(2b^2\ln\frac{b}{a} - b^2 + a^2\right),$$

k being the yield stress in shear. For $b = 2a$, $M_E = 0.258ka^2$.

Initially Curved Plates: Elastic-Plastic State

The state of stress in an initially curved plate or wide beam in plane strain under a bending moment $M > M_E$ was studied by Shaffer and House [1955]. At first the plastic zone spans $a \le r \le c_1$, while $c_1 \le r \le b$ is elastic. In the plastic zone the equilibrium equation (4.3.18), together with the Tresca yield condition $\sigma_\theta - \sigma_r = 2k$ and the boundary condition $\sigma_r(a) = 0$, gives the stresses

$$\sigma_r = 2k\ln\frac{r}{a}, \quad \sigma_\theta = 2k\left(\ln\frac{r}{a} + 1\right).$$

In the elastic region the stresses are still given by (4.4.17), but the constants A, B, and C must be determined to satisfy continuity with the plastic stresses at $r = c_1$, in addition to $\sigma_r(b) = 0$.

As the moment is increased further, at a certain value, say M_1, $\sigma_\theta(b)$ attains the value $-2k$. For $b = 2a$, $M_1 = 0.387ka^2$. When $M > M_1$, a second plastic zone forms in $c_2 \le r \le b$, in which stresses are given by

$$\sigma_r = 2k\ln\frac{b}{r}, \quad \sigma_\theta = -2k\left(1 - \ln\frac{b}{r}\right).$$

The stresses in the elastic core $c_1 \le r \le c_2$ are still given by (4.4.17). The four continuity conditions for σ_r and σ_θ at $r = c_1$ and $r = c_2$ furnish the three constants A, B, and C, and in addition, a relation between c_1 and c_2, namely,

$$c_1^2\left(1 + \ln\frac{c_2^2}{ab}\right) = c_2^2\left(1 - \ln\frac{ab}{c_1^2}\right).$$

The fully plastic state is attained when $c_1 = c_2 = \sqrt{ab}$. The ultimate moment is

$$M_U = 2k\left[-\int_a^{\sqrt{ab}}\left(\ln\frac{r}{a} + 1\right)r\,dr + \int_{\sqrt{ab}}^b\left(1 - \ln\frac{b}{r}\right)r\,dr\right] = \frac{k}{2}(b-a)^2.$$

For $b = 2a$, $M_U = 0.5\kappa^2$.

As in the thick-walled tube in plane strain, both the elastic and plastic parts of ε_z vanish as long as σ_z is the intermediate principal stress, so that $\sigma_z = \nu(\sigma_r + \sigma_\theta)$ in the plastic regions as well. In the inner plastic zone, the condition that $\sigma_r < \sigma_z < \sigma_\theta$ is $(1 - 2\nu)\ln(c_1/a) < \nu$, while in the outer plastic zone the corresponding condition $\sigma_\theta < \sigma_z < \sigma_r$ is $(1 - 2\nu)\ln(b/c_2) < 1 - \nu$. Clearly, these conditions are satisfied at all stages if they are satisfied when $c_1 = c_2 = \sqrt{ab}$, and the second condition is satisfied whenever the first one is, since $\nu \leq 1 - \nu$. The results are consequently valid provided $ln(b/a) \leq 2\nu/(1 - 2\nu)$. For $\nu = 0.3$, the limiting value of b/a is $e^{1.5} = 4.48$.

The displacement field in the elastic-plastic pure bending of wide bars was derived by Shaffer and House [1957] for an incompressible material and by Eason [1960a] for compressible materials. The overall change in geometry can be measured by the fractional change in the angle subtended by the bar, $\Delta\alpha/\alpha$, and this is found to remain of the elastic order of magnitude until M reaches about $0.95M_U$, similar to the case of initially straight bars. The corresponding plane-stress problem was treated by Eason [1960b].

Exercises: Section 4.4

1. Find the shape factor M_U/M_E for the elastic–perfectly plastic I-beam having the cross-section shown in Figure 6.2.2.

2. Find the moment-curvature relation of an elastic–perfectly plastic beam of square cross-section that is bent about a diagonal.

3. Derive Equation (4.4.10).

4. Find the moment-curvature relation analogous to (4.4.10) for a rectangular beam made of a linearly work-hardening material with initial yield stress σ_E and work-hardening modulus H.

5. Find the shape factor for an elastic–perfectly plastic T-beam made of two equal rectangles of length d and thickness t. Calculate for $d/t = 8$ and for the limiting case as $d/t \to \infty$.

6. Find the equations of the interaction diagram for pure bending of an elastic–perfectly plastic beam of ideal angle section composed of two narrow rectangles of length a and b, respectively, having the same thickness t. Sketch the diagram for $b = 0.6a$.

7. Derive an equation analogous to (4.4.13) for the residual curvature in the beam of Exercise 4.

8. Discuss the behavior of an elastic–perfectly plastic rectangular beam that is bent by means of a uniform bending moment to a curvature $\kappa_0 > \kappa_E$, and then unloaded and reloaded by a bending moment in the

opposite direction until plastic deformation again takes place. Plot the moment-curvature diagram.

9. Discuss the development of a secondary plastic zone due to shear stresses in a uniformly loaded cantilever beam and a uniformly loaded simply supported beam of rectangular cross-section.

10. Discuss the plane-stress elastic-plastic solution, including the secondary plastic zone, for a simply supported beam of rectangular cross-section carrying a concentrated load at $x = \alpha L$, with $0 < \alpha < \frac{1}{2}$.

Section 4.5 Numerical Methods

The problems discussed in the preceding sections of this chapter, for which solutions were often presented in closed form, have been extremely simple: highly regular shapes, simple boundary conditions, and idealized material behavior. Real bodies whose mechanical behavior we wish to study are rarely characterized by such simplicity. But models that approximate the behavior of real bodies involve complex computations which have become feasible only with the development of high-speed digital computers over the past thirty years.

A computer can perform only a finite number of calculations in a finite time. The description of the mechanical state of a body, if it is to be achieved numerically, must therefore be defined by a finite number of variables. In the simple problems discussed in Section 4.1, stress and strain were assumed to be essentially constant throughout the body, and therefore each such problem is governed by the rate equations alone, constituting a set of one or more ordinary differential equations. For example, Equations (4.1.2) and (4.1.3) each represent a set of two coupled nonlinear ordinary differential equations.

In the more general problems studied in Sections 4.2 to 4.4, stress and displacement are unknown functions of position, with the former required to satisfy the equilibrium equations and the latter related to them through constitutive relations. Knowledge of the values of these functions at all points of the body amounts to an infinity of data. For the problem to be numerically tractable, this infinity must be reduced to a finite amount. That is, the stress and displacement fields must be determined, at least approximately, by a finite number of parameters. The process whereby this reduction is accomplished is the discretization of the problem, which was discussed in general terms in 1.3.5. The discretized problem is again governed by a coupled set of nonlinear ordinary differential equations. Their number, however, can be quite large, and therefore special techniques are required in order to economize on computer time.

Numerical techniques of time integration of the differential equations arising in elastic-plastic problems are discussed in 4.5.1. In 4.5.2 we present an overview of the most usual discretization scheme, namely, the finite-element method (other methods are mentioned in passing). Finally, in 4.5.3 we discuss, in summary form, the combination of the two topics, that is, the formulation of finite-element methods for elastic-plastic and elastic-viscoplastic continua.

Matrix notation is used throughout this section, but with **boldface** rather than *underline* notation. For a column-matrix-valued function of a column-matrix-valued variable — say $\boldsymbol{\phi}(\boldsymbol{\xi})$ — the rectangular matrix $[\partial\phi_i/\partial\xi_j]$ is denoted $\partial\boldsymbol{\phi}/\partial\boldsymbol{\xi}$, $\partial_{\boldsymbol{\xi}}\boldsymbol{\phi}$ or $\boldsymbol{\phi}_{\boldsymbol{\xi}}$. In particular, if $\phi(\boldsymbol{\xi})$ is scalar-valued, then $\partial\phi/\partial\boldsymbol{\xi} = \phi_{\boldsymbol{\xi}} = \partial_{\boldsymbol{\xi}}\phi$ is a row matrix.

4.5.1. Integration of Rate Equations

Viscoplasticity

For viscoplasticity, including viscoplasticity without a yield surface as given by the "unified" models (i.e., rate-dependent behavior with internal variables), the governing equations are

$$\boldsymbol{\sigma} = \mathsf{C}(\boldsymbol{\varepsilon} - \boldsymbol{\varepsilon}^i) \tag{4.5.1}$$

and

$$\dot{\boldsymbol{\xi}} = \phi\mathbf{h}, \tag{4.5.2}$$

where ϕ and $\mathbf{h}$ are functions of $(\boldsymbol{\sigma},\, \boldsymbol{\xi},\, T)$ or of $(\boldsymbol{\varepsilon},\, \boldsymbol{\xi},\, T)$, and the internal-variable matrix $\boldsymbol{\xi}$ includes $\boldsymbol{\varepsilon}^i$.

When the temperature and the stress *or* strain are prescribed functions of time, the viscoplastic rate equations (4.5.2) are just a set of coupled first-order ordinary differential equations, which may be written in matrix notation as

$$\dot{\boldsymbol{\xi}} = \boldsymbol{\phi}(\boldsymbol{\xi},\, t). \tag{4.5.3}$$

A rather general method of integrating Equation (4.5.3) is the **generalized Euler method**, according to which, if $\boldsymbol{\xi}(t)$ is known and $\Delta\boldsymbol{\xi} = \boldsymbol{\xi}(t + \Delta t) - \boldsymbol{\xi}(t)$, then, approximately,

$$\Delta\boldsymbol{\xi} = \Delta t\,[(1 - \beta)\boldsymbol{\phi}(\boldsymbol{\xi},\, t) + \beta\boldsymbol{\phi}(\boldsymbol{\xi} + \Delta\boldsymbol{\xi},\, t + \Delta t)], \tag{4.5.4}$$

where $\boldsymbol{\xi} = \boldsymbol{\xi}(t)$, and β is a parameter between 0 and 1 to be chosen in accordance with the solution algorithm followed. The choice $\beta = 0$ represents the **forward Euler method** (sometimes called simply **Euler's method**), one that is fully *explicit* in the sense that the unknown quantity $\Delta\boldsymbol{\xi}$ appears only on the left-hand side of the equation. While this method is simple, the error accumulates rather rapidly unless very small time increments are used. With $0 < \beta \leq 1$, the procedure is *implicit*. In particular, the choice $\beta = 1$ represents the **backward Euler method**. A common choice is $\beta = \frac{1}{2}$, representing the **Crank–Nicholson method**.

With $\beta > 0$, Equation (4.5.4) constitutes a set of coupled nonlinear equations for the $\Delta\xi_\alpha$. An iterative scheme is required, as a rule, in order to achieve a solution. In the **direct iteration method** or **method of successive approximations**, an initial guess $\Delta\boldsymbol{\xi}^{(0)}$ (e.g., $\Delta\boldsymbol{\xi}^{(0)} = \mathbf{0}$) is made and substituted for $\Delta\boldsymbol{\xi}$ in the right-hand side of (4.5.4). The result of computing the right-hand side is then called $\Delta\boldsymbol{\xi}^{(1)}$, is substituted again, and so forth. The iteration stops when two successive approximations are sufficiently close, that is, when an appropriately defined magnitude (or *norm*) of $\Delta\boldsymbol{\xi}^{(k+1)} - \Delta\boldsymbol{\xi}^{(k)}$, denoted $||\Delta\boldsymbol{\xi}^{(k+1)} - \Delta\boldsymbol{\xi}^{(k)}||$, is less than some prescribed error tolerance. The magnitude $||\boldsymbol{\xi}||$ may be defined in a variety of ways, for example, $\max_\alpha |\xi_\alpha|$, $\sum_\alpha |\xi_\alpha|$, or $\sqrt{\sum_\alpha \xi_\alpha^2}$.

An iteration method that usually produces faster convergence than the direct iteration method is the **Newton–Raphson method**. The right-hand side of Equation (4.5.4) is subtracted from its left-hand side, and the result is rewritten as

$$\boldsymbol{\psi}(\Delta\boldsymbol{\xi}) = 0,$$

it being understood that $\boldsymbol{\psi}$ also depends on $\boldsymbol{\xi}$, t and Δt. The initial guess $\Delta\boldsymbol{\xi}^{(0)}$ (which may again be zero, or which may be calculated from an explicit scheme) is introduced; this is the *predictor* phase of the solution. Next, $\boldsymbol{\psi}$ at $\Delta\boldsymbol{\xi}^{(1)}$ is evaluated by the approximation

$$\boldsymbol{\psi}(\Delta\boldsymbol{\xi}^{(1)}) \doteq \boldsymbol{\psi}(\Delta\boldsymbol{\xi}^{(0)}) + \mathbf{J}(\Delta\boldsymbol{\xi}^{(1)} - \Delta\boldsymbol{\xi}^{(0)}),$$

where

$$\mathbf{J} = (\partial\boldsymbol{\psi}/\partial\Delta\boldsymbol{\xi})|_{\Delta\boldsymbol{\xi}^{(0)}} = \mathbf{I} - \beta\,\Delta t\ (\partial\boldsymbol{\phi}/\partial\boldsymbol{\xi})|_{\boldsymbol{\xi}+\Delta\boldsymbol{\xi}^{(0)}}\,.$$

By treating the approximation as an equality and setting $\boldsymbol{\psi}(\Delta\boldsymbol{\xi}^{(1)}) = 0$, we obtain

$$\Delta\boldsymbol{\xi}^{(1)} = \Delta\boldsymbol{\xi}^{(0)} + \mathbf{J}^{-1}\boldsymbol{\psi}(\Delta\boldsymbol{\xi}^{(0)});$$

this is the *corrector* phase. The process may be continued with $\Delta\boldsymbol{\xi}^{(1)}$ replacing $\Delta\boldsymbol{\xi}^{(0)}$ in order to produce $\Delta\boldsymbol{\xi}^{(2)}$, and so on, until $||\boldsymbol{\psi}(\Delta\boldsymbol{\xi}^{(k)})||$ is sufficiently small. With a reasonably good guess for $\Delta\boldsymbol{\xi}^{(0)}$, it is usually not necessary to recalculate $\mathbf{J}$ at each iteration

Rate-Independent Plasticity

The incremental stress-strain relations of rate-independent plasticity are

$$\dot{\boldsymbol{\sigma}} = \mathsf{C}(\dot{\boldsymbol{\varepsilon}} - \dot{\boldsymbol{\varepsilon}}^p). \qquad (4.5.5)$$

The flow equations (3.2.1) are written in matrix notation as

$$\dot{\varepsilon}^p = \dot{\lambda}\mathbf{g}, \tag{4.5.6}$$

where $\mathbf{g}$ is the column-matrix representation of the flow tensor with components h_{ij}, defined in 3.1.1. If $f(\boldsymbol{\sigma}, \boldsymbol{\xi}) = 0$ is the equation of the yield surface, then, in accordance with Equation (3.2.8),

$$\dot{\lambda} = \begin{cases} \frac{1}{L} < f_{\sigma}\mathsf{C}\,\dot{\varepsilon} >, & f = 0, \\ 0, & f < 0, \end{cases} \tag{4.5.7}$$

where L is given by Equation (3.2.9), or, in matrix notation,

$$L = H + f_{\sigma}\mathsf{C}\mathbf{g},$$

with H denoting the work-hardening modulus as before.

Equations (4.5.5)–4.5.7) can be combined symbolically in the form

$$\dot{\boldsymbol{\sigma}} = \mathsf{C}_{ep}\dot{\varepsilon}, \tag{4.5.8}$$

where it must be understood that in view of (4.5.7), the *elastic-plastic modulus matrix* C_{ep} is a nonlinear operator, since it depends on the direction of $\dot{\varepsilon}$ in addition to depending on the current state as specified by $(\boldsymbol{\sigma}, \boldsymbol{\xi}, T)$ or by $(\varepsilon, \boldsymbol{\xi}, T)$ (recall that ε^p is included in $\boldsymbol{\xi}$). The rate equations for $\boldsymbol{\xi}$,

$$\dot{\boldsymbol{\xi}} = \dot{\lambda}\mathbf{h}, \tag{4.5.9}$$

when combined with (4.5.7) will symbolically be written in the form

$$\dot{\boldsymbol{\xi}} = \boldsymbol{\Lambda}\dot{\varepsilon}, \tag{4.5.10}$$

where $\boldsymbol{\Lambda}$ is a nonlinear operator like C_{ep}.

Note that the governing equations, (4.5.5)–4.5.7) and (4.5.9), have here been chosen in their strain-space form, while the statically determinate problems studied in Section 4.1, in particular those described by Equations (4.1.2) and (4.1.3), were formulated under stress control. The reason for the choice of strain control as the basis for the development of numerical methods is twofold: first, because strain control, not being limited to work-hardening materials, is more general than stress control; and second, because the strain-space formulation is naturally associated with a displacement-based spatial discretization scheme, such as the most usual versions of the finite-element method.

The predictor/corrector scheme described for the viscoplastic rate equations can be applied to rate-independent plasticity as well. An effective method consists of an elastic predictor followed by a plastic corrector. It is supposed that $\boldsymbol{\sigma}$ and $\boldsymbol{\xi}$ (which includes $\boldsymbol{\varepsilon}^p$) have been determined with sufficient accuracy, with $f(\boldsymbol{\sigma}, \boldsymbol{\xi}) \le 0$ satisfied, at a certain value of the strain tensor $\boldsymbol{\varepsilon}$. The elastic predictor/plastic corrector scheme is used to determine the effect of a small strain increment $\Delta\boldsymbol{\varepsilon}$: it is initially guessed that the effect is purely elastic, so that the stress will change to $\boldsymbol{\sigma}^{(1)} = \boldsymbol{\sigma} + \mathsf{C}\,\Delta\boldsymbol{\varepsilon}$, and the internal-variable matrix will remain as $\boldsymbol{\xi} = \boldsymbol{\xi}^{(1)}$. For simplicity, the superscript (k) is applied to quantities evaluated at $\boldsymbol{\sigma}^{(k)}$, $\boldsymbol{\xi}^{(k)}$. If $f^{(1)} \le 0$, then the elastic prediction is correct, and the process is repeated for the next strain increment. If, however, $f^{(1)} > 0$, then the strain increment must include some plastic strain, and a correction must be applied to $\boldsymbol{\sigma}^{(1)}$ and $\boldsymbol{\xi}^{(1)}$.

The stress correction is $\Delta\boldsymbol{\sigma}^{(2)} = -\mathsf{C}\,\Delta\boldsymbol{\varepsilon}^{p(2)}$, and $\Delta\boldsymbol{\varepsilon}^{p(2)}$ is given, in principle, by integrating 4.5.6) over the increment. In practice, it will be defined by an approximation:

$$\Delta\boldsymbol{\varepsilon}^{p(2)} = \Delta\lambda\,\mathbf{g}^{(1)}.$$

Likewise,

$$\Delta\boldsymbol{\xi}^{(2)} = \Delta\lambda\,\mathbf{h}^{(1)},$$

and $\Delta\lambda$ is defined so that the state defined by $\boldsymbol{\sigma}^{(2)} = \boldsymbol{\sigma}^{(1)} + \Delta\boldsymbol{\sigma}^{(2)}$ and $\boldsymbol{\xi}^{(2)} = \boldsymbol{\xi}^{(1)} + \Delta\boldsymbol{\xi}^{(2)}$ lies on the yield surface, at least in the first approximation. To within an error that is of an order higher than the first in $\Delta\lambda$,

$$f^{(2)} \doteq f^{(1)} + (\partial f/\partial\boldsymbol{\sigma})^{(1)}\,\Delta\boldsymbol{\sigma}^{(2)} + (\partial f/\partial\boldsymbol{\xi})^{(1)}\,\Delta\boldsymbol{\xi}^{(2)} = f^{(1)} - L^{(1)}\,\Delta\lambda,$$

where $L^{(1)}$ is the value at $\boldsymbol{\sigma}^{(1)}$, $\boldsymbol{\xi}^{(1)}$ of the quantity L defined by (3.2.9). The satisfaction of the yield criterion at $\boldsymbol{\sigma}^{(2)}$, $\boldsymbol{\xi}^{(2)}$, to this approximation, requires that

$$\Delta\lambda = \frac{f^{(1)}}{L^{(1)}}.$$

In fact, it can be shown that the state at (2) is on or outside the yield surface if the yield function f is a convex function of its arguments,[1] since convexity implies [see Equation (1.5.11)] that

$$f^{(2)} - f^{(1)} - (\partial f/\partial\boldsymbol{\sigma})^{(1)}\,\Delta\boldsymbol{\sigma}^{(2)} - (\partial f/\partial\boldsymbol{\xi})^{(1)}\,\Delta\boldsymbol{\xi}^{(2)} \ge 0,$$

and the definitions of $\Delta\boldsymbol{\sigma}^{(2)}$, $\Delta\boldsymbol{\xi}^{(2)}$, $\Delta\lambda$ and L reduce the left-hand side of the inequality to its first term. Usually, $f^{(2)}$ is close enough to zero so that this correction is sufficient. Otherwise, the plastic correction process can be repeated until the state obtained is on the yield surface to a sufficient degree of accuracy (for fairly simple yield surfaces and flow rules, the final

[1]A yield function f is convex if it is, for example, of the combined-hardening type (3.3.7), with F and $-k$ convex functions of their arguments. The convexity of F is necessary for the yield surface to be convex, while the convexity of $-k$ means essentially that the isotropic hardening rate is at most linear.

state may be obtained by scaling). The next strain increment can then be applied.

Under stress control, the effect of a stress increment $\Delta\boldsymbol{\sigma}$ on the yield function is examined. If $f^{(1)} = f(\boldsymbol{\sigma} + \Delta\boldsymbol{\sigma}, \boldsymbol{\xi}) \le 0$, then $\Delta\boldsymbol{\varepsilon} = \mathsf{C}^{-1}\Delta\boldsymbol{\sigma}$, and the next increment can be applied. If $f^{(1)} > 0$, then $\boldsymbol{\xi}$ is corrected by $\Delta\boldsymbol{\xi}^{(2)} = \Delta\lambda\,\mathbf{h}^{(1)}$, where $\Delta\lambda = f^{(1)}/H^{(1)}$ in order that $f^{(2)} \doteq 0$. As under strain control, the correction is repeated as needed.

The preceding scheme, called the **return-mapping algorithm**, is a generalization of the **radial-return algorithm**, so called because, according to the Mises flow rule, $\mathsf{C}\mathbf{g}$ is proportional to the stress deviator $\mathbf{s}$, and therefore $\Delta\boldsymbol{\sigma}^{(2)} \propto -\mathbf{s}^{(1)}$, that is, the stress correction is directed toward the origin of the stress-deviator space. Other predictor/corrector algorithms have been proposed, for example the **initial-stress algorithm** of Zienkiewicz, Valliappan, and King [1969].

More generally, the elastic predictor/plastic corrector method may be thought of as a split of the elastic-plastic problem into an elastic problem governed by

$$\Delta\varepsilon_{ij} = \frac{1}{2}(\Delta u_{i,j} + \Delta u_{j,i}), \qquad \Delta\varepsilon^p_{ij} = 0, \qquad \Delta\xi_\alpha = 0,$$

and a plastic problem governed by the differential equations

$$\frac{d}{d\lambda}\boldsymbol{\sigma}(\lambda) = -\mathsf{C}\mathbf{g}(\boldsymbol{\sigma}(\lambda), \boldsymbol{\xi}(\lambda)), \qquad \frac{d}{d\lambda}\boldsymbol{\xi}(\lambda) = \mathbf{h}(\boldsymbol{\sigma}(\lambda), \boldsymbol{\xi}(\lambda)), \tag{4.5.11}$$

constrained by the yield criterion

$$f(\boldsymbol{\sigma}(\lambda), \boldsymbol{\xi}(\lambda)) = 0, \tag{4.5.12}$$

and subject to the initial condition

$$\{\boldsymbol{\sigma}(\lambda), \boldsymbol{\xi}(\lambda)\}|_{\lambda=0} = \{\boldsymbol{\sigma}^{(1)}, \boldsymbol{\xi}\},$$

where $\boldsymbol{\sigma}^{(1)}$ is, as defined above, the stress tensor in the trial elastic state and $\boldsymbol{\xi}$ is the internal-variable array at the end of the preceding time step.

The problem formed by Equations (4.5.11)–4.5.12) may be solved either by direct iteration or by the Newton–Raphson method. Some general methods, applicable to any yield criterion and hardening rule, have been developed, including the general closest-point projection method and the cutting-plane algorithm (see Simo and Hughes [1988]). For example, in the closest-point projection method, the equations of the backward Euler method for perfect plasticity with an associated flow rule, which may be written as

$$f(\boldsymbol{\sigma} + \Delta\boldsymbol{\sigma}) = 0,$$

$$\boldsymbol{\psi}(\Delta\boldsymbol{\sigma}, \Delta\boldsymbol{\xi}) \stackrel{\text{def}}{=} \Delta\boldsymbol{\varepsilon}^p - \Delta\lambda\,\partial_{\boldsymbol{\sigma}} f(\boldsymbol{\sigma} + \Delta\boldsymbol{\sigma})^T = 0,$$

are linearized by means of a Newton–Raphson scheme with $\Delta\varepsilon^p = -\mathsf{C}^{-1}\,\Delta\boldsymbol{\sigma}$ (since $\Delta\varepsilon = 0$ in the corrector phase), yielding

$$\psi^{(0)} - [\mathsf{C}^{-1} + \Delta\lambda^{(0)}\,\partial_{\sigma\sigma} f(\boldsymbol{\sigma} + \Delta\boldsymbol{\sigma}^{(0)})](\Delta\boldsymbol{\sigma}^{(1)} - \Delta\boldsymbol{\sigma}^{(0)})$$
$$- (\Delta\lambda^{(1)} - \Delta\lambda^{(0)})\,\partial_{\sigma} f(\boldsymbol{\sigma} + \Delta\boldsymbol{\sigma}^{(0)})^T = 0$$

and

$$f(\boldsymbol{\sigma} + \Delta\boldsymbol{\sigma}^{(0)}) + \partial_{\sigma} f(\boldsymbol{\sigma} + \Delta\boldsymbol{\sigma}^{(0)})(\Delta\boldsymbol{\sigma}^{(1)} - \Delta\boldsymbol{\sigma}^{(0)}) = 0.$$

These equations may be solved for $\Delta\lambda^{(1)}$ and $\Delta\boldsymbol{\sigma}^{(1)}$, and hence for $\Delta\varepsilon^{p(1)}$.

The accuracy and stability of integration algorithms for rate equations are discussed by Ortiz and Popov [1985].

4.5.2. The Finite-Element Method

In statically indeterminate problems, the equilibrium equations must be solved simultaneously with the constitutive relations. When the latter are represented by nonlinear rate equations, the resulting problem is one of non-linear partial differential equations, and must be solved numerically, that is, the differential equations must be replaced by algebraic equations, and this is accomplished by means of a spatially discretized model of the continuum, as discussed in 1.3.5.

Generalities

The most commonly used spatial discretization method nowadays is the **finite-element method**, originally developed as a method of analysis of complex framed structures (such as aircraft structures) in which the "elements" are simple elastic members such as truss bars, straight and curved beams, and the like. The method was soon extended to inelastic behavior and to bodies modeled by one-, two-, and three-dimensional domains of arbitrary shape, the elements being small subdomains having a relatively simple geometry. In the case of one-dimensional domain (which may be straight or curved), finite elements are necessarily line segments. In two-dimensional and three-dimensional domains, a particular division into subdomains is known as a *finite-element mesh*, and the elements are typically polygons and polyhedra, respectively. Some meshes are shown in Figure 4.5.1. In order to maintain the simplicity of the element geometry, it may be necessary to approximate the boundary of the domain — typically, a curved boundary by one that is piecewise straight, as in Figure 4.5.1(b). Occasionally, elements with curved boundaries may be used advantageously, as for example in a domain with circular geometry, in which the elements may be partly bounded by arcs of circles [see Figure 4.5.1(c)].

Associated with each element are a number of points known as *nodes*, usually located on the boundary (but occasionally in the interior) of the

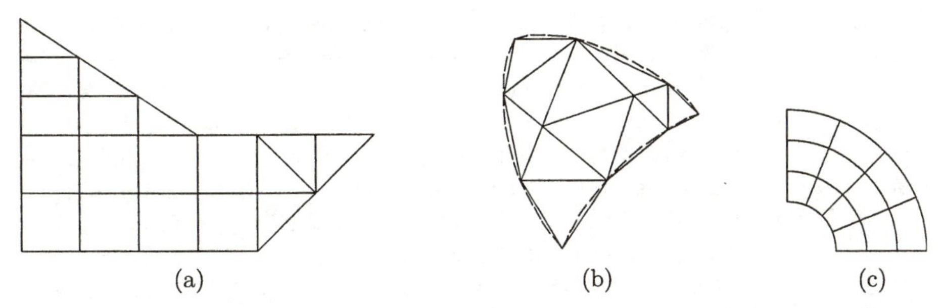

Figure 4.5.1. Finite-element meshes: (a) general mesh; (b) approximation of curved boundary; (c) domain with circular geometry using curvilinear elements.

element. The nodes usually include, at the very least, the vertices of the polygon. It is generally most convenient to use, as the generalized coordinates of the model, the values *at the nodes* of the functions describing the displacement field; these may simply be the nodal displacements, although in finite-element models for beams, plates and shells the nodal rotations may be included as well. It is often convenient to include, among the nodal displacements, initially even those that are prescribed, and to eliminate them after the global formulation. A finite element is characterized, then, by its shape, by the nodes associated with it, and by the assumed variation of displacement within the element, which is described by means of *shape functions* or *interpolation functions*.[1] These functions are ordinarily taken as polynomial in terms of a local set of rectilinear coordinates.

The simplest finite element is one having the shape of a triangle in two dimensions, or of a tetrahedron in three dimensions, and in which the nodes are the vertices only [see Figure 4.5.2(a-b)]. The variation of displacement is

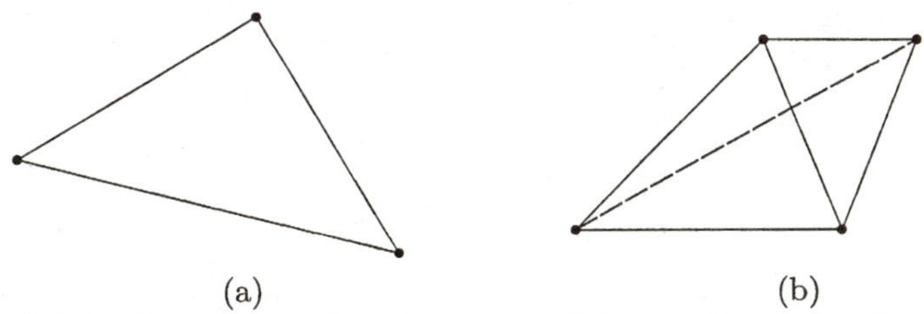

Figure 4.5.2. Constant-strain elements: (a) two-dimensional; (b) three-dimensional.

taken as linear. It is easy to see that the number of parameters necessary to

[1] If the shape functions describing the variation of displacement in terms of the nodal displacements are the same as those describing the geometry of the element in terms of the global coordinates, then the element is called *isoparametric*.

describe the variation of each displacement component in such an element is just equal to the number of vertices. Since the partial derivatives of the displacement field are constant within the element, the strain is likewise constant, and the element is called a *constant-strain element*. Since the displacement in a constant-strain element varies linearly along the edges (or sides), it is continuous between adjacent elements. Furthermore, a rigid-body displacement of all the nodes produces a rigid-body displacement field in each element and therefore in the whole domain. The linear variation of the displacement also means that the strain within each element is constant, so that a combination of nodal displacements representing a global state of constant strain does in fact produce such a state. An element that meets these criteria is known as a *conforming* element. It has been shown that the approximate solutions obtained by successive refinements of conforming finite-element meshes converge monotonically to the exact solution of the corresponding elasticity problem.

The constant-strain element has, as its principal advantage, its simplicity. On the other hand, since the displacement field is represented by a piecewise-linear approximation, it is clear that a rather fine mesh is required in order to obtain reasonably accurate results. More refined elements permit the use of coarser meshes and thus reduce computational time and storage. Some two-dimensional examples are shown in Figure 4.5.3 (while they are shown as rectilinear, they may also be curvilinear):

(a) The six-node triangular element ($T6$), with a general quadratic variation of the displacement.

(b) The four-node quadrilateral element ($Q4$), with a bilinear variation of the displacement; that is, when one local coordinate is held fixed, then the variation with respect to the other coordinate is linear.

(c) The nine-node quadrilateral element ($Q9$), with a biquadratic variation of the displacement.

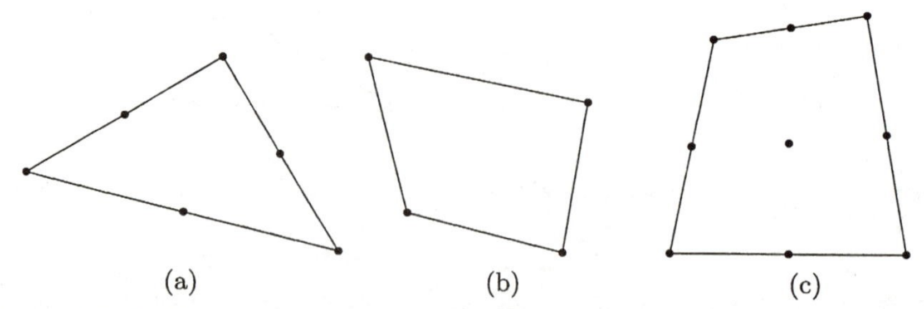

Figure 4.5.3. Higher-order two-dimensional elements: (a) six-node triangle ($T6$); (b) four-node quadrilateral ($Q4$); (c) nine-node quadrilateral ($Q9$).

Other elements in use include the seven-node triangular element (*T7*), like *T6* but with an additional node at the center, and the eight-node quadrangular element (*Q8*), like *Q9* but without the central node.

Each element of Figure 4.5.3 has a three-dimensional counterpart: (a) the ten-node tetrahedron, (b) the eight-node hexahedron, and (c) the twenty-seven-node hexahedron. In addition, wedge-type elements are sometimes used, as in Figure 4.5.4. Elements used for the analysis of plates and shells generally have the same shapes as the two-dimensional elements of Figure 4.5.2 (flat for plates, curved for shells), but as mentioned above, the generalized coordinates may include nodal rotations in addition to nodal displacements.

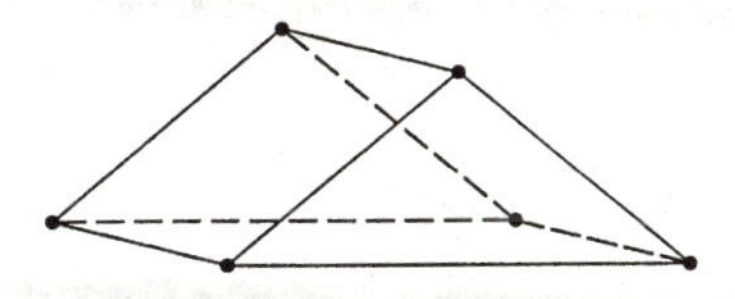

Figure 4.5.4. Wedge element.

The finite-element method is not the only technique for the numerical solution of boundary-value problems in solid mechanics. An older technique is the **finite-difference method**; see, for example, Mendelson [1968] for its application to plasticity problems. A relatively new technique is the **boundary-element method**, based on certain fundamental solutions in elasticity theory; for its application to elastic-plastic bodies, see Telles and Brebbia [1979] and Maier [1983]. For torsion problems, a "cellular analogy," in which solid shafts are idealized as multi-celled structures, was developed by Johnson [1988].

Displacement Formulation

The most common formulation of the finite-element method is as a displacement-based discretization of the type discussed in 1.3.5. In order to simplify the discussion, it will be assumed that all constrained generalized coordinates are zero. Let the number of unknown nodal displacements associated with a given element be N^e (in general in this section, the superscript e denotes a reference to the given element). The column matrix of these displacements will be denoted $\mathbf{q}^e$, and may be derived from the global nodal-displacement matrix $\mathbf{q}$ by means of the relation

$$\mathbf{q}^e = \mathbf{A}^e\mathbf{q}, \tag{4.5.13}$$

where $\mathbf{A}^e$ is an $N^e \times N$ matrix. If the local description of the nodal displacements is in the same frame as the global one, then $\mathbf{A}^e$ is what is known as a *Boolean* matrix: $A^e_{mn} = 1$ if the mth elemental degree of freedom corresponds to the nth global degree of freedom, and $A^e_{mn} = 0$ otherwise.

For a given element occupying the subdomain Ω^e, the interpolation functions ϕ_n of Equation (1.3.17), when assembled in a $3 \times N^e$ matrix, will be denoted $\mathbf{N}^e$. Similarly, the functions $(\nabla\phi_n)^S$ of Equation (1.3.18), when assembled in a $6 \times N^e$ matrix, will be denoted $\mathbf{B}^e$; this matrix is constant in a constant-strain element. The finite-element analogues of Equations (1.3.17)–(1.3.18) accordingly are

$$\mathbf{u}^h(\mathbf{x})\Big|_{\Omega^e} = \mathbf{N}^e(\mathbf{x})\mathbf{q}^e, \qquad \boldsymbol{\varepsilon}^h(\mathbf{x})\Big|_{\Omega^e} = \mathbf{B}^e(\mathbf{x})\mathbf{q}^e.$$

Here the superscript h designates the discretization, and $|_{\Omega^e}$ denotes the restriction of a function to Ω^e.

The element internal-force matrix $\mathbf{Q}^e$ is analogously obtained by the virtual-work method discussed in 1.3.5 as

$$\mathbf{Q}^e = \int_{\Omega^e} \mathbf{B}^{eT} \boldsymbol{\sigma}\, d\Omega,$$

where $d\Omega$ denotes an infinitesimal element of area or volume as appropriate. In a constant-strain element, with constant material properties assumed, the stress and internal variables are also constant, so that the preceding integral reduces to a product. In a higher-order element the values of the state variables must be specified at certain points within the element known as *Gauss points*, and numerical integration must be carried out in terms of these values.

By the principle of virtual work,

$$\mathbf{Q}^T \delta\mathbf{q} = \sum_e \mathbf{Q}^{eT} \delta\mathbf{q}^e,$$

where $\sum_e$ denotes summation over all the elements. In view of (4.5.13), the global internal-force matrix is immediately obtained as

$$\mathbf{Q} = \sum_e \mathbf{A}^{eT}\mathbf{Q}^e.$$

Let $\mathbf{F}$ denote the column matrix representing the global discretized external loads, that is,

$$\mathbf{F} = \sum_e \mathbf{A}^{eT} \left[\int_{\Omega^e} \mathbf{N}^{eT}\mathbf{f}\, d\Omega + \int_{\partial\Omega_t^e} \mathbf{N}^{eT}\mathbf{t}^a\, d\Gamma \right],$$

where $\partial\Omega_t^e$ is that portion (if there is any) of the boundary of Ω^e that forms a part of ∂R_t, and $d\Gamma$ is an infinitesimal (surface or line) element of such a boundary. The equilibrium equation is then

$$\mathbf{Q} = \mathbf{F}.$$

For an elastic body, this equation becomes

$$\mathbf{K}\mathbf{q} = \mathbf{F},$$

where

$$\mathbf{K} = \sum_e \mathbf{A}^{eT}\mathbf{K}^e\mathbf{A}^e,$$

is the global stiffness matrix, with

$$\mathbf{K}^e = \int_{\Omega^e} \mathbf{B}^{eT}\mathsf{C}\mathbf{B}^e \, d\Omega,$$

being the element stiffness matrix. The corresponding equations for nonlinear continua with rate-type constitutive equations are discussed in 4.5.3.

Mixed Formulation

In inelastic (and some elastic) problems it is sometimes advantageous to base finite-element methods on a mixed formulation, in which the approximations for displacement, strain and stress are by means of distinct functions; one instance of the advantage is the treatment of incompressibility. According to this point of view, the strain-displacement relations and the constitutive relations are constraints, with the stress and strain, respectively, as the corresponding Lagrange multipliers. Consequently, these relations are to be satisfied in a weak way, rather than pointwise, and the corresponding weak forms may be derived from the Hu–Washizu principle (see Exercise 14 of Section 1.4 and Exercise 2 of Section 3.4). Since the variations $\delta\mathbf{u}$, $\delta\boldsymbol{\varepsilon}$, and $\delta\boldsymbol{\sigma}$ are independent of one another, the principle furnishes three separate variational equations, written (for consistency with the rest of the present section) in matrix notation:

$$\int_R (\boldsymbol{\sigma}^T \nabla^S \delta\mathbf{u} - \mathbf{f}^T \delta\mathbf{u})\, dV - \int_{\partial R_t} \mathbf{t}^{aT}\, \delta\mathbf{u}\, dS = 0, \tag{4.5.14}$$

$$\int_R [\boldsymbol{\sigma} - \mathsf{C}(\boldsymbol{\varepsilon} - \boldsymbol{\varepsilon}^i)]^T\, \delta\boldsymbol{\varepsilon}\, dV = 0, \tag{4.5.15}$$

$$\int_R (\nabla^S \mathbf{u}_- \boldsymbol{\varepsilon})^T\, \delta\boldsymbol{\sigma}\, dV = 0, \tag{4.5.16}$$

where $\nabla^S\mathbf{u}$ is the appropriate column-matrix representation of $(\nabla\mathbf{u})^S$, and Equation 4.5.1) is assumed as the constitutive relation ($\boldsymbol{\varepsilon}^i$ may be replaced by $\boldsymbol{\varepsilon}^p$).

Equations (4.5.14)–4.5.16) are the weak forms, respectively, of the equilibrium equations, of the stress-strain relations and of the strain-displacement relations. When the strain-displacement and stress-strain relations are enforced pointwise, the displacement formulation is recovered.

In the mixed finite-element formulation, stress and strain have distinct discretizations:

$$\boldsymbol{\sigma}^h(\mathbf{x})\Big|_{\Omega^e} = \mathbf{S}^e(\mathbf{x})\mathbf{c}^e, \tag{4.5.17}$$

$$\boldsymbol{\varepsilon}^h(\mathbf{x})\Big|_{\Omega^e} = \mathbf{E}^e(\mathbf{x})\mathbf{a}^e, \tag{4.5.18}$$

where in general $\mathbf{S}^e != \mathbf{E}^e$. Since the interpolations of stress and strain are in general assumed discontinuous between the elements, the variational equations (4.5.15)–4.5.16) hold for each element Ω^e. Substituting (4.5.17)–4.5.18) into (4.5.15) and 4.5.16), respectively, gives

$$\boldsymbol{\sigma}^h(\mathbf{x})\Big|_{\Omega^e} = \mathbf{S}^e(\mathbf{x})\mathbf{H}^{eT-1}\int_{\Omega^e} \mathbf{E}^{eT}\mathsf{C}(\boldsymbol{\varepsilon} - \boldsymbol{\varepsilon}^i)\, d\Omega, \tag{4.5.19}$$

$$\boldsymbol{\varepsilon}^h(\mathbf{x})\Big|_{\Omega^e} = \mathbf{E}^e(\mathbf{x})\mathbf{H}^{e-1}\int_{\Omega^e} \mathbf{S}^{eT}\,\nabla^S\mathbf{u}\, d\Omega \overset{\text{def}}{=} \bar{\nabla}^S\mathbf{u}, \tag{4.5.20}$$

where

$$\mathbf{H}^e = \int_{\Omega^e} \mathbf{S}^{eT}\mathbf{E}^e\, d\Omega,$$

and $\bar{\nabla}^S$ is a discrete approximation to the operator ∇^S. Equation (4.5.14), with the aid of Equations 4.5.19)–4.5.20), now yields

$$G(\mathbf{u},\,\delta\mathbf{u}) \overset{\text{def}}{=} \sum_e \left[\int_{\Omega^e} (\bar{\nabla}^S\mathbf{u} - \boldsymbol{\varepsilon}^i)^T\mathsf{C}\,\bar{\nabla}^S\delta\mathbf{u}\, d\Omega - G^e_{ext}\right] = 0, \tag{4.5.21}$$

where

$$G^e_{ext} = \int_{\Omega^e} \mathbf{f}^T\,\delta\mathbf{u}\, d\Omega + \int_{\partial\Omega^e_t} \mathbf{t}^{aT}\,\delta\mathbf{u}\, d\Gamma.$$

Equation (4.5.21) is a variational equation in the displacements, to be used in place of the virtual-work equation in the discretization process discussed in 1.3.5.

4.5.3. Finite-Element Methods for Nonlinear Continua

Nonlinear Problems: Incremental Solution

As long as inertia effects can be ignored and the deformation remains infinitesimal, the discretization procedure described above can be applied in nonlinear problems when the displacement, strain, and stress fields are replaced by their respective time derivatives, that is, the velocity, strain-rate, and stress-rate fields $\mathbf{v}$, $\dot{\boldsymbol{\varepsilon}}$, and $\dot{\boldsymbol{\sigma}}$.

We consider a problem in which the discretized external-load matrix $\mathbf{F}$ is a given function of time. One approach is to insert $\mathbf{F}$ in the time-differentiated equilibrium equation

$$\dot{\mathbf{Q}} = \dot{\mathbf{F}}.$$

This equation can be integrated over a time interval Δt to yield

$$\Delta\mathbf{R} = \Delta\mathbf{Q} - \Delta\mathbf{F} = 0.$$

Viscoplastic Problems

As has already been said, as the temperature goes up, the behavior of metals becomes significantly rate-dependent and is therefore much better described by viscoplasticity theory (whether with or without a yield surface) than by rate-independent plasticity. In addition, it was pointed out in 3.1.2 that the solutions of problems in classical viscoplasticity under constant loads attain asymptotically the equivalent rate-independent plasticity solutions, and are often easier to achieve.

To simplify the writing, the factor ϕ in Equation (4.5.2) is taken as unity, and the yield criterion, if any, is assumed to be embodied directly in $\mathbf{h}$. The rate equations are thus

$$\dot{\boldsymbol{\xi}} = \mathbf{h}(\boldsymbol{\sigma}, \boldsymbol{\xi}); \tag{4.5.22}$$

these equations include the flow equation

$$\dot{\boldsymbol{\varepsilon}}^i = \mathbf{g}(\boldsymbol{\sigma}, \boldsymbol{\xi}). \tag{4.5.23}$$

When Equation (4.5.1) is differentiated with respect to time and combined with (4.5.23), the displacement-based discretization leads to

$$\dot{\mathbf{Q}} = \mathbf{K}\dot{\mathbf{q}} - \mathbf{G}, \tag{4.5.24}$$

where

$$\mathbf{G} = \sum_e \mathbf{A}^{eT} \int_{\Omega^e} \mathbf{B}^{eT}\mathsf{C}\mathbf{g}\, d\Omega,$$

and $\mathbf{K}$ is the previously defined elastic global stiffness matrix. The problem, then, is to integrate Equations (4.5.22) and (4.5.24) for $\boldsymbol{\xi}$ and $\mathbf{q}$, with $\boldsymbol{\sigma}$ given by (4.5.1) together with $\boldsymbol{\varepsilon}^h\big|_{\Omega^e} = \mathbf{B}^e\mathbf{q}^e$, subject to the equilibrium equation $\mathbf{Q} = \mathbf{F}$.

As with the previously considered integration of the rate equation alone, the simplest technique is the Euler method, expressed by

$$\Delta\boldsymbol{\xi} = \Delta t\, \mathbf{h} \tag{4.5.25}$$

and

$$\Delta\boldsymbol{\sigma} = \mathsf{C}(\Delta\boldsymbol{\varepsilon} - \Delta t\, \mathbf{g}), \tag{4.5.26}$$

where $\mathbf{h}$ (which includes $\mathbf{g}$) is evaluated at the beginning of the time step. Applying the discretization process to Equation (4.5.26) gives

$$\Delta\mathbf{q} = \mathbf{K}^{-1}(\Delta\mathbf{F} + \Delta t\, \mathbf{G}). \tag{4.5.27}$$

The method is effective only with small time increments, not only because of the accumulation of discretization error, but because roundoff error may produce unstable results — that is, small changes in the initial conditions may produce large differences in the solution after a certain time. The Euler method is *conditionally stable* in the sense that stability holds only if the time increment Δt is less than some critical value Δt_{cr}.

The generalized Euler method discussed previously has been shown to be unconditionally stable when $\beta \geq \frac{1}{2}$. The initial values of $\mathbf{h}$ and $\mathbf{g}$ in Equations (4.5.25–26) are replaced by linear approximations to their values at $t + \beta\,\Delta t$, where t denotes the beginning of the time step. These approximations are

$$\mathbf{h} + \beta(\mathbf{h}_\sigma\,\Delta\sigma + \mathbf{h}_\xi\,\Delta\boldsymbol{\xi}) \quad \text{and} \quad \mathbf{g} + \beta(\mathbf{g}_\sigma\,\Delta\sigma + \mathbf{g}_\xi\,\Delta\boldsymbol{\xi}),$$

respectively. Hence Equations (4.5.25–26) are replaced by

$$(\mathbf{I} - \beta\,\Delta t\,\mathbf{h}_\xi)\,\Delta\boldsymbol{\xi} = DT\,(\mathbf{h} + \beta\mathbf{h}_\sigma\,\Delta\sigma)$$

and

$$(\mathbf{I} + \beta\,\Delta t\,\mathsf{C}\mathbf{g}_\sigma)\,\Delta\sigma = \mathsf{C}[\Delta\varepsilon - DT\,(\mathbf{g} + \beta\mathbf{g}_\xi)\,\Delta\boldsymbol{\xi}]$$

(note that the dimension of the identity matrices $\mathbf{I}$ in these equations is not in general the same). Eliminating $\Delta\boldsymbol{\xi}$ between the two equations, we obtain

$$\Delta\sigma = \bar{\mathsf{C}}(\Delta\varepsilon - \Delta t\,\bar{\mathbf{g}}),$$

where

$$\bar{\mathsf{C}} = [\mathsf{C}^{-1} + \beta\,\Delta t\,\mathbf{g}_\sigma + \beta^2\,DT\,\mathbf{g}_\xi(\mathbf{I} - \beta\,\Delta t\,\mathbf{h}_\xi)^{-1}\mathbf{h}_\sigma]^{-1}$$

and

$$\bar{\mathbf{g}} = \mathbf{g} + \beta\mathbf{g}_\xi(\mathbf{I} - \beta\,\Delta t\,\mathbf{h}_\xi)^{-1}\mathbf{h}.$$

By analogy with (4.5.27), we obtain the discretized equation

$$\Delta\mathbf{q} = \bar{\mathbf{K}}^{-1}(\Delta\mathbf{F} + \Delta t\,\bar{\mathbf{G}}),$$

where

$$\bar{\mathbf{K}} = \sum_e \mathbf{A}^{eT}\left(\int_{\Omega^e} \mathbf{B}^{eT}\bar{\mathsf{C}}\mathbf{B}^e\,d\Omega\right)\mathbf{A}^e$$

and

$$\bar{\mathbf{G}} = \sum_e \mathbf{A}^{eT}\int_{\Omega^e} \mathbf{B}^{eT}\mathsf{C}\bar{\mathbf{g}}\,d\Omega.$$

The improved accuracy and stability of the generalized Euler method is achieved at the cost of computing and inverting a new stiffness matrix at each time step. If desired, several iterations may be performed in each time step in order to improve the results.

Elastic-Plastic Problems

The finite-element solution of elastic-plastic problems has been the subject of intense development since the 1960s, and a number of different approaches have been proposed. Differences persist over such issues as the use of many constant-strain elements against fewer higher-order elements, the use of local constitutive equations in rate form against that of variational inequalities in the derivation of the discrete equations, and others. In this section a brief outline of some of the most common approaches is presented. Much of the theory can be found in textbooks such as Oden [1972] and Zienkiewicz [1977] (see also Zienkiewicz and Taylor [1989]), while computer implementation is treated by Owen and Hinton [1980]. More recent developments based on a mixed formulation are discussed by Simo and Hughes [1988].

A finite-element model may be regarded as successful if it achieves convergence of the solutions obtained under successive refinements of the mesh. It is especially so if, in those cases where a closed-form solution (or another reliable solution) is available, the convergence is to this solution. The strain-space formulation of the rate equations of plasticity permits, in principle, the application of any solution method to problems involving bodies that may be work-hardening, perfectly plastic, or strain-softening. However, the finite-element method has been successful only for work-hardening and perfectly plastic bodies. Attempts to solve problems involving strain-softening behavior have almost invariably led to *mesh-sensitivity*, that is, lack of convergence under mesh refinement. This result should not be too surprising. Our study of uniqueness criteria (Section 3.4) showed that no such criteria exist for softening bodies (if a body is nonstandard, then it must even have some hardening). Furthermore, our discussion of the physical nature of softening behavior in 2.3.2 showed that such behavior may be caused by a localization of strain rather than by pointwise strain-softening. Several schemes that account for localization (or a *size effect*) have been formulated in order to permit successful finite-element solutions of problems involving strain-softening bodies. Such schemes are beyond the scope of the present introductory treatment, which is consequently confined to work-hardening and perfectly plastic bodies.

We begin with an introduction to displacement-based methods. We suppose that the current global displacement matrix $\mathbf{q}$ and the current local state $(\boldsymbol{\sigma}, \boldsymbol{\xi})$ at each point are known.[1] The constitutive relation can be taken in the form (4.5.8), and, if the sign of $\dot{\lambda}$ is assumed to be known at all

[1] *Points* in the present context means elements if these are constant-strain, and Gauss points in the case of higher-order elements.

points, then the tangent stiffness C_ε is

$$\mathsf{C}_\varepsilon = \begin{cases} \mathsf{C}, & f < 0 \text{ or } \dot{\lambda} \le 0, \\ \mathsf{C} - \frac{1}{L}\mathsf{C}\mathbf{g} f_\sigma \mathsf{C}, & f = 0 \text{ and } \dot{\lambda} > 0. \end{cases}$$

A possible initial assumption is that $\dot{\lambda}$ has the same sign at the current state as it did at the preceding state. The global tangent stiffness $\mathbf{K}_t$ may be defined in terms of C_ε in the same way as $\mathbf{K}$ in terms of C, that is,

$$\mathbf{K}_t = \sum_e \mathbf{A}^{eT} \left(\int_{\Omega^e} \mathbf{B}^{eT} \mathsf{C}_{ep} \mathbf{B}^e \, d\Omega \right) \mathbf{A}^e.$$

The problem of determining the state at successive discrete points in time when the external-load history is prescribed[2] may be broken down, at each time point, into two phases. The first phase constitutes the *rate problem*: the generalized velocities are obtained from

$$\dot{\mathbf{q}} = \mathbf{K}_t^{-1} \dot{\mathbf{F}}, \tag{4.5.28}$$

and, with $\dot{\mathbf{q}}$ known, $\dot{\boldsymbol{\varepsilon}}$ may be determined from the interpolation $\dot{\boldsymbol{\varepsilon}}^h\big|_{\Omega^e} = \mathbf{B}^e \dot{\mathbf{q}}^e$, and the assumption on $\dot{\lambda}$ may be checked. If it is verified, then the rate problem has been solved, since $\dot{\mathbf{q}}$ and $\dot{\boldsymbol{\varepsilon}}$ are now known, as are $\dot{\boldsymbol{\xi}}$ through Equation (4.5.9) or (4.5.10), and $\dot{\boldsymbol{\sigma}}$ through (4.5.5) or (4.5.8). Otherwise, an iteration scheme may be used: If $\mathbf{K}_t^{(0)}$ denotes $\mathbf{K}_t$ evaluated on the basis of the initial assumption on $\dot{\lambda}$, then the solution

$$\dot{\mathbf{q}}^{(k)} = \mathbf{K}_t^{(k-1)^{-1}} \dot{\mathbf{F}}, \quad k = 1, 2, \ldots,$$

permitting the evaluation of a new distribution of $\dot{\lambda}$ and hence the calculation of a new tangent stiffness matrix $\mathbf{K}_t^{(k)}$. The process normally converges after a small number of iterations.

Once the rate problem is solved, the second phase of the solution is the determination of the state resulting from the imposition of an incremental load $\Delta\mathbf{F}$. An explicit method begins with

$$\Delta\mathbf{q}^{(0)} = \mathbf{K}_t^{-1} \Delta\mathbf{F},$$

from which $\Delta\boldsymbol{\varepsilon}^{(0)}$ is computed, and hence $\Delta\boldsymbol{\xi}^{(1)}$ and $\Delta\boldsymbol{\sigma}^{(1)}$ as in 4.5.1, with a single implementation of the return-mapping algorithm. In general, the incremental nodal forces $\Delta\mathbf{Q}^{(1)}$ resulting from $\Delta\boldsymbol{\sigma}^{(1)}$ do not equal $\Delta\mathbf{F}$, creating residual forces $\Delta\mathbf{R}^{(1)}$ that may be added to $\Delta\mathbf{F}$ in the next step of the iteration, which yields $\Delta\mathbf{q}^{(1)}$ from the general algorithm

$$\Delta\mathbf{q}^{(k)} = \Delta\mathbf{q}^{(k-1)} + \mathbf{K}_t^{-1} \Delta\mathbf{R}^{(k)}, \tag{4.5.29}$$

[2]It is assumed, for simplicity, that any prescribed boundary velocities are zero.

where the stiffness matrix $\mathbf{K}_t$ may or may not be modified at each iteration according to the new values of $\boldsymbol{\sigma}$ and $\boldsymbol{\xi}$. In the **tangent-stiffness method**, which is equivalent to the Newton–Raphson method, the stiffness matrix $\mathbf{K}_t$ is recomputed at each iteration, providing faster convergence of the iteration process at the expense of more computation at each iteration. These computation costs are reduced in the so-called **modified tangent-stiffness methods**, in which $\mathbf{K}_t$ is computed only once for each load increment, for example in the first or in the second iteration. In the **initial-stiffness method** the elastic stiffness $\mathbf{K}$ is used throughout the process, greatly decreasing the need for matrix inversion but increasing the number of iterations. Convergence occurs when the residual forces are sufficiently small, based on some appropriate norm. The next load increment can then be applied, and the iteration procedure can begin again.

Algorithmic Tangent Moduli

When the return-mapping algorithm is used in conjunction with the tangent stiffness defined by C_ε, the result is a loss of the quadratic rate of asymptotic convergence, particularly important for large time steps (Nagtegaal [1982], Simo and Taylor [1985]). A procedure that preserves the quadratic rate of asymptotic convergence replaces C_{ep} in the expression for $\mathbf{K}_t$ by the so-called *algorithmic* (or *consistent*) tangent moduli, defined as follows: given an algorithm that produces $\Delta\boldsymbol{\sigma}$ when $\boldsymbol{\sigma}$, $\boldsymbol{\varepsilon}$, $\boldsymbol{\xi}$ and $\Delta\boldsymbol{\varepsilon}$ are given, or formally, if

$$\Delta\boldsymbol{\sigma} = \Delta\boldsymbol{\sigma}(\boldsymbol{\sigma},\, \boldsymbol{\varepsilon},\, \boldsymbol{\xi},\, \Delta\boldsymbol{\varepsilon}),$$

then

$$\mathsf{C}_{alg} = \partial\Delta\boldsymbol{\sigma}/\partial\Delta\boldsymbol{\varepsilon}|_{t+\Delta t}\,.$$

While C_{alg} coincides with the elastic-plastic tangent modulus tensor C_ε in problems with only one independent stress component, it does not do so in general. For more details, see Simo and Taylor [1985] and Simo and Hughes [1988].

The definition of C_{alg} given above is consistent with the incremental form of the equilibrium equations, $\Delta\mathbf{Q} = \Delta\mathbf{F}$. Other forms may be associated with the total form, $\mathbf{Q} = \mathbf{F}$, and with the mixed formulation, given by Equation (4.5.21). The derivation for the case of the mixed formulation is shown next.

Application of Mixed Formulation

In the mixed formulation, the finite-element discretization may be written as

$$\begin{aligned} \mathbf{u}^h(\mathbf{x})\Big|_{\Omega^e} &= \mathbf{N}^e(\mathbf{x})\mathbf{q}^e, \\ \nabla^S\mathbf{u}^h(\mathbf{x})\Big|_{\Omega^e} &= \mathbf{B}^e(\mathbf{x})\mathbf{q}^e, \\ \boldsymbol{\varepsilon}^h(\mathbf{x})\Big|_{\Omega^e} = \bar{\nabla}^S\mathbf{u}^h(\mathbf{x})\Big|_{\Omega^e} &= \bar{\mathbf{B}}^e(\mathbf{x})\mathbf{q}^e, \end{aligned} \qquad (4.5.30)$$

where[1]

$$\bar{\mathbf{B}}^e(\mathbf{x}) = \mathbf{E}^e(\mathbf{x})\mathbf{H}^{e-1}\int_{\Omega^e} \mathbf{S}^{eT}\mathbf{B}^e\,d\Omega.$$

Substitution of (4.5.30) into Equation (4.5.21), the reduced weak form of the equilibrium equation, gives

$$\mathbf{Q} = \sum_e \mathbf{A}^{eT}\int_{\Omega^e} \bar{\mathbf{B}}^{eT}\mathsf{C}(\bar{\nabla}^S\mathbf{u} - \varepsilon^p)\,d\Omega$$

for the global internal-force matrix, where ε^p is evaluated by integration of the rate equations. An application of the Newton–Raphson method again yield Equation (4.5.29), with the tangent stiffness matrix now given by

$$\begin{aligned}\mathbf{K}_t &= \partial\mathbf{Q}/\partial\mathbf{q}\\ &= \sum_e \mathbf{A}^{eT}\int_{\Omega^e} \bar{\mathbf{B}}^{eT}\{\partial[\mathsf{C}(\bar{\nabla}^S\mathbf{u} - \varepsilon^p)]/\partial(\bar{\nabla}^S\mathbf{u})\}\,[\partial(\bar{\nabla}^S\mathbf{u})/\partial\mathbf{q}]\,d\Omega\\ &= \sum_e \mathbf{A}^{eT}\left(\int_{\Omega^e} \bar{\mathbf{B}}^{eT}\mathsf{C}_{alg}\bar{\mathbf{B}}^e\,d\Omega\right)\mathbf{A}^e,\end{aligned}$$

since

$$\partial(\bar{\nabla}^S\mathbf{u})/\partial\mathbf{q} = [\partial(\bar{\nabla}^S\mathbf{u})/\partial\mathbf{q}^e]\,(\partial\mathbf{q}^e/\partial\mathbf{q}) = \bar{\mathbf{B}}^e\mathbf{A}^e,$$

and

$$\mathsf{C}_{alg} = \partial[\mathsf{C}(\bar{\nabla}^S\mathbf{u} - \varepsilon^p)]/\partial(\bar{\nabla}^S\mathbf{u})\Big|_{t+\Delta t}$$

is the tensor of algorithmic (consistent) tangent moduli obtained by differentiation of the algorithmic (discrete) flow law.

Examples

As one example of the application of the finite-element method to the solution of elastic-plastic problems, we examine the thick-walled elastic–perfectly plastic cylindrical tube in plane strain, with the material assumed to obey the Mises criterion and its associated flow rules; the corresponding problem for the Tresca criterion was treated analytically in 4.3.5. If the material is plastically *and elastically* incompressible (Poisson's ratio equal to one-half), then the Mises and Tresca results coincide.

Two cases are presented. In both, the ratio b/a of the outer to the inner radius is 2, and $k/G = 0.003$. The finite elements used are concentric rings (toroids) of rectangular cross-section, so that the mesh in any rz-plane is a rectangular one; four-noded ($Q4$) elements are chosen for their simplicity. It was shown by Nagtegaal, Parks and Rice [1974] that in a nearly incompressible body in plane strain represented by $Q4$ elements, an unrealistically stiff response ("locking") is obtained unless special precautions are taken.

[1] The use of the matrix function $\bar{\mathbf{B}}^e$ has given the procedure the name "B-bar".

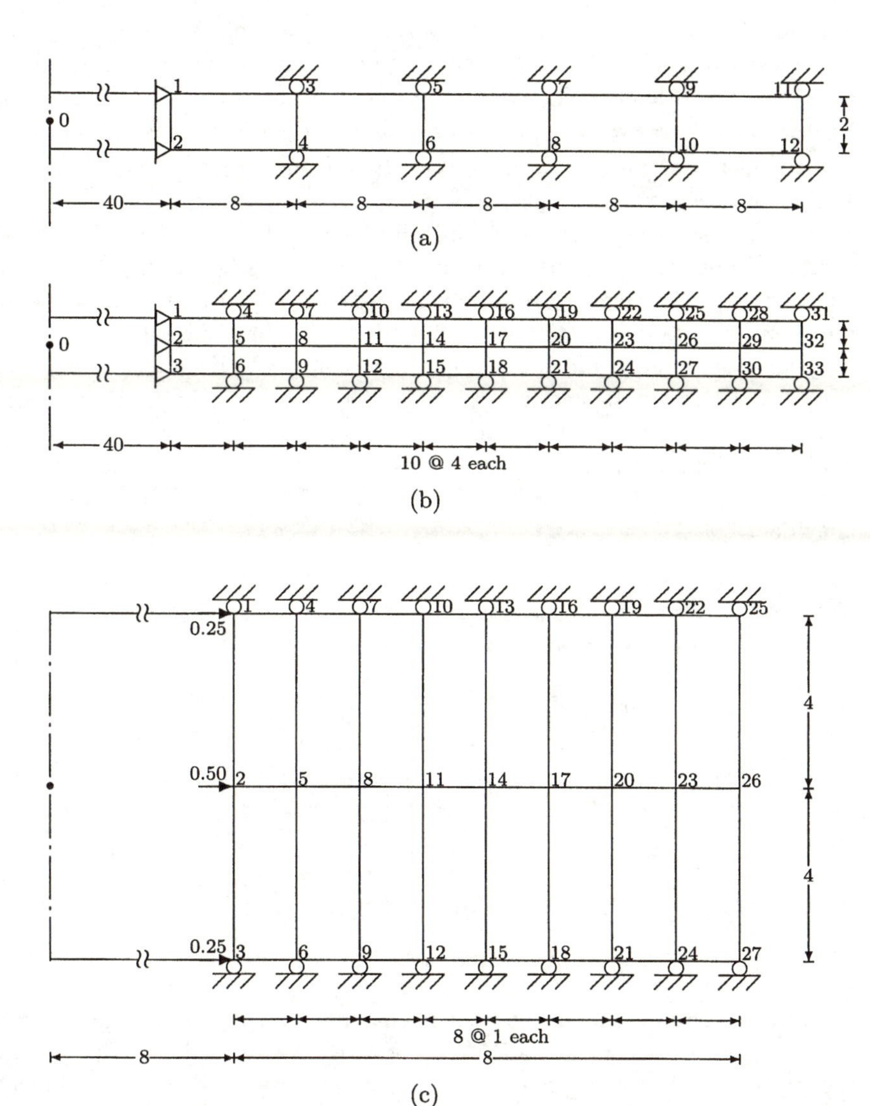

Figure 4.5.5. Meshes used in analysis of elastic-plastic cylinder: (a) coarse mesh, displacement control, equal radial displacements applied at nodes 1 and 2; (b) fine mesh, displacement control, equal radial displacements applied at nodes 1, 2, and 3; (c) fine mesh, force control, forces applied at nodes 1, 2, and 3 in the proportions shown.

The previously discussed B-bar procedure, in which the volume strain is independently approximated by a discontinuous interpolation, is one way out, and such a procedure is used in the present analysis. The program used is FEAP (Finite Element Analysis Program), developed by R. L. Taylor (for an introductory treatment, see Zienkiewicz [1977], Chapter 24).

In the first case the cylinder is virtually incompressible ($\nu = 0.4999$) and is analyzed under displacement control, with the radial displacement of the inner surface, $u(a)$, monotonically increased up to $4ka/3G$ and then decreased until the pressure turns negative. Two meshes are used, a coarser and a finer one, shown respectively in Figure 4.5.5(a) and (b). The mesh represents a 1-radian sector of the cylinder, and therefore, at each value of $u(a)$ the pressure is calculated according to $p = (\sum R)/ah$, where $\sum R$ is the sum of the radial reactions at nodes 1 and 2 in mesh (a) and at nodes 1, 2, and 3 in mesh (b). The results of the computations are shown in Table 4.5.1 as $p_{(a)}$ and $p_{(b)}$, respectively, and they are compared with the exact values p_{ex} obtained according to the theory of Section 4.3. The convergence is evident.

Table 4.5.1. Pressure-Displacement Relation for Incompressible Cylinder: Computed and Exact Values

$3G\,u(a)/ka$	$p_{(a)}/k$	$p_{(b)}/k$	p_{ex}/k
1	0.5010	0.5000	0.4996
2	0.9578	0.9562	0.9539
3	1.1973	1.1937	1.1928
4	1.3192	1.3156	1.3139
3	0.8184	0.8162	0.8143
2	0.3177	0.3160	0.3147

In the second case, the cylinder is analyzed under pressure control, with the mesh shown in Figure 4.5.5(c), and both compressible ($\nu = 0.3$) and incompressible behavior are examined. For the former, the computed results for the stress distributions σ_r, σ_θ, and σ_z, and for the displacements $u(a)$ and $u(b)$, are shown for various values of c/a (where c is the radius of the elastic-plastic boundary) in Figure 4.5.6, where they are compared with the numerical solution of Hodge and White [1950];[1] the graph of p/k against c/a in Figure 4.5.6(d) was used to obtain c/a for the imposed values of p/k.

Finally, the distributions of σ_z for both compressible and incompressible behavior when $c/a = 1.5$, and a plot of p/k against $2Gu(b)/ka$ for the incompressible case, are shown in Figure 4.5.7.

[1]The graphs are taken from Prager and Hodge [1951].

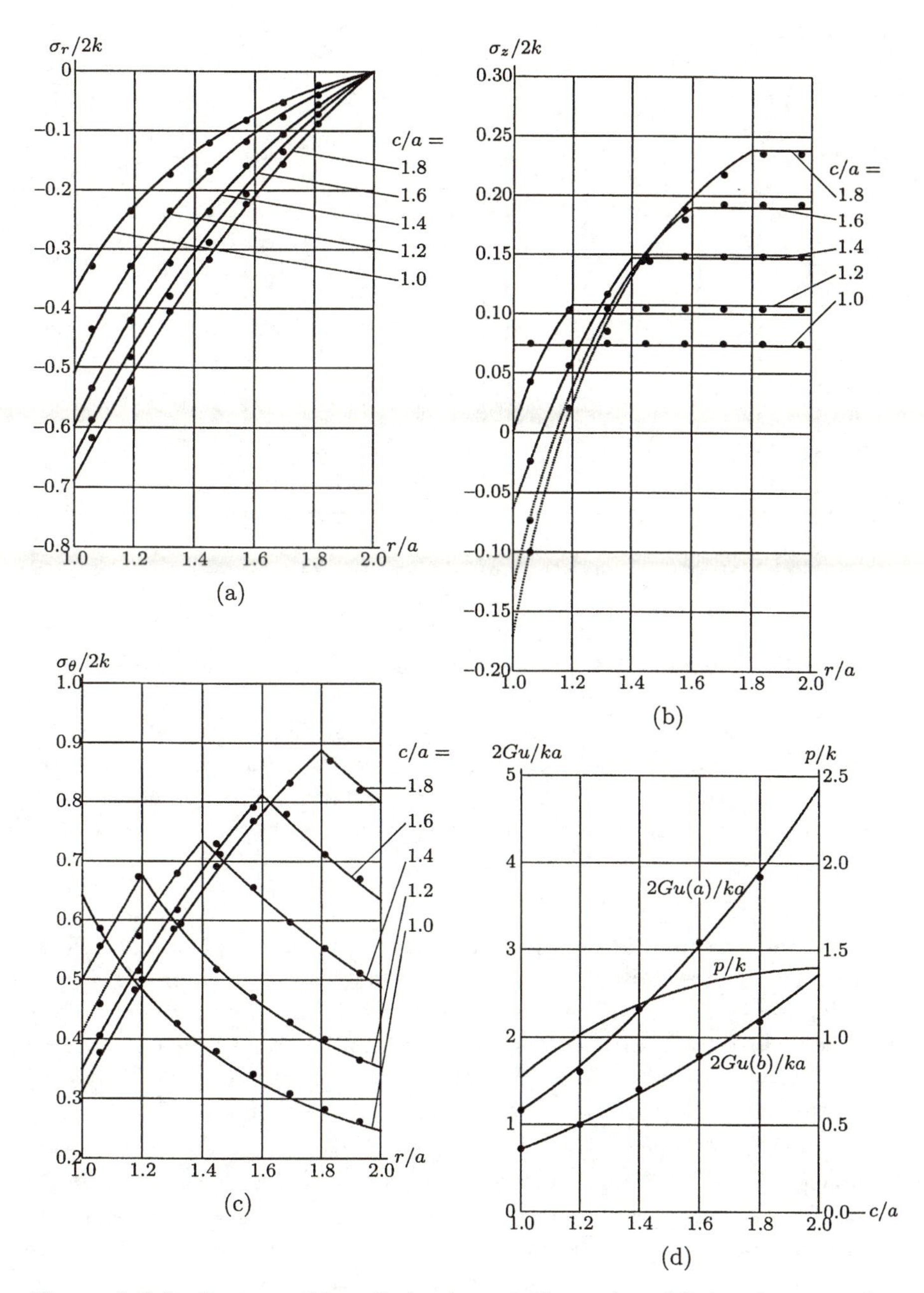

Figure 4.5.6. Compressible cylinder ($\nu = 0.3$), results of finite-element calculations: (a)–(c) stress distributions at various values of c/a; (d) displacements of inner and outer curfaces, and internal pressure, against c/a.

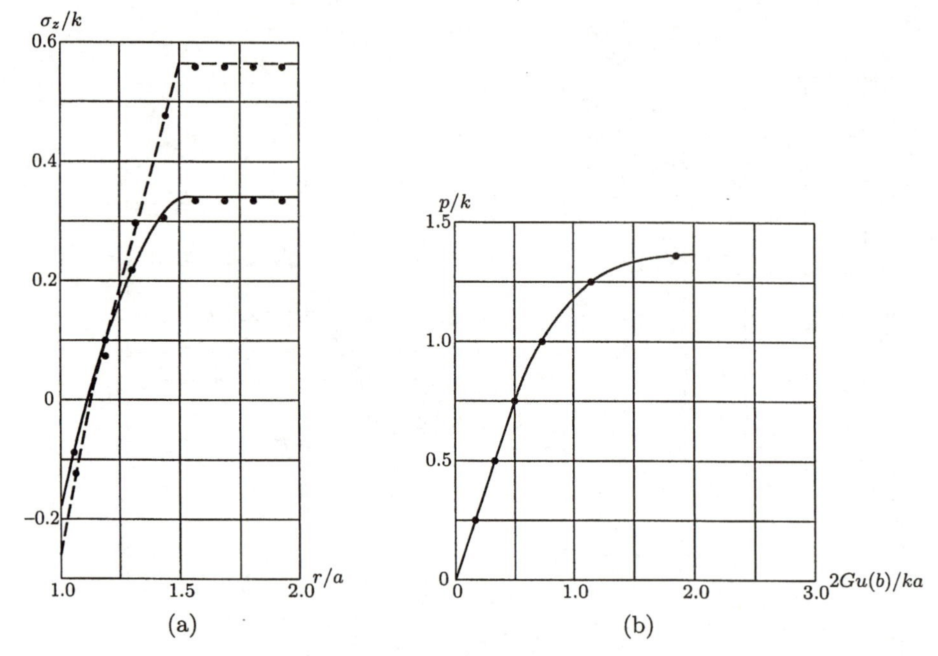

Figure 4.5.7. Compressible and incompressible cylinder, results of finite-element analysis (dots) and of Hodge and White [1950] or exact analysis (curves): (a) axial stress distributions at $c/a = 1.5$ (dashed line, incompressible [exact]; solid line, compressible [Hodge and White]; (b) pressure against displacement of outer surface (incompressible only).

Another example is the end-loaded elastic–perfectly plastic cantilever beam in plane stress, discussed in 4.4.2. Computations, based on the return-mapping algorithm for plane stress formulated by Simo and Taylor [1986], were performed with a 4×40 and an 8×80 mesh of $Q4$ elements, covering half the beam (above or below the middle line). Numerical and analytical results for the load–deflection relation are shown in Figure 4.5.8.

Lastly, an example is given of a problem for which no analytical solution is known. Figure 4.5.9 shows the torque-twist relation for an elastic–perfectly plastic I-beam having the cross-section shown in the figure. The computed results, due to Baba and Kajita [1982], are shown along with the ultimate torque calculated by means of the sand-heap analogy (see 4.2.3). As we saw in 4.2.1, in a displacement formulation of the torsion problem the warping function $\psi(x_1, x_2)$ is the only unknown; four-noded rectangular elements are used, with a cubic interpolation for ψ, and with the values of ψ, $\psi_{,1}$, and $\psi_{,2}$ at the nodes as the nodal variables. A comparison of the results with those obtained by the "cellular analogy" was made by Johnson [1988].

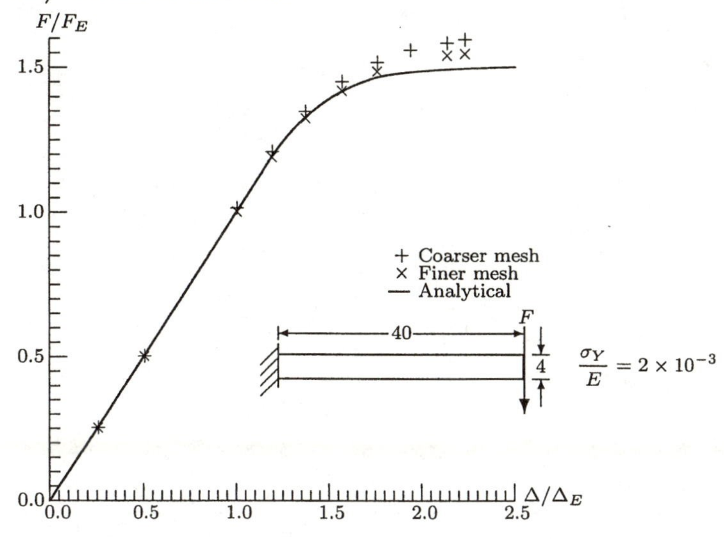

Figure 4.5.8. End-loaded rectangular cantilever beam in plane stress: finite-element and analytical results for the force-deflection diagram.

Exercises: Section 4.5

1. Find the elastic-plastic tangent-modulus tensor C_{ep} for the solid obeying the associated flow rule with the yield function

$$f(\boldsymbol{\sigma}, \boldsymbol{\rho}, \bar{\varepsilon}^p) = \sqrt{2\bar{J}_2} - \sqrt{2}/3\,(\sigma_E + \beta H \bar{\varepsilon}^p),$$

where σ_E, β, and H are constants,

$$\bar{J}_2 = \frac{1}{2}(s_{ij} - \rho_{ij})(s_{ij} - \rho_{ij}),$$

and $\boldsymbol{\rho}$ and $\bar{\varepsilon}^p$ are governed respectively by

$$\dot{\rho}_{ij} = \frac{2}{3}(1 - \beta)H\dot{\varepsilon}^p_{ij}, \qquad \dot{\bar{\varepsilon}}^p = \sqrt{\tfrac{2}{3}\dot{\varepsilon}^p_{ij}\dot{\varepsilon}^p_{ij}}.$$

2. (a) Formulate the radial-return algorithm for the elastic-plastic solid of Exercise 1.

   (b) By differentiating the algorithm, find the consistent tangent-modulus tensor, and compare with the result of Exercise 1.

3. Provide the details of the derivation of Equation (4.5.21) from Equation (4.5.14) by using the relations (4.5.19–20).

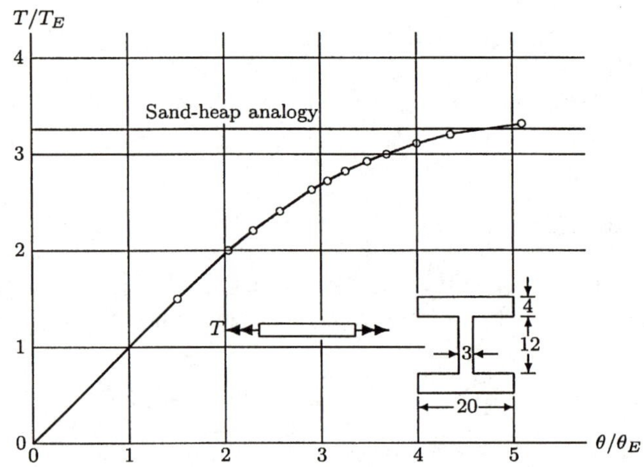

Figure 4.5.9. Torsion of an I-beam: finite-element results for the torque-twist diagram (Baba and Kajita [1982]). The ultimate load obtained by the sand-heap analogy is shown for comparison.

Chapter 5

Problems in Plastic Flow and Collapse I

Theories and "Exact" Solutions

Introduction

In Chapter 4 the concepts of *plastic flow* and *plastic collapse* were regarded as essentially equivalent, representing a state in which a body continues to deform under constant applied forces. In practical applications, however, the two concepts have quite different meanings. Plastic collapse describes undesirably large deformations of an already formed body (a *structure*) that result from excessive forces; the calculation of *collapse loads* of simple structures was studied in Section 4.1. The concept of plastic flow, on the other hand, is usually applied to the deliberate forming of a mass of solid (such as metal or clay) into a desired shape through the application of appropriate forces.[1] It is remarkable that these two large classes of problems, of fundamental importance in mechanical and civil engineering, can be attacked by the same methodology — the theory of rigid–perfectly plastic bodies, with the help of the theorems of limit analysis. A particularly extensive body of theory, filling entire books, exists for problems of plane strain; a summary of the theory, with some applications, is presented in Section 5.1. In Section 5.2 we deal with the plastic collapse of circular plates.

Apart from plastic collapse, collapse of a elastic-plastic body may also be due to structural instability. Such collapse (e.g., the buckling of a column) may begin when the body is still fully elastic, and plastic deformation occurs as a part of *post-buckling behavior*. Buckling that follows yield is covered by the theory of plastic instability, which is treated in Section 5.3.

[1] The plastic flow of soil constitutes an exception, since substantial movement of a soil mass supporting a building or forming an earth dam is generally regarded as failure.

Section 5.1 Plane Problems

In Section 4.3 we found the stress field in a pressurized elastic–perfectly plastic hollow cylinder in a state of plane strain by solving the equilibrium equations together with, on the one hand, the compatibility condition in the elastic region and, on the other hand, the yield criterion in the plastic region, and by satisfying the boundary conditions on the outer and inner surfaces and continuity conditions at the elastic-plastic boundary. The solution is valid for all pressures up to the ultimate pressure p_U.

When the ultimate pressure is reached, the tube becomes fully plastic. The compatibility and continuity conditions then become irrelevant. The equilibrium equation (4.3.18) and the yield criterion now constitute equations for the two unknown stress components σ_r and σ_θ, which may be solved so that the traction boundary conditions are satisfied. For both the Mises and Tresca yield criteria (which are equivalent in plane plastic flow), the solution produces the stresses given by Equations $(4.3.30)_{1,2}$ with $c = b$.

Since the fully plastic tube problem is formulated entirely in terms of the stresses, it is often said to be *statically determinate*, though in a looser sense than that of Section 4.1, since it is not equilibrium ("statics") alone that determines the stress field, but the yield criterion as well. The fully plastic torsion problem of 4.2.3 can similarly be characterized as statically determinate, since once again there is one nontrivial equilibrium equation which, together with the yield criterion, can be used to determine the two unknown stress components subject to the traction boundary conditions.

The same notion of static determinacy can be applied, in principle, to a body of arbitrary shape that is assumed to be undergoing plane plastic deformation, or to be in a state of plane stress: there are, in general, three unknown stress components (σ_{11}, σ_{22}, σ_{12} in Cartesian coordinates) and two plane equilibrium equations, plus the yield criterion. However, static determinacy in this sense is effective in producing a unique stress field only in special cases, namely those in which (a) only traction boundary conditions are relevant, and (b) the entire body must become plastic for unrestricted plastic flow to occur. As we saw in the beam problems studied in Section 4.5, it is in general possible to have unrestricted flow in a plastic region that occupies only a part of the body, the rest of the body remaining in the elastic range and hence behaving as though it were rigid (recall, from 3.5.1, the vanishing of the elastic strain rates at incipient plastic flow).

Even when the aforementioned conditions (a) and (b) are satisfied, there may occur situations in which the traction boundary are not by themselves sufficient to choose between two possible stress tensors at a point (see Figure 5.1.4, which is discussed later).

Since, in the general case, some of the boundary conditions may be kinematic (velocity boundary conditions), it becomes necessary to find a kinematically admissible velocity field such that the strain rates derived from it obey the flow rule. Consequently, the stress and velocity problems are coupled. In the special cases where a unique stress field can be found directly, then the velocity field can be found afterwards, as was done with the warping of a fully plastic rectangular shaft (see 4.2.3). In the axisymmetric plane-strain problem, the only velocity component is the radial velocity v, and the flow rule is equivalent to the incompressibility constraint

$$\frac{dv}{dr} + \frac{v}{r} = 0,$$

which may be solved to give $v(r) = v(a)(a/r)$.

If, in addition, elastic regions remain in the course of plastic flow, then a solution of the problem requires the determination of the elastic-plastic boundary and of a stress field in the elastic regions which is continuous with the plastic stress field at the boundary. While this was accomplished for the beam problems of Section 4.5, it is in general an exceedingly difficult task. In many problems, the main objective is to find the load that produces plastic flow or collapse, and a complete elastic-plastic solution is not necessary to achieve this objective: we know from the theorems of limit analysis (Section 3.5) that the correct critical load is obtained from a plastically and statically admissible stress field and a kinematically admissible velocity field that is associated with it in the plastic region. Thus a rigid-plastic boundary may be established on a purely kinematic basis, and the plastic stress field needs to be extended into the rigid region only so that is statically and plastically admissible. With this extension, the solution becomes a complete rigid–plastic solution, usually known simply as a *complete solution* (Bishop [1953]).

A systematic method of determining stress fields and associated velocity fields in perfectly plastic bodies obeying the Mises (or Tresca) yield criterion in plane strain was developed in the 1920s by Prandtl, Hencky, Mises and others, and generalized by Mandel [1962] to include other yield criteria and plane stress. This method, generally known as **slip-line theory**, is discussed in the next subsection. Some applications are presented in succeeding subsections.

5.1.1. Slip-Line Theory

Shear Directions

A convenient way to establish the necessary relations for the stress field in a plastic region is with the help of the definitions $n = \frac{1}{2}(\sigma_1 + \sigma_2)$ and $r = \frac{1}{2}|\sigma_1 - \sigma_2|$, as given in 3.3.4, in conjunction with the yield condition (3.3.6). We introduce the Mohr's circle relations

$$\sigma_{11} = n + r\sin 2\theta, \quad \sigma_{22} = n - r\sin 2\theta, \quad \sigma_{12} = -r\cos 2\theta, \qquad (5.1.1)$$

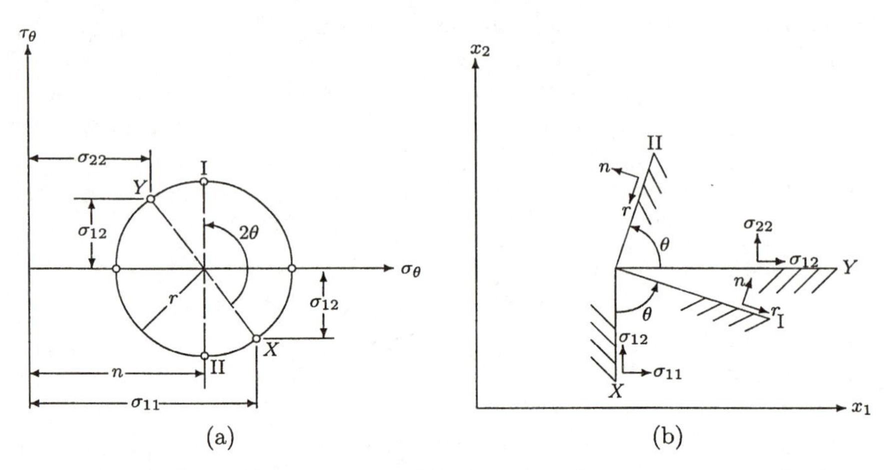

Figure 5.1.1. First (I) and second (II) shear-line directions in the (a) Mohr's-circle and (b) physical planes.

where θ is the angle from the x_1 axis to one of the principal shear directions, namely the one along which the maximum shear stress r is directed to the left when one is facing the outer normal (see Figure 5.1.1). This direction will be called the *second shear direction*, and a line having this direction locally everywhere will be called a *second shear line*. The other shear direction (shear line) will be called the first.

When equations (5.1.1) are substituted into the equilibrium equations

$$\sigma_{11,1} + \sigma_{12,2} = 0, \quad \sigma_{12,1} + \sigma_{22,2} = 0,$$

the resulting equations are, upon substitution of $r = h(n)$,

$$[1 + h'(n) \sin 2\theta] n_{,1} - h'(n) \cos 2\theta \, n_{,2} + 2r(\theta_{,1} \cos 2\theta + \theta_{,2} \sin 2\theta) = 0,$$
$$-h'(n) \cos 2\theta \, n_{,1} + [1 - h'(n) \sin 2\theta] n_{,2} + 2r(\theta_{,1} \sin 2\theta - \theta_{,2} \cos 2\theta) = 0. \tag{5.1.2}$$

Equations (5.1.2) constitute a pair of nonlinear partial differential equations for n, θ. A useful method of numerical solution for these equations is the **method of characteristics**.

Method of Characteristics

In order to understand how this method can be used to solve the system, it is simpler to consider first a single first-order partial differential equation of the form

$$Av_{,1} + Bv_{,2} = C,$$

where A, B and C are functions of x_1, x_2 and v. The equation can be multiplied by an infinitesimal increment dx_1 and rewritten as

$$A\Big(\Big(v_{,1}\,dx_1 + \frac{B}{A}v_{,2}\,dx_1\Big) = C\,dx_1.$$

The quantity in parentheses becomes a perfect differential $dv = (v_{,1})\,dx_1 + (v_{,2})\,dx_2$ if $dx_2 = m\,dx_1$, where $m = B/A$. The direction defined by $dx_2/dx_1 = m$ is called a *characteristic direction*, and a curve that is everywhere tangent to a characteristic direction is known as a *characteristic curve* or simply a *characteristic*. Along such a curve, then,

$$\frac{dx_1}{A} = \frac{dx_2}{B} = \frac{dv}{C}.$$

If v is known at one point of a characteristic curve, then dv can be calculated for a neighboring point on this curve, and continuing this operation allows v to be determined at all points on the curve. If v is known at all points of a curve Σ that is nowhere tangent to a characteristic, then its values may be calculated along all the characteristics that intersect Σ.

If two characteristics emanating from different points of Σ should intersect at some point Q of the x_1x_2-plane, then in general two different values of v will be obtained there. The point Q is therefore the locus of a discontinuity in v.

Consider, now, a curve Σ that is nowhere tangent to a characteristic, and suppose that a discontinuity (jump) in $v_{,1}$ or $v_{,2}$ occurs at a point P of Σ. Since the only information necessary to determine the characteristic curve through P and the values of v along this curve is the value of v at P, the directional derivative of v *along* the characteristic will have the same value on either side of it. Consequently, if any jump in $v_{,2}$ or $v_{,1}$ is propagated through the x_1x_2-plane, it must occur *across* the characteristic through P.

In order to apply the method of characteristics to a system of several first-order partial differential equations, the number of real characteristic directions at each point must equal the number of unknown variables. The system is then called *hyperbolic*. The system (5.1.2) is consequently hyperbolic if it has two real characteristic directions. It is called *parabolic* if there is one such direction, and *elliptic* if there are none.

Characteristics of Equations (5.1.2)

In order to determine whether Equations (5.1.2) constitute a hyperbolic system, we begin by solving them for $n_{,1}$, $n_{,2}$ in terms of $\theta_{,1}$, $\theta_{,2}$:

$$n_{,1} = \frac{-2r}{1-h'^2}(\theta_{,1}\cos 2\theta + \theta_{,2}\sin 2\theta - h'\theta_{,2}),$$

$$n_{,2} = \frac{-2r}{1-h'^2}(\theta_{,1}\sin 2\theta - \theta_{,2}\cos 2\theta + h'\theta_{,1}).$$

Note that the solutions break down when $|h'(n)| = 1$. Assuming that this breakdown does not occur, we attempt to determine the characteristics by assuming that, along a characteristic, $dn = \lambda\, d\theta$. Now

$$\begin{aligned} dn &= n_{,1}\, dx_1 + n_{,2}\, dx_2 \\ &= \frac{-2r}{1-h'^2}\{\theta_{,1}\,[\cos 2\theta\, dx_1 + (\sin 2\theta + h')\, dx_2] + \theta_{,2}\,[(\sin 2\theta - h')\, dx_1 - \cos 2\theta\, dx_2]\} \\ &= \lambda\, d\theta \\ &= \theta_{,1}\,(\lambda\, dx_1) + \theta_{,2}\,(\lambda\, dx_2). \end{aligned}$$

Equating the coefficients of $\theta_{,1}$ and $\theta_{,2}$, respectively, in the second and fourth lines leads to the system of equations

$$[(1-h'^2)\lambda + 2r\cos 2\theta]\, dx_1 + 2r(\sin 2\theta + h')\, dx_2 = 0,$$

$$2r(\sin 2\theta - h')\, dx_1 + [(1-h'^2)\lambda - 2r\cos 2\theta]\, dx_2 = 0,$$

which constitute a second-order eigenvalue problem. The characteristic equation is

$$(1-h'^2)^2\lambda^2 - 4r^2\cos^2 2\theta - 4r^2\sin^2 2\theta + 4r^2h'^2,$$

yielding the eigenvalues

$$\lambda = \pm\frac{2r}{\sqrt{1-h'^2}}.$$

The roots are real if and only if $|h'| \le 1$; the problem is hyperbolic if $|h'| < 1$, parabolic when $|h'| = 1$, and elliptic when $|h'| > 1$.

Referring to the examples of the yield condition (3.3.6) discussed following its formulation, we note that the plane-stress problem for the Mises criterion is hyperbolic only when $|n| < 3k/2$, parabolic when $|n| = 3k/2$, and elliptic when $|n| > 3k/2$; this last condition occurs when

$$\frac{1}{2} < \frac{\sigma_1}{\sigma_2} < 2.$$

For the Tresca criterion, the problem is hyperbolic when $|n| < k$ and parabolic when $|n| \ge k$; the elliptic case does not arise.

In plane strain, on the other hand, $h' \equiv 0$ for both criteria with the associated flow rule when the elastic strains can be neglected, and therefore the problem is hyperbolic throughout the plastic domain. The same is true for the Mohr–Coulomb criterion with the appropriate (not necessarily associated) flow rule, since $h' = -\sin\phi$, where ϕ is the angle of internal friction. More generally, for any yield criterion given by Equation (3.3.6) with $|h'| < 1$, it is convenient to define a variable angle of internal friction, $\phi(n)$, by $\sin\phi(n) = -h'(n)$; this is just the inclination of the Mohr envelope with respect to the σ-axis at the point of tangency. The eigenvalues λ are correspondingly given by $\pm 2r\sec\phi$.

From the trigonometric identity

$$\tan(a+b) = \frac{\sin 2a + \sin 2b}{\cos 2a + \cos 2b}$$

we may derive the directions of the eigenvectors. The eigenvector corresponding to $\lambda = 2r \sec\phi$ is given by $dx_2/dx_1 = -\cot(\theta - \frac{1}{2}\phi)$, and will be called the first or α characteristic, while the eigenvector corresponding to $\lambda = -2r\sec\phi$ is given by $dx_2/dx_1 = \tan(\theta + \frac{1}{2}\phi)$, and will be called the second or β characteristic. In the case of the Mises and Tresca criteria in plane strain (the classical case), the characteristic directions coincide with the shear directions and are orthogonal. In general the characteristics of the two families intersect at an angle $\frac{1}{2}\pi \pm \phi$; they coalesce into a single family when $\phi = \frac{1}{2}\pi$, that is, in the parabolic case.

Defining the dimensionless variable ω by

$$\omega = \int \frac{\cos\phi(n)}{2h(n)} dn,$$

we may write the characteristic relations, following Mandel [1962], as

$$\begin{aligned} d\omega &= d\theta \quad \text{along a first characteristic,} \\ d\omega &= -d\theta \quad \text{along a second characteristic.} \end{aligned} \tag{5.1.3}$$

Note that $\omega = n/2k$ in the classical case, and

$$\omega = -\frac{\cot\phi}{2}\ln\left(1 - \frac{n}{c}\tan\phi\right)$$

for the Mohr–Coulomb material, yielding the preceding value in the limit as $\phi \to 0$, with $c = k$.

If we introduce a curvilinear coordinate system α, β such that first and second characteristics are given respectively by $\beta =$ constant and $\alpha =$ constant, then we can write the canonical equations

$$\frac{\partial}{\partial\alpha}(\omega - \theta) = 0, \quad \frac{\partial}{\partial\beta}(\omega + \theta) = 0,$$

which have the general solution

$$\omega = \xi(\alpha) + \eta(\beta), \quad \theta = \xi(\alpha) - \eta(\beta),$$

ξ and η being arbitrary functions.

In the following discussion of the geometric properties of the characteristic network we shall limit ourselves to the case with ϕ constant — that is, the Mohr–Coulomb criterion — so that a change in θ is also a change in the direction of the characteristic. This case includes the classical case, $\phi = 0$, and the characteristic network is then called a *Hencky–Prandtl network*. Among its properties are the following:

1. Suppose that one α characteristic is straight, that is, for a given β, $\partial\theta/\partial\alpha = 0$. It follows that $\xi'(\alpha) = 0$, and consequently all the α characteristics are straight. Obviously, the same result holds for β characteristics.

2. Consider a pair of α characteristics, defined by β and β', respectively, and a pair of β characteristics defined by α and α', the points of intersection being labeled $A = (\alpha, \beta)$, $B = (\alpha', \beta)$, $C = (\alpha, \beta')$, and $D = (\alpha', \beta')$, as in Figure 5.1.2. It follows from the solution above that

$$\chi_{AC} = \theta(\alpha, \beta) - \theta(\alpha, \beta') = \eta(\beta') - \eta(\beta),$$
$$\chi_{BD} = \theta(\alpha', \beta) - \theta(\alpha', \beta') = \eta(\beta') - \eta(\beta),$$

and consequently the two angles χ_{AC} and χ_{BD} are equal. It can similarly be shown that χ_{AB} equals χ_{CD}. This result is due to Hencky [1923] and is known as **Hencky's theorem.** In words: the angle formed by the tangents of two given characteristics of one family at their points of intersection with a characteristic of the other family does not depend on the choice of the intersecting characteristic of the other family.

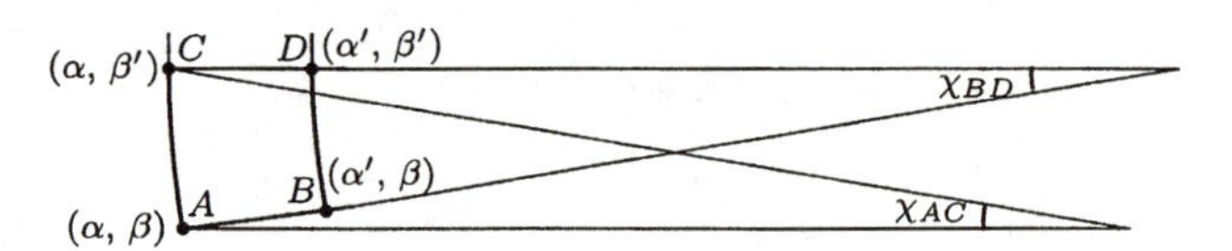

Figure 5.1.2. Characteristics of Equations (5.1.2).

3. Now take (α', β') in Figure 5.1.2 infinitesimally close to (α, β). If $R_\beta(\alpha, \beta)$ denotes the radius of curvature of the β characteristics at (α, β) and if ds_α and ds_β denote infinitesimal arc lengths along the α and β characteristics, respectively, then

$$ds_{\beta(AC)} = R_\beta(\alpha, \beta)\chi_{AC} = [R_\beta(\alpha', \beta) + ds_\alpha]\chi_{BD}.$$

Since $R_\beta(\alpha', \beta) = R_\beta(\alpha, \beta) + dR_\beta$, it follows that $dR_\beta = -ds_\alpha$ along an α characteristic. Similarly, $dR_\alpha = ds_\beta$ along a β characteristic. This result is due to Prandtl [1923].

Traction Boundary Conditions

Traction boundary-value problems may be of three types, with the construction of characteristics corresponding to each type shown in Figure 5.1.3.

Problem 1. The boundary is nowhere tangent to a characteristic.

Problem 2. The boundary is composed of characteristics of both families.

Problem 3. The boundary is of mixed type.

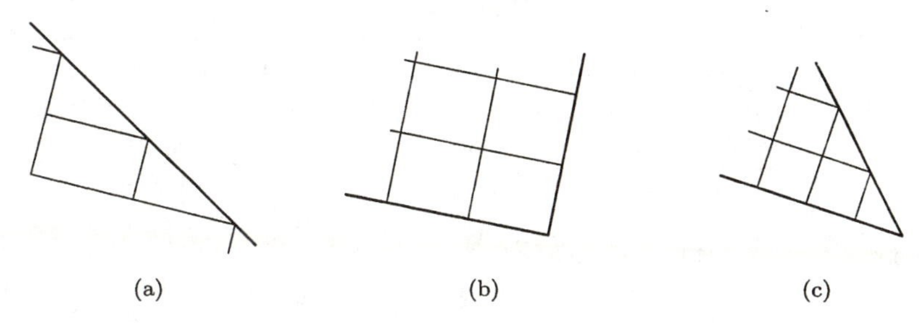

Figure 5.1.3. Traction boundary-value problems: (a) Problem 1; (b) Problem 2; (c) Problem 3.

If the relation between ω and n is invertible, then the state of stress at a point is determined by $(\omega,\ \theta)$. Referring to Figure 5.1.4, we note that, along an arc whose normal forms an angle χ with the x_1-axis, the normal stress, shear stress and transverse (interior) normal stress are respectively given by

$$\sigma = n + r\sin 2(\theta - \chi), \quad \tau = r\cos 2(\theta - \chi), \quad \sigma' = n - r\sin 2(\theta - \chi), \tag{5.1.4}$$

where n and r are determined by ω. If the arc forms a part of the boundary, however, then at most σ and τ will be given there (*traction boundary conditions*); σ' can then have either of the values $\sigma \pm 2\sqrt{r^2 - \tau^2}$. Usually the right value of σ' can be chosen by physical intuition.

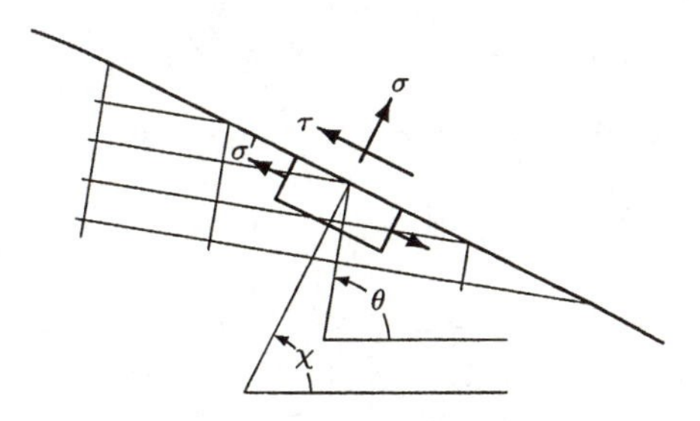

Figure 5.1.4. Stresses at a boundary.

In the classical case, the two choices for σ' give the respective explicit expressions for θ and ω

$$\theta = \chi \pm \frac{1}{2}\cos^{-1}\frac{\tau}{k}, \quad \omega = \frac{\sigma}{2k} \pm \frac{1}{2}\sqrt{1 - \frac{\tau^2}{k^2}}.$$

Stress Discontinuities

An arc such as that of Figure 5.1.4 may also be located in the interior of the plastic domain and be part of a *line of stress discontinuity*. To satisfy equilibrium, σ and τ must be continuous across such a line, but ω (and therefore n) and θ are discontinuous, so that the directions of the characteristics of each family change abruptly. In the general case r is also discontinuous, but in the classical case r equals k and is therefore continuous; σ' then takes each of the two possible values $\pm 2\sqrt{k^2 - \tau^2}$ on either side of the discontinuity line. Note that there is no discontinuity if and only if $|\tau| = k$, that is, if the arc is along a characteristic (which in the classical case is a shear line). When $|\tau| < k$, θ changes by $\cos^{-1}(\tau/k)$. It can be seen from a Mohr's-circle construction that the line of stress discontinuity must bisect the angles formed by the characteristics of each family on either side of it. It can also be seen that if the discontinuity line is thought of as the limit of a narrow zone of continuous but rapid change in σ' while σ and τ remain constant, all the intermediate Mohr's circles must be of radius less than k, showing that this zone is elastic and that the discontinuity line is therefore the remnant of an elastic zone (just like the ridge lines in the torsion problem). As a result of this property, it was shown by Lee [1950] that a line of stress discontinuity acts like an inextensible but perfectly flexible filament.

If the two regions separated by a line of stress discontinuity are denoted 1 and 2, and if the inclination of the line is χ, then σ and τ as given by the first two Equations (5.1.4) are continuous across this line. Specializing to the classical case, with $n = 2\omega k$ and $r = k$, leads to

$$2\omega_1 + \sin 2(\theta_1 - \chi) = 2\omega_2 + \sin 2(\theta_2 - \chi),$$

$$\cos 2(\theta_1 - \chi) = \cos 2(\theta_2 - \chi).$$

These equations may be solved for θ_2 and ω_2 in terms of θ_1 and ω_1, yielding the **jump conditions** due to Prager [1948]:

$$\theta_2 = 2\chi - \theta_1 \pm n\pi, \quad \omega_2 = \omega_1 \pm \sin 2(\theta_1 - \chi), \tag{5.1.5}$$

n being an integer, and the appropriate sign being taken as indicated by the problem.

It was shown by Winzer and Carrier [1948] that if several straight lines of stress discontinuity separating domains of constant stress meet at a point, then these lines must number at least four. Winzer and Carrier [1949] also discussed stress discontinuities between fields of variable stress.

The preceding arguments can be carried over from the classical to the general case (see Salençon [1977]). In particular, the jump conditions may be written directly in terms of the variables in Equations (5.1.4),

$$\begin{aligned} n_1 + r_1 \sin 2(\theta_1 - \chi) &= n_2 + r_2 \sin 2(\theta_2 - \chi), \\ r_1 \cos 2(\theta_1 - \chi) &= r_2 \cos 2(\theta_2 - \chi). \end{aligned} \tag{5.1.6}$$

Velocity Fields

If a traction boundary-value problem is solved by constructing a characteristic network, as described above, over a part of the region representing the body, then the loading forms an upper bound to that under which plastic flow becomes possible, since, as will be shown below, a kinematically admissible velocity field can then be found. As mentioned before, the exact flow load is found when the stress field can be extended in a statically and plastically admissible manner into the rigid region. The following discussion will be limited to the classical case; for a more general discussion, see, for example, Salençon [1977].

The equations governing the velocity components v_1, v_2 are found by combining the associated flow rule,

$$\frac{\dot{\varepsilon}_{11}}{\sigma_{11}-\sigma_{22}} = \frac{\dot{\varepsilon}_{22}}{\sigma_{22}-\sigma_{11}} = \frac{\dot{\varepsilon}_{12}}{2\sigma_{12}},$$

with the strain-rate–velocity relations

$$\dot{\varepsilon}_{11} = v_{1,1}\,, \quad \dot{\varepsilon}_{22} = v_{2,2}\,, \quad \dot{\varepsilon}_{12} = \frac{1}{2}(v_{1,2} + v_{2,1})$$

and with Equations (5.1.1) to obtain

$$v_{1,1} + v_{2,2} = 0, \quad v_{1,2} + v_{2,1} - 2\cot 2\theta\, v_{2,2} = 0. \tag{5.1.7}$$

The characteristics of Equations (5.1.7) are found by assuming $dv_1 = \lambda\, dv_2$. Thus

$$\begin{aligned} dv_1 = v_{1,1}\, dx_1 + v_{1,2}\, dx_2 &= v_{2,1}\,(-dx_2) + v_{2,2}\,(2\cot 2\theta\, dx_2 - dx_1) \\ = \lambda\, dv_2 &= v_{2,1}\,(\lambda\, dx_1) + v_{2,2}\,(\lambda\, dx_2), \end{aligned}$$

so that the characteristic directions are the eigenvectors of the system

$$\lambda\, dx_1 + dx_2 = 0,$$

$$dx_1 + (\lambda - 2\cot 2\theta)\, dx_2 = 0.$$

The eigenvalues λ are the roots of

$$\lambda^2 - 2\lambda\cot 2\theta - 1 = 0,$$

namely,

$$\lambda = \cot 2\theta \pm \csc 2\theta = \begin{cases} \cot\theta \\ -\tan\theta. \end{cases}$$

For $\lambda = \cot\theta$ we have $dx_2/dx_1 = -\cot\theta$ (i.e., the first shear direction) while for $\lambda = -\tan\theta$ we have $dx_2/dx_1 = \tan\theta$ (i.e., the second shear direction).

We see therefore that *the characteristics of the velocity equations are the same as those of the stress equations*. The characteristic relations are thus

$$\begin{aligned} dv_1 &= \cot\theta\, dv_2 && \text{along an } \alpha \text{ characteristic,} \\ dv_1 &= -\tan\theta\, dv_2 && \text{along a } \beta \text{ characteristic.} \end{aligned}$$

A plane whose coordinates are v_1 and v_2 is known as the *hodograph plane*, and a diagram in this plane showing the velocity distribution is called a *hodograph*. If P and Q are two neighboring points in the x_1x_2-plane (the *physical plane*) lying on the same shear line, and if P' and Q' are the points in the hodograph plane representing the respective velocities, then as shown by Geiringer [1951] and Green [1951], it follows from the characteristic relations that the line element $\bar{P'Q'}$ is perpendicular to $\bar{PQ}$. Consequently the Hencky–Prandtl properties apply to the hodograph as well. A rigid region, if it does not rotate, is represented by a single point in the hodograph plane.

It is also instructive to express the characteristic relations in terms of the velocity components along the characteristic directions, given respectively by $v_\alpha = v_1 \sin\theta - v_2 \cos\theta$ and $v_\beta = v_1 \cos\theta + v_2 \sin\theta$. The relations were derived by Geiringer [1931] and are known as the **Geiringer equations**:

$$\begin{aligned} dv_\alpha &= v_\beta\, d\theta && \text{along an } \alpha \text{ characteristic,} \\ dv_\beta &= -v_\alpha\, d\theta && \text{along a } \beta \text{ characteristic.} \end{aligned} \qquad (5.1.8)$$

It can be seen that these relations express the condition that the longitudinal strain rates $\dot{\varepsilon}_{11}$ and $\dot{\varepsilon}_{22}$ vanish when the x_1 and x_2 axes coincide locally with the characteristic directions, in other words, that the shear lines are *inextensible*. This result also follows directly from the flow rule, since with respect to such axes we have $\sigma_{11} - \sigma_{22} = 0$.

It must be remembered, however, that the flow rule actually gives only the ratios among the strain rates, and the preceding result must strictly be written as $\dot{\varepsilon}_{11}/\dot{\varepsilon}_{12} = \dot{\varepsilon}_{22}/\dot{\varepsilon}_{12} = 0$. Another interpretation of this result is that $\dot{\varepsilon}_{12}$ is infinite, meaning that either $v_{1,2}$ or $v_{2,1}$ is infinite. This happens if the tangential velocity component is discontinuous across a characteristic (the normal component must be continuous for material continuity), that is, if *slip* occurs. The characteristics are thus the potential loci of slip and are therefore also called *slip lines*. Kinematically admissible velocity fields with discontinuities across slip lines are often used in the construction of solutions. In particular, slip may occur along a characteristic forming the boundary between the plastic and rigid regions.

If slip occurs along an α characteristic, with the tangential velocity having the values v_α and v_α^* on either side of it, then, since v_β has the same value on both sides, an application of Equation $(5.1.8)_1$ along the two sides of the slip line gives $dv_\alpha = dv_\alpha^*$, or $d(v_\alpha^* - v_\alpha) = 0$. An analogous result applies to a β characteristic. Thus *the discontinuity in the tangential velocity* *remains constant along a slip line.* It follows further that the curves in the hodograph plane that form the images of the two sides of a line of velocity discontinuity are parallel.

5.1.2. Simple Slip-Line Fields

A great many practical problems can be solved by means of slip-line fields containing straight slip lines. As we have seen, if one slip line (characteristic) of a given family is straight, then all the slip lines of that family must be straight. Families of straight slip lines, as can be seen in Figure 5.1.5, may be of three types: (a) parallel, (b) meeting at a point, and (c) forming an envelope.

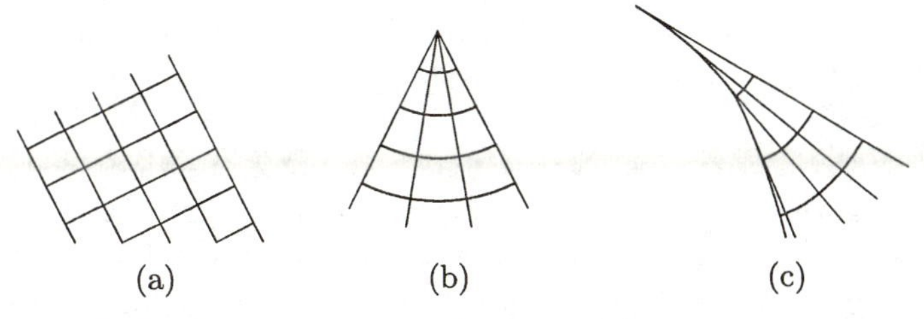

Figure 5.1.5. Families of straight slip lines: (a) constant-state field; (b) centered fan; (c) noncentered fan with envelope.

(a) If all the slip lines of one family are straight and parallel, then those of the other family must be likewise. Since θ is constant, it follows from Equations (5.1.3) that ω is constant as well, and therefore that the state of stress is uniform. A region in which the slip lines are of this type is called a region of *constant state*; this term is taken from wave-propagation theory and is not strictly applicable here, because, while the stress components are constant, the velocity components are not necessarily so; Equations (5.1.8), with $d\theta = 0$, yield the solutions $v_\alpha = f(\beta)$, $v_\beta = g(\alpha)$. The functions f and g are arbitrary except as constrained by boundary conditions.

(b) If the slip lines of one family are straight and meet at a point, then those of the other family must be concentric circular arcs. Such a system of slip lines is called a *centered fan*. A number of problems may be solved by inserting a centered fan between two regions of constant state, in such a way that the bounding radial lines of the fan are also the bounding parallel lines of the constant-state regions. ω is constant along all the straight lines, while along the circular arcs of the fan, $d\omega = \pm d\theta$, depending on whether the arcs are α or β characteristics. If the θ difference between the two constant-state regions is $\Delta\theta$, then this is just the angle subtended by the bounding lines of the fan, and $\Delta\omega = \pm\Delta\theta$.

(c) An envelope of slip lines is also called a *limiting line*, and a family of straight slip lines forming an envelope is called a noncentered fan; the envelope is called the *base curve* of the fan. A limiting line cannot be in the interior of the plastic region, and therefore must form a part of either the boundary of the body or of the rigid-plastic boundary. Other properties of limiting lines are discussed by Prager and Hodge [1951], Section 25.

Some Applications

We now consider some simple applications of slip-line fields consisting of constant-state regions and centered fans, illustrated in Figure 5.1.6. These results are of considerable importance in soil mechanics, where they are used to study the stability of slopes and the carrying capacity of foundations made of clays for which the hypothesis of constant shear strength (the undrained strength discussed in 2.3.1) can be justified.

(a) Consider, first, a wedge of angle 2γ with a uniform pressure on one side and no traction on the other. A possible shear-line net consists of two regions of constant state, separated by a line of stress discontinuity bisecting the wedge. With the principal stresses $2k - p$, $-p$ on one side and 0, $-2k$ on the other, continuity of the normal stress across the discontinuity line is possible only if $p = 4k \sin^2 \gamma$. This value is therefore a lower bound to the pressure causing incipient plastic flow. As can be seen in Figure 5.1.6(a), when the wedge is acute ($\gamma < \frac{1}{4}\pi$) a velocity field may be constructed such that regions ABC and AEF slip along the slip lines AB and AF, respectively, while $ACDE$ flows perpendicular to the stress-discontinuity line AD, so that slip also occurs along AC and AE. The material below BAF may be rigid.

(b) When the wedge is obtuse ($\gamma > \frac{1}{4}\pi$) no such velocity field is possible. On the other hand, it is now possible to insert a centered fan of angle $2\gamma - \frac{1}{2}\pi$ between two constant-state regions, producing the pressure $p = 2k(1 + 2\gamma - \frac{1}{2}\pi)$, which exceeds the previously obtained lower bound for all $\gamma > \frac{1}{4}\pi$.

(c) Now consider a truncated wedge with pressure on the top face. At each corner we can construct a plastic zone consisting of a centered fan between two triangular regions of constant state, and plastic flow can occur when the two plastic regions meet, so that the top face can have a downward velocity. The angle subtended by each fan is just γ, so that $p = 2k(1 + \gamma)$.

(d) The limit of the preceding case as $\gamma \to \frac{1}{2}\pi$ represents a half-plane carrying a rigid block, and therefore the limiting pressure on the interface between the half-plane and the block is $k(2 + \pi)$. This result was obtained by Prandtl [1920] by assuming a single triangular region of constant state under the block, with a centered fan on either side and another constant-state region outside each fan [see Figure 5.1.6(f)]. Prandtl's solution was criticized by Hill, who pointed out that, since the elastic solution of the problem leads to infinite stresses at the corners of the block, plastic zones must be there from the outset. As the pressure is increased, these zones will

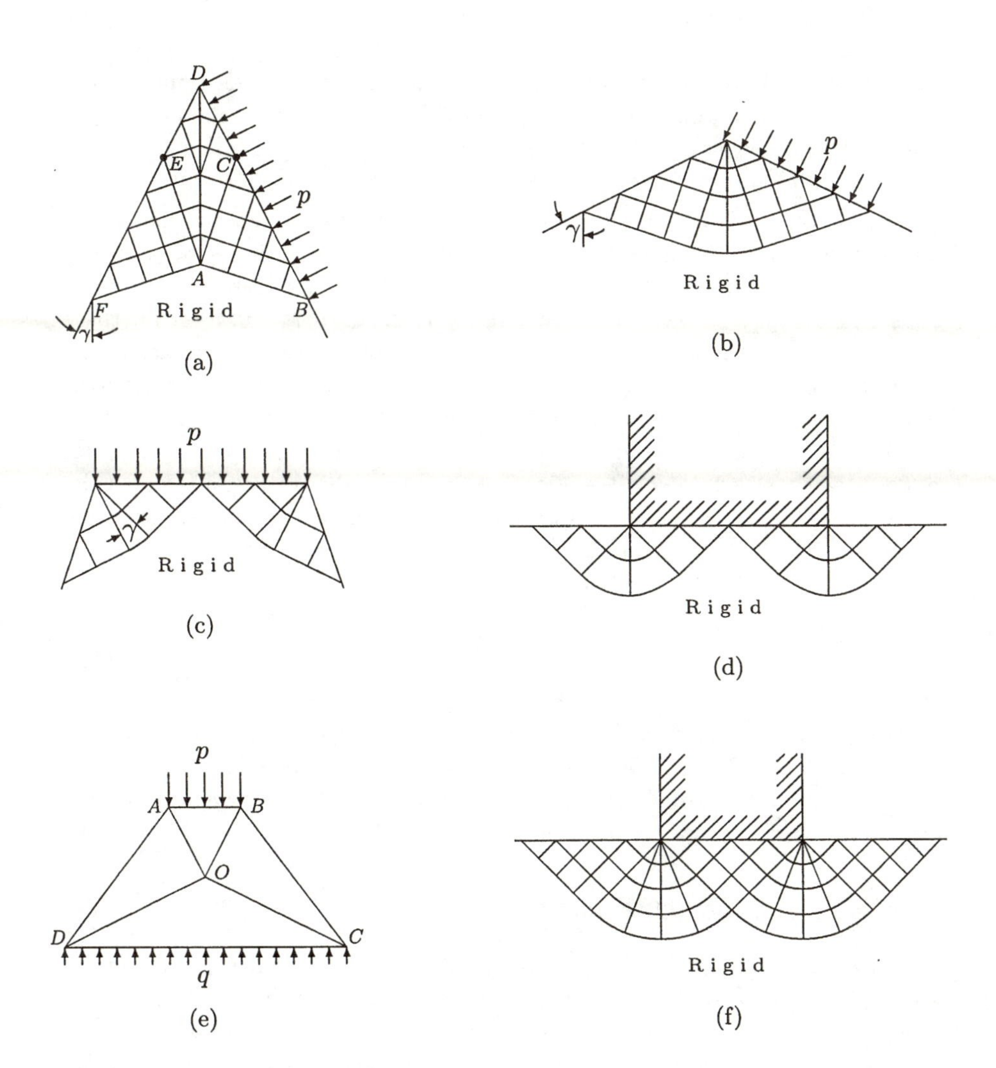

Figure 5.1.6. Simple slip-line fields: (a) acute wedge; (b) obtuse wedge; (c) truncated wedge loaded on the top edge; (d) rigid block on a half-plane, Hill solution; (e) finite truncated wedge, discontinuous stress field; (f) rigid block on a half-plane, Prandtl solution.

grow until they meet, as in case (c).

In the last two examples, no stress field outside the plastic regions has been presented, so that the resulting pressures must be regarded as upper bounds only. For example (c), Drucker and Chen [1968] have shown how to construct a statically admissible stress field, leading to a lower bound equal to the upper bound on the pressure. The following example shows that for a *finite* truncated wedge, a lower pressure than that obtained in (c) above can be found.

(e) A solution for a finite truncated wedge, uniformly loaded on its top and bottom faces, is based on a fully plastic stress distribution, with the four triangular constant-state regions separated by the stress-discontinuity lines OA, ..., OD, which bisect the angles at the corners. In order for these bisectors to meet at one point, the ratio of the bottom to the top face of the trapezoid must be $(1+\sin\gamma)/(1-\sin\gamma)$. Because of symmetry, only the right half of the wedge need be considered. Since the flank BC is traction-free, region 2 is in a state of simple compression parallel to BC, and the values of θ and ω there are $\theta_2 = \gamma + \pi/4$ and $\omega_2 = -\frac{1}{2}$, respectively. In region 1, by symmetry, $\theta = \theta_1 = \pi/4$. By Equation $(5.1.5)_2$, $\omega_1 = -\frac{1}{2} - \sin\gamma$. Equation $(5.1.1)_2$ then gives, in region 1,

$$\sigma_y = -2k(1 + \sin\gamma) = -p,$$

where p is the pressure on the top face; note that this value is less than that obtained in (c) above. The pressure on the bottom face can similarly be found as $q = 2k(1 - \sin\gamma)$. It can be seen that the aforementioned geometric restriction is necessary for equilibrium.

Problems with Circular Symmetry

In classical problems with axial symmetry, the slip lines are along the shear directions and therefore at 45° to the radial and tangential directions. They are therefore given by logarithmic spirals, $r \propto e^{\pm\theta}$. These slip lines can be used to construct velocity fields in hollow prisms with a circular bore under internal pressure. In Figure 5.1.7, for example, an axisymmetric stress field — the same as in the hollow cylinder under internal pressure — is assumed in the region inside the largest circle, with a vanishing stress field outside this circle. Plastic flow is assumed to occur only in the curved triangular regions bounded by the slip lines that meet the sides of the square at their midpoints. The remaining regions move diagonally outward as rigid bodies. A statically admissible stress field and an associated kinematically admissible velocity field are thus found, and the pressure must be that for the cylinder, namely, $p = 2k \ln b/a$.

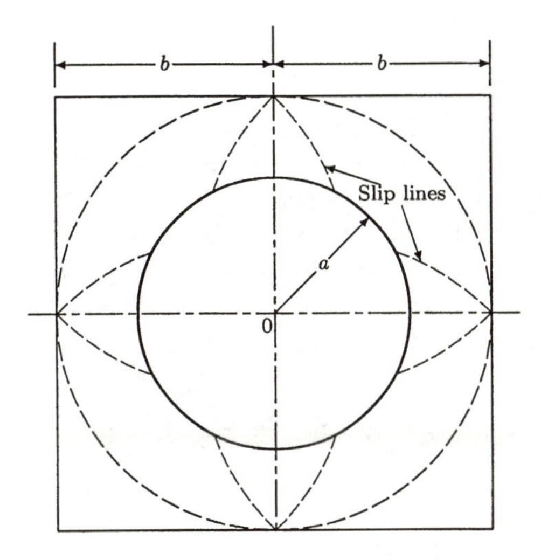

Figure 5.1.7. Hollow prism with a circular cutout.

5.1.3. Metal-Forming Problems

A number of problems representing metal-forming processes can be solved approximately by means of the theory of plane plastic strain, if the metal is idealized as a material that is rate-independent and rigid–perfectly plastic, and if the thermal stresses that result from the temperature gradients induced by the forming process can be neglected. In a metal-forming problem — unlike a structural problem — unrestricted plastic flow is the desired condition. The solution is intended to furnish the smallest applied force under which the metal will flow, rather than the largest load under which the structure will not collapse. For this reason an upper bound to the force is a "safe" answer.

Metal-forming processes that closely approximate plane-strain conditions (and for which slip-line theory can be used to generate upper bounds on the forming forces) include forging, indentation, and cutting of wide strips, as well as continuous processes such as extrusion, drawing and rolling. In problems representing the latter category of processes, it is not the initiation of the process that is studied, but a state of *steady plastic flow* in which a large amount of plastic deformation has already taken place, and the stress and velocity fields are taken as constant in time in an Eulerian sense, much as in steady-flow problems of fluid mechanics.

An extensive bibliography of slip-line fields for metal-forming processes can be found in Johnson, Sowerby and Venter [1982]. Many examples are also to be found in Johnson and Mellor [1973], Chapters 11 and 14, and Chakrabarty [1987], Chapters 7 and 8. Only a few selected problems will be treated here.

Indentation

The problem of the half-plane carrying a rigid block, illustrated in Figure 5.1.6(d), can also be interpreted as describing the beginning of indentation of a half-plane by a flat punch. The Hill solution, which requires slip between the plastic zones and the punch, implies smooth contact. In the Prandtl solution, on the other hand, the triangular region directly under the punch moves rigidly downward with it, corresponding to rough contact. This material in this region is often called *dead metal.*

The solution shown in Figure 5.1.6(c) for the truncated wedge of half-angle γ with uniform pressure on its top face can be adapted to the problem of indentation by a flat indenter at the bottom of a flat trench (Figure 5.1.8) if γ is replaced by $\pi - \gamma$. The indentation pressure is thus $p = 2k(1 + \pi - \gamma)$.

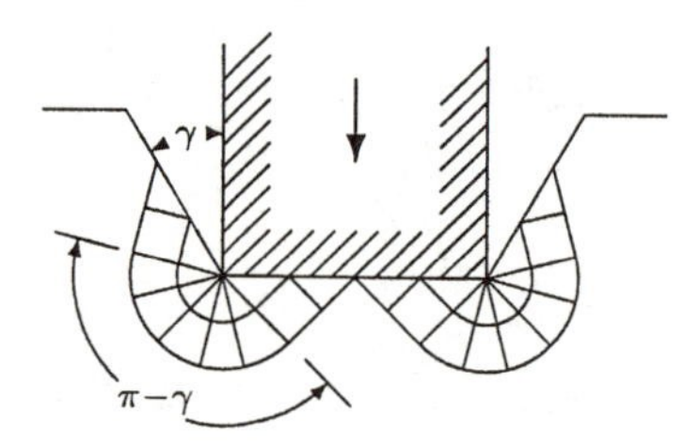

Figure 5.1.8. Flat indenter at the bottom of a flat trench.

A solution for the frictionless indentation of a half-plane by an acute wedge-shaped indenter of half-angle α ($\alpha < \pi/4$) was proposed by Hill, Lee and Tupper [1947]. This problem is one of *pseudo-steady plastic flow*, in which the geometry of the slip-line field (and therefore the stress and velocity fields) changes as penetration proceeds, but in a geometrically similar manner — that is, the angles remain the same, and only the scale changes. As shown in Figure 5.1.9, the slip-line field covers Zone 1. Zone 3 is that in which the material is elastic and therefore treated as rigid, while the intermediate Zone 2 contains material that has yielded but is restrained from moving. As can be seen, the solution allows for the piling up of material (the formation of a "coronet") about the indenter, although in practice such piling up is observed only in work-hardening materials. The angle θ subtended by the centered fan between the two constant-state triangles can be determined from the wedge half-angle α by means of the condition that the volume of piled-up material (shown as shaded in the figure) equals the volume of that portion of the indenter that has penetrated the work (shown crosshatched). The slip-line field is equivalent to that in Figure 5.1.6(b), with $\theta = 2\gamma - \pi/2$, and therefore the contact pressure q is

$$q = 2k(1 + \theta).$$

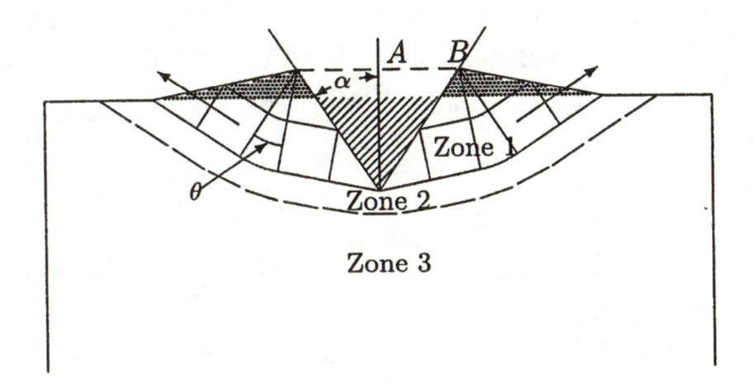

Figure 5.1.9. Indentation of a half-plane by an acute wedge-shaped indenter.

The indenter force, at a given state of indentation, is thus

$$F = 4k(1+\theta)\overline{AB}.$$

Because of geometric similarity, the indenter force is consequently proportional to indentation depth.

Forging and Cutting

The simultaneous application of identical flat punches to a strip of finite thickness, Figure 5.1.10(a), may be used to model forging, with the bottom punch representing the anvil and the top punch the forging tool. Similarly, the cutting of a strip of metal with a wirecutter-like tool can be described as the simultaneous indentation by a pair of identical wedge-shaped indenters located opposite each other [Figure 5.1.10(b)]. Because of symmetry, in each case only the top half of the strip needs to be considered, and the middle plane may be regarded as a frictionless foundation. The solution of both problems was studied by Hill [1953].

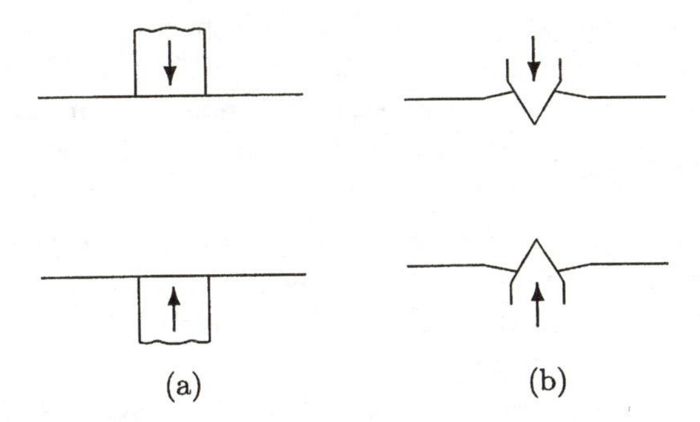

Figure 5.1.10. Forging and cutting: (a) forging; (b) cutting.

For the cutting problem Hill showed that when the plastic region has not yet reached the foundation, the slip-line field is the same as for the semi-infinite domain. When, however, the plastic region extends through the thickness of the strip, a different mode of deformation takes over: piling up ceases, and the material on either side of the plastic region moves rigidly outward; the slip-line field is shown in Figure 5.1.11(a).

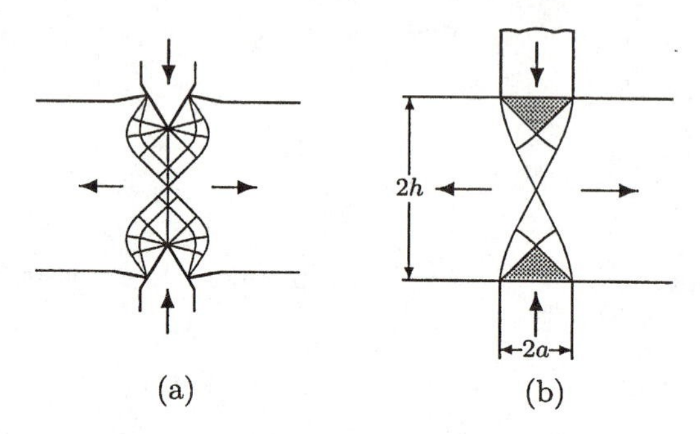

Figure 5.1.11. Slip-line fields: (a) forging; (b) cutting.

For the forging problem, the slip-line field is shown in Figure 5.1.11(b). A triangular dead-metal region attaches itself to the punch, and indentation proceeds as in the cutting problem, provided that $h < 8.74a$, where h is the half-thickness of the strip and a the half-width of the punch. When $h = 8.74a$, it can be shown that the punch pressure is $p = 2k(1 + \pi/2)$, as for the semi-infinite domain. It follows that for $h > 8.74a$ the zone of plastic deformation does not go through the strip and the pressure remains at this value.

Drawing and Extrusion

Drawing and extrusion are processes in which a billet of material is forced to flow through a die shaped to produce the required cross-section. In drawing, as the name suggests, the material is pulled. Extrusion involves pushing. In *direct* extrusion the die is stationary with respect to the container holding the billet, and a ram moves in the container, pushing the billet outward with the help of a pressure pad. In *reverse* extrusion the container is closed at one end, and the die is pushed inside the container. The three processes are shown in Figure 5.1.12.

The technologically important applications of these processes are predominantly three-dimensional — drawing produces wire, and extrusion is used to make lightweight structural shapes, trim and the like. In the absence of three-dimensional solutions, however, the solutions of the corresponding plane problems provide qualitative information on the nature of the plastic regions and hence allow estimates for the required forces.

Figure 5.1.13(a) illustrates a solution due to Hill [1948c] describing frictionless extrusion through a square die with 50% reduction. The slip-line field consists the two centered fans OAB and $OA'B'$; because of symmetry,

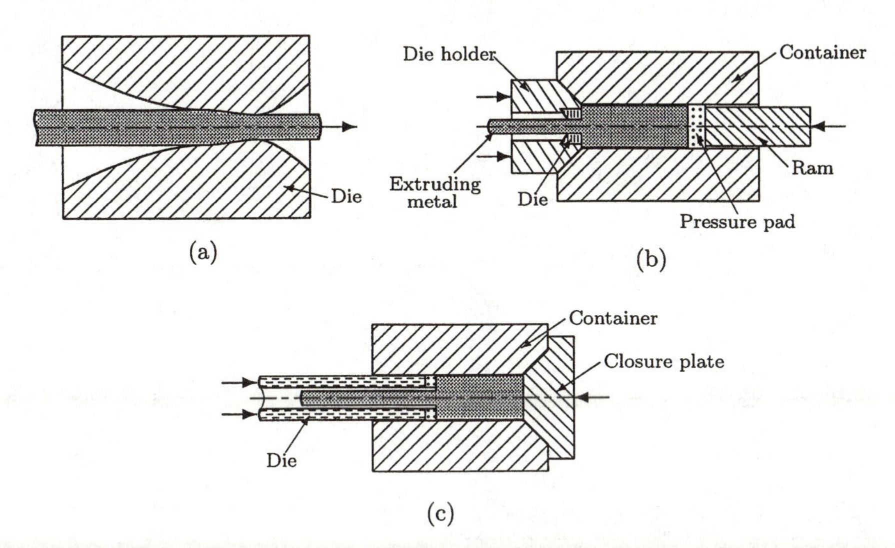

Figure 5.1.12. Drawing and extrusion: (a) drawing; (b) direct extrusion; (c) reverse extrusion.

only OAB need be considered. Since the *exit slip line* OA is a line of constant stress, σ_{11} must vanish identically on it in order for the extruded metal to its left to be in equilibrium, and $\sigma_{12} = 0$ because the line forms an angle of 45° with the x-axis. The yield criterion requires $|\sigma_{22}| = 2k$, and, since the sheet is being compressed, it follows that $\sigma_{22} = -2k$. Equations (5.1.1), with $n = 2k\omega$ and $r = k$, accordingly require that $\omega = -\frac{1}{2}$ and $\theta = \pi/4$ on OA.

The characteristic relations can now be used to determine the state along AB. Since the fan subtends 90°, $\theta = 3\pi/4$, and Equation $(5.1.3)_2$ shows that $\omega = \frac{1}{2}(1+\pi)$ there. By Equations (5.1.1), then, we have $\sigma_{11} = -(2+\pi)k$, $\sigma_{22} = -\pi k$, and $\sigma_{12} = 0$ on AB. The average value of $-\sigma_{11}$ along OAB is thus equal to the extrusion pressure,

$$p = \left(1 + \frac{\pi}{2}\right) k.$$

A statically admissible extension of the stress field into the rigid region due to Alexander [1961] shows this to be the exact pressure, not merely an upper bound. The stress field on OB is extended into the dead-metal region ABC, while that on OA is extended into OAP; the extruded metal to the left of AP is stress-free, so that AP is a line of stress discontinuity. The extension to the right of the arc OB is achieved analytically.

A simple slip-line field for a reduction of $\frac{2}{3}$ is shown in Figure 5.1.13(b),

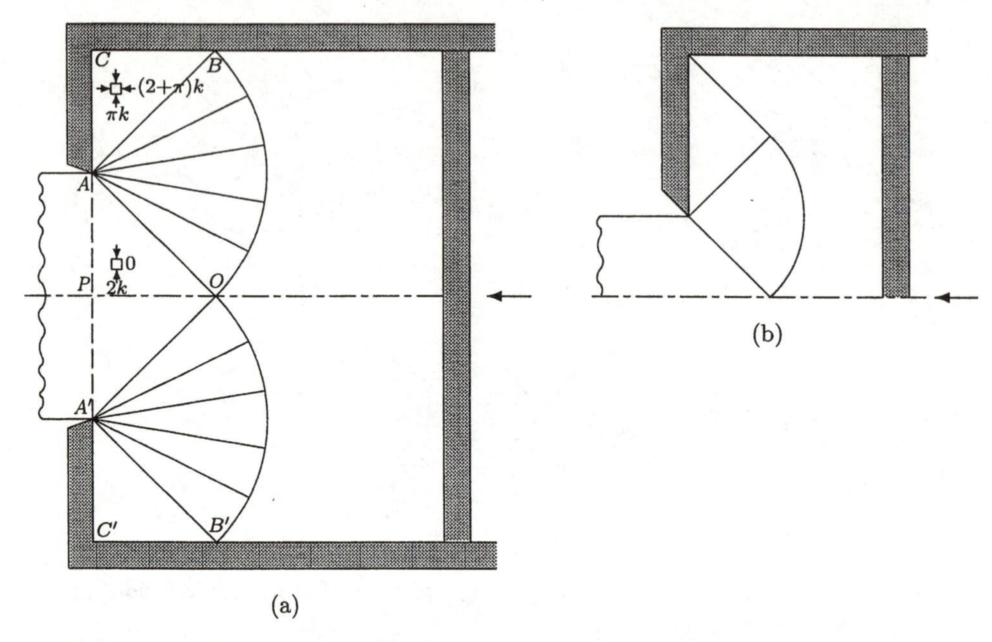

Figure 5.1.13. Extrusion: (a) frictionless extrusion through a square die with 50% reduction; (b) slip-line field for two-thirds reduction.

and leads to an extrusion pressure of

$$p = \frac{4}{3}\left(1 + \frac{\pi}{2}\right) k.$$

Figures 5.1.14(a) and (b) illustrate both drawing and extrusion through a tapered die. If the container walls in the extrusion problem are smooth, then the slip-line fields are identical if the die angle α and the reduction ratio r are the same; the stress fields in the regions covered by the slip-line field differ only by a hydrostatic stress.

A particularly simple slip-line field due to Hill and Tupper [1948], valid for a smooth die when $r = 2\sin\alpha/(1 + 2\sin\alpha)$, is shown in Figure 5.1.14(c). In the extrusion problem, ω and θ on the exit slip line OA are the same as in the preceding problem, and therefore their values in the constant-state region ABC are $\omega = -\frac{1}{2}(1 + 2\alpha)$, $\theta = \pi/4 + \alpha$. The normal pressure on AC is

$$q = (1 - 2\omega)k = 2(1 + \alpha)k,$$

and the tangential stress there is zero. For equilibrium, the extrusion pressure p must be

$$p = rq = \frac{4(1 + \alpha)\sin 2\alpha}{1 + 2\sin 2\alpha} k.$$

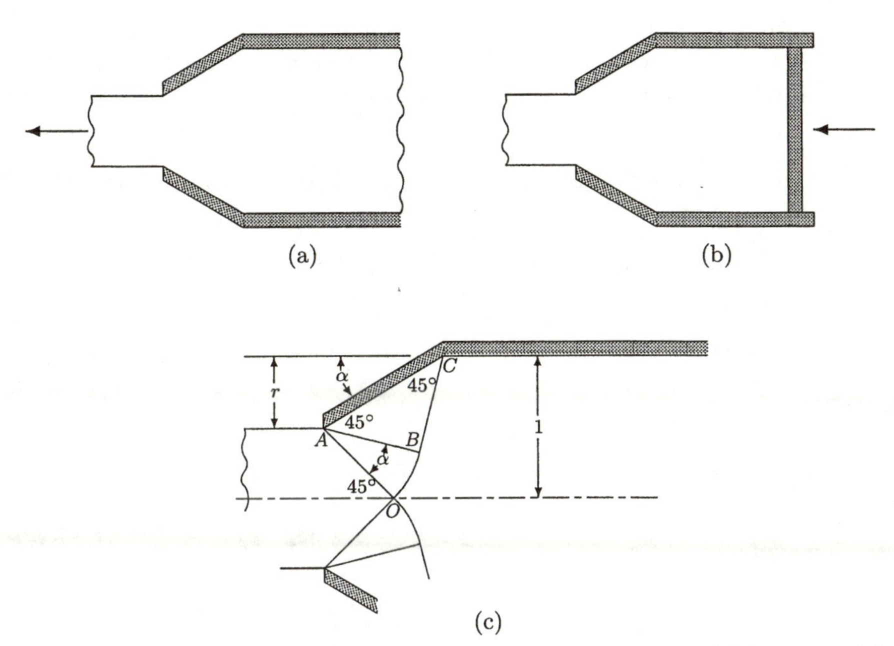

Figure 5.1.14. Drawing and extrusion through a tapered die: (a) drawing; (b) extrusion; (c) slip-line field for drawing and extrusion through a smooth tapered die with $r = 2\sin\alpha/(1 + 2\sin\alpha)$ (Hill and Tupper [1948].

The corresponding drawing problem is solved by superposing a hydrostatic tension equal in magnitude to this pressure. The drawing stress therefore has the same value as the extrusion pressure.

For further solutions of metal-forming problems using slip-line theory, see the aforementioned references by Johnson and Mellor [1973], Johnson, Sowerby and Venter [1982], and Chakrabarty [1987].[1] Additional plane problems, in which complete solutions are not available, are discussed in Section 6.1 in the context of limit analysis.

Exercises: Section 5.1

1. For a Hencky–Prandtl network in which the values of $\omega - \theta$ on any two neighboring first characteristics and the values of $\omega + \theta$ on any two neighboring second characteristics differ by the same small constant, show that the diagonals of the network are lines of constant ω or of constant θ (use Figure 5.1.2).

[1] Finite-element methods for metal-forming problems are treated by Kobayashi, Oh, and Altan [1989].

2. Show that when four straight lines of stress discontinuity meet on an axis of symmetry of stress field and separate four regions of constant stress, as in Figure 5.1.6(e), the angles AOD and BOC must be right angles.

3. Show that any velocity field that is associated with the stress field of the preceding exercise represents rigid-body motion.

4. Derive the Geiringer equations (5.1.8).

5. Find the differential equations for the velocity field in plane plastic flow in a standard Mohr–Coulomb material. Determine the characteristics of the velocity field.

6. Show that in classical problems of plane plastic flow with axial symmetry, the slip lines are given in polar coordinates by $r \propto e^{\pm\theta}$.

7. Show that the relation between θ and α in the slip-line field of Figure 5.1.9 is
$$\alpha = \frac{1}{2}\left[\theta + \cos^{-1}\tan\left(\frac{\pi}{4} - \frac{\theta}{2}\right)\right].$$

8. The slip-line field of Figure 5.1.6(c) may be regarded as representing a stage in the squashing of an originally pointed wedge of half-angle α (greater than γ) by a lubricated flat plate.

   (a) Using geometry and volume constancy, find the relation between α and γ.

   (b) Find the relation between the applied force and the distance moved by the plate.

   (c) Determine the smallest value of α for which the solution is valid.

9. Discuss how the slip-line field of Figure 5.1.6(c) can be used to study the necking of a symmetrically notched tension specimen.

10. Discuss the velocity fields associated with the slip-line fields of Figures 5.1.13(a) and (b).

Section 5.2 Collapse of Circular Plates

The goal of this section is derivation of collapse loads for axisymmetrically loaded circular plates made of a perfectly plastic material obeying the Tresca criterion. For such plates, complete solutions exist in closed form, and they are treated in 5.2.3. For the collapse loads of plates without circular symmetry, limit analysis must be used to obtain estimates, and this is done in

Chapter 6. The introduction to plate theory given in 5.2.1 is general, and not limited to circular plates. Similarly, the presentation of elastic relations and yield criteria at the beginning of 5.2.2 and 5.2.3, respectively, is general, but solutions will be given for axisymmetric problems only.

5.2.1. Introduction to Plate Theory

Derivation of Plate Equilibrium Equations

A plate may be defined as a solid body occupying in the undeformed configuration the region $A \times [-h/2, h/2]$, that is, the set of points $\{(x_1, x_2, x_3) \mid (x_1, x_2) \in A, -h/2 \le x_3 \le h/2\}$, where A is a closed domain in the x_1x_2-plane bounded by a simple closed curve C (we are assuming no holes in the plate), with h considerably smaller than the typical dimension of A. The plane $x_3 = 0$ is called the *middle plane* of the plate. The outward normal unit vector to C in the x_1x_2-plane has components ν_α ($\alpha = 1, 2$), and the counterclockwise tangential unit vector has components t_α.

We shall approach the study of the mechanics of plates by a combined use of the three-dimensional equilibrium equations and of virtual work. The plate equilibrium equations will be derived directly from the former. First, we define the stress resultants as follows:

$$
\begin{aligned}
N_{\alpha\beta} &= \int_{-h/2}^{h/2} \sigma_{\alpha\beta}\, dx_3 && \text{(membrane forces)},\\
Q_\alpha &= \int_{-h/2}^{h/2} \sigma_{\alpha 3}\, dx_3 && \text{(shear forces)},\\
M_{\alpha\beta} &= -\int_{-h/2}^{h/2} x_3 \sigma_{\alpha\beta}\, dx_3 && \text{(moments)}.
\end{aligned}
$$

The surface loads are

$$
\begin{aligned}
p_\alpha &= \sigma_{\alpha 3}\,\Big|_{-h/2}^{h/2} + \int_{-h/2}^{h/2} f_\alpha\, dx_3,\\
q &= \sigma_{33}\,\Big|_{-h/2}^{h/2} + \int_{-h/2}^{h/2} f_3\, dx_3,\\
m_\alpha &= -x_3\sigma_{\alpha 3}\,\Big|_{-h/2}^{h/2} - \int_{-h/2}^{h/2} x_3 f_\alpha\, dx_3,
\end{aligned}
$$

where $\mathbf{f}$ is the body force per unit volume.

Distinguishing the x_3-coordinate from the x_α ($\alpha = 1, 2$), we write the local equilibrium equations

$$\sigma_{\alpha\beta,\beta} + \sigma_{\alpha 3,3} + f_\alpha = 0, \tag{5.2.1}$$

$$\sigma_{\alpha 3,\alpha} + \sigma_{33,3} + f_3 = 0. \tag{5.2.2}$$

Integrating these equations through the thickness and performing integration by parts where necessary yields

$$N_{\alpha\beta,\beta} + p_\alpha = 0 \tag{5.2.3}$$

and

$$Q_{\alpha,\alpha} + q = 0. \tag{5.2.4}$$

When Equations (5.2.1) are multiplied by x_3 and then integrated through the thickness, the result is

$$M_{\alpha\beta,\beta} + Q_\alpha + m_\alpha = 0. \tag{5.2.5}$$

We can eliminate Q_α between Equations (5.2.4) and (5.2.5), and obtain

$$M_{\alpha\beta,\alpha\beta} = q - m_{\alpha,\alpha}\,. \tag{5.2.6}$$

Equations (5.2.3) and (5.2.4–5) or (5.2.6) are the plate equilibrium equations, the former for in-plane or membrane forces and the other for bending forces. Note that the two modes of behavior — in-plane deformation and bending — are statically uncoupled. In the elementary theory they are also kinematically uncoupled, and therefore can be studied separately.

Displacement Assumptions and Virtual Work

The elementary displacement model for plate behavior is described by the following displacement field:

$$u_\alpha(x_1, x_2, x_3) = \bar{u}_\alpha(x_1, x_2) - x_3 w_{,\alpha}(x_1, x_2), \qquad u_3(x_1, x_2, x_3) = w(x_1, x_2);$$

here the $\bar{u}_\alpha$ are the in-plane displacements and w is the deflection. It follows that $\varepsilon_{i3} = 0$, $i = 1, 2, 3$, so that $\sigma_{ij}\,\delta\varepsilon_{ij} = \sigma_{\alpha\beta}\,\delta\varepsilon_{\alpha\beta}$, and $\varepsilon_{\alpha\beta} = \bar{\varepsilon}_{\alpha\beta} - x_3 w_{,\alpha\beta}$, where $\bar{\varepsilon}_{\alpha\beta} = \frac{1}{2}(\bar{u}_{\alpha,\beta} + \bar{u}_{\beta,\alpha})$.

It is important to check whether the assumption of infinitesimal strain is valid. The Green–Saint-Venant strain tensor (Section 1.2) has the in-plane components

$$E_{\alpha\beta} = \varepsilon_{\alpha\beta} + \frac{1}{2}u_{\gamma,\alpha}\,u_{\gamma,\beta} + \frac{1}{2}w_{,\alpha}\,w_{,\beta}\,.$$

If we neglect the contributions of the in-plane displacements $\bar{u}_\alpha$, then the right-hand side reduces to

$$-x_3 w_{,\alpha\beta} + \frac{1}{2}x_3^2 w_{,\alpha\gamma}\,w_{,\beta\gamma} + \frac{1}{2}w_{,\alpha}\,w_{,\beta}\,.$$

If δ is a typical deflection and l a typical dimension of A, then the first term is of order $h\delta/l^2$, the second of order $(h\delta/l^2)^2$, and the third of order $(\delta/l)^2$. While the second term is negligible in comparison to the third whenever h/l is sufficiently small, as is normal in plate theory, for the third term to be negligible in comparison to the first term *it is necessary for the deflection to be small compared to the plate thickness.* Otherwise, the Green–Saint-Venant strain tensor must be used, given in general by

$$E_{\alpha\beta} = \bar{E}_{\alpha\beta} - x_3 w_{,\alpha\beta},$$

with

$$\bar{E}_{\alpha\beta} = \bar{\varepsilon}_{\alpha\beta} + \frac{1}{2} w_{,\alpha} w_{,\beta}. \tag{5.2.7}$$

Equation (5.2.7) will be used in the next section when the buckling of plates is studied.

Under the hypothesis of infinitesimal strain, the internal virtual work becomes

$$\overline{\delta W}_{int} = \int_A (N_{\alpha\beta}\, \delta\bar{\varepsilon}_{\alpha\beta} + M_{\alpha\beta}\, \delta w_{,\alpha\beta})\, dA.$$

This equation may be rewritten as

$$\overline{\delta W}_{int} - \overline{\delta W}^{(1)}_{ext} = \overline{\delta W}^{(2)}_{ext},$$

where $\overline{\delta W}^{(1)}_{ext}$ denotes the part of the external virtual work due to the body force and the surface tractions on the planes $x_3 = +-h/2$, and $\overline{\delta W}^{(2)}_{ext}$ is that due to applied forces and moments along the edge. The first part accordingly is given by

$$\overline{\delta W}^{(1)}_{ext} = \int_A \left[\int_{-h/2}^{h/2} f_i\, \delta u_i\, dx_3 + (\sigma_{i3}\, \delta u_i)\, \Big|_{-h/2}^{h/2} \right] dA.$$

Now

$$f_i\, \delta u_i = f_\alpha\, \delta\bar{u}_\alpha - x_3 f_\alpha \delta w_{,\alpha} + f_3\, \delta w$$

and

$$(\sigma_{i3}\, \delta u_i)\, \Big|_{-h/2}^{h/2} = \sigma_{\alpha 3}\, \Big|_{-h/2}^{h/2} \delta\bar{u}_\alpha - (x_3 \sigma_{\alpha 3})\, \Big|_{-h/2}^{h/2} \delta w_{,\alpha} + \sigma_{33}\, \Big|_{-h/2}^{h/2} \delta w,$$

so that

$$\overline{\delta W}^{(1)}_{ext} = \int_A (p_\alpha\, \delta\bar{u}_\alpha + q\, \delta w + m_\alpha\, \delta w_{,\alpha})\, dA.$$

Since $N_{\alpha\beta}$ is symmetric, it follows that $N_{\alpha\beta}\, \delta\bar{\varepsilon}_{\alpha\beta} = N_{\alpha\beta}\, \delta\bar{u}_{\alpha,\beta}$, and

$$\overline{\delta W}_{int} - \overline{\delta W}^{(1)}_{ext} = \int_A [(N_{\alpha\beta}\, \delta\bar{u}_{\alpha,\beta} - p_\alpha\, \delta\bar{u}_\alpha) + (M_{\alpha\beta} \delta w_{,\alpha\beta} - m_\alpha\, \delta w_{,\alpha} - q\, \delta w)]\, dA.$$

Now

$$N_{\alpha\beta}\, \delta\bar{u}_{\alpha,\beta} = (N_{\alpha\beta}\, \delta\bar{u}_\alpha)_{,\beta} - N_{\alpha\beta,\beta}\, \delta\bar{u}_\alpha,$$

and

$$\begin{aligned} M_{\alpha\beta}\, \delta w_{,\alpha\beta} &- m_\alpha\, \delta w_{,\alpha} \\ &= (M_{\alpha\beta}\, \delta w_{,\alpha})_{,\beta} - [(M_{\alpha\beta,\beta} + m_\alpha)\, \delta w]_{,\alpha} + (M_{\alpha\beta,\alpha\beta} + m_{\alpha,\alpha})\, \delta w, \end{aligned}$$

so that upon applying the two-dimensional divergence theorem we obtain

$$\overline{\delta W}_{int} - \overline{\delta W}^{(1)}_{ext} = \oint_C [\nu_\beta N_{\alpha\beta}\, \delta\bar{u}_\alpha + \nu_\beta M_{\alpha\beta}\, \delta w_{,\alpha} + \nu_\alpha (M_{\alpha\beta,\beta} + m_\alpha)\, \delta w] ds$$
$$- \int_A [(N_{\alpha\beta,\beta} + p_\alpha)\, \delta\bar{u}_\alpha + (M_{\alpha\beta,\alpha\beta} + m_{\alpha,\alpha} - q)\, \delta w]\, dA. \quad (5.2.8)$$

However, by the equilibrium equations (5.2.3) and (5.2.6) the area integral vanishes. In the remaining contour integral, the expression in parentheses may be replaced by $-Q_\alpha$ as a result of Equation (5.2.5). Moreover, the three functions w, $w_{,1}$ and $w_{,2}$ are not independent on C, because, if $w_{,\alpha}$ is decomposed as $w_{,\alpha} = \nu_\alpha \partial w/\partial n + t_\alpha \partial w/\partial s$, then the normal derivative $\partial w/\partial n$ (which represents the local rotation of the plate) can be prescribed independently of w, but the tangential derivative $\partial w/\partial s$ is entirely determined by w. Performing the aforementioned decomposition and defining $M_n = \nu_\alpha \nu_\beta M_{\alpha\beta}$ (the normal bending moment), $M_{nt} = \nu_\alpha t_\beta M_{\alpha\beta}$ (the twisting moment), and $Q_n = \nu_\alpha Q_\alpha$, we obtain

$$\overline{\delta W}_{int} - \overline{\delta W}^{(1)}_{ext} = \oint_C \left(\nu_\beta N_{\alpha\beta}\, \delta\bar{u}_\alpha + M_n \frac{\partial \delta w}{\partial n} + M_{nt} \frac{\partial \delta w}{\partial s} + Q_n\, \delta w \right) ds.$$

By integration by parts,

$$\oint_C M_{nt} \frac{\partial\, \delta w}{\partial s} ds = - \oint_C \frac{\partial M_{nt}}{\partial s}\, \delta w\, ds,$$

and therefore

$$\overline{\delta W}_{int} - \overline{\delta W}^{(1)}_{ext} = \oint_C \left(\nu_\beta N_{\alpha\beta}\, \delta\bar{u}_\alpha + M_n\, \delta \frac{\partial w}{\partial n} + V_n\, \delta w \right) ds,$$

where $V_n = Q_n - \partial M_{nt}/\partial s$ is the *effective shear force* along the edge. A graphic illustration of the equivalence between a *varying* twisting moment and a distributed transverse force may be seen in Figure 5.2.1.

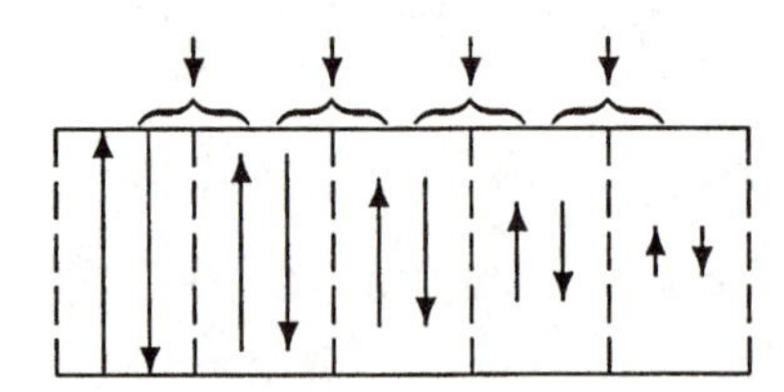

Figure 5.2.1. Effective shear force along a plate edge.

Finally, let the applied in-plane forces along the edge be F^a_α, the applied bending moment (acting about the tangent to C) M^a_n, and the applied transverse force V^a_n, all per unit length. Then

$$\overline{\delta W}^{(2)}_{ext} = \oint_C \left(F^a_\alpha\, \delta\bar{u}_\alpha + M^a_n\, \delta \frac{\partial w}{\partial n} + V^a_n\, \delta w \right) ds,$$

so that the boundary conditions are

$$\begin{aligned}
&\text{either } \nu_\beta N_{\alpha\beta} = F_\alpha^a && \text{or } \bar{u}_\alpha \text{ prescribed,}\\
&\text{either } M_n = M_n^a && \text{or } \frac{\partial w}{\partial n} \text{ prescribed,}\\
&\text{either } V_n = V_n^a && \text{or } w \text{ prescribed.}
\end{aligned}$$

Leaving aside the in-plane forces and displacements, we see that at every point of the edge two conditions must be specified. For example, along a *clamped* edge the conditions are $w = 0$ and $\partial w/\partial n = 0$; along a *simply supported* edge, $w = 0$ and $M_n = 0$; and along a *free* edge, $V_n = 0$ and $M_n = 0$. The condition $V_n = 0$ was first derived by Kirchhoff, and consequently the theory of plates that has thus far been outlined is known as **Kirchhoff plate theory**. In the original plate theory formulated by Sophie Germain, a free edge was assumed to be subject to the *three* boundary conditions $Q_n = 0$, $M_{nt} = 0$, and $M_n = 0$, resulting in an improperly posed boundary-value problem for elastic plates.

In the present treatment the principle of virtual work was used to derive the boundary conditions that are consistent with the displacement model adopted, while the equilibrium equations (5.2.3)–(5.2.6) were derived from the three-dimensional ones — that is, they were shown to be necessary, but not sufficient. However, it can easily be seen that the principle of virtual work also implies the equilibrium equations (5.2.3) and (5.2.8), and as the only necessary ones: since the displacement components $\bar{u}_\alpha$, w can vary independently in A, and since the area integral in Equation (5.2.8) must vanish, it follows that the coefficients of $\delta\bar{u}_\alpha$ and δw must vanish. Equation (5.2.5) may then be used as the definition of Q_α.

Before introducing constitutive equations, it must be noted that although the displacement model is one in which $\varepsilon_{33} = 0$, this constraint is not realistic. Actually, it is the stress σ_{33} which is very nearly zero, or at least, its maximum value is very small in comparison to those of the stresses $\sigma_{\alpha\beta}$ (α, $\beta = 1, 2$). Similarly, the shear stresses $\sigma_{3\alpha}$, though important in the equilibrium equations, are generally of small magnitude. Consequently, most points of the plate are nearly in a state of plane stress. The elastic behavior of isotropic plates should therefore be described by Equations (1.4.13), and plasticity by a plane-stress yield criterion.

5.2.2. Elastic Plates

Elastic Relations

Equations (1.4.13),

$$\sigma_{\alpha\beta} = \frac{E}{1-\nu^2}[(1-\nu)\varepsilon_{\alpha\beta} + \nu\varepsilon_{\gamma\gamma}\delta_{\alpha\beta}].$$

lead to

$$N_{\alpha\beta} = \frac{Eh}{1-\nu^2}[(1-\nu)\bar{\varepsilon}_{\alpha\beta} + \nu\bar{\varepsilon}_{\gamma\gamma}\,\delta_{\alpha\beta}]$$

and

$$M_{\alpha\beta} = D[(1-\nu)\kappa_{\alpha\beta} + \nu\kappa_{\gamma\gamma}\,\delta_{\alpha\beta}],$$

where $D = Eh^3/[12(1-\nu^2)]$ is the *plate bending modulus*, and $\kappa_{\alpha\beta} = w_{,\alpha\beta}$ is the *curvature tensor*.

It can be seen that the problem of the in-plane forces is identical with the plane-stress problem, with $N_{\alpha\beta}$, F_α^a, $\bar{u}_\alpha$ and $\bar{\varepsilon}_{\alpha\beta}$ corresponding to $\sigma_{\alpha\beta}$, T_α^a, u_α, and $\varepsilon_{\alpha\beta}$, respectively. For the flexure problem, the equilibrium equation, when combined with the moment-curvature and curvature-deflection relations, becomes

$$\nabla^4 w = \frac{\bar{q}}{D},$$

where $\bar{q} = q - m_{\alpha,\alpha}$ is the effective transverse load per unit area. In what follows we shall assume, as is the case in most problems, that $m_\alpha = 0$ and therefore $\bar{q}$ will be replaced by q.

Axisymmetrically Loaded Circular Plates

Given a circular plate of radius a, if the load q is a function (in polar coordinates) of r only and if the edge conditions are uniform, then the deflection w can likewise be assumed to be a function of r only, the only nonzero shear force is $Q_r = Q$, and the only moments are M_r and M_θ. Equation (5.2.4) then reduces to

$$\frac{1}{r}\frac{d}{dr}(rQ) + q = 0,$$

which can be integrated to yield

$$Q = -\frac{1}{r}\int_0^r rq\,dr,$$

and Equation (5.2.5) becomes

$$\frac{dM_r}{dr} + \frac{M_r - M_\theta}{r} + Q = 0.$$

The curvature tensor components are

$$\kappa_r = \frac{d^2w}{dr^2}, \qquad \kappa_\theta = \frac{1}{r}\frac{dw}{dr},$$

and therefore the elastic relations take the form

$$M_r = D\left(\frac{d^2w}{dr^2} + \nu\frac{1}{r}\frac{dw}{dr}\right), \quad M_\theta = D\left(\frac{1}{r}\frac{dw}{dr} + \nu\frac{d^2w}{dr^2}\right).$$

Substituting this result in the moment-shear equations results in

$$D\left(\frac{d^3w}{dr^3}+\frac{1}{r}\frac{d^2w}{dr^2}-\frac{1}{r^2}\frac{dw}{dr}\right)=-Q.$$

But the left-hand side is just

$$D\frac{d}{dr}\left[\frac{1}{r}\frac{d}{dr}\left(r\frac{dw}{dr}\right)\right],$$

so that the differential equation can be solved by integration.

The simplest problem is the one where the load is uniform, that is, $q =$ constant. Then $Q = -qr/2$, and the integration for w results in

$$w(r)=\frac{qr^4}{64D}+Ar^2+B\ln r+C,$$

where A, B, C are constants of integration. For the deflection to be finite at the center we must have $B = 0$, and it is convenient to set $C = 0$, that is, measure the deflection relative to the center rather than relative to the edge. The remaining constant, A, is then determined from the edge condition.

Clamped Edge. Here the edge condition is $w'(a) = 0$, leading to $A = -qa^2/32D$. The deflection of the edge relative to the center is thus $qa^4/64D$, or equivalently, the center deflection relative to the edge is $qa^4/64D$. The moments are

$$M_r(r)=\frac{q}{16}[(3+\nu)r^2-(1+\nu)a^2],\qquad M_\theta(r)=\frac{q}{16}[(1+3\nu)r^2-(1+\nu)a^2].$$

Simply Supported Edge. We have

$$M_r(r)=\frac{(3+\nu)qr^2}{16}+2(1+\nu)DA,$$

and therefore the condition $M_r(a) = 0$ leads to $A = -(3+\nu)qa^2/32(1+\nu)D$. The deflection is therefore

$$w(r)=\frac{q}{64(1+\nu)D}[(1+\nu)r^4-2(3+\nu)a^2r^2],$$

the maximum deflection being $(5+\nu)qa^4/64(1+\nu)D$, or, with $\nu = 0.3$, about four times as large as for the clamped plate. The moments are

$$M_r(r)=\frac{(3+\nu)q}{16}(r^2-a^2),\qquad M_\theta(r)=\frac{q}{16}[(1+3\nu)r^2-(3+\nu)a^2].$$

As a preliminary step to determining the deflection due to an arbitrary axisymmetric load $q(r)$, we consider the case of a force F concentrated on a circle of radius b. This may be viewed as the limit as $c \to b$ of the annular loading

$$q(r) = \begin{cases} 0, & 0 < r < b, \\ \dfrac{F}{\pi(c^2 - b^2)}, & b < r < c, \\ 0, & c < r < a. \end{cases}$$

It follows that rQ is constant for $r < b$ and for $r > c$ (with $Q = 0$ in the former region), and that

$$rQ|_b^c = -\int_b^c q(r) r\, dr = -\frac{F}{2\pi}.$$

Consequently, in the limit as $c \to b$ we have

$$\frac{d}{dr}\left[\frac{1}{r}\frac{d}{dr}\left(r\frac{dw}{dr}\right)\right] = \begin{cases} 0, & r < b, \\ \dfrac{F}{2\pi D r}, & r > b. \end{cases}$$

Denoting the deflections in $r < b$ and $r > b$ by w_1 and w_2, respectively, and setting $w_1(0) = 0$ for convenience, we have

$$w_1(r) = A_1 r^2,$$

$$w_2(r) = A_2 r^2 + B_2 b^2 \ln\frac{r}{b} + C_2 b^2 + \frac{F}{8\pi D} r^2 \ln\frac{r}{b}.$$

The continuity conditions $w_1(b) = w_2(b)$, $w_1'(b) = w_2'(b)$, and $w_1''(b) = w_2''(b)$ yield

$$A_2 = A_1 - \frac{F}{8\pi D}, \quad B_2 = C_2 = \frac{F}{8\pi D}.$$

Thus

$$w_2(r) = A_1 r^2 + \frac{F}{8\pi D}\left[(r^2 + b^2)\ln\frac{r}{b} + b^2 - r^2\right],$$

$$w_2'(r) = 2A_1 r + \frac{F}{8\pi D}\left(2r\ln\frac{r}{b} + \frac{b^2}{r} - r\right),$$

and in $r > b$,

$$M_r(r) = 2(1+\nu)DA_1 + \frac{F}{8\pi}\left[2(1+\nu)\ln\frac{r}{b} + (1-\nu)\left(1 - \frac{b^2}{r^2}\right)\right].$$

If the edge $r = a$ is clamped, then $w_2'(a) = 0$ and therefore

$$A_1 = -\frac{F}{8\pi D}\left[\ln\frac{a}{b} - \frac{1}{2}\left(1 - \frac{b^2}{a^2}\right)\right],$$

while if the edge is simply supported, then $M_r(a) = 0$ and

$$A_1 = -\frac{F}{8\pi D}\left[\ln\frac{a}{b} + \frac{1-\nu}{2(1+\nu)}\left(1 - \frac{b^2}{a^2}\right)\right].$$

These results may be combined by writing

$$A_1 = -\frac{F}{8\pi D}\left[\ln\frac{a}{b} + \lambda\left(1 - \frac{b^2}{a^2}\right)\right],$$

where λ equals $-\frac{1}{2}$ for the clamped plate and $(1-\nu)/2(1+\nu)$ for the simply supported plate. Since we have set $w(0) = 0$, the center deflection *relative to the edge* is

$$-w(a) = \frac{F}{8\pi D}\left[b^2\ln\frac{a}{b} - (1+\lambda)(a^2 - b^2)\right].$$

Let the solution derived above be written as

$$w(r) = -\frac{F}{8\pi D}g(r,\, b;\, \lambda),$$

where, by definition,

$$g(r,\, \rho;\, \lambda) = \begin{cases} \left[\ln\dfrac{a}{\rho} + \lambda\left(1 - \dfrac{\rho^2}{a^2}\right)\right] r^2, & r < \rho, \\ r^2\ln\dfrac{a}{r} - \rho^2\ln\dfrac{r}{\rho} + (1+\lambda)r^2 - \left(1 + \lambda\dfrac{r^2}{a^2}\right)\rho^2, & r > \rho. \end{cases}$$

Now consider an arbitrary load distribution $q(r)$. The total load contained in the infinitesimal annulus $\rho < r < \rho + d\rho$ is $2\pi q(\rho)\rho\, d\rho$, and consequently the deflection due to this load alone is $-(1/4D)q(\rho)g(r,\, \rho;\, \lambda)\rho\, d\rho$. The deflection due to the entire load is obtained by superposition:

$$w(r) = -\frac{1}{4D}\int_0^a q(\rho)g(r,\, \rho;\, \lambda)\rho\, d\rho,$$

and for the center deflection, in particular, we have

$$-w(a) = \frac{1}{4D}\int_0^a q(\rho)\left[(1+\lambda)(a^2 - \rho^2) - \rho^2\ln\frac{a}{\rho}\right]\rho\, d\rho.$$

Consider, for example, a *downward* load F uniformly distributed over an inner circle of radius b; then

$$\begin{aligned} w(a) &= \frac{F}{4\pi b^2 D}\int_0^b \left[(1+\lambda)(a^2 - \rho^2) - \rho^2\ln\frac{a}{\rho}\right]\rho\, d\rho \\ &= \frac{F}{4\pi D}\left[(1+\lambda)\left(\frac{a^2}{2} - \frac{b^2}{4}\right) - \frac{b^2}{4}\ln\frac{a}{b} + \frac{b^2}{16}\right]. \end{aligned}$$

When $b = a$, we obtain

$$w(a) = [4(1+\lambda) - 1]\frac{Fa^2}{64\pi D},$$

from which we obtain the previous results for the clamped and simply supported cases (with $q = -F/\pi a^2$) by inserting the appropriate values of λ.

With $b = 0$ we obtain the solution for a load concentrated at the center. The deflection at any r is obtained by evaluating $w(r)$ by superposition, with $r > b$, and then taking the limit as $b \to 0$:

$$w(r) = \lim_{b\to 0} \frac{1}{4D} \cdot \frac{F}{\pi b^2} \int_0^b \left[r^2 \ln\frac{a}{r} - \rho^2 \ln\frac{r}{\rho} + (1+\lambda)r^2 - \left(1 + \lambda\frac{r^2}{a^2}\right)\rho^2 \right] \rho\, d\rho$$

$$= \frac{F}{8\pi D}\left[r^2 \ln\frac{a}{r} + (1+\lambda)r^2 \right].$$

The maximum deflection is

$$w(a) = \frac{Fa^2}{16\pi D}\begin{cases} 1, & \text{clamped,} \\ \dfrac{3+\nu}{1+\nu}, & \text{simply supported.} \end{cases}$$

The bending moments are

$$M_r = \frac{F}{8\pi}\left[2(1+\nu)\ln\frac{a}{r} - (1-\nu) + 2(1+\nu)\lambda \right],$$

$$M_\theta = \frac{F}{8\pi}\left[2(1+\nu)\ln\frac{a}{r} + (1-\nu) + 2(1+\nu)\lambda \right].$$

The logarithmic singularity at the center indicates that the assumptions of elementary plate theory do not hold near the point of application of a concentrated load. For more details, see Timoshenko and Woinowsky-Krieger [1959], Section 5.1.

5.2.3. Yielding of Plates

Plate Yield Criteria

A plate will be said to yield in bending at a point (x_1, x_2) if the stress tensor there obeys the yield criterion at every x_3 except $x_3 = 0$; points in the middle plane are regarded as remnants of the elastic core. If the stresses σ_{i3} are assumed negligible in magnitude next to the $\sigma_{\alpha\beta}$ (this does not mean that they can be neglected in the equilibrium equations, because derivatives occur there), then we may apply a plane-stress yield criterion, say

$$f\left(\frac{\sigma_{\alpha\beta}}{\sigma_y}\right) = 0,$$

in every plane $x_3 \neq 0$. Equilibrium is satisfied if

$$\sigma_{\alpha\beta} = -\frac{4}{h^2} M_{\alpha\beta} \operatorname{sgn} x_3. \tag{5.2.9}$$

If the **ultimate moment** is defined as $M_U = \sigma_y h^2/4$, then the plate yield criterion is given by

$$f(m_{\alpha\beta}) = 0,$$

where $m_{\alpha\beta} = M_{\alpha\beta}/M_U$; the Mises and Tresca criteria become, respectively,

$$\begin{aligned} m_{11}^2 - m_{11}m_{22} + m_{22}^2 + 3m_{12}^2 &= 1 \quad \text{(Mises)}, \\ \max(|m_1|, |m_2|, |m_1 - m_2|) &= 1 \quad \text{(Tresca)}. \end{aligned}$$

A yield criterion that approximates the behavior of doubly reinforced concrete slabs is the **Johansen criterion**,

$$\max(|m_1|, |m_2|) = 1.$$

Problems in contained plastic bending of plates may be studied by means of numerical methods analogous to those of Section 4.5.[1] The plastic collapse of perfectly plastic plates may in general be investigated by means of limit analysis, as is done in Section 6.4. If such plates are, however, circular and axisymmetrically loaded and supported, then exact solutions may be obtained for the collapse state.

Fully Plastic Axisymmetrically Loaded Circular Plates

If the loading and support are axisymmetric, then the only nonvanishing moments are M_r and M_θ, and the equilibrium equation is

$$(rM_r)' - M_\theta = \int_0^r qr\, dr. \tag{5.2.10}$$

This equation and the yield condition constitute two equations for M_r and M_θ. Equivalently, if the yield condition is solved for M_θ in terms of M_r and the resulting expression for M_θ is substituted in (5.2.10), the result is a nonlinear first-order differential equation for $M_r(r)$. At the center of the plate, $M_r = M_\theta$ and consequently, if the curvature there is positive (concave upward), $M_r(0) = M_U$ constitutes an initial condition with which the differential equation may be solved. In addition, a boundary condition at the edge $r = a$ must be satisfied; this yields the ultimate load. Let us recall that for a simply supported plate, the edge conditions are $w = M_r = 0$; thus $M_r = 0$ is a boundary condition with which the differential equation may be solved. For a clamped plate, the edge must form a **hinge circle**,

[1] For an introduction to finite-element methods for plates, see Zienkiewicz [1977], Chapter 10.

that is, a locus of slope discontinuity (a special case of the hinge curve discussed in Section 6.2). As we shall see, the edge condition there becomes $M_r(a) = -M_U$ or $M_r(a) = -2M_U/\sqrt{3}$ for the Tresca or Mises material, respectively.

Figure 5.2.2 shows the Mises and Tresca yield criteria for axisymmetrically loaded circular plates. It follows from the preceding discussion that the center of the plate is in the moment state corresponding to point B, and that a simply supported edge corresponds to point C. A simply supported plate may thus be assumed to be entirely in the regime BC. For the the Tresca material, this means that $M_\theta = M_U$ everywhere, and the problem to be solved is therefore linear.

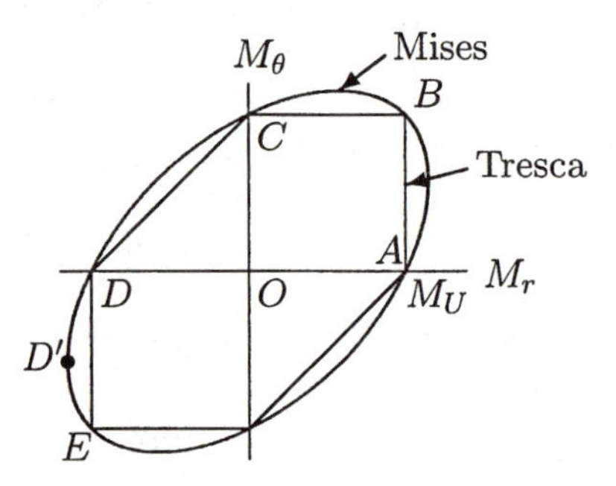

Figure 5.2.2. Mises and Tresca yield criteria for axisymmetrically loaded circular plates

Solution for Tresca Plate

Let us consider, for example, a downward load F uniformly distributed over a circle of radius b, the plate being unloaded outside this circle. The equilibrium equation is then

$$(rM_r)' - M_U = \begin{cases} -\dfrac{Fr^2}{2\pi b^2}, & r < b, \\ -\dfrac{F}{2\pi}, & r > b. \end{cases}$$

The solution for $r < b$ satisfying the condition at $r = 0$ is

$$M_r = M_U - \frac{Fr^2}{6\pi b^2},$$

while the solution for $r > b$ satisfying the condition at $r = a$ is

$$M_r = \left(\frac{F}{2\pi} - M_U\right)\left(\frac{a}{r} - 1\right).$$

Continuity at $r = b$ requires that

$$F = 2\pi \frac{M_U}{1 - 2b/3a}.$$

This result includes the extreme cases $F = 6\pi M_U$ for the uniformly loaded plate ($b = a$) and $F = 2\pi M_U$ for a plate with concentrated load. This last case could not have been treated directly because the moments would have to go to infinity at the center —a condition incompatible with plasticity.

On segment BC of the Tresca hexagon, the flow rule yields $\dot{\kappa}_r/\dot{\kappa}_\theta = 0$, that is, $\dot{w}''/(\dot{w}'/r) = 0$. Consequently the velocity field obeying the edge condition must be $\dot{w}(r) = (1 - r/a)v_0$, where v_0 is the center velocity (i.e., the plate deforms in the shape of a cone). It is shown in Chapter 6 that the upper-bound load obtained with this velocity field equals the one obtained here, as indeed it must, since the solution is complete.

In order to study the clamped plate, we must know the velocity fields associated with the other sides of the Tresca hexagon. On CD we have $\dot{\kappa}_r + \dot{\kappa}_\theta = 0$, that is, $\dot{w}'' + \dot{w}'/r = 0$, a differential equation whose solution is $\dot{w} = C + D \ln r$. Neither this velocity field nor the preceding (conical) one can possibly meet a condition of zero slope at a clamped edge, and this is why a hinge circle is necessary there. Finally, on AB and DE the flow rule gives $\dot{\kappa}_\theta/\dot{\kappa}_r = 0$. If this is interpreted as $\dot{\kappa}_\theta = \dot{w}'/r = 0$, then the velocity field is $\dot{w} =$ constant (i.e., rigid-body motion). Alternatively, we may interpret it as $\dot{\kappa}_r = inf$; this would be the state at a hinge circle, and from this follows the clamped-edge condition $M_r = -M_U$. On the Mises ellipse, the only relevant point where $\dot{\kappa}_\theta/\dot{\kappa}_r = 0$ is D', where $M_r = -2M_U/\sqrt{3}$.

It follows from these considerations that a clamped plate must be in regime BC near the center and in CD near the edge; point C gives the state at $r = c$ for some c such that $0 < c < a$. With the same loading as assumed for the simply supported plate above, we have to solve the problem both for $c > b$ and $c < b$.

Case 1 : $c > b$. The equilibrium equation is

$$\begin{aligned}
(rM_r)' &= M_U - \frac{Fr^2}{2\pi b^2}, \quad && 0 < r < b, \\
(rM_r)' &= M_U - \frac{F}{2\pi}, && b < r < c, \\
rM_r' &= M_U - \frac{F}{2\pi}, && c < r < a.
\end{aligned}$$

Let $m = M_r/M_U$ and $p = F/2\pi M_U$; then the solution satisfying the conditions at $r = 0$ and $r = c$ is

$$\begin{aligned} m(r) &= 1 - \frac{pr^2}{3b^2} && 0 < r < b, \\ m(r) &= (p-1)\left(\frac{c}{r} - 1\right), && b < r < c, \\ m(r) &= -(p-1)\ln\frac{r}{c}, && c < r < a. \end{aligned}$$

The additional conditions to be met are continuity at $r = b$ and $m = -1$ at $r = a$, producing the two equations

$$1 - \frac{1}{3}p = (p-1)\left(\frac{c}{b} - 1\right),$$

$$(p-1)\ln\frac{a}{c} = 1.$$

The assumption $c > b$ requires $1 < p < 3$, and the limiting case $c = b$, $p = 3$ corresponds to $b/a = e^{-1/2}$. Consequently, the present case represents the range $0 < b/a < e^{-1/2}$. Eliminating c between the two equations, we obtain

$$\frac{b}{a} = \frac{3(p-1)}{2p}e^{-\frac{1}{p-1}}, \quad p < 3.$$

For example, $p = 2$ corresponds to $b/a = 0.276$, $p = 1.5$ to $b/a = 0.068$, and $p = 1.286$ to $b/a = 0.01$. The ultimate concentrated load is $2\pi M_U$ ($p = 1$, $b = 0$), the same as for the simply supported plate. This limit is approached, however, only at extremely small values of b/a.

Case 2 : $b > c$. The equilibrium equation is

$$\begin{aligned} (rM_r)' &= M_U - \frac{Fr^2}{2\pi b^2} && 0 < r < c, \\ rM_r' &= M_U - \frac{Fr^2}{2\pi b^2} && c < r < b, \\ rM_r' &= M_U - \frac{F}{2\pi} && b < r < a. \end{aligned}$$

An analysis similar to that in Case 1 leads to

$$\frac{b}{a} = e^{-\frac{5-p+\ln(p/3)}{2(p-1)}}, \quad p > 3.$$

The extreme case $b = a$ (uniformly loaded plate) corresponds to $p = 5.63$, an increase of 88% over the ultimate load of the simply supported case.

Much of the preceding theory is due to H. G. Hopkins and various collaborators. For more details and other solutions, see, for example, Hopkins and Prager [1953], Hopkins and Wang [1954], and Drucker and Hopkins [1955].

For plates without circular symmetry, complete solutions are not available. Estimates of collapse loads will be found with the help of the upper-bound theorem of limit analysis — and exceptionally of the lower-bound theorem — in 6.4.1.

The effect of large deflections on the collapse load of a simply supported circular plate was studied by Onat and Haythornthwaite [1956]; the analysis is based on the deformation of the plate into a conical shell (see also Hodge [1959], Section 11-7). Approximate methods for determining collapse loads of shells are studied in 6.4.3.

Exercises: Section 5.2

1. Derive the plate equilibrium equations (5.2.6) by means of virtual work. Discuss the significance of the shear forces Q_α in this formulation.

2. A uniformly loaded circular plate made of an elastic–perfectly plastic material with yield stress σ_Y is assumed to obey the Tresca criterion. If the total load is F, determine the value F_E at which yielding begins when the plate is (a) simply supported and (b) clamped. Compare with the corresponding ultimate loads.

3. Determine the velocity field in a fully plastic clamped circular Tresca plate of radius a carrying a load F that is uniformly distributed over an inner circle of radius b. Use the upper-bound theorem to determine the relation between the load and the ratio b/a. Compare with the results in the text.

4. A simply supported circular Tresca plate of radius a carries a load F that is uniformly distributed over the perimeter of the circle $r = b$. Find the value of F and the moment distribution when the plate is fully plastic.

5. A simply supported circular Tresca plate of radius a carries a load F that is uniformly distributed over the entire plate. The ultimate moment is M_U inside the circle $b = a/2$ and $\frac{1}{2}M_U$ outside this circle. Find the value of F and the moment distribution when the plate is fully plastic.

Section 5.3 Plastic Buckling

While Sections 5.1 and 5.2 dealt with the plastic collapse of bodies whose material behavior may be idealized as perfectly plastic, the present section is devoted to the study of the buckling collapse of bodies made of work-hardening material. The elementary theory of elastic column buckling is part of the knowledge of all students of mechanics. It is worthwhile, however, to begin this section, in 5.3.1, with a general introduction to stability theory and to use the buckling of bars as an illustration of the theory. In 5.3.2 we discuss theories of the modulus that must be used in the determination of critical loads. Finally, in 5.3.3 the plastic buckling of plates and shells is studied.

5.3.1. Introduction to Stability Theory

Elastic Stability

In 1.4.3 it was shown that an elastic body under conservative loads is in equilibrium if and only if the total potential energy Π is stationary with respect to virtual displacements, a condition expressed by Equation (1.4.18):

$$\delta\Pi = 0.$$

It was further stated without proof that the equilibrium is stable only if Π is a *minimum*. An intuitive, though not strictly rigorous proof can be based on the following observation: if the configuration of the body is to change slightly from the one at equilibrium, and if the potential energy at the altered configuration is greater than at equilibrium, then additional work must be done on the body in order to effect the change, and therefore the change cannot take place spontaneously. The proof can be easily made rigorous for discrete systems (those with a finite number of degrees of freedom), and the result is known as the **Lagrange–Dirichlet theorem**. The proof for continua runs into technical difficulties, but these will be ignored here, and the result will be accepted.

Mathematically, the condition that Π is a minimum at equilibrium can be expressed as follows: let Π denote the potential energy evaluated at the displacement field $\mathbf{u}$, and $\Pi + \Delta\Pi$ the potential energy evaluated at the varied displacement field $\mathbf{u} + \delta\mathbf{u}$. Assuming the dependence of Π on $\mathbf{u}$ to be smooth, we can write

$$\Delta\Pi = \delta\Pi + \frac{1}{2}\delta^2\Pi + \ldots,$$

where $\delta\Pi$ (the first variation defined in 1.4.3) is linear in $\delta\mathbf{u}$ (and/or in its derivatives, and therefore also in $\delta\varepsilon$), $\delta^2\Pi$ is quadratic, and so on. We may limit ourselves to virtual displacements that are small enough so that terms beyond the quadratic can be neglected. Since $\delta\Pi$ vanishes if Π is stationary at $\mathbf{u}$, clearly Π is a minimum only if $\delta^2\Pi$ is nonnegative for all $\delta\mathbf{u}$ that are compatible with the constraints (i.e., for all virtual displacements).

We may therefore say that the equilibrium is stable only if $\delta^2\Pi > 0$ for all virtual displacements. The criterion for the onset of instability, known as the **Trefftz criterion**, is thus

$$\delta^2\Pi = 0.$$

Since $\Pi = \Pi_{int} + \Pi_{ext}$, we have $\delta^2\Pi = \delta^2\Pi_{int} + \delta^2\Pi_{ext}$. An equivalent statement of the Trefftz criterion is therefore

$$\delta^2\Pi_{int} + \delta^2\Pi_{ext} = 0.$$

In a linearly elastic body,

$$\Pi_{int} = \frac{1}{2}\int_R C_{ijkl}\varepsilon_{ij}\varepsilon_{kl}\, dV,$$

and therefore

$$\begin{aligned}\Delta\Pi_{int} &= \frac{1}{2}\int_R C_{ijkl}[(\varepsilon_{ij} + \delta\varepsilon_{ij})(\varepsilon_{kl} + \delta\varepsilon_{kl}) - \varepsilon_{ij}\varepsilon_{kl}]\, dV \\ &= \int_R C_{ijkl}\varepsilon_{ij}\delta\varepsilon_{kl}\, dV + \tfrac{1}{2}\int_R C_{ijkl}\delta\varepsilon_{ij}\delta\varepsilon_{kl}\, dV.\end{aligned}$$

As a result of the definitions of the first and second variations, the first integral in the last expression is $\delta\Pi_{int}$, and the second integral is $\delta^2\Pi_{int}$, that is,

$$\delta^2\Pi_{int} = \int_R C_{ijkl}\delta\varepsilon_{ij}\delta\varepsilon_{kl}\, dV. \tag{5.3.1}$$

Generalization to Quasi-Elastic Materials

A general theory of stability in elastic–plastic solids capable of large deformations was formulated by Hill [1958]. Here, we shall limit our consideration to infinitesimal deformations. We can then immediately generalize the preceding result for linearly elastic bodies to nonlinearly elastic ones if we interpret C as the *tangent modulus* tensor defined by $C_{ijkl} = \partial\sigma_{ij}/\partial\varepsilon_{kl}$. Since $\delta\sigma_{ij} = C_{ijkl}\,\delta\varepsilon_{kl}$, the integrand of (5.3.1) can be rewritten as $\delta\sigma_{ij}\,\delta\varepsilon_{ij}$.

Since it is a given equilibrium state that is examined as to its stability, the past history of the body is irrelevant to this examination, and therefore the result can be further extended to materials that are *quasi-elastic* in the sense that, *at a given state*, a small stress variation $\delta\boldsymbol{\sigma}$ can be uniquely associated with a small strain variation $\delta\boldsymbol{\varepsilon}$. As we have seen, rate-independent plastic materials have this property. We may therefore state, as a generalization to quasi-elastic bodies of the Trefftz criterion, the **energy criterion** for the stability of such bodies subject to conservative loads:

$$\int_R \delta\sigma_{ij}\,\delta\varepsilon_{ij}\, dV + \delta^2\Pi_{ext} = 0. \tag{5.3.2}$$

Furthermore, the first term on the left-hand side will be written as $\delta^2\Pi_{int}$ without thereby implying the existence of an internal potential energy Π_{int}. In other words, we shall *define*

$$\delta^2\Pi_{int} \stackrel{\text{def}}{=} \int_R \delta\sigma_{ij}\,\delta\varepsilon_{ij}\,dV.$$

In the absence of internal constraints, the quantity $\delta\sigma_{ij}\,\delta\varepsilon_{ij}$ is positive for all nonvanishing $\delta\varepsilon$ if the material is linearly elastic. It is so likewise for rate-independent plastic materials that are stable in the sense of Drucker (see 3.2.1), that is, work-hardening and obeying an associated flow rule. For bodies made of such materials, then, instability can occur only if it is possible for $\delta^2\Pi_{ext}$ to be negative.

If the loads $\mathbf{f}$ and $\mathbf{t}^a$ are independent of displacement, the external potential energy is

$$\Pi_{ext} = -\int_R f_i u_i\,dV - \int_{\partial R} t_i^a u_i\,dS,$$

as given in 1.4.3. We see, then, that $\Delta\Pi_{ext} = \delta\Pi_{ext}$ (i.e. $\delta^2\Pi_{ext} = 0$) *unless the virtual displacement significantly alters the region occupied by the body,* thereby introducing terms that are quadratic in $\delta\mathbf{u}$ into $\Delta\Pi$.

Buckling of Quasi-Elastic Bars

The classic example of this occurrence is the buckling of a bar (column, strut) under a compressive axial load P acting through the neutral axis. If the bar is initially straight, with length L, and then bends, with the deflection of the neutral axis given by $v(x)$ $(0 < x < L)$, then the chord spanned by the neutral axis becomes $\int_0^L \sqrt{1-v'^2}\,dx$, so that the work done by the load P is

$$P\left(L - \int_0^L \sqrt{1-v'^2}\,dx\right) \doteq \frac{1}{2}P\int_0^L v'^2\,dx.$$

If the only other load is a distributed transverse load q, then the external potential energy is

$$\Pi_{ext} = -\int_0^L qv\,dx - \frac{1}{2}P\int_0^L v'^2\,dx, \tag{5.3.3}$$

and the first and second variations are, respectively,

$$\begin{aligned}\delta\Pi_e &= -\int_0^L q\,\delta v\,dx - P\int_0^L v'\,\delta v'\,dx\\ &= \int_0^L (Pv'' - q)\,\delta v\,dx - Pv'\,\delta v\big|_0^L\end{aligned}$$

(the last form having been obtained by integration by parts), and

$$\delta^2\Pi_{ext} = -P\int_0^L (\delta v')^2\,dx.$$

If the bar is subject to a distributed axial compressive load of intensity p per unit length in addition to end loads P_0 and P_L at $x = 0$ and $x = L$, respectively, then the external potential energy can be written as

$$\Pi_{ext} = -\int_0^L qv\,dx - \int_0^L ps\,dx + P_0 s(0) - P_L s(L),$$

where $s(x)$ is the shortening of the bar due to bending at point x, given by

$$s(x) = s(0) + \frac{1}{2}\int_0^x v'^2\,dx.$$

If $P(x)$ now denotes the internal axial force at x (positive in compression), then equilibrium requires

$$P' + p = 0,$$

and therefore

$$-\int_0^L ps\,dx = \int_0^L P's\,dx = Ps|_0^L + \frac{1}{2}\int_0^L Pv'^2\,dx.$$

Equation (5.3.3) for Π_{ext} needs to be changed only by placing P under the integral sign, and

$$\delta^2\Pi_{ext} = -\int_0^L P(\delta v')^2\,dx.$$

In accordance with elementary beam theory, the state of stress at each point will be approximated as uniaxial, so that $\delta\sigma_{ij}\,\delta\varepsilon_{ij} = \delta\sigma_x\delta\varepsilon_x$. Furthermore, $\delta\varepsilon_x = -y\,\delta v''$, so that

$$\delta^2\Pi_{int} = \int_0^L \left[\int_A (-y)\,\delta\sigma_x\,dA\right]\delta v''\,dx = \int_0^L \delta M\,\delta v''\,dx,$$

where M is the bending moment. The energy criterion for bars therefore takes the form

$$\int_0^L \delta M\,\delta v''\,dx - \int_0^L P(\delta v')^2\,dx = 0. \tag{5.3.4}$$

Integration by parts of the first integral leads to

$$\int_0^L (\delta M' + P\,\delta v')\,\delta v'\,dx = \delta M\,\delta v'|_0^L.$$

Since the end conditions on beams are usually such that either the rotation v' or the bending moment M cannot be varied, the right-hand side of this equation vanishes.

Finally, we assume an *effective modulus* $\bar{E}$ such that $\delta M = \bar{E}I\,\delta v''$; in a linearly elastic material, of course, this is just the Young's modulus E. In nonlinear materials $\bar{E}$ may be assumed to be determined by the average stress in the bar, $\sigma = P/A$. Theories of the effective modulus are discussed in 5.3.2.

In problems in which constraints on the deflection v itself can be disregarded, the virtual rotation $\delta v'$ may be taken as the unknown variable. Writing this variable as θ, we obtain, upon observing that the rotation is not constrained in the interior of the bar, the differential equation

$$(\bar{E}I\theta')' + P\theta = 0. \tag{5.3.5}$$

More generally, the first integral in (5.3.4) must be integrated by parts twice, so that the energy criterion becomes

$$\int_0^L (\delta M' + P\,\delta v')'\,\delta v\,dx = (\delta M' + P\,\delta v')\,\delta v\big|_0^L. \tag{5.3.6}$$

The quantity in parentheses may be interpreted as the virtual shear force, and therefore either it or the virtual displacement can be expected to vanish at each end. The differential equation expressing the energy criterion is therefore

$$(\bar{E}I\theta')'' + (P\theta)' = 0. \tag{5.3.7}$$

End-Loaded Prismatic Bars

If the bar is prismatic then A and I are constant. If, moreover, it is subject to an axial end load only, then P is constant. In such a bar, then, the average stress P/A and therefore, by hypothesis, the effective modulus are constant. Equation (5.3.7) thus becomes

$$\bar{E}I\theta''' + P\theta' = 0, \tag{5.3.8}$$

and can be immediately solved as

$$\theta = B\cos\lambda x + C\sin\lambda x + D,$$

where B, C, and D are constants, and λ is defined by

$$\lambda^2 = \frac{P}{\bar{E}I}.$$

If the bar is pinned at both ends, then $\theta'(0) = 0$ and $\theta'(L) = 0$. The first condition requires $C = 0$, and the second

$$\lambda B\sin\lambda L = 0.$$

For a nontrivial solution,[1] λ and B must both be different from zero, and instability occurs only if

$$\sin\lambda L = 0,$$

[1]Note that D is irrelevant since it represents a rigid rotation.

that is, if $\lambda = n\pi/L$, where n is a positive integer. The *fundamental mode* of buckling corresponds to $n = 1$, that is, $\lambda = \pi/L$, and the load producing it (the *critical load*) obeys

$$\frac{P}{\bar{E}} = \pi^2 \frac{I}{L^2}.$$

If the bar is elastic, $\bar{E} = E$, and

$$P = \pi^2 \frac{EI}{L^2} \stackrel{\text{def}}{=} P_E,$$

where P_E is known as the *Euler load.*

For other end conditions, the equation governing the critical load can be written as

$$\frac{P}{\bar{E}} = \pi^2 \frac{I}{L_e^2}, \tag{5.3.9}$$

where L_e is known as the *effective length* of the bar, given by $L_e = \pi/\lambda_1$, λ_1 being the smallest nonzero value of λ. For a cantilever column, for example, the characteristic equation is $\cos \lambda L = 0$, leading to a fundamental mode descibed by $\lambda_1 = \pi/2L$, and hence $L_e = 2L$. For a bar that is clamped at $x = 0$ and pinned at $x = L$, the end conditions are $\theta(0) = \theta'(L) = 0$, in addition to the constraint $\int_0^L \theta\, dx = 0$, describing zero deflection of the pinned end relative to the clamped end. The three conditions lead to the characteristic equation $\lambda L = \tan \lambda L$, whose lowest root is $\lambda_1 L = 4.4934$. Hence $L_e/L = \pi/4.4934 = 0.699$.

The solution of Equation (5.3.9) for the critical load for inelastic bars will be postponed until after the discussion of the effective modulus.

5.3.2. Theories of the Effective Modulus

Tangent-Modulus (Engesser–Shanley) Theory

The first analysis of inelastic column buckling is due to Engesser [1889], who based the calculation of the critical load on the incremental relation $\delta\sigma_x = E_t\, \delta\varepsilon_x$, where E_t is the tangent modulus, defined as the slope $d\sigma/d\varepsilon$ of the uniaxial compression curve at the current value of $\sigma = P/A$. The essential assumption is that, while a bending moment requires a nonuniform stress distribution, the deviation of the stress from the average is, at least initially, sufficiently small so that the stress-strain curve can be locally approximated by straight line with slope E_t. Consequently, the effective modulus $\bar{E}$ is just E_t. Since this is a function of σ, it is convenient to rewrite Equation (5.3.9) in terms of σ rather than P, and to designate the solution as the *critical stress* $\sigma_{cr} = P_{cr}/A$. It is conventional to define the "radius of gyration" $r \stackrel{\text{def}}{=} \sqrt{I/A}$, and to designate L_e/r as the *slenderness ratio*. Equation (5.3.9) now reads

$$\frac{\sigma}{E_t(\sigma)} = \frac{\pi^2}{(L_e/r)^2}. \tag{5.3.10}$$

A plot of the solution of this equation, σ_{cr} against L_e/r, is often called a *column curve*. When σ_{cr} is given explicitly as a function of L_e/r, the relation is called a *column formula*.

If the material has a definite elastic-limit stress σ_E such that $E_t(\sigma) = E$ for $\sigma < \sigma_E$, then the slenderness ratio at which $\sigma = \sigma_E$ is called the *critical slenderness ratio*, defined by

$$\left(\frac{L}{r}\right)_{cr} = \pi\sqrt{\frac{E}{\sigma_E}}.$$

The critical slenderness ratio is clearly a material property (for mild steel it is about 90), and for supercritically slender bars the critical load is given by the elastic solution. The portion of the column curve for $L_e/r > (L/r)_{cr}$ is thus the *Euler hyperbola* given by

$$\sigma_{cr} = \frac{\pi^2 E}{(L_e/r)^2}.$$

If the material is perfectly plastic, so that $\sigma_Y = \sigma_E$, then a stress greater than σ_E is not possible, and therefore $\sigma_{cr} = \sigma_E$ for $L_e/r \le (L/r)_{cr}$. For materials without a definite elastic limit, the column curve approaches the Euler hyperbola asymptotically as $L_e/r \to \infty$. For a material whose uniaxial stress-strain relation is described by the Ramberg–Osgood equation (2.1.2), the tangent modulus is easily obtained as

$$E_t = \frac{E}{1 + \alpha m(\sigma/\sigma_R)^{m-1}}, \tag{5.3.11}$$

and the column curve is obtained from

$$\frac{\sigma}{\sigma_R} + \alpha m\left(\frac{\sigma}{\sigma_R}\right)^m = \frac{\pi^2 E/\sigma_R}{(L_e/r)^2}.$$

Since such a material hardens indefinitely, no cutoff stress exists for short bars.

A formula describing a stress-strain relation with no definite elastic limit that approaches perfect plasticity asymptotically, with an ultimate stress σ_∞, was proposed by Prager [1942] in the form

$$\sigma = \sigma_\infty \tanh\frac{E\varepsilon}{\sigma_\infty}.$$

The asymptote is approached quite fast: when the total strain equals twice the elastic strain ($\varepsilon = 2\sigma/E$), the stress is already given by $\sigma = 0.9575\sigma_\infty$. The tangent modulus can be readily obtained as

$$E_t = E\left[1 - \left(\frac{\sigma}{\sigma_\infty}\right)^2\right].$$

The preceding formula can be easily generalized to

$$E_t = E\left[1 - \left(\frac{\sigma}{\sigma_\infty}\right)^n\right], \tag{5.3.12}$$

where $n = 2$ corresponds to the Prager formula. The case $n = 1$ describes an exponential stress-strain curve, $\sigma = \sigma_\infty[1 - \exp(-E\varepsilon/\sigma_\infty)]$. The greater the value of n, the more rapid the approach to perfect plasticity. The column curve for the generalized Prager formula may be plotted from

$$\frac{\sigma/\sigma_\infty}{1 - (\sigma/\sigma_\infty)^n} = \frac{\pi^2 E/\sigma_\infty}{(L_e/r)^2}.$$

Explicit column formulas can be obtained from this equation for $n = 1$ and $n = 2$.

The Reduced-Modulus (Kármán) Theory

Soon after the publication of Engesser's theory, it was recognized by engineers, beginning with Considère [1891] and eventually including Engesser himself, that the tangent-modulus theory was in contradiction with the elastic–plastic behavior of metals; a formal theory was proposed by von Kármán [1910], based on the following reasoning.

When an initially straight bar under a compressive axial load begins to buckle, the fibers on the concave side undergo additional compression, but in those on the convex side the compressive strain, and hence the stress, is reduced. The stress change in the latter fibers is consequently elastic. Writing, for convenience, stress and strain as positive in compression, the incremental stress-strain relation is accordingly

$$\delta\sigma = \left\{ \begin{array}{ll} E_t\,\delta\varepsilon, & \delta\varepsilon > 0, \\ E\,\delta\varepsilon, & \delta\varepsilon < 0, \end{array} \right\}$$

where E_t is, as before, the elastic–plastic tangent modulus at the stress σ. Let $y = y_0$ give the location of the neutral fibers, with $y > y_0$ and $y < y_0$ being the areas of additional and reduced compression, respectively. Since $\delta\varepsilon = (y - y_0)\,\delta v''$, the additional bending moment is

$$\delta M = \int_A (y - y_0)\,\delta\sigma\,dA$$

or

$$\delta M = E_r I \, \delta v'',$$

where

$$E_r = \frac{1}{I}\left[E\int_{y<y_0}(y-y_0)^2\,dA + E_t\int_{y>y_0}(y-y_0)^2\,dA\right] \tag{5.3.13}$$

is the *reduced modulus*. The value of y_0 is determined, as usual, by the constancy of the axial load:

$$0 = \delta P = \int_A \delta\sigma\,dA,$$

or

$$E\int_{y<y_0}(y-y_0)\,dA + E_t\int_{y>y_0}(y-y_0)\,dA = 0. \tag{5.3.14}$$

Elimination of y_0 permits the expression of E_r as a function of σ and of bar geometry.

If the bar is rectangular, with width b and depth h, then Equation (5.3.13), with the help of $I = bh^3/12$, gives

$$E_r = \frac{4}{h^3}(E_t h_1^3 + E h_2^3).$$

where

$$h_1 = \frac{h}{2} - y_0, \quad h_2 = \frac{h}{2} + y_0.$$

Equation (5.3.14), with the factor b omitted, then becomes

$$E\int_{-h/2}^{y_0}(y-y_0)\,dy + E_t\int_{y_0}^{h/2}(y-y_0)\,dy = \frac{1}{2}(E_t h_1^2 - E h_2^2) = 0,$$

so that $h_1/h_2 = \sqrt{E/E_t}$, and

$$\frac{h_1}{h} = \frac{\sqrt{E}}{\sqrt{E}+\sqrt{E_t}}, \quad \frac{h_2}{h} = \frac{\sqrt{E_t}}{\sqrt{E}+\sqrt{E_t}}.$$

Finally, then,

$$E_r = \frac{4EE_t}{\left(\sqrt{E}+\sqrt{E_t}\right)^2}.$$

The reduced modulus is seen to be greater than the tangent modulus by a factor between 1 and 4, and therefore gives a correspondingly greater critical load.

Figure 5.3.1 shows column curves based on both the tangent-modulus and reduced-modulus theories, for the Ramberg–Osgood formula with $\alpha = 0.1$ and $m = 10$, and for the Prager formula.

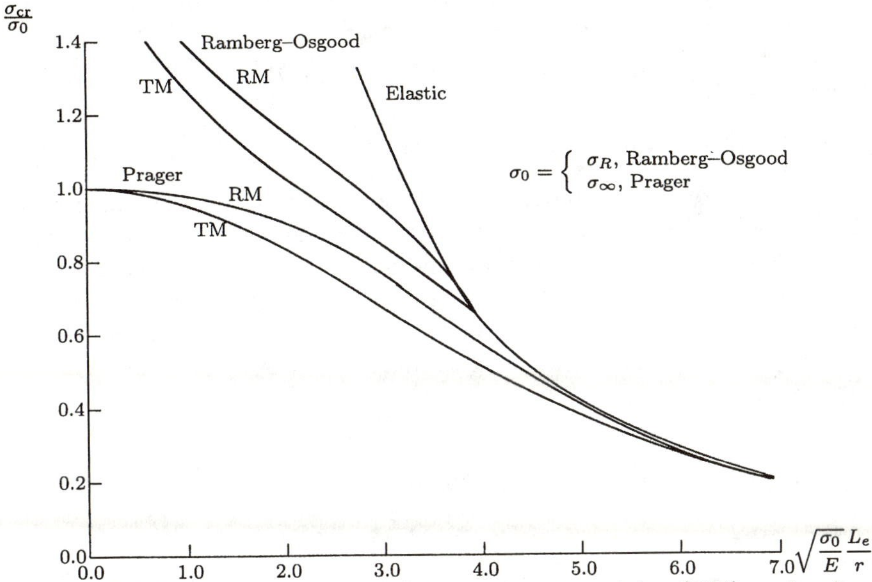

Figure 5.3.1. Column curves based on tangent-modulus (TM) and reduced-modulus (RM) theories for the Ramberg–Osgood formula with $\alpha = 0.1$ and $m = 10$, and for the Prager formula. The elastic column curve is included for comparison.

Comparison of the Two Theories

The reduced-modulus theory dominated engineering practice for most of the first half of the twentieth century, since it was rigorously based on elastic–plastic theory. However, experiments in which bars are subjected to increasing axial loads until they buckle have consistently shown the results to be in much better agreement with the tangent-modulus theory. The first explanation of this discrepancy was given by Shanley [1947].

The reduced-modulus theory is based on the assumption that buckling occurs with no first-order change in the axial load, $\delta P = 0$. But, as Shanley wrote, "upon reaching the critical tangent-modulus load, there is nothing to prevent the column from bending simultaneously with increasing axial load." It is thus possible for the neutral fibers to coincide initially with the extreme fibers on the concave side, all the other fibers undergoing additional compression. Any other possible location of the neutral fibers would correspond to a higher axial load, and therefore the tangent-modulus load P_{tm} is a *lower bound* to the elastic–plastic buckling load.

The tangent-modulus load is thus not a load at which instability occurs, but *bifurcation*: when this load is exceeded the bar may be in one of several configurations — it may remain straight, or it may be bent in either direction (buckling in one plane only is assumed). If the bar buckles at P_{tm}, then, as the load is increased, the neutral axis moves inward, tending asymptotically to the position corresponding to the reduced-modulus load. The latter load is therefore an *upper bound* to the buckling load.

Effect of Imperfections

In tests on real bars, imperfections such as initial curvature or eccentricity of the load are inevitable. Some bending moment, however slight, must therefore be present as soon as any load is applied, and consequently bending proceeds from the beginning of loading. Consider a pinned elastic column with the load applied at a small distance e from the centroidal axis. The deflection v can easily be shown to be governed by

$$EIv'' + Pv = Pe,$$

and the maximum deflection is found to be

$$v_{max} = e\left[\sec\left(\frac{\pi}{2}\sqrt{\frac{P}{P_E}}\right) - 1\right].$$

The deflection remains of the order of e until the load gets close to the Euler load, when it begins to grow large.

A similar conclusion holds for a column with an initial deflection v_0 of amplitude e. The equation governing the total deflection v under an axial load P is

$$EI(v'' - v_0'') + Pv = 0,$$

and if the column is pinned while v_0 is assumed as $v_0(x) = e\sin(\pi x/L)$, then the maximum deflection is

$$v_{max} = \frac{e}{1 - P/P_E}.$$

For an imperfect column, then, the buckling load may be interpreted as the load in the vicinity of which the imperfections become significantly amplified. It is for this reason that the tangent-modulus load must necessarily appear as the buckling load of work-hardening elastic–plastic columns, since the possibility of remaining straight when this load is exceeded is open only to perfect columns.

In columns made of an elastic–perfectly plastic material, both the reduced and tangent moduli are zero, and therefore a perfect column, as discussed previously, will buckle elastically or yield in direct compression at supercritical and subcritical slenderness ratios, respectively. Such a sharp transition is, in fact, hardly ever observed in real columns made of structural steel, a material that is fairly well represented as elastic–perfectly plastic.

The deviation from theoretical behavior can also be ascribed to imperfections. An approximate theory, developed by several nineteenth-century authors and recently reviewed by Mortelhand [1987], results in the formula

$$\frac{(1-\sigma/\sigma_Y)\sigma/\sigma_Y}{1-(1+\eta)\sigma/\sigma_Y} = \frac{\pi^2 E}{\sigma_Y (L_e/r)^2},$$

where $\eta = Ahe/2I$, h being the beam depth and e, as before, the amplitude of the initial deflection.

The effect of imperfections is of even greater significance in the analysis of post-bifurcation behavior. This topic has been extensively reviewed by Calladine [1973] and Hutchinson [1974].

Other Uniaxial Buckling Problems: Rings and Arches

From the preceding arguments we can infer that in all buckling problems in which uniaxial stress is assumed, the elastic solution can be used to give the buckling load with E replaced by E_t, provided that E_t can be taken as constant. Take, for example, a circular ring under an external pressure q per unit length of center line, the radius of the center line being R. The compressive force in the ring is thus qR, and the stress is $\sigma = qR/A$, where A is the cross-sectional area. The well-known solution is due to Bresse (see Timoshenko and Gere [1961]). The fundamental buckling mode is shown in Figure 5.3.2(a), with the ring deformed into an ellipse, and the critical pressure is

$$q_{cr} = \frac{3EI}{R^3},$$

where I is the second moment of area of the cross-section for bending in the plane of the ring.

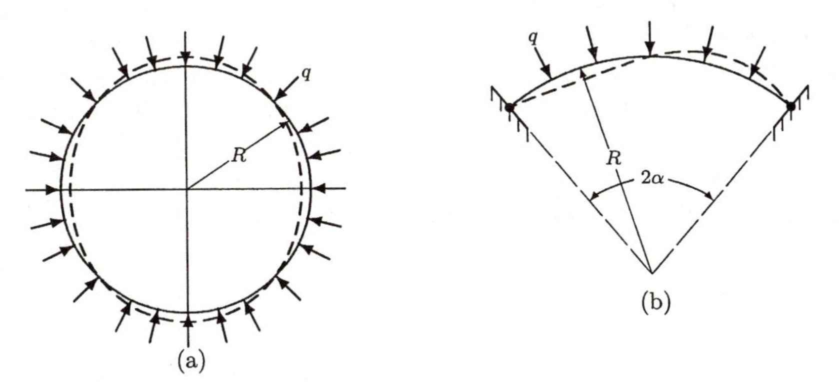

Figure 5.3.2. Buckling of a ring or arch: (a) complete ring; (b) hinge-ended arch.

The four points on the buckled ring whose radial displacement is zero are the nodes of the buckling modes. Any half of the ring between two opposite nodes is equivalent to a semicircular arch that is hinged at both ends, and therefore Bresse's formula furnishes the critical buckling pressure for such an arch. More generally, for a hinged-ended arch subtending an angle 2α the fundamental buckling mode can be expected to be as shown in Figure 5.3.2(b), and the result for the critical pressure is

$$q_{cr} = \left(\frac{\pi^2}{\alpha^2} - 1\right)\frac{EI}{R^3}.$$

These results may be immediately converted to the inelastic case. Limiting ourselves to the ring, we obtain

$$\frac{\sigma}{E_t(\sigma)} = \frac{3}{(R/r)^2},$$

where $r = \sqrt{I/A}$ as before. If the stress-strain relation is given by the generalized Prager formula with $n = 1$, an explicit formula for the stress, and hence for the critical pressure, is obtained:

$$q_{cr} = \frac{\sigma_\infty A/R}{1 + (\sigma_\infty/3E)(R/r)^2}.$$

For a ring of rectangular cross-section, with depth h (in the radial direction) and width b (in the axial direction), we have $A = bh$ and $r^2 = h^2/12$. Defining $p = q/b$ as the pressure in the usual sense (per unit area), we obtain

$$p_{cr} = \frac{\sigma_\infty h/R}{1 + 4(\sigma_\infty/E)(R/h)^2},$$

a formula that coincides with that derived by Southwell [1915] when σ_∞ is identified with the yield stress. Southwell's result was actually intended for cylindrical tubes rather than rings, but based on an assumption of uniaxial stress — a highly questionable assumption for a shell, as shown in the next subsection.

5.3.3. Plastic Buckling of Plates and Shells

Introduction

Consider a flat plate subject to an applied in-plane force per unit length F_α^a around its edge and a distributed in-plane force per unit area p_a. The membrane-force field $N_{\alpha\beta}$ obeys the equilibrium equation (5.2.3) and the boundary condition $\nu_\beta N_{\alpha\beta} = F_\alpha^a$. The equilibrium becomes unstable if it is possible for the plate to undergo a deflection $w(x_1, x_2)$ with no additional forces applied. The membrane forces $N_{\alpha\beta}$ can be assumed to be related to the average Green–Saint-Venant strains $\bar{E}_{\alpha\beta}$ given by Equation (5.2.7), and if the former do not change, then neither do the latter. The middle plane will therefore undergo a second-order displacement $\bar{u}_\alpha$ such that

$$2\bar{\varepsilon}_{\alpha\beta} = \bar{u}_{\alpha,\beta} + \bar{u}_{\beta,\alpha} = -w_{,\alpha}\, w_{,\beta}\,.$$

The work done by the applied forces on this displacement is

$$\oint_C F^a_\alpha \bar{u}_\alpha\, ds + \int_A p_\alpha \bar{u}_\alpha\, dA = \oint_C \nu_\beta N_{\alpha\beta} \bar{u}_\alpha ds + \int_A p_\alpha \bar{u}_\alpha\, dA$$
$$= \int_A [(N_{\alpha\beta,\beta} + p_\alpha)\bar{u}_\alpha + N_{\alpha\beta}\bar{u}_{\alpha,\beta}]\, dA$$

with the help of the two-dimensional divergence theorem. The quantity in parentheses vanishes as a result of (5.2.3). Since $N_{\alpha\beta}$ is symmetric, $N_{\alpha\beta}\bar{u}_{\alpha,\beta} = N_{\alpha\beta}\bar{\varepsilon}_{\alpha\beta} = -\frac{1}{2}N_{\alpha\beta}w_{,\alpha}\, w_{,\beta}$, and the second variation of the external potential energy can finally be obtained as

$$\delta^2\Pi_{ext} = \int_A N_{\alpha\beta}\, \delta w_{,\alpha}\, \delta w_{,\beta}\, dA. \tag{5.3.15}$$

For $\delta^2\Pi_{int}$ we have

$$\delta^2\Pi_{int} = \int_A \delta M_{\alpha\beta}\, \delta w_{,\alpha\beta}\, dA. \tag{5.3.16}$$

Since

$$\delta M_{\alpha\beta}\, \delta w_{,\alpha\beta} = (\delta M_{\alpha\beta}\, \delta w_{,\alpha})_{,\beta} - \delta M_{\alpha\beta,\beta}\, \delta w_{,\alpha}\,,$$

and since the edge conditions are usually such that $\nu_\beta\, \delta M_{\alpha\beta}\, \delta w_{,\alpha} = 0$ on C, the equation expressing the energy criterion becomes

$$\int_A (\delta M_{\alpha\beta,\beta} - N_{\alpha\beta}\, \delta w_{,\beta})\, \delta w_{,\alpha}\, dA = 0.$$

Support conditions on a plate are rarely such that the deflection itself is unconstrained; another integration by parts is usually necessary, leading to the differential equation

$$\delta M_{\alpha\beta,\alpha\beta} - (N_{\alpha\beta}\, \delta w_{,\beta})_{,\alpha} = 0. \tag{5.3.17}$$

Let the incremental relation between strain and stress for an isotropic material in a state of plane stress be written in the form

$$\dot{\varepsilon}_{\alpha\beta} = \frac{1}{\bar{E}}[(1+\bar{\nu})\dot{\sigma}_{\alpha\beta} - \bar{\nu}\dot{\sigma}_{\gamma\gamma}\, \delta_{\alpha\beta}],$$

where $\bar{\nu}$ is the instantaneous contraction ratio, and $\bar{E}$ is the instantaneous modulus, in general *not* equal to the uniaxial tangent modulus discussed before. If the tangent-modulus theory is applied to the plate problem, then an incremental moment-curvature relation can be written in the form

$$\delta M_{\alpha\beta} = \bar{D}[(1-\bar{\nu})\,\delta w_{,\alpha\beta} + \bar{\nu}\,\delta w_{,\gamma\gamma}\,\delta_{\alpha\beta}], \tag{5.3.18}$$

where $\bar{D} = \bar{E}h^3/12(1-\bar{\nu}^2)$ is the effective plate modulus. The parameters $\bar{E}$ and $\bar{\nu}$, and hence also $\bar{D}$, are functions of $N_{\alpha\beta}$, and the general form of (5.3.17) when the relations (5.3.18) are substituted is complicated. In what follows, only examples with a uniform membrane-force field will be considered.

Circular Plate Under Radial Load

A simple example is that of a circular plate under a uniformly distributed compressive radial load applied around its edge. Let a be the radius of the plate, h its thickness, and N the magnitude of the applied load per unit length of circumference. It is easy to see that a uniform state of plane stress, $\sigma_r = \sigma_\theta = -N/h$, is in equilibrium, satisfies the compatibility conditions if the plate is elastic, and obeys the yield criterion everywhere if it obeys it anywhere when the plate is plastic.

If the buckling is assumed axisymmetric, the deflection of the middle plane being $w(r)$, then the second integration by parts leading to (5.3.17) may be dispensed with, and the energy criterion can be expressed by the differential equation

$$\frac{1}{r}[(r\,\delta M_r)' - \delta M_\theta] + N\phi = 0,$$

where $\phi = \delta w'$ is the virtual rotation of the radial lines, with the prime denoting differentiation with respect to r. The axisymmetric form of (5.3.18) is

$$\delta M_r = \bar{D}\left(\phi' + \bar{\nu}\frac{\phi}{r}\right), \quad \delta M_\theta = \bar{D}\left(\frac{\phi}{r} + \bar{\nu}\phi'\right),$$

and the equation governing ϕ is thus

$$\phi'' + \frac{1}{r}\phi' + \left(\frac{N}{\bar{D}} - \frac{1}{r^2}\right)\phi = 0, \tag{5.3.19}$$

a Bessel equation of order 1. The general solution that is regular at $r = 0$ is $\phi(r) = J_1(\lambda r)$, where $\lambda = \sqrt{N/\bar{D}}$ and J_1 is the Bessel function of the first kind of order 1. Let $\lambda = k/a$ be the smallest nonzero root for which ϕ satisfies the boundary condition. The critical load is then $N_{cr} = \sigma_{cr}h$, where σ_{cr} is, in view of the definition of $\bar{D}$, the solution of the nonlinear equation

$$\sigma = \frac{k^2}{12}\frac{\bar{E}}{1-\bar{\nu}^2}\left(\frac{h}{a}\right)^2, \tag{5.3.20}$$

$\bar{E}$ and $\bar{\nu}$ being, as noted above, functions of σ.

If the edge of the plate is clamped, then the edge condition $\phi(a) = 0$ leads to the characteristic equation $J_1(k) = 0$, whose smallest nonzero root is $k = 3.832$. If the edge of the plate is free to rotate, then M_r must vanish there, so that the edge condition is $\phi'(a) + \bar{\nu}\phi(a)/a = 0$. The characteristic equation then becomes

$$J_0(k) - \frac{1-\bar{\nu}}{k} J_1(k) = 0.$$

Except in the case of the elastic plate, for which $\bar{\nu} = \nu$ (the Poisson's ratio), this equation must be solved simultaneously with (5.3.20) in order to find σ_{cr}.

It is now necessary to evaluate $\bar{E}$ and $\bar{\nu}$ as functions of σ. We assume the plate material to be work-hardening and governed by the Mises criterion and its associated flow rule. The plastic strain rate can then be written as

$$\dot{\varepsilon}_{ij}^p = \frac{9}{4} \frac{s_{kl}\dot{s}_{kl}}{H\sigma_Y^2} s_{ij},$$

where σ_Y is the current value of the uniaxial yield stress and H is the uniaxial plastic modulus (related to the tangent modulus by $Hinv = E_t inv - Einv$), as can easily be verified by the substitutions $s_{11} = \frac{2}{3}\sigma$, $s_{22} = s_{33} = -\frac{1}{3}\sigma$, and $|\sigma| = \sigma_Y$. In a state of plane stress,

$$s_{kl}\dot{s}_{kl} = \sigma_{\alpha\beta}\dot{\sigma}_{\alpha\beta} - \frac{1}{3}\sigma_{\alpha\alpha}\dot{\sigma}_{\beta\beta},$$

and therefore, if currently $\sigma_1 = \sigma_2 = \sigma$,

$$\dot{\varepsilon}_1^p = \dot{\varepsilon}_2^p = \frac{1}{4H}(\dot{\sigma}_1 + \dot{\sigma}_2).$$

The complete incremental stress-strain relations are therefore

$$\dot{\varepsilon}_1 = \left(\frac{1}{E} + \frac{1}{4H}\right)\dot{\sigma}_1 - \left(\frac{\nu}{E} - \frac{1}{4H}\right)\dot{\sigma}_2,$$

$$\dot{\varepsilon}_2 = \left(\frac{1}{E} + \frac{1}{4H}\right)\dot{\sigma}_2 - \left(\frac{\nu}{E} - \frac{1}{4H}\right)\dot{\sigma}_1.$$

Thus

$$\frac{1}{\bar{E}} = \frac{1}{E} + \frac{1}{4H} = \frac{1}{4E_t} + \frac{3}{4E}$$

and

$$\bar{\nu} = \frac{1 + 4\nu - E/E_t}{3 + E/E_t}.$$

It should be noticed that for small values of E_t/E, $\bar{\nu} \doteq -1 + 4(1+\nu)E_t/E$ and $\bar{E} \doteq 4E_t$, so that the factor $\bar{E}/(1-\bar{\nu}^2)$ in Equation (5.3.20) tends to a constant fraction, $\frac{1}{2}(1-\nu)$, of the elastic value. This result, which is similar to what occurs in other plate and shell buckling problems, is quite unreasonable when compared with the uniaxial case, and indeed with experimental data. Furthermore, the result is not limited to the Mises criterion but would also be produced by any isotropic yield criterion that is smooth at $\sigma_1 = \sigma_2 = \pm\sigma_Y$, since all such yield loci must be tangent there. We are left with the conclusion that incremental plasticity with a smooth yield surface may not be applicable to the analysis of multiaxial instability problems.

Considerable improvement is obtained when the deformation theory of plasticity discussed in 3.2.1 is used. The Hencky theory, in particular, is based on the Mises criterion, and gives the plastic strain as

$$\varepsilon_{ij}^p = \frac{3\bar{\varepsilon}^p}{2\bar{\sigma}} s_{ij}, \tag{5.3.21}$$

where

$$\bar{\sigma} = \sqrt{\frac{3}{2}\sigma_{ij}\sigma_{ij}}, \quad \bar{\varepsilon}^p = \sqrt{\frac{2}{3}\varepsilon_{ij}^p\varepsilon_{ij}^p}$$

are the equivalent stress and plastic strain, related to each other by the uniaxial relation. The incremental form of (5.3.21) is

$$d\varepsilon_{ij}^p = \frac{3}{2\bar{\sigma}}\left[\left(d\bar{\varepsilon}^p - \frac{\bar{\varepsilon}^p}{\bar{\sigma}}d\bar{\sigma}\right)s_{ij} + \bar{\varepsilon}^p\, ds_{ij}\right].$$

Upon introducing the uniaxial secant and tangent moduli E_s and E_t, defined by

$$\frac{1}{E_s} = \frac{1}{E} + \frac{\bar{\varepsilon}^p}{\bar{\sigma}}, \quad \frac{1}{E_t} = \frac{1}{E} + \frac{d\bar{\varepsilon}^p}{d\bar{\sigma}},$$

the incremental relation may be written as

$$d\varepsilon_{ij}^p = \frac{3}{2}\left[\left(\frac{1}{E_t} - \frac{1}{E_s}\right)\frac{d\bar{\sigma}}{\bar{\sigma}}s_{ij} + \left(\frac{1}{E_s} - \frac{1}{E}\right)ds_{ij}\right].$$

Applying the relation to the plane-stress case with $\sigma_1 = \sigma_2$ leads to the instantaneous modulus and contraction ratio

$$\frac{1}{\bar{E}} = \frac{1}{4E_t} + \frac{3}{4E_s}$$

and

$$\bar{\nu} = -\frac{E/E_t + 2(1-2\nu) - 3E/E_s}{E/E_t + 3E/E_s}.$$

For a gradually flattening uniaxial stress-strain curve, when $E_t \ll E_s \ll E$ we have $\bar{E} \doteq 4E_t$ as in the incremental theory, but $\bar{\nu} \doteq -(1 - 6E_t/E_s)$, and therefore $\bar{E}/(1 - \bar{\nu}^2) \doteq E_s/3$. To the first approximation, then, it is the secant modulus, rather than the elastic modulus, that governs the buckling, resulting in a much smaller critical load than that given by the incremental theory.

An explanation of the failure of incremental theory based on an isotropic yield criterion to predict a reasonable buckling load is due to Phillips [1972], who points out that the yield criteria of work-hardening materials become anisotropic almost immediately upon plastic loading.

Torsional Buckling of a Cruciform Column

If a column is made up of thin plate sections that do not form a closed tube and is sufficiently short, then under the action of a compressive axial load it will buckle by twisting rather than bending. Consider, for example, the cross-shaped column shown in Figure 5.3.3, and in particular the flange whose middle plane is the xy-plane with y positive. If the virtual twist angle per unit length of the cross-section at x is $\phi(x)$, then the virtual deflection at (x, y) is $\delta w(x, y) = y\phi(x)$. The axial load P may be assumed to be uniformly distributed with intensity $P/4b$ per unit width of flange. The second-order external potential energy on a fiber of width dy is therefore

$$d(\delta^2\Pi_{ext}) = -\frac{P}{4b}dy\, y^2 \int_0^L \phi'^2\, dx,$$

and, for the whole flange,

$$\delta^2\Pi_{ext} = \int_{y=0}^{y=b} d(\delta^2\Pi_{ext}) = -\frac{Pb^2}{12}\int_0^L \phi'^2\, dx.$$

The energy criterion may be expressed by adding this quantity to $\delta^2\Pi_{int}$ as given by (5.3.16) and equating the result to zero. Now

$$\delta w_{,11} = y\phi'', \quad \delta w_{,12} = \phi', \quad \delta w_{,22} = 0,$$

so that

$$\delta M_{\alpha\beta}\, \delta w_{,\alpha\beta} = \bar{D}[y^2\phi''^2 + (1-\bar{\nu})\phi'^2].$$

However, ϕ'' is of order ϕ'/L, and, if $L \gg b$, then the first term in brackets may be neglected in comparison with the second. Consequently, independently of the form taken by ϕ',

$$P_{cr} = \frac{12\bar{D}(1-\bar{\nu})}{b},$$

or, because of the definition of $\bar{D}$, we can define $\sigma_{cr} = P_{cr}/4bh$ as the solution of

$$\sigma = \bar{G}\left(\frac{h}{b}\right)^2,$$

where $\bar{G} = \bar{E}/2(1+\bar{\nu})$ is the instantaneous shear modulus, $d\tau/d\gamma$. Note that the result is independent of the column length L.

In any incremental theory with a yield criterion of the form (3.3.5) and an associated flow rule, when $\tau = 0$ the normal to the yield locus is directed

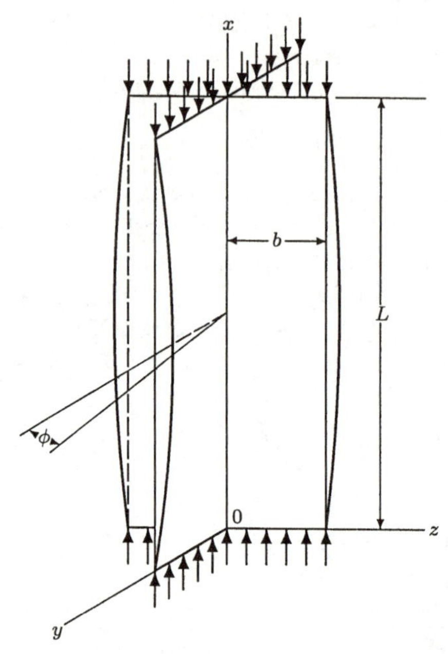

Figure 5.3.3. Cruciform (cross-shaped) column: initial and buckled geometry.

in the σ-direction, and therefore $d\gamma^p = 0$. Consequently $\bar{G} = G$, so that the buckling load is unaffected by plasticity — again an untenable result.

On the other hand, in the Hencky theory we have

$$d\gamma^p = 3\left(\frac{1}{E_s} - \frac{1}{E}\right) d\tau,$$

so that

$$\bar{G} = \frac{E_s}{3 + (1 - 2\nu)E_s/E}.$$

Once again, it is primarily the secant modulus that determines the critical load. Experiments by Gerard and Becker [1957] on aluminum columns show very good agreement with the prediction of deformation theory for $\nu = \frac{1}{2}$ (see Figure 5.3.4).

Shell Under External Pressure

The behavior of shells is in general much more complicated than that of plates. However, the buckling behavior of certain thin-walled shells may be studied by means of a simplified theory, known as the Donnell–Mushtari–Vlasov (DMV) theory, whose structure is essentially the same as that of plate theory (see Niordson [1985], Chapter 15). It must be kept in mind,

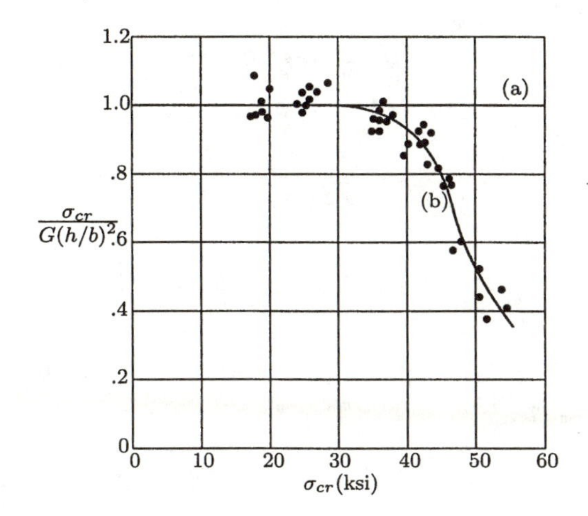

Figure 5.3.4. Torsional buckling of a cross-shaped column: predictions of (a) incremental and (b) deformation theories with $\nu = 1/2$. Dots represent experimental results (Gerard and Becker [1957]).

though, that shells are far more imperfection-sensitive than bars or plates; even the slightest imperfections can greatly reduce the buckling load.

For an elastic thin-walled cylindrical tube of mean radius R and thickness h, subject to an external pressure p, the critical pressure can be found from the Bresse formula for the ring by substituting D for EI, resulting in

$$p_{cr} = \frac{E}{4(1-\nu^2)}\left(\frac{h}{R}\right)^3.$$

The substitution is equivalent to assuming that the axial strain is zero. In other words the tube, when viewed axially, is in a state of plane strain as opposed to the plane-stress state of the ring. The axial stress is thus $\sigma_z = \nu\sigma_\theta$, where $\sigma_\theta = -pR/h$ is the circumferential stress, and the circumferential stress-strain relation is $\sigma_\theta = E\varepsilon_\theta/(1-\nu^2)$.

In an elastic–plastic tube governed by incremental theory, the plane-strain condition can only be enforced incrementally, that is, $d\sigma_z = \bar{\nu}\, d\sigma_\theta$. If buckling occurs after yielding, then the state of stress just before buckling is not known *a priori*, but must be determined by integrating the incremental relation, and checking at each step whether the buckling criterion is met.

The situation is somewhat simpler with deformation theory. The condition $\varepsilon_z = 0$ results in

$$\sigma_z = \frac{1}{2}\left[1 - (1-2\nu)\frac{E_s}{E}\right]\sigma_\theta.$$

The incremental relation is

$$d\sigma_z = \frac{1}{2}\left[1 - (1-2\nu)\frac{E_s}{E}\right] d\sigma_\theta - \frac{1-2\nu}{2E}\frac{dE_s}{d\bar{\sigma}}\, d\bar{\sigma}.$$

But

$$\frac{dE_s}{d\bar{\sigma}} = -\frac{E_s}{\bar{\sigma}}\left(\frac{E_s}{E_t} - 1\right)$$

and

$$d\bar{\sigma} = d\sqrt{\sigma_\theta^2 - \sigma_\theta\sigma_z + \sigma_z^2} = \frac{(2\sigma_\theta - \sigma_z)\, d\sigma_\theta + (2\sigma_z - \sigma_\theta)\, d\sigma_z}{2\bar{\sigma}}.$$

Thus both $d\bar{\sigma}$ and $d\sigma_z$ can be expressed as multiples of $d\sigma_\theta$, and the expressions can be substituted in

$$d\varepsilon_\theta = \frac{1}{E}(d\sigma_\theta - \nu\, d\sigma_z) + \frac{1}{2}\left(\frac{1}{E_s} - \frac{1}{E}\right)\frac{2\sigma_\theta - \sigma_z}{\bar{\sigma}}\, d\bar{\sigma},$$

resulting in

$$d\bar{\varepsilon}_\theta = \frac{1}{\tilde{E}}\, d\sigma_\theta,$$

where $\tilde{E}$ is in general a complicated function of σ_θ; it is this $\tilde{E}$ that replaces E in the Bresse formula. When $\nu = \frac{1}{2}$, however, $\tilde{E}$ turns out to be just $4E_t/3$, where E_t is the uniaxial tangent modulus. In this case the same result is given by incremental plasticity theory.

The problem of the *spherical* shell is easier, since the state of stress before buckling is $\sigma_\theta = \sigma_\phi = -\sigma$, where $\sigma = pR/2h$. The critical value of σ for elastic buckling is given by

$$\sigma_{cr} = \frac{E}{\sqrt{3(1-\nu^2)}}\frac{h}{R}.$$

The value for plastic buckling can be obtained by solving

$$\sigma = \frac{\bar{E}}{\sqrt{3(1-\bar{\nu}^2)}}\frac{h}{R},$$

where $\bar{E}$ and $\bar{\nu}$ are the same functions of σ as are used for the circular plate under radial force. Figure 5.3.5 shows the critical stress for the buckling of a spherical shell, based on both the incremental and deformation theories, using the Ramberg–Osgood formula with $\alpha = 0.1$ and $m = 6$. For comparison, the stress-strain curve is also shown.

Exercises: Section 5.3

1. Find the reduced modulus E_r in terms of the elastic modulus E and the tangent modulus E_t for (a) an ideal sandwich section and (b) for a thin-walled square tube section. Compare with the result for a rectangular section when $E_t/E = 0.1$ and when $E_t/E = 0.02$.

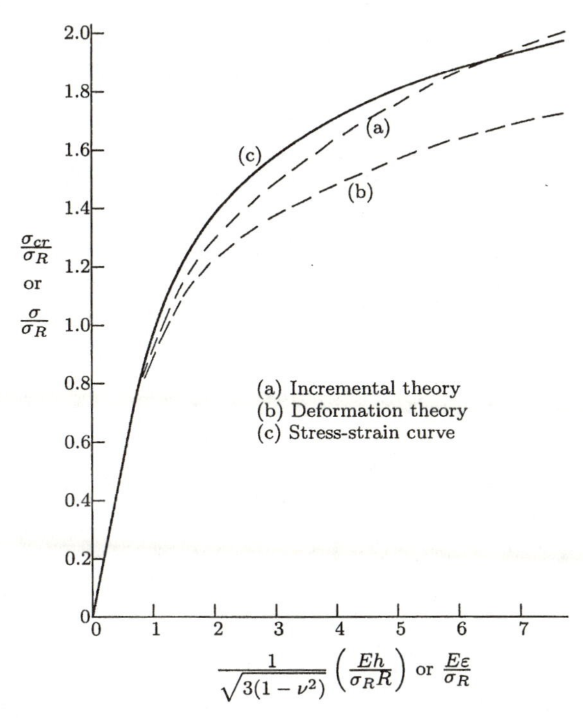

Figure 5.3.5. Buckling of a spherical shell under internal pressure: critical stress based on (a) incremental and (b) deformation theories, with (c) stress-strain curve included (from Hutchinson [1972]).

2. Plot column curves based on both the tangent-modulus and the reduced-modulus theories (for a rectangular section) for (a) the generalized Prager formula with $n = 1$, (b) the generalized Prager formula with $n = 4$, (c) the Ramberg–Osgood formula with $\alpha = 0.3$ and $m = 4$.

3. For a ring under radial pressure, plot suitably nondimensionalized buckling curves (q_{cr} against R/r) on the basis of (a) the Prager formula and (b) the Ramberg–Osgood formula with $\alpha = 0.2$ and $m = 6$.

4. Is it possible to find an instantaneous plate modulus $\bar{D}$ and a contraction ratio $\bar{\nu}$ for the buckling problem of a circular plate under a compressive radial load when the plate material obeys the incremental theory of plasticity with the Tresca criterion and its associated flow rule? Explain.

5. Perform the analysis leading to the results shown in Figure 5.3.5.

Chapter 6

Problems in Plastic Flow and Collapse II

Applications of Limit Analysis

Introduction

The theorems of limit analysis for standard elastic–perfectly plastic three-dimensional continua in arbitrary states of deformation were proved in 3.5.1. The proof, due to Drucker, Prager, and Greenberg [1952],[1] was the final link in a chain of development of the theory, which began with proofs for beams and frames by Gvozdev [1938], Horne [1950], and Greenberg and Prager [1951], followed by a proof for bodies in plane strain by Drucker, Greenberg, and Prager [1951].

In this chapter applications of the theorems are presented. Section 6.1 deals with plane problems in both plane strain and plane stress. Section 6.2 deals with beams under combined loading (including arches), Section 6.3 with trusses and frames, and Section 6.4 with plates and shells.

It is shown that as a rule, plausible velocity fields are easier to guess than stress fields, and therefore in many cases only upper-bound estimates are available. Of particular importance are velocity fields called *mechanisms*, in which deformation is concentrated at points, lines, or planes, with the remaining parts of the system moving as rigid bodies. The use of mechanisms for estimating collapse loads antedates the development of plasticity theory. Examples include Coulomb's method of slip planes for studying the collapse strength of soil, the plastic-hinge mechanism due to Kazinczy [1914] for steel frames, and the yield-line theory of Johansen [1932] for reinforced-concrete slabs, later extended to plates in general.

[1]But already given in the book by Prager and Hodge [1951].

Section 6.1 Limit Analysis of Plane Problems

As pointed out in Section 5.1, slip-line theory as a rule gives only upper bounds, unless an admissible extension of the stress field from the region covered by the slip-line field to the rest of the body is found. A convenient method for finding lower bounds is by means of discontinuous stress fields, some examples of which were studied in Section 5.1.

6.1.1. Blocks and Slabs with Grooves or Cutouts

Among the earliest applications of limit analysis to plane problems are studies of the effects of cutouts on the yielding of rectangular slabs or blocks subject to tension perpendicular to a side. In the absence of cutouts, the slab may be assumed to collapse under a uniform state of tensile stress $\sigma = \sigma_Y = 2k$, the Tresca criterion being assumed in plane stress. The cutouts may be expected to reduce the average or nominal tensile stress σ required for collapse to $2\rho k$ $(0 < \rho < 1)$, where ρ is called the *cutout factor*. Limit analysis may be used to find bounds on the cutout factor.

Tension of a Grooved Rectangular Block

The following example, illustrated in Figure 6.1.1, is discussed by Prager and Hodge [1951]. A statically admissible stress field is shown in Figure 6.1.1(a). The trapezoidal regions on either side of the line of symmetry OO correspond to the truncated wedge of Figure 5.1.6(e) (page 289), except that the stresses are tensile rather than compressive. The rectangular regions beyond the trapezoids are in simple tension with stress $\sigma = 2k(1 - \sin\gamma)$, where γ is the wedge semi-angle. In order to produce the best lower bound, γ should be as small as possible, subject to the geometric restriction discussed in connection with Figure 5.1.6(e) and the evident additional restriction that the flanks of the wedge be tangent to the cutout circles. It can be shown, the details being left to an exercise, that these restrictions require that $2\sin 2\gamma = 1 + \sin\gamma$. The smallest angle obeying this equation is 0.379 radian, so that $\sin\gamma = 0.370$, and the lower bound to the cutout factor is 0.630.

A simple kinematically admissible velocity field is shown in Figure 6.1.1(b). The portion above the line AB slides rigidly along this line with respect to the portion below. If the sliding speed is v, then the plastic dissipation per unit area along AB is kv, while the length of the line is $3a\sqrt{2}$. The external rate of work per unit thickness is $\sigma \cdot 4a \cdot v/\sqrt{2}$. Equating this rate to the total internal dissipation per unit thickness, $3\sqrt{2}kva$, yields the upper bound $\sigma = 3k/2$, or $\rho = 3/4$.

Figure 6.1.1(c) shows a slip-line field in which the region bounded by the slip lines AB and $A'B$ (as well as its mirror image) is in a state of

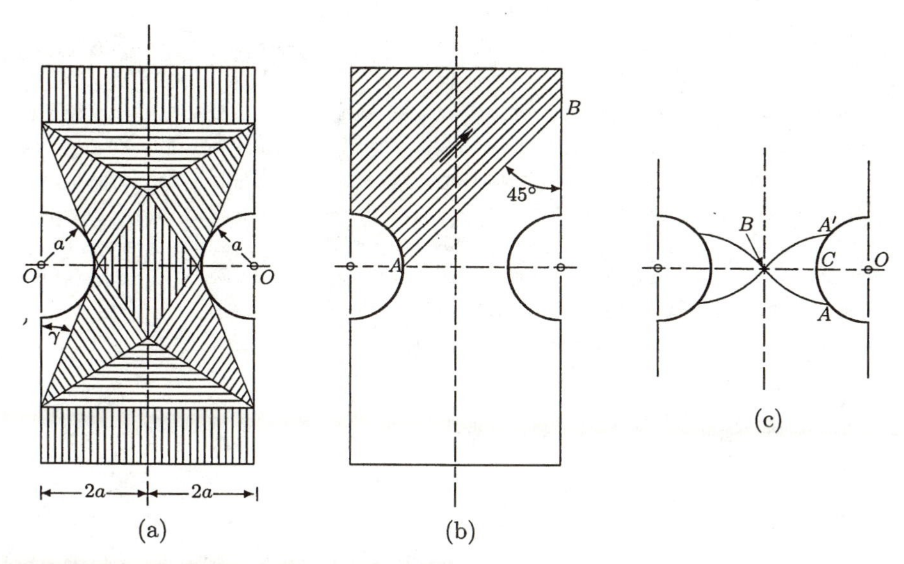

Figure 6.1.1. Tension of a grooved rectangular block: (a) statically admissible stress field; (b) kinematically admissible velocity field; (c) slip-line field.

axisymmetric stress as studied in Section 4.4. The stress σ_θ in this region is

$$\sigma_\theta = 2k\left(1 + \ln\frac{r}{a}\right),$$

where r is measured from the center O of the circle. The resultant axial force per unit thickness is therefore

$$F = 2\int_a^{2a} \sigma_\theta\, dr = 4k\int_a^{2a}\left(1 + \ln\frac{r}{a}\right) dr$$
$$= 8ka\ln 2.$$

The nominal stress is thus $\sigma = F/4a = 2k\ln 2$, corresponding to a cutout factor of $\ln 2 = 0.693$. Since the stress field has not been extended to the regions outside the slip-line field, this value cannot be a lower bound. On the other hand, the slip-line field implies a solution for the velocity field, which, together with the rigid axial motion of the regions outside it, forms a kinematically admissible velocity field for the body as a whole, and therefore the result gives an improved upper bound for the cutout factor. The bounds on the cutout factor are therefore

$$0.630 \le \rho \le 0.693.$$

Tension of a Square Slab With a Slit

A thin square slab with a narrow slit perpendicular to the direction of the load is shown in Figure 6.1.2(a). The slit width is assumed to be of the same order of magnitude as the slab thickness, both being much smaller than the sides of the square. A kinematically admissible velocity field may be based on a shearing failure mode as shown in Figure 6.1.2(b). If the shearing plane makes an angle α with the load direction then the total area of the surface of sliding is $(1-\beta)ah\csc\alpha$, where h is the slab thickness; and if the relative velocity of motion of the two parts of the slab is v, then the plastic dissipation per unit area is $kv\sec\alpha$. The total internal dissipation is therefore $2k(1-\beta)ahv\csc 2\alpha$. The external work rate is σahv, and therefore an upper bound to the cutout factor is $(1-\beta)\csc 2\alpha$. The best upper bound is obtained for $\alpha = \pi/4$, and gives $\rho = 1-\beta$.

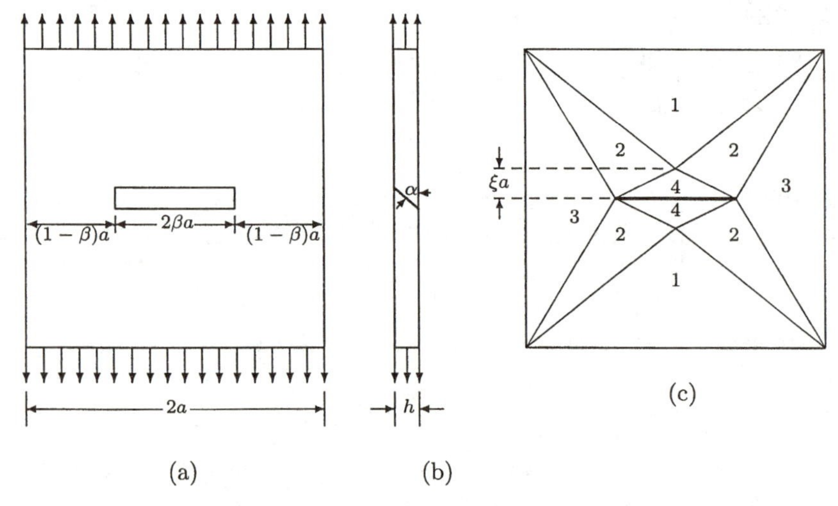

Figure 6.1.2. Tension of a square slab with a narrow slit: (a) geometry and loading; (b) kinematically admissible velocity field; (c) statically admissible stress field.

A discontinuous stress field for the determination of a lower bound was constructed by Hodge [1953] (see Hodge [1959], Section 12-2), and is shown in Figure 6.1.2(c). Since the problem is one of plane stress and not plane strain, it is not the Prager jump conditions (5.1.5) but the more general jump conditions (5.1.6), together with the yield criterion, that must be used. There being four distinct regions, the total number of unknown stress variables (n_i, r_i, θ_i) $(i = 1, 2, 3, 4)$ is 12. The equations for these variables are furnished by, first, two traction boundary conditions each on two sides of the square and on the face of the slit, and second, by two jump conditions each on the boundaries between regions 1 and 2, 2 and 3, and 2 and 4. In particular, the fact that τ_{xy} vanishes on the external boundaries of regions 1, 3 and 4 means that it vanishes throughout these regions. Regions 3 and 4 can reasonably be assumed to be in a state of simple tension and compression, respectively, so that $\theta_3 = \theta_4 = 3\pi/4$. In region 1, θ is either $3\pi/4$ or

$\pi/4$; it will be assumed that $\sigma_y \geq \sigma_x$ there, so that $\theta_1 = \pi/4$. The remaining boundary conditions accordingly give

$$n_1 - r_1 = \sigma, \quad n_3 - r_3 = 0, \quad n_4 + r_4 = 0,$$

so that n_1, n_3 and n_4 can be eliminated. The jump conditions are

$$n_2 + r_2 \sin 2(\theta_2 - \chi_{12}) = \sigma + r_1 \cos 2\chi_{12},$$

$$n_2 + r_2 \sin 2(\theta_2 - \chi_{23}) = r_3(1 - \cos 2\chi_{23}),$$

$$n_2 + r_2 \sin 2(\theta_2 - \chi_{24}) = -r_4(1 + \cos 2\chi_{24}),$$

$$r_2 \cos 2(\theta_2 - \chi_{12}) = r_1 \sin 2\chi_{12},$$

$$r_2 \cos 2(\theta_2 - \chi_{23}) = -r_3 \sin 2\chi_{23},$$

$$r_2 \cos 2(\theta_2 - \chi_{24}) = -r_4 \sin 2\chi_{24},$$

where the angles χ_{12}, χ_{23} and χ_{24}, giving the inclinations of the normals to the boundary lines with the x-axis, can be expressed in terms of β and ξ. Once the equations are solved, satisfaction of the yield inequality

$$2r + |n - r| + |n - r| \leq 4k$$

in each region produces four inequalities on the cutout factor $\rho = \sigma/2k$:

$$\rho\beta \leq 1 - \xi, \quad \rho \leq 1 - \beta, \quad \rho\beta \leq \xi,$$

$$\rho\sqrt{[\xi + \beta(1 - \beta)]^2 + 4\beta^2} \leq \beta + \xi(1 - \beta).$$

If the second inequality were satisfied as an equality, then the result $\beta = 1 - \rho$ would coincide with the upper bound. It turns out that if the equality is assumed, then the remaining inequalities are obeyed if $\xi = 1 - \beta + \beta^2$. In this problem, then, the exact cutout factor has been found. Other problems involving slabs with cutouts were treated by Hodge and various coworkers (see Hodge [1959], Chapter 12, for a survey).

6.1.2. Problems in Bending

Pure Bending of a Notched Bar

The notched bar shown in Figure 6.1.3(a) is of rectangular cross-section and subject to equal and opposite couples M applied at its ends. The discontinuous stress field shown is statically admissible, and plastically admissible in classical plane strain as well as for the Tresca criterion in plane stress. It is, of course, equivalent to the limiting stress distribution in a perfect beam, limited to the material below the notch, and gives a lower bound of $\frac{1}{2}kba^2$ for the ultimate moment, where b is the width of the beam.

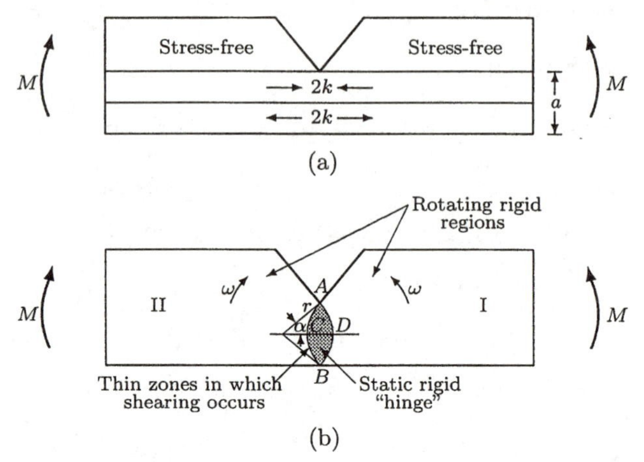

Figure 6.1.3. Pure bending of a notched bar: (a) statically admissible stress field; (b) mechanism.

An upper bound can be found on the basis of a mechanism, shown in Figure 6.1.3(b), in which the outer portions of the bar rotate rigidly by sliding along the arcs ACB and ADB, the inner region $ACBD$ remaining stationary; the arcs must accordingly be circular. The mechanism resembles the plastic hinge discussed in 4.4.2.

If the angular velocity of rotation is ω and the radius of the arcs is r, then the plastic dissipation per unit area of sliding surface is $kr\omega$. If the angle subtended by the arcs is 2α, then the total area of the sliding surfaces is $4br\alpha$. But $r = \frac{1}{2}a\csc\alpha$, so that the total internal dissipation is $k\omega ba^2\alpha\csc^2\alpha$. Equating this to the external work rate $2M\omega$ gives the upper bound $M = \frac{1}{2}kba^2\alpha\csc^2\alpha$. The smallest value of this occurs when $\tan\alpha = 2\alpha$, and gives the upper bound of $M = 0.69kba^2$. For a bar that is wide enough to be regarded as bending in plane strain, A. P. Green [1953] found a slip-line field that gives the improved upper bound of $M = 0.63kba^2$

End-Loaded Cantilever in Plane Strain and Plane Stress

Another problem studied by Green [1954a] by means of slip-line theory concerns the impending collapse of an end-loaded wide tapered cantilever, shown in Figure 6.1.4.[1] If the taper is not extreme and if the ratio L/h of the length to the least depth is sufficiently great, it is reasonable to suppose that the collapse mechanism is of the plastic-hinge type. The slip-line fields shown in Figures 6.1.4(a) and (b) produce deformations similar to those

[1]For the prismatic beam, the problem was also studied by Onat and Shield [1955].

corresponding to a hinge mechanism. In (a), the rigid portion to the right of the slip-line field rotates about point Y; in (b) it slides over the circular arc PYQ, as in the preceding problem, and $APYQB$ is a continuous slip line that is a locus of velocity discontinuity.

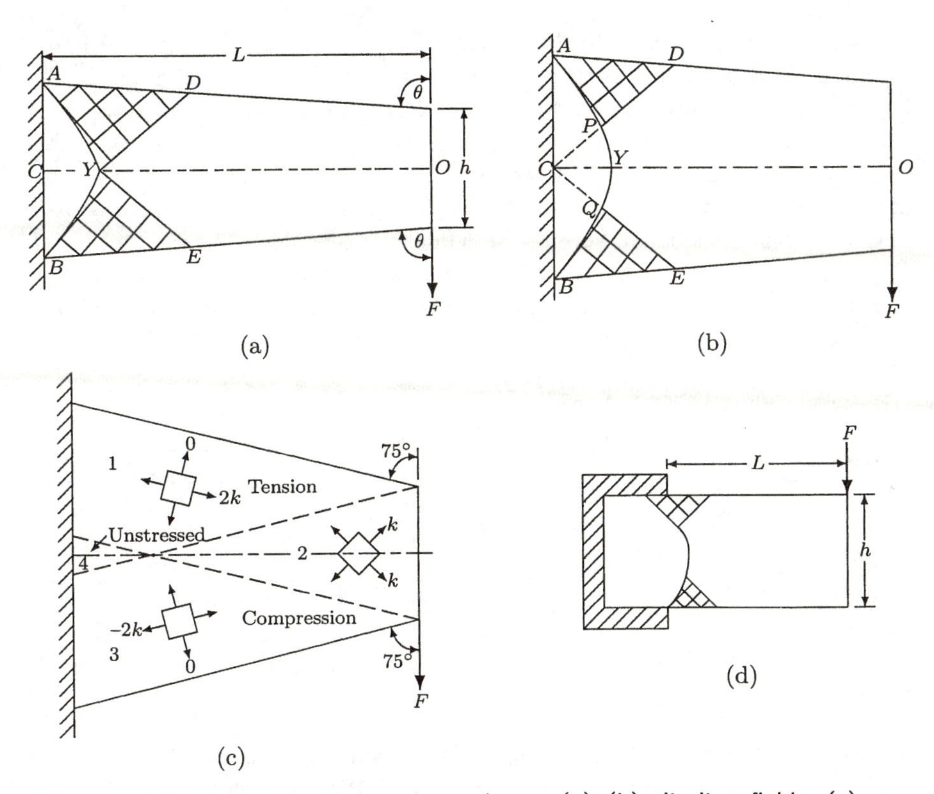

Figure 6.1.4. End-loaded tapered cantilever: (a)–(b) slip-line fields; (c) statically admissible stress field. (d) Weakly supported prismatic cantilever.

A complete solution is found, with the help of the discontinuous but statically and plastically admissible stress field shown in Figure 6.1.4(c), for the beam with 15° taper ($\theta = 75°$), in which case $F_U = kbh$ for all values of L/h. For greater values of θ, no extension of the stress field into the rigid regions is given and therefore the calculated value of F is an upper bound to F_U, though, as Green argues, it is likely to be very close, since the proposed slip-line field shows remarkable similarity with the plastically deformed region observed in experiments. For $\theta \geq 75°$ the load is given, in accordance with the slip-line field in (a), by

$$F = 2kbh \sin 2\theta \quad \text{for} \quad \frac{L}{h} \geq \frac{1}{2} \tan\theta(\sin 2\theta - \cos 2\theta).$$

For a given θ, as L/h is decreased below the limiting value, F increases.

For prismatic beams, Green [1954a,b] also constructed solutions for uniformly distributed loading, and for other boundary conditions, including "weakly" supported cantilevers (see Figure 6.1.4(d)) as well as beams fixed at both ends.

A much simpler velocity field for a prismatic beam is shown in Figure 6.1.5(a). In this picture region 1 undergoes simple shearing, regions 2 and 3 rotate rigidly, and regions 4 and 5 undergo tension or compression. In the last-named regions the vertical velocity does not vanish, so that the condition of no motion at the wall is not satisfied. A modification of the mechanism that satisfies this condition, shown in Figure 6.1.5(b), was proposed by Drucker [1956a]; here region 6 does not displace. The mechanism of Figure 6.1.5(a) is equivalent (with the direction of the force reversed) to that of Figure 6.1.5(c) for a center-loaded simply supported beam. Calculations are done here only for mechanism (a).

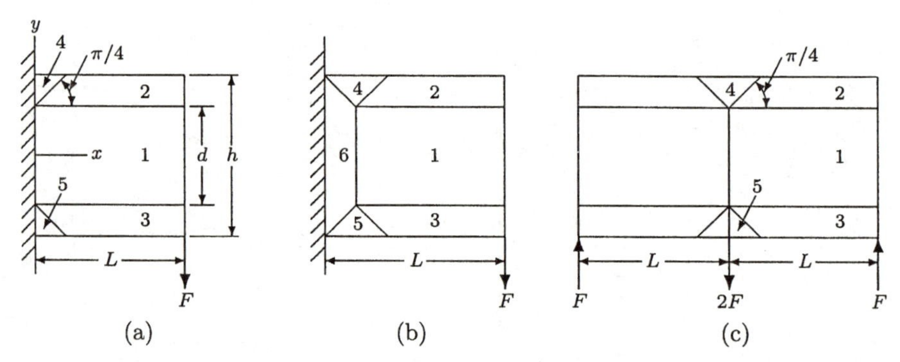

Figure 6.1.5. Velocity fields for a prismatic beam: (a) simple velocity field for a cantilever; (b) Drucker's modification; (c) center-loaded simply supported beam.

The velocity components v_x and v_y will be denoted $\dot{u}$ and $\dot{v}$, respectively, with the subscripts referring to the regions. The mechanism implies the following velocity field:

$$\dot{u}_1 = 0, \qquad \dot{v}_1 = \dot{\Delta}\left(1 - \frac{x}{L}\right),$$
$$\dot{u}_2 = \frac{\dot{\Delta}}{L}\left(y - \frac{d}{2}\right), \quad \dot{v}_2 = \dot{\Delta}\left(1 - \frac{x}{L}\right),$$
$$\dot{u}_4 = \dot{\Delta}\left(1 - \frac{x}{L}\right), \quad \dot{v}_4 = \frac{\dot{\Delta}}{L}\left(y - \frac{d}{2}\right).$$

The velocities in regions 3 and 5 are analogous to those in 2 and 4, respectively. Note that the velocity field is continuous. Note further that $\dot{\varepsilon}_x + \dot{\varepsilon}_y = 0$ in every region, so that the mechanism applies to plane strain as well as to plain stress. For either the Mises or the Tresca criterion, the dissipation $D_p(\dot{\varepsilon})$ equals $\tau_Y\dot{\Delta}/L$ in region 1 and $2\tau_Y\dot{\Delta}/L$ in regions 4 and 5. Equating the external work rate to the total internal dissipation,

$$F\dot{\Delta} = \tau_Y\frac{\dot{\Delta}}{L}\cdot bLd + 2\tau_Y\frac{\dot{\Delta}}{L}\cdot b\left(\frac{h-d}{2}\right)^2,$$

leads to the upper bound

$$F = \tau_Y bh\left[\frac{d}{h} + \frac{h}{2L}\left(1-\frac{d}{h}\right)^2\right]. \tag{6.1.1}$$

The upper bound can be optimized by minimizing with respect to d. The minimum occurs at $d = h - L$ and leads to $F = \tau_Y bh(1 - L/2h)$; but this result is valid only for $L < h$. For $L \geq h$, $d = 0$, that is, the mechanism is a plastic hinge consisting of two deforming triangles, and $F = \tau_Y bh^2/2L$ — a result that, in plane stress, agrees with that of elementary beam theory for the Tresca criterion.

End-Loaded Cantilever in Plane Stress: Lower Bound

Drucker [1956a] also constructed a lower bound for the problem by means of the following statically admissible stress field:

$$\sigma_x = \frac{F}{bh}\frac{x}{h}\frac{2\alpha^2}{1-\cos\alpha}\cos\left[\alpha\left(1-\frac{2|y|}{h}\right)\right]\operatorname{sgn} y,$$

$$\tau_{xy} = \frac{F}{bh}\frac{\alpha}{1-\cos\alpha}\sin\left[\alpha\left(1-\frac{2|y|}{h}\right)\right], \qquad \sigma_y = 0,$$

with $\alpha \leq \pi/2$. The yield criterion (3.3.5)[1] is met at $x = L$ if

$$\alpha = \frac{h\sigma_Y}{2L\tau_Y} \quad \text{and} \quad F = \tau_Y bh\frac{1-\cos\alpha}{\alpha}. \tag{6.1.2}$$

In accordance with the limitation on α, the result is limited to $h/L \leq \pi\tau_Y/\sigma_Y$. It can easily be ascertained that for small values of h/L, the result for F approaches $\sigma_Y bh^2/4L$ as in the beam-theory approach.

For $h/L > \pi\tau_Y/\sigma_Y$, the stress field is

$$\sigma_x = 0, \quad \sigma_y = 0, \quad \tau_{xy} = \tau_Y \quad \text{for} \quad |y| < \frac{d}{2},$$

and

$$\sigma_x = \sigma_Y\frac{x}{L}\cos\left[\frac{\pi}{2}\left(1-\frac{2|y|-d}{h-d}\right)\right]\operatorname{sgn} y, \qquad \sigma_y = 0,$$

[1] Drucker [1956a] considered a Tresca material only (i.e., $\sigma_Y = 2\tau_Y$). The extension to the more general yield criterion (3.3.5) is straightforward.

$$\tau_{xy} = \tau_Y \sin\left[\frac{\pi}{2}\left(1 - \frac{2|y| - d}{h - d}\right)\right] \quad \text{for } |y| > \frac{d}{2}.$$

The stress field is in equilibrium if $d = h - \pi\tau_Y L/\sigma_Y$. The load is then

$$F = \frac{2}{\pi}\tau_Y bh\left[1 + \frac{d}{h}\left(\frac{\pi}{2} - 1\right)\right] = \tau_Y bh\left[1 - (\pi - 2)\frac{\tau_Y L}{\sigma_Y h}\right]. \tag{6.1.3}$$

The upper and lower bounds are compared in Figure 6.1.6.

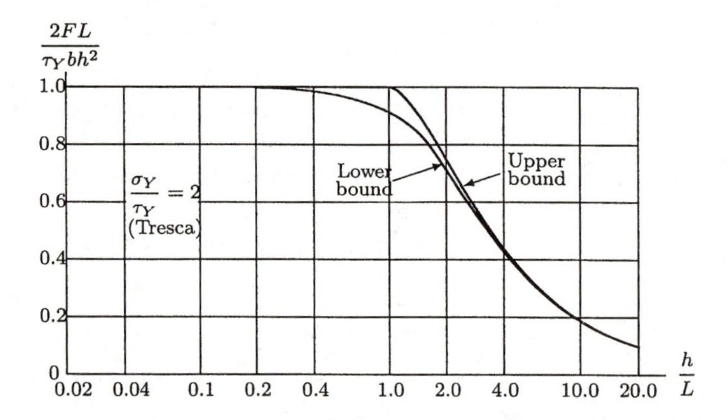

Figure 6.1.6. Upper and lower bounds for an end-loaded prismatic cantilever beam

I-Beams

Green [1954b] extended his results for prismatic beams to I-beams by assuming that the slip-line fields derived for rectangular beams in plane stress prevail in the web, while the flanges are in pure tension or compression. The shear force is thus carried entirely by the web, while the bending moment is the sum of that provided by the web and the couple formed by the flange forces.

The simple velocity fields shown in Figure 6.1.5 were proposed for short I-beams by Leth [1954] with d as the actual web depth. The upper-bound load based on the field in (a) or (c) is

$$F = \tau_Y\left(A_w + 2A_f\frac{h - d}{L}\right),$$

where $A_w = t_w d$ is the web area and $A_f = b_f(h - d)$ is the flange area, t_w and b_f being respectively the web thickness and flange width. Since A_f and A_w are typically of the same order of magnitude, the second term in parentheses may usually be neglected, and the upper-bound loaded may be approximated as $\tau_Y A_w$.

. For sufficiently long I-beams, however, it is reasonable to assume that failure is by a plastic-hinge mechanism, with $F \doteq M_U/L$ as given by elementary beam theory. For an I-beam,

$$M_U = \frac{\sigma_Y}{4}[A_w d + 2A_f(h+d)] \doteq \frac{\sigma_Y h}{4}(A_w + 4A_f),$$

the approximate expression being based on the assumption that $h - d \ll h$. Comparing the two upper bounds, then, we find that the plastic-hinge mechanism furnishes the lower one for

$$\frac{L}{h} \geq \frac{\sigma_Y}{4\tau_Y}\left(1 + 4\frac{A_f}{A_w}\right),$$

approximately, or about 5/2 for a beam made of Tresca material with $A_f = A_w$. Additional results in the limit analysis of beams, derived on the basis of *local* behavior, are discussed in Section 6.2 (see 6.2.4).

6.1.3. Problems in Soil Mechanics

Yield Criterion and Flow Rule

The most commonly used yield criterion for soils is the Mohr–Coulomb criterion discussed in 3.3.3,

$$\sigma_{max} - \sigma_{min} + (\sigma_{max} + \sigma_{min})\sin\phi = 2c\cos\phi,$$

where c is the cohesion and ϕ is the angle of internal friction. The Mohr–Coulomb criterion includes, as limiting cases, (1) the Tresca criterion (with $\phi = 0$ and $c = k$), used to describe, for example, saturated clays, and (2) the cohesionless friction model ($c = 0$) for cohesionless soils (dry sands and gravels).

If the material is taken as standard, then the flow rule at a regular point of the yield surface is

$$\dot{\varepsilon}^p_{max} = \dot{\lambda}(1 + \sin\phi), \quad \dot{\varepsilon}^p_{min} = -\dot{\lambda}(1 - \sin\phi), \quad \dot{\varepsilon}^p_{int} = 0.$$

The flow rule implies a constant *dilatancy ratio*, defined as $(\dot{\varepsilon}^p_1 + \dot{\varepsilon}^p_2 + \dot{\varepsilon}^p_3)/\dot{\gamma}^p_{max}$, which at a regular point is given by $(\dot{\varepsilon}^p_{max} + \dot{\varepsilon}^p_{min})/(\dot{\varepsilon}^p_{max} - \dot{\varepsilon}^p_{min})$ and is therefore equal to $\sin\phi$. The measured dilatancy of most soils (as well as rocks and concrete) is usually significantly less than this,[1] except in the case of undrained clays in which both internal friction and dilatancy are negligible. Most such materials, therefore, cannot be modeled as standard. In a nonstandard model, the flow rule may be taken in the same form as above, but with the dilatancy angle ψ replacing ϕ.

[1]The dilatancy ratio rarely exceeds 0.1, while the angle of internal friction may be as high as 45° in dense, well-graded soils with angular particles.

The plastic dissipation in a standard Mohr–Coulomb material was shown by Drucker [1953] to be

$$D_p(\dot{\varepsilon}^p) = c \cot \phi (\dot{\varepsilon}_1^p + \dot{\varepsilon}_2^p + \dot{\varepsilon}_3^p)$$

at any point of the yield surface, including the corners.

In plane plastic flow with $\dot{\varepsilon}_3 = 0$, as was pointed out in 3.3.4, σ_3 is the intermediate principal stress, even if $\psi \neq \phi$. The criterion therefore takes the form

$$|\sigma_1 - \sigma_2| + (\sigma_1 + \sigma_2) \sin \phi = 2c \cos \phi.$$

Mechanisms

A *Coulomb mechanism* in plane strain is one in which polygonal blocks of material move rigidly relative to one another. The interfaces between the blocks may be regarded as very narrow zones in which the strain rates are very large. In a nondilatant material, only shearing takes place in such a zone, so that the movement is one of sliding, and the interfaces are just slip lines, as discussed in Section 5.1. In the presence of dilatancy, dilatation as well as shearing takes place, and the movement involves separation in addition to sliding.[1] In fact, it can easily be shown (the details are left to an exercise) that the velocity discontinuity forms an angle equal to ψ with the interface – that is, if the magnitude of the discontinuity is v then the sliding speed is $v \cos \psi$ and the separation speed is $v \sin \psi$. If the thickness of the zone is h, then the average longitudinal strain rate perpendicular to the interface is $(v \sin \psi)/h$ and the average shearing rate is $(v \cos \psi)/h$, so that the principal strain rates are $\frac{1}{2}(\pm 1 + \sin \psi)v/h$. In the standard material, then, the plastic dissipation per unit area of interface is $cv \cos \phi$.

In another type of mechanism, introduced by Petterson and developed by Fellenius, the velocity discontinuities are along circular arcs, with the material inside an arc rotating rigidly about the center of the circle and thus sliding past the remaining material. This mechanism, known as a *slip circle*, is clearly appropriate only for a nondilatant (e.g., Tresca) material. Nevertheless, like the Coulomb mechanism with pure sliding, it is often used regardless of material properties. Both the location of the center of the circle and its radius can be chosen so as to minimize the upper-bound load predicted by the mechanism.

In a standard Mohr–Coulomb material, a velocity discontinuity such that the mass on one side rotates rigidly while that on the other side remains stationary takes the form of a logarithmic spiral rather than a circle. As can be seen from Figure 6.1.7, $dr/(r\, d\theta) = \tan \phi$, which can be integrated to give

[1] In practice, soil mechanicians often use the Coulomb mechanism with pure sliding regardless of the yield criterion or flow rule assumed. Also, the mechanism is usually analyzed by means of statics rather than kinematics. For the Tresca material, the results are equivalent.

$r = r_0 \exp[\tan\phi(\theta - \theta_0)]$. Clearly, for $\phi = 0$, the curve becomes the circle $r = r_0$.

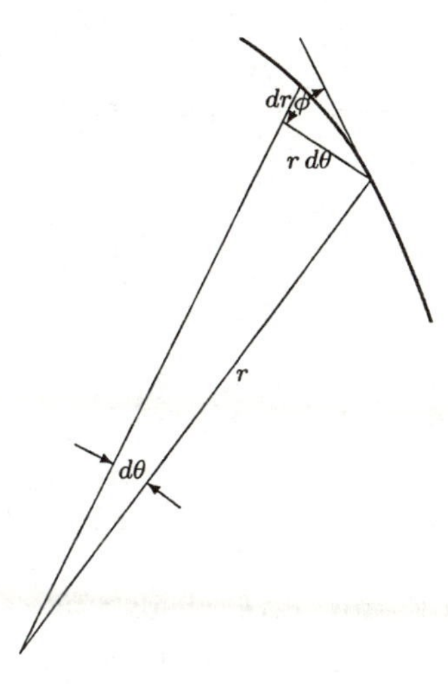

Figure 6.1.7. Velocity discontinuity in a standard Mohr–Coulomb material.

In accordance with Radenkovic's theorems (see 3.5.1), any upper bound found for a standard Mohr–Coulomb material is also an upper bound for a nonstandard material with the same yield criterion, while any lower bound for a standard Mohr–Coulomb material whose angle of internal friction equals the dilatancy angle of the nonstandard material is also a lower bound for the latter. The following examples illustrate the procedure. The examples are limited to soils that can be modeled as homogeneous, a condition rarely encountered in real soil masses. For an extensive survey of applications of limit analysis to soil mechanics, including numerical results, see Chen [1975].

Stability of a Vertical Bank

A vertical bank of height h occupies the half-strip $0 \le y \le h$, $x \ge 0$. We wish to determine the maximum height so that the bank does not collapse. Since the weight of the bank per unit horizontal area is wh, where w is the specific weight, we may regard h as the equivalent of a load, and denote its greatest safe value by h_U. A lower bound to h_U may be found by assuming, with Drucker and Prager [1952], the stress field $\sigma_x = \tau_{xy} = 0$, $\sigma_y = -w(h - y)$, which satisfies the equilibrium equations and leaves the vertical and horizontal surfaces of the bank traction-free. The greatest numerical value of σ_y, equal to wh, must not exceed the yield stress in uniaxial compression for the standard Mohr–Coulomb material with internal-friction angle ψ, namely

$$\sigma'_C = \frac{2c\cos\psi}{1-\sin\psi} = 2c\tan\left(\frac{\pi}{4}+\frac{\psi}{2}\right).$$

If necessary — that is, if the bank is not situated atop a hard stratum — the stress field may be extended into the half-space $y \le 0$ without violating the yield criterion as follows:

$$\sigma_x = \alpha wy, \quad \tau_{xy} = 0, \quad \sigma_y = \left\{ \begin{array}{ll} -w(h-y), & x > 0, \\ wy, & x < 0, \end{array} \right\}$$

where $\alpha = (1-\sin\psi)/(1+\sin\psi)$. The stress field thus contains admissible discontinuities that separate it into three zones.

For the standard Tresca material, the lower bound may be written as $2k/w$. An improvement to $2\sqrt{2}k/w$ is achieved by means of an admissible stress field consisting of seven zones, proposed by Heyman [1973], who also discusses incomplete stress fields leading to somewhat higher lower bounds. A numerical solution by Pastor [1976] yields a lower bound of $3.1k/w$.

However, the stress fields producing these improvements include tensile stresses in some regions, while the Drucker–Prager stress field does not. Indeed, for a material that cannot take tension it was shown by Drucker [1953], on the basis of a mechanism including a *tension crack*, that $2k/w$ is an upper bound as well.

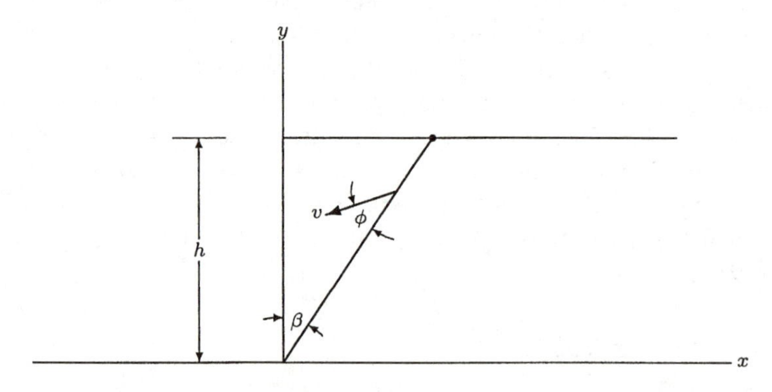

Figure 6.1.8. Coulomb mechanism in a vertical bank.

An upper bound for a Mohr–Coulomb material may be found with the help of the Coulomb mechanism shown in Figure 6.1.8, in which a wedge of angle β separates from the remainder of the bank. The weight (per unit thickness perpendicular to the page) of the wedge is $W = \frac{1}{2}wh^2\tan\beta$, and if the magnitude of the velocity of the wedge is v, then its downward component is $v\cos(\beta+\phi)$. Equating the external work rate to the internal dissipation,

$$\frac{1}{2}wh^2\tan\beta\cos(\beta+\phi)v = chv\cos\phi\sec\beta, \tag{6.1.4}$$

produces the upper bound

$$h = \frac{2c\cos\phi}{w\sin\beta\cos(\beta+\phi)}.$$

The least upper bound is obtained by maximizing the denominator with respect to β:

$$\frac{d}{d\beta}[\sin\beta\cos(\beta+\phi)] = \cos\beta\cos(\beta+\phi) - \sin\beta\sin(\beta+\phi) = 0,$$

or

$$\tan(\beta+\phi) = \cot\beta = \tan\left(\frac{\pi}{2}-\beta\right).$$

The equation is obeyed if $\beta = \frac{1}{4}\pi - \frac{1}{2}\phi$. But

$$\sin\left(\frac{\pi}{4}-\frac{\phi}{2}\right)\cos\left(\frac{\pi}{4}+\frac{\phi}{2}\right) = \frac{1}{2}(1-\sin\phi),$$

and therefore $h_U \le 2\sigma_C/w$ ($= 4k/w$ for a Tresca material). We are thus left with the wide limits,

$$\frac{2c}{w}\tan\left(\frac{\pi}{4}+\frac{\psi}{2}\right) \le h_U \le \frac{4c}{w}\tan\left(\frac{\pi}{4}+\frac{\phi}{2}\right).$$

The mechanism based on the logarithmic spiral produces a slight improvement over that of Coulomb: the optimal spiral reduces the factor in the upper bound from 4 to 3.83.

Stability of a Simple Slope

The improvement in the upper bound obtained by using a curved rather than a planar slip surface is significantly greater for inclined banks, known as *simple slopes*. It is conventional in soil mechanics to define the *stability factor*[1] $N_s = wh_U/c$, a function of the slope α as well as of the internal-friction angle ϕ. The optimal Coulomb wedge gives

$$N_s = \frac{4\sin\alpha\cos\phi}{1-\cos(\alpha-\phi)}.$$

It was shown by Taylor [1937] that limit-equilibrium calculations based on the logarithmic spiral are so close to those based on the slip circle as to be undistinguishable. Chen [1975] showed further that there is no significant difference between the results of the limit-equilibrium method and upper-bound limit analysis. Some examples are shown in Table 6.1.1.

Very little work has to date been done on lower bounds in the limit analysis of slope stability.

---

[1]Some authorities, including D. W. Taylor [1948], use the *stability number* $m = 1/N_s$ instead.

Table 6.1.1. Stability Factors of Homogeneous Simple Slopes

α	ϕ	Failure Surface	
		Plane	Curved
90°	0°	4.00	3.83
	25°	6.28	6.03
75°	0°	5.21	4.56
	25°	9.80	6.03
60°	0°	6.93	5.24
	25°	17.36	12.74

Thrust on Retaining Walls

If a vertical bank of soil is too high for stability — and in a cohesionless soil, any height is too great — then it must be held back by a retaining wall. The soil and the wall exert on each other a mutual thrust, equal to the resultant of the horizontal stress σ_x in the soil. Failure of the soil may occur when it is in one of two states, *active* and *passive*.

In a passive failure, the wall moves into the soil, increasing the horizontal pressure until the yield criterion is reached in the soil. The thrust P is thus an increasing load on the soil mass, doing positive work, and its limiting value, known as the *passive thrust* and denoted P_p, is an ultimate load in the usual sense.

In an active failure, the wall is pushed outward as a result of the horizontal pressure, and this pressure is reduced until the soil yields. In the process the thrust decreases and does negative work. If the limiting thrust (the *active thrust*) is denoted P_a, then the upper and lower bounds are bounds on $-P_a$, and the usual nomenclature of limit analysis must be reversed: the upper-bound theorem gives a lower bound to P_a, and the lower-bound theorem gives an upper bound.

The static analysis is due to Rankine. The geometry is taken as the same as in the preceding problem, Figure 6.1.8, and the wall is assumed smooth. Equilibrium is satisfied if $\sigma_y = -w(h - y)$, $\tau_{xy} = 0$, and σ_x depends on y only. The Mohr–Coulomb criterion is assumed to be met everywhere, that is,

$$|\sigma_x - \sigma_y| + \sin\phi\,(\sigma_x + \sigma_y) = 2c\cos\phi.$$

Two solutions exist for σ_x:

$$\begin{aligned}\sigma_x &= \sigma_y \frac{1 - \sin\phi}{1 + \sin\phi} - 2c\frac{\cos\phi}{1 + \sin\phi} \\ &= \sigma_y \tan^2\left(\frac{\pi}{4} - \frac{\phi}{2}\right) - 2c\tan\left(\frac{\pi}{4} - \frac{\phi}{2}\right),\end{aligned}$$

representing active failure, and

$$\sigma_x = \sigma_y \frac{1+\sin\phi}{1-\sin\phi} + 2c\frac{\cos\phi}{1-\sin\phi}$$
$$= \sigma_y \tan^2\left(\frac{\pi}{4}+\frac{\phi}{2}\right) + 2c\tan\left(\frac{\pi}{4}+\frac{\phi}{2}\right),$$

representing passive failure. Introducing $\sigma_y = -w(h-y)$ and integrating over $0 \leq\leq h$ gives the limiting thrusts,

$$P_a = \frac{1}{2}wh^2\tan^2\left(\frac{\pi}{4}-\frac{\phi}{2}\right) - 2ch\tan\left(\frac{\pi}{4}-\frac{\phi}{2}\right)$$

and

$$P_p = \frac{1}{2}wh^2\tan^2\left(\frac{\pi}{4}+\frac{\phi}{2}\right) + 2ch\tan\left(\frac{\pi}{4}+\frac{\phi}{2}\right).$$

The preceding formulas, known as **Rankine's formulas**, are widely used in soil mechanics. In view of Radenkovic's second theorem, however, it must be recognized that they are not true lower bounds (in the usual sense for P_p, in the reverse sense for P_a) unless the friction angle ϕ is replaced by the dilatancy angle ψ. Upper bounds can be obtained by means of the Coulomb mechanism, following Coulomb's own analysis of 1776, but making sure that a kinematic approach with an associated flow rule is taken; Coulomb assumed pure sliding, and analyzed the wedge statically.

For active failure, the mechanism is the same as for the free-standing vertical bank. With the wall again assumed to be smooth, the external rate of work is given by the left-hand side of (6.1.4) with the additional term $-Pv\sin(\beta+\phi)$, and the internal dissipation equals the right-hand side. Consequently,

$$P = \frac{1}{2}wh^2\tan\beta\cot(\beta+\phi) - ch\cos\phi\sec\beta\csc(\beta+\phi).$$

Both terms on the right-hand side can be shown to be stationary at $\beta = \frac{1}{4}\pi - \frac{1}{2}\phi$, so that the largest P (corresponding to the smallest upper bound on $-P$) is given by

$$P_a = \frac{1}{2}wh^2\tan^2\left(\frac{\pi}{4}-\frac{\phi}{2}\right) - ch\cos\phi\sec^2\left(\frac{\pi}{4}-\frac{\phi}{2}\right).$$

But

$$\cos\phi = \sin\left(\frac{\pi}{2}-\phi\right) = \sin 2\left(\frac{\pi}{4}-\frac{\phi}{2}\right) = 2\sin\left(\frac{\pi}{4}-\frac{\phi}{2}\right)\cos\left(\frac{\pi}{4}-\frac{\phi}{2}\right),$$

so that

$$\cos\phi\sec^2\left(\frac{\pi}{4}-\frac{\phi}{2}\right) = 2\tan\left(\frac{\pi}{4}-\frac{\phi}{2}\right),$$

and the Rankine formula for the active thrust is recovered.

An analogous result is obtained for the passive thrust. Here the wedge moves upward and to the right, the velocity forming an angle $\beta - \phi$ with the vertical (details are left to an exercise). We thus see that the Rankine formulas give the correct limiting thrusts on a smooth wall for the standard Mohr–Coulomb material. For the nonstandard material, they furnish the Radenkovic bounds.

The kinematic approach may be extended to obtain upper bounds in the presence of friction between the soil and the wall, by adding the term $\mu P|v_y|$ to the internal dissipation, μ being an average coefficient of friction and v_y the vertical component of velocity; it is assumed that the wall moves horizontally. While no analytical solutions exist, the wedge angle β giving the lowest upper bound can easily be found numerically for given μ, ϕ, and wh/c.

Exercises: Section 6.1

1. Find the best value of the wedge angle γ for the stress field of Figure 6.1.1(a).

2. Assume that the velocity discontinuity in Figure 6.1.1(b) is inclined at an arbitrary angle α. Find the upper bound to the cutout factor, and show that $\alpha = 45°$ gives the best upper bound.

3. Show that for a rectangular slab of sides $2a$ and $2b$ ($b > a$) with a slit of length $2\beta a$ parallel to the shorter side, the cutout factor for simple tension perpendicular to the slit is still $1 - \beta$.

4. Find lower and upper bounds on the cutout factor for the slab in Exercise 1 when $b < a$.

5. Find the value of the load F corresponding to the slip-line fields of Figure 6.1.4(a) for $\theta \geq 75°$.

6. Determine the velocity fields corresponding to the mechanisms of Figure 6.1.5(b) and (c) and the corresponding upper bounds to the collapse load F.

7. Derive Equation (6.1.2) for beams with $h/L \leq \pi\tau_Y/\sigma_Y$.

8. Derive Equation (6.1.3) for beams with $h/L \geq \pi\tau_Y/\sigma_Y$.

9. Find an upper bound to the critical height h of the bank of Figure 6.1.8 when the straight velocity-discontinuity line is replaced by a logarithmic spiral (Figure 6.1.7).

10. Derive the equations governing the upper bound to the passive thrust and the lower bound to the active thrust on a retaining wall in the presence of friction between the soil and the wall.

11. Assuming plane strain, use a stress field like that of Figure 5.1.5(e) to find a lower bound to the ultimate tensile force carried by the symmetrically notched tension specimen of Exercise 9, Section 5.1. Compare with the result of that exercise.

Section 6.2 Beams Under Combined Stresses

6.2.1. Generalized Stress

Introduction

A concept of great usefulness in the limit analysis of beams, arches, frames, plates and shells was introduced by Prager [1955b, 1956b, 1959]. It is that of *generalized stress and strain.*

Consider the ideal sandwich beam as shown in Figure 3.5.1(a) (page 155), but subject to distributed loading so that the axial force P and bending moment M vary along its length. While M and P can no longer be regarded as generalized loads, they may be regarded as generalized stresses in the following sense. At any section of the beam, the stresses in the flanges are $\sigma = P/2A \pm M/Ah$. The *local* values of the elongation and rotation obey the relations

$$\frac{d\Delta}{dx} = \varepsilon, \quad \frac{d\theta}{dx} = \kappa,$$

where ε is the mean longitudinal strain and κ is the curvature. The strains in the flanges are $\varepsilon \pm h\kappa/2$, and the internal virtual work can easily be shown to be

$$\overline{\delta W}_{int} = \int_0^L (P\,\delta\varepsilon + M\delta\kappa)dx,$$

the span of the beam being $0 < x < L$. The local axial force P and bending moment M may now be regarded as the *generalized stresses*, with ε and κ, respectively, as their conjugate *generalized strains*.

Figure 3.5.1(a), in addition to representing the limit-load locus for the beam under external axial force and moment, thus also represents the *yield locus* for the ideal sandwich beam. Such yield loci are also called interaction diagrams; an example was already studied in 4.4.1 (see Figure 4.4.5(a), page 224).

The ideal sandwich beam is statically determinate in the sense of 4.1.1, since it has no range of contained plastic deformation: if the material is perfectly plastic, then the beam can undergo unlimited deformation as soon as either flange yields. In any real beam, as we already know, the ultimate moment M_U is greater than the elastic-limit moment M_E, and therefore two distinct yield loci exist: the elastic-limit (or initial yield) locus and the ultimate yield locus [again, see Figure 4.4.5(a)]. Under the hypothesis of rigid–plastic behavior, however, only the ultimate yield locus is relevant.

Generalized Stress and Strain: Definitions

Generalized stresses may coincide with the actual stresses, or they may be local stress resultants integrated over one or (as in the present example) two dimensions, or even over a whole finite element of the body (such as a bar in a truss). If the generalized stresses are denoted Q_j ($j = 1, ..., n$), then the conjugate generalized strains q_j are in general defined by

$$\overline{\delta W}_{int} = \int_\Omega \sum_{j=1}^{n} Q_j \, \delta q_j \, d\Omega,$$

where $\int_\Omega (\cdot) \, d\Omega$ describes integration over the entire body with respect to volume, area, or length, as appropriate, or summation over all finite elements.[1]

Let $\mathbf{Q}$ and $\mathbf{q}$ denote the generalized stress and strain vectors, respectively. As illustrated by the ideal sandwich beam, a yield locus in terms of generalized stresses, say $\Phi(\mathbf{Q}) = 0$, may be derived in exactly the same way as the limit locus in terms of generalized loads was derived in 3.5.1. For rigid–plastic materials, the generalized plastic dissipation is thus

$$\bar{D}_p = \mathbf{Q} \cdot \dot{\mathbf{q}}, \tag{6.2.1}$$

and the principle of maximum plastic dissipation may be written as

$$(\mathbf{Q} - \mathbf{Q}^*) \cdot \dot{\mathbf{q}} \geq 0$$

or

$$\bar{D}_p(\dot{\mathbf{q}}) \geq \mathbf{Q}^* \cdot \dot{\mathbf{q}} \tag{6.2.2}$$

for any $\mathbf{Q}^*$ such that $\Phi(\mathbf{Q}^*) \leq 0$.

Finally, the theorems of limit analysis may be restated as follows.

Lower-Bound Theorem. A load point $\mathbf{P}$ is on or inside the limit locus if a generalized stress field $\mathbf{Q}^*$ can be found that is in equilibrium with $\mathbf{P}$ and obeys $\Phi(\mathbf{Q}^*) \leq 0$ everywhere.

Upper-Bound Theorem. A load point $\mathbf{P}$ is on or outside the limit locus if a kinematically admissible velocity field, yielding the generalized velocity

[1] In technical mathematical language, $d\Omega$ is a *measure* in a space of three, two, one or zero dimensions.

vector $\dot{\mathbf{p}}^*$ conjugate to $\mathbf{P}$ and the generalized strain-rate field $\dot{\mathbf{q}}^*$, can be found so that

$$\mathbf{P} \cdot \dot{\mathbf{p}}^* = \int_\Omega \bar{D}_p(\dot{\mathbf{q}}^*)\, d\Omega. \tag{6.2.3}$$

Elastic and Plastic Generalized Strain

When it is desired to describe elastic–plastic behavior in terms of generalized stress and strain, then it is necessary to decompose the generalized strain into elastic and plastic parts:

$$\mathbf{q} = \mathbf{q}^e + \mathbf{q}^p.$$

But with the exception of some simple cases, there is in general no one-to-one correspondence between $\mathbf{q}^e$ and $\boldsymbol{\varepsilon}^e$ or between $\mathbf{q}^p$ and $\boldsymbol{\varepsilon}^p$. Consider, for example, a real (as distinct from ideal) beam subject to symmetric bending only; the moment M is the only generalized stress, and the curvature κ is the only generalized strain. The actual strain at a point is given by

$$\varepsilon = -\kappa y.$$

In the elastic range, the moment-curvature relation is $M = EI\kappa$, and therefore the elastic part of the curvature is

$$\kappa^e = \frac{M}{EI}.$$

The plastic strain is

$$\begin{aligned} \varepsilon^p = \varepsilon - \varepsilon^e &= -(\kappa^e + \kappa^p)y - \frac{\sigma}{E} \\ &= -\kappa^p y - \frac{1}{E}\left(\sigma + \frac{My}{I}\right). \end{aligned}$$

Thus, while

$$\kappa^p = -\frac{1}{I}\int_A y\varepsilon^p\, dA,$$

there is no inverse relation by which κ^p determines ε^p. The quantity $\sigma + My/I$ does not vanish in the range of contained plastic deformation, and neither does its time derivative. Consequently,

$$\int_A \sigma\dot{\varepsilon}^p\, dA \neq M\dot{\kappa}^p,$$

except when $\dot{\sigma} = 0$, a condition that implies that $\dot{M} = 0$ and hence $\dot{\kappa}^e = 0$, and therefore holds only on the ultimate yield locus.

It follows that the principle of maximum plastic dissipation in terms of generalized stress and generalized *plastic* strain is in general valid only under unrestricted plastic flow, with $\dot{\mathbf{q}}^p = \dot{\mathbf{q}}$. The exceptional cases are those in which no contained plastic deformation occurs *locally*, as at a section of an ideal beam, or in a bar carrying axial force only (truss member).

6.2.2. Extension and Bending

Introduction

The theory of symmetric pure bending of elastic–plastic beams, discussed in 4.4.1, can easily be extended to beams subject to an axial force P in addition to the bending moment M, provided the deflection is so small that the additional bending moment resulting from the axial force acting over the deflection (the so-called **P-Δ effect**) is negligible. With this proviso satisfied, and with the uniaxial stress-strain relation given by $\sigma = f(\varepsilon)$, the first of Equations (4.4.7) needs only to have its right-hand side changed from 0 to P.

As seen in 4.4.1, the problem of asymmetric bending is difficult even in the absence of axial force. An analysis of rectangular beams subject to axial force and bending moments about both axes was carried out by Shakir-Khalil and Tadros [1973]. In the present subsection, only symmetric bending is considered.

Consider a beam whose centroidal fiber coincides with the z-axis, whose cross-section is symmetric about the y-axis, and which is made of an elastic–perfectly plastic material. In bending in the yz-plane, the strain is given as by Equation (4.4.2) with $\alpha = 0$ (but with the axes renamed): $\varepsilon = -\kappa(y - y_0)$, so that $\kappa y_0 = \varepsilon_0$ is the strain of the centroidal fibers or, equivalently, the average strain over the cross-section.

In the elastic range, this strain is elastically related to the average stress P/A, that is, $\kappa y_0 = P/AE$. Since $\kappa = M/EI$ regardless of the value of P, it follows that the coordinate y_0 of the neutral axis is given by $y_0 = PI/MA$, and the stress distribution is

$$\sigma = \frac{P}{A} - \frac{My}{I}.$$

Let the y-coordinates of of the extreme bottom and top fibers be $y = -c_1$ and $y = c_2$, respectively. The magnitudes of the extreme values of the stresses are

$$\sigma_1 = \left|\frac{P}{A} + \frac{Mc_1}{I}\right|, \quad \sigma_2 = \left|\frac{P}{A} - \frac{Mc_2}{I}\right|,$$

and the elastic limit corresponds to $\max(\sigma_1, \sigma_2) = \sigma_Y$, that is,

$$\max\left(\left|\frac{P}{P_U} + \frac{M}{M_{E1}}\right|, \left|\frac{M}{M_{E2}} - \frac{P}{P_U}\right|\right) = 1,$$

where

$$P_U = \sigma_Y A, \qquad M_{E1} = \sigma_Y \frac{I}{c_1}, \qquad M_{E2} = \sigma_Y \frac{I}{c_2}.$$

If the section is doubly symmetric, then $c_1 = c_2 = h/2$ and $M_{E1} = M_{E2} = M_E = 2\sigma_Y I/h$. The elastic-limit locus is then simply $|P/P_U| +$

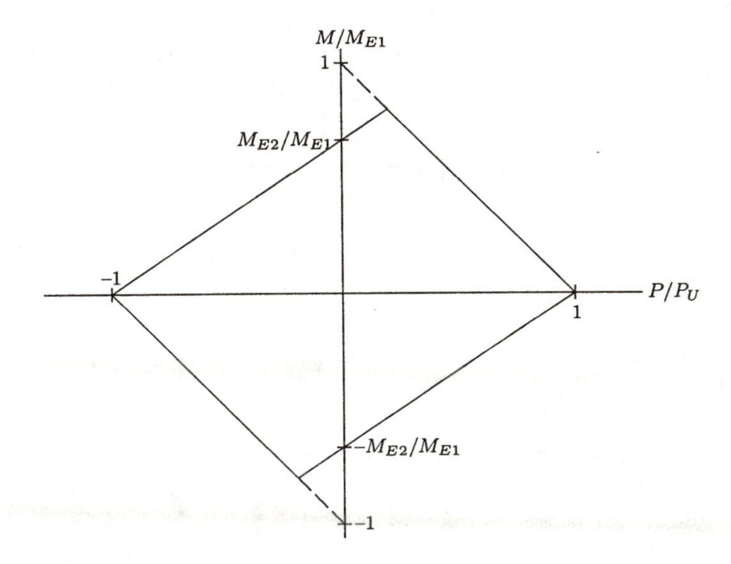

Figure 6.2.1. Elastic-limit locus for an asymmetric beam under combined bending and extension.

$|M/M_E| = 1$, that is, it has exactly the same form as the yield locus of the ideal sandwich beam [Figure 3.5.1(b)] but with the elastic-limit moment M_E replacing the ultimate moment M_U.[1] For a section without double symmetry, the elastic-limit locus is as shown in Figure 6.2.1 if $c_2 > c_1$ (so that $M_{E1} > M_{E2}$).

Note that for a certain ratio of P to M, namely $P/M = (c_2 - c_1)A/2I$, yielding occurs simultaneously at the top and bottom fibers, and two plastic zones form as the generalized stress (M, P) moves outside the elastic-limit locus. Otherwise only one plastic zone forms at first. As the point (M, P) moves farther from the origin, the stress at the extreme fiber opposite the plastic zone increases until it, too, reaches the yield-stress value, creating a second plastic zone. Further loading results in the shrinking of the elastic core, exactly as in pure bending, until it becomes negligibly thin, indicating that the ultimate yield locus has been reached.

Yield Locus for Symmetric Bending

In order to determine the ultimate yield locus, let us assume for the sake of definiteness that $M > 0$, so that the bottom fibers are in tension, with $\sigma = \sigma_Y$, and the top fibers are in compression, with $\sigma = -\sigma_Y$. If y_0 is the

[1]For the ideal sandwich beam, of course, $M_E = M_U$.

y-coordinate of the vanishing elastic core, then

$$M = -\sigma_Y \left[\int_{y=-c_1}^{y=y_0} y\, dA - \int_{y=y_0}^{y=c_2} y\, dA \right] = 2\sigma_Y \int_{y=y_0}^{y=c_2} y\, dA \qquad (6.2.4)$$

and

$$P = \sigma_Y \left[\int_{y=-c_1}^{y=y_0} dA - \int_{y=y_0}^{y=c_2} dA \right] \qquad (6.2.5)$$

since $\int_{y=-c_1}^{y=c_2} y\, dA = 0$. For a given cross-section, the integrals in Equations (6.2.4)–(6.2.5) can be evaluated in terms of y_0, so that these equations furnish a parametric representation of the ultimate yield locus in the MP-plane in terms of the parameter y_0, the range of y_0 being $-c_1 \le y_0 \le c_2$.

Associated Flow Rule

The principle of maximum plastic dissipation holds in states of uniaxial stress regardless of the yield criterion and flow rule obeyed by the material under multiaxial stress. Consequently the generalized strain rate ($\dot{\varepsilon}_0$, $\dot{\kappa}$) must be perpendicular to the yield locus in the MP-plane, that is,

$$\dot{\varepsilon}_0\, dP + \dot{\theta}\, dM = 0.$$

Now consider Equations (6.2.4)–(6.2.5). If $b(y)$ denotes the width of the beam at the level y, then $dA = b(y)dy$, and therefore

$$dM = -2\sigma_Y b(y_0) y_0 dy_0, \qquad dP = 2\sigma_Y b(y_0) dy_0,$$

so that normality is equivalent to $\dot{\varepsilon}_0 - y_0\dot{\kappa} = 0$ — precisely the definition of $\dot{\varepsilon}_0$.

Examples of Yield Loci

In order to compare the behavior of different cross-sections, it is convenient to describe the yield locus in terms of the dimensionless generalized stresses $m = M/M_U$ and $p = P/P_U$, where $P_U = \sigma_Y A$ (A being the total cross-sectional area) and M_U is given by Equation (4.4.11). The parameter can also be made dimensionless by defining, for example, $\eta = 2y_0/h$, where $h = c_1 + c_2$ is the depth of the beam; the range of η is thus $-\eta_1 \le \eta \le \eta_2$, where $\eta_i = 2c_i/h$, $i = 1$, 2. Equations (6.2.4)–(6.2.5) can therefore be written symbolically as

$$m = \bar{m}(\eta), \qquad p = \bar{p}(\eta). \qquad (6.2.6)$$

As an example, consider a rectangular beam of width b and depth h. It is easy to see that Equations (6.2.4)–(6.2.5) become

$$M = \sigma_Y b \left(\frac{h^2}{4} - y_0^2 \right)$$

and

$$P = 2\sigma_Y b y_0$$

In dimensionless form,

$$m = 1 - \eta^2, \qquad p = \eta,$$

so that in this case the yield locus can be described in explicit form,

$$m = 1 - p^2,$$

and forms a parabola.

For a circular bar of radius a, it is more convenient to define η by $\eta = \sin^{-1}(y_0/a)$, so that its range is $-\frac{1}{2}\pi \le \eta \le \frac{1}{2}\pi$. The yield locus is given by

$$p = \frac{1}{\pi}(2\eta + \sin 2\eta), \qquad m = \cos^3 \eta.$$

If M is negative, it need only be replaced by $-M$ (and m by $-m$) in all the results for doubly symmetric sections. Without double symmetry, P must also be replaced by $-P$ (and p by $-p$).

The yield loci for the rectangular and circular bars are shown as curves 1 and 2 in Figure 6.2.2. It is seen that they differ only slightly over the entire range. The yield locus for an I-beam, on the other hand, can be expected to lie between the loci for the rectangular beam and for the ideal sandwich beam, and closer to the latter. Figure 6.2.2 also shows, as curve 3, the locus for the I-beam shown in the adjacent picture.

In practice, curved yield loci such as the one for the rectangular bar are often replaced, following Onat and Prager [1954], by piecewise linear approximations such as the one shown with dashed lines in Figure 6.2.2.[1] Such approximations may also be regarded as the exact yield loci for certain idealized sections, which are in turn approximations to the true sections. The advantage of this point of view (see Hodge [1959], Section 7-3) is that a velocity field that gives generalized strain rates associated with the approximate yield locus may be easily visualized in the context of the idealized section. A frequently used yield locus for wide-flange steel sections is given by

$$\begin{aligned} 0.85|m| + |p| &= 1, \quad |p| \ge 0.15, \\ |m| &= 1, \quad |p| \le 0.15. \end{aligned} \tag{6.2.7}$$

Application: Collapse of a Semicircular Arch

As an illustration of the use of the yield locus in terms of moment and axial force, we consider the pinned-ended semicircular arch of radius a, loaded by a concentrated vertical force $2F$ at midspan, shown in Figure 6.2.3(a).

[1]A general theory of plasticity with piecewise linear yield loci is due to Hodge [1957a].

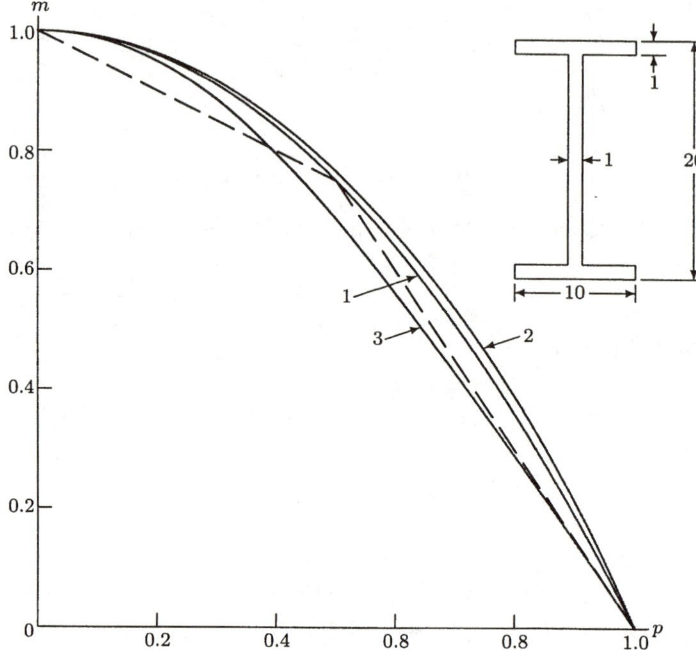

Figure 6.2.2. Yield loci under combined bending and extension for rectangular beams (curve 1), circular beams (curve 2), and the I-beam shown (curve 3). Dashed lines represent a piecewise linear approximation for rectangular beams.

The vertical reactions at the supports are each equal to F by symmetry, but the horizontal reactions $\pm H$ are unknown. Let χ denote the angle between the resultant reaction and the vertical, so that $\tan\chi = H/F$. A free-body diagram of a segment of the arch is shown in Figure 6.2.3(b), and equilibrium shows that the axial force and moment are

$$
\begin{aligned}
P(\phi) &= -F\sin\phi - H\cos\phi = -F(\sin\phi + \tan\chi\sin\phi) \\
&= -\frac{F}{\cos\chi}\sin(\chi+\phi), \\
M(\phi) &= Fa(1-\sin\phi) - Ha\cos\phi \\
&= \frac{Fa}{\cos\chi}[\cos\chi - \sin(\chi+\phi)].
\end{aligned}
$$

With the moment and axial force varying along the arch, it must be assumed that the yield criterion (whichever is chosen) is met only at certain critical sections. As in the case of transversely loaded beams discussed in 4.4.2, *plastic hinges* form at those sections. If a single hinge were to form in the arch, it would necessarily, because of symmetry, be at midspan. A three-hinged arch, however, is a stable structure. Consequently, plastic collapse of the arch requires the formation of two additional plastic hinges, located symmetrically about the midpoint of the arch. The collapse mechanism is shown by the dashed curves in Figure 6.2.3(a); the arch segments between the hinges move as rigid bodies.

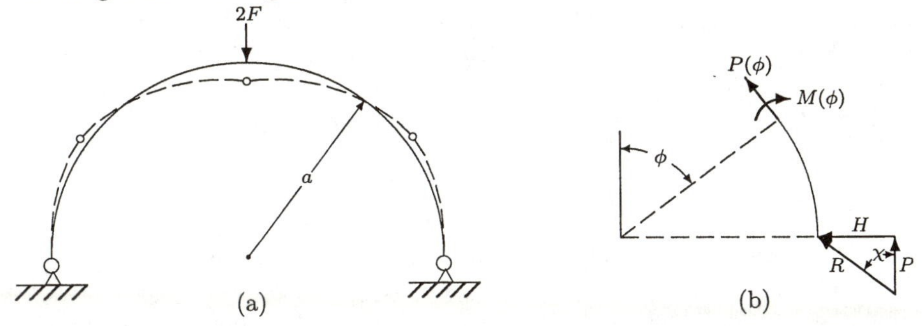

Figure 6.2.3. Pin-ended circular arch: (a) initial geometry and loading (solid line) and collapse mechanism (dashed line); (b) free-body diagram of a segment.

The location of the off-center hinges can easily be determined by noting that both moment and axial force have local maxima only at $\phi = \pi/2 - \chi$, and therefore any convex combination of them also has a local maximum there. Collapse is thus determined by the requirement that the yield criterion $\Phi(M, P) = 0$ is met at $\phi = 0$ and $\phi = \pi/2 - \chi$. Assuming a section with double symmetry so that $\Phi(M, P) = \Phi(|M|, |P|)$, we can find χ and the value of F at collapse by solving simultaneously

$$\Phi(F \tan\chi, Fa|1 - \tan\chi|) = 0 \quad \text{and} \quad \Phi(F \sec\chi, Fa(\sec\chi - 1)) = 0.$$

The choice of a nonlinear yield criterion necessitates the simultaneous solution of two nonlinear equations, a task that may be unpleasant. With a piecewise linear yield criterion, on the other hand, once the side of the polygon is chosen that is appropriate for each equation, the equations are linear in F. Eliminating F, a quadratic equation in $\cos\chi$ or $\sin\chi$ is obtained.

For simplicity, the square yield locus corresponding to the ideal sandwich section will be used. With the help of the dimensionless quantity $\eta = M_U/P_U a$, the equations to be solved are

$$[|1 - \tan\chi| + \eta \tan\chi]Fa = M_U,$$

$$[(1 + \eta) \sec\chi - 1]Fa = M_U.$$

Consequently χ satisfies

$$|1 - \tan\chi| + \eta \tan\chi = (1 + \eta) \sec\chi - 1.$$

It can immediately be seen that if $\tan\chi > 1$, then $\tan\chi = sec\chi$, or $\chi = \pi/2$, so that the additional plastic hinges coincide with the central hinge, and no collapse is possible. It is necessary, then, that $\tan\chi < 1$. For η negligibly small, $\tan\chi$ is very nearly 3/4 and therefore $F_U \doteq 4M_U/a$ — a result equivalent to neglecting the influence of axial forces. For η small but not negligible, a first-order approximation in η may be effected, leading to $\tan\theta \doteq 3/4 - 5\eta/16$ and $F_U \doteq (4M_U/a)/(1+4.25\eta)$. Other examples of arch collapse based on piecewise linear yield loci were studied by Onat and Prager [1953, 1954]; see also Hodge [1959], Section 7-4.

6.2.3. Combined Extension, Bending and Torsion

In a bar under a combination of axial force, bending moment, and torque, the nonvanishing stress components are σ_z, τ_{xz}, and τ_{yz}, with $\tau_{xz} = \partial\phi/\partial y$ and $\tau_{yz} = -\partial\phi/\partial x$, ϕ being the stress function. Any point of the bar is in a state of plane stress in the plane that is perpendicular to $\nabla\phi$; the state of stress in that plane can be given by $\begin{bmatrix} \sigma & \tau \\ \tau & 0 \end{bmatrix}$, where $\sigma = \sigma_z$ and $\tau = |\nabla\phi| = \sqrt{\tau_{xz}^2 + \tau_{yz}^2}$. Both the Mises and the Tresca yield criteria are given by Equation (3.3.5):

$$\left(\frac{\sigma}{\sigma_Y}\right)^2 + \left(\frac{\tau}{\tau_Y}\right)^2 = 1.$$

Since $\sigma_{ij}\dot{\varepsilon}_{ij} = \sigma\dot{\varepsilon} + \tau\dot{\gamma}$, the associated flow rule may be obtained from the maximum-plastic-dissipation principle as

$$\dot{\varepsilon}^p = \dot{\lambda}\frac{\sigma}{\sigma_Y^2}, \quad \dot{\gamma}^p = \dot{\lambda}\frac{\tau}{\tau_Y^2}, \tag{6.2.8}$$

and the plastic dissipation is

$$D_p(\dot{\varepsilon}) = \sqrt{(\sigma_Y\dot{\varepsilon})^2 + (\tau_Y\dot{\gamma})^2}. \tag{6.2.9}$$

Lower Bound

A lower-bound yield locus for combined extension, bending, and torsion can be found by generalizing an approach proposed by Hill and Siebel [1951] for combined bending and torsion only. The approach is based on assuming, on the one hand, the same distribution of normal stress as in 6.2.2, but with $|\sigma| = \alpha\sigma_Y$, where $0 \le \alpha \le 1$; and on the other hand, the same distribution of shear stress as in fully plastic torsion, but with $\tau = \sqrt{1-\alpha^2}\tau_Y$. The second assumption leads immediately to

$$t \stackrel{\text{def}}{=} \frac{T}{T_U} = \sqrt{1-\alpha^2}.$$

The first assumption means that the m-p relation would be given by (6.2.6) if m and p were defined as $M/\alpha M_U$ and $P/\alpha P_U$, respectively. With the standard definitions, the relation is therefore

$$\frac{p}{\alpha} = \bar{p}(\eta), \qquad \frac{m}{\alpha} = \bar{m}(\eta).$$

Eliminating α, we can describe the yield surface in *mpt*-space in terms of the single parameter η:

$$p = \sqrt{1-t^2}\bar{p}(\eta), \qquad m = \sqrt{1-t^2}\bar{m}(\eta).$$

For the rectangular beam, we have the explicit description

$$m = \sqrt{1-t^2} - \frac{p^2}{\sqrt{1-t^2}}.$$

The projections of this surface on both the *mt*- and *pt*-planes are unit circles.

Upper Bound

An upper-bound yield curve for combined extension, bending and torsion can be found following the method of Hill and Siebel [1953]. The assumed velocity field is taken so that it results only in an extension rate $\dot{\varepsilon}_0$ of the centroidal fiber, a pure curvature rate $\dot{\kappa}$ about the centroidal axis, and a rate of twist $\dot{\theta}$ about the centroid; warping is neglected.[1] With symmetric bending in the yz-plane assumed, the strain rates are

$$\dot{\varepsilon}_z = \dot{\varepsilon}_0 - \dot{\kappa}y, \qquad \dot{\gamma}_{\theta z} = r\dot{\theta},$$

where $r = \sqrt{x^2+y^2}$. The plastic dissipation is therefore

$$D_p(\dot{\varepsilon}^*) = \tau_Y|\dot{\theta}|\sqrt{r^2 + (\alpha y - \beta)^2},$$

where $\alpha = (\sigma_Y/\tau_Y)(\dot{\kappa}/\dot{\theta})$ and $\beta = (\sigma_Y/\tau_Y)(\dot{\varepsilon}_0/\dot{\theta})$.

A simple way of finding the upper-bound generalized stresses corresponding to the assumed velocity field is to determine the stresses that are related to it by the associated flow rule and that obey the yield criterion — but that do not, in general, form a statically admissible stress field — and then to calculate their resultants. The associated flow rule (6.2.8) produces the stresses

$$\sigma = \frac{\sigma_Y^2}{\dot{\lambda}}(\dot{\varepsilon}_0 - \dot{\kappa}y) = -\mu\sigma_Y(\alpha y - \beta)$$

and

$$\tau = \mu\tau_Y r,$$

[1] A correction for warping, resulting in better upper bounds for noncircular sections, is due to Gaydon and Nuttall [1957].

where $\tau = \tau_{z\theta}$ (so that $\tau_{rz} \equiv 0$),[1] and $\mu = \tau_Y\dot{\theta}/\dot{\lambda}$ is a function of position that can be determined by requiring that the yield criterion (3.3.5) be obeyed everywhere, resulting in

$$\mu = \frac{1}{\sqrt{r^2 + (\alpha y - \beta)^2}}.$$

The stress resultants M, P and T are therefore given by

$$\begin{aligned} M &= -\int_A y\sigma\, dA = \alpha\sigma_Y \int_A \frac{y^2}{\sqrt{r^2 + (\alpha y - \beta)^2}}\, dA, \\ P &= \int_A \sigma\, dA = \beta\sigma_Y \int_A \frac{1}{\sqrt{r^2 + (\alpha y - \beta)^2}}\, dA, \\ T &= \int_A r\tau\, dA = \tau_Y \int_A \frac{r^2}{\sqrt{r^2 + (\alpha y - \beta)^2}}\, dA. \end{aligned} \tag{6.2.10}$$

The integrations in Equations (6.2.10) may be performed, numerically if necessary, to yield a parametric representation of the upper-bound yield surface in *mpt*-space in terms of the parameters α and β. Computed yield curves representing the projections of the lower-bound and upper-bound yield surfaces in the *mt*- and *pt*-planes, are shown in Figure 6.2.4(a) and (b) for a circular and a square bar, respectively.

Extension and Torsion of a Circular Bar

For a circular bar, the upper-bound solution presented above is, in fact, a complete solution, since an axisymmetric shear-stress distribution $\tau_{\theta z} = \tau(r)$, $\tau_{rz} = 0$ is statically admissible. A closed-form result can be obtained for extension and torsion alone, that is, for $\alpha = 0$.

We define the dimensionless parameter $\zeta = \beta/a$, where a is the radius of the bar. Equations $(6.2.10)_{2,3}$ give

$$p = \frac{2\pi}{\sigma_Y \pi a^2}\zeta a\sigma_Y \int_0^a \frac{r\, dr}{\sqrt{\zeta^2 a^2 + r^2}} = \zeta \int_0^a \frac{dx}{\sqrt{x + \zeta^2}}$$

and

$$t = \frac{2\pi}{\frac{2}{3}\tau_Y \pi a^3}\tau_Y \int_0^a \frac{r^3\, dr}{\sqrt{\zeta^2 a^2 + r^2}} = \frac{3}{2}\int_0^1 \frac{x\, dx}{\sqrt{x + \zeta^2}},$$

or

$$p = 2\zeta\left(\sqrt{\zeta^2 + 1} - \zeta\right), \quad t = 2\zeta^3 - (2\zeta^2 - 1)\sqrt{\zeta^2 + 1}. \tag{6.2.11}$$

The first of Equations (6.2.11) can be rewritten as

$$p + 2\zeta^2 = 2\zeta\sqrt{\zeta^2 + 1}.$$

[1]Thus the traction boundary conditions are not satisfied for any but a circular bar.

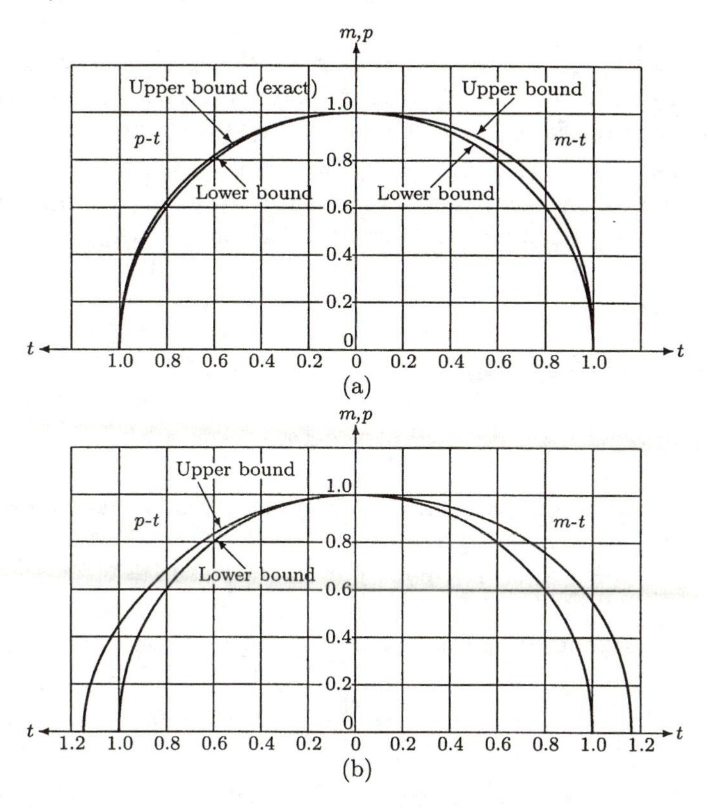

Figure 6.2.4. Lower-bound and upper-bound yield loci under bending, torsion and extension: (a) circular bar; (b) square bar.

Squaring both sides, we find

$$4\zeta^4 + 4\zeta^2 = 4\zeta^4 + 4\zeta^2 p + p^2,$$

or $4\zeta^2(1-p) = p^2$, which can immediately be solved for ζ:

$$\zeta = \frac{p}{2\sqrt{1-p}}.$$

Substituting in the equation for t, we obtain the explicit relation

$$t = \frac{1}{2}(2+p)\sqrt{1-p}.$$

Squaring both sides of this equation, we may rewrite it as

$$t^2 + p^2 = 1 + \frac{p^2 - p^3}{4},$$

a form that is convenient for determining the extent to which the present upper-bound yield curve differs from the previously found lower bound, described by the unit circle. The right-hand side of the last equation has its maximum at $p = 2/3$, its value there being 28/27. Consequently the distance of the points on the dimensionless yield curve from the origin lies between 1 and $\sqrt{28/27} = 1.018$, and the curve differs only slightly from the circle.

The stress distribution giving rise to the resultants just obtained can also be determined as the solution of a constrained extremum problem: find the stresses satisfying (3.3.5) everywhere such that the torque,

$$T = \int_A r\tau \, dA,$$

is maximum for a given axial force,

$$P = \int_A \sigma \, dA.$$

Substituting for τ by solving (3.3.5), the problem may be written as

$$\tau_Y \, \delta \int_A r\sqrt{1 - (\sigma/\sigma_Y)^2} \, dA + \nu \, \delta \int_A \sigma \, dA = 0,$$

where ν is a Lagrangian multiplier, or

$$\int_A \left[\beta - \frac{(\sigma/\sigma_Y)r}{\sqrt{1 - (\sigma/\sigma_Y)^2}}\right] \delta\sigma \, dA = 0,$$

where $\beta = \nu\sigma_Y/\tau_Y$. There being no further constraint on the distribution of σ, the quantity in brackets under the integral sign must be zero, that is,

$$\frac{\sigma}{\sigma_Y} = \frac{\beta}{\sqrt{\beta^2 + r^2}},$$

and the yield criterion is satisfied everywhere if and only if

$$\frac{\tau}{\tau_Y} = \frac{r}{\sqrt{\beta^2 + r^2}}.$$

Integration leads immediately to p and t given by Equations (6.2.11), with $\zeta = \beta/a$ as before.

Bending and Torsion of a Bar of Arbitrary Cross-Section

A complete solution for combined bending and torsion of a fully plastic bar of arbitrary cross-section may be obtained by integrating numerically a nonlinear partial differential equation first derived by Handelman [1944]. The stress function $\phi(x, y)$ is now regarded as unknown, and it is to be found so that it maximizes the bending moment for a given torque (or vice versa). For doubly symmetric bending in the yz-plane, we have

$$M = \frac{\sigma_Y}{\tau_Y} \int_A |y| \sqrt{\tau_Y^2 - (\partial\phi/\partial x)^2 - (\partial\phi/\partial y)^2} \, dA.$$

The constrained extremum problem may therefore be written as

$$\int_A \left[|y|\,\delta\sqrt{\tau_Y^2 - (\partial\phi/\partial x)^2 - (\partial\phi/\partial y)^2} - \nu\,\delta\phi \right] dA = 0,$$

where ν is again a Lagrangian multiplier. Since $\phi = 0$ and hence $\delta\phi = 0$ on the boundary, integration by parts leads to

$$\int_A \left\{ \frac{\partial}{\partial x}\left[\frac{y\partial\phi/\partial x}{\sqrt{\tau_Y^2 - (\partial\phi/\partial x)^2 - (\partial\phi/\partial y)^2}} \right] \right. $$
$$\left. + \frac{\partial}{\partial y}\left[\frac{y\partial\phi/\partial y}{\sqrt{\tau_Y^2 - (\partial\phi/\partial x)^2 - (\partial\phi/\partial y)^2}} \right] + \nu \right\} \delta\phi\, dA = 0.$$

Since ϕ is unconstrained in the interior of A, the contents of the curly brackets must vanish everywhere, and this furnishes the required partial differential equation for ϕ, an elliptic equation subject to the boundary condition $\phi = 0$. Different values of ν give different ratios of M to T. The equation was solved numerically for a square cross-section by Steele [1954], and for other cross-sections by Imegwu [1960], who showed that the interaction diagram in the mt-plane is remarkably insensitive to cross-sectional shape.

6.2.4. Bending and Shear

In the problems involving combinations of bending moments, axial force and torque that have been considered so far in this section, these stress resultants are independent of one another as far as the equilibrium of the beam is concerned, and therefore their interaction may be rigorously studied on a purely local basis. In particular, for the purpose of lower-bound analysis, if a statically admissible stress distribution is found for each such resultant separately, then a linear combination of such stress distributions is also statically admissible.

If a beam is subject to transverse loading, then this loading determines both the shear force V and the bending moment M, which are therefore related to each other by the equilibrium equation

$$\frac{dM}{dz} = V.$$

Strictly speaking, then, M and V are not generalized stresses that can be specified independently of each other.

Indeed, each case of a transversely loaded beam presents a distinct problem of limit analysis. As discussed in 6.1.2, even such closely related cases as the end-loaded cantilever and the center-loaded simply supported beam are different. If the span-to-depth ratio is sufficiently high, however, the limit loads tend to those predicted by the elementary plastic-hinge mechanism — that is, the overall collapse of the beam is almost completely determined by local behavior at a critical section. It thus becomes reasonable to formulate an approximate local yield locus, in terms of M and V, that governs local behavior, analogous to the local yield loci found above. The problem of limit analysis of an arbitrarily loaded beam then reduces to the determination of the critical section or sections; this problem is studied in 6.3.2.

The particular problem of the end-loaded cantilever may be used as the test problem for studying the local yield locus, as discussed by Drucker [1956]. Here, the problem has already been investigated from two points of view: in 4.5.2 as the limit of elastic–plastic bending, and in 6.1.2 as a problem in plane limit analysis. The latter approach, as was indicated there, gives both upper and lower bounds for the collapse load of the beam. The former approach, as is shown next, furnishes a lower bound.

Lower Bound as Limit of Elastic-Plastic Solution

We saw in 4.5.2 that as the elastic core shrinks under the influence of an increasing moment, the maximum shear stress there grows until it also reaches the yield value, after which a secondary plastic zone forms in which $|\tau| = \tau_Y$, $\sigma = 0$. Between this central plastic zone and each of the two outer plastic zones is an elastic zone in which, going from inside to outside, the normal stress increases linearly in magnitude from 0 to σ_Y, and the shear stress decreases (parabolically in a rectangular beam or in the web of an I-beam) from τ_Y to zero. The limiting state is, of course, one in which the elastic zones shrink to vanishing thickness. This limit does not represent, strictly speaking, a statically admissible stress field, since it includes a discontinuity in τ. Since, however, the discontinuity occurs at two isolated points, equilibrium of any finite element is not violated and the stress distribution is acceptable. If the width of the plastic zone is c, then $V = \tau_Y bc$ and $M = \sigma_Y b(h^2 - c^2)/4$. Defining $v = V/V_U$, where $V_U = \tau_Y A$, we immediately obtain the dimensionless interaction curve given by

$$m = 1 - v^2.$$

This curve has clearly the same form as the one previously found for the interaction between moment and axial force, with v replacing p. In fact, it can be shown that for any doubly symmetric section the yield locus in the mv-plane is given parametrically by

$$m = \bar{m}(\eta), \qquad v = \bar{p}(\eta),$$

where $\bar{m}$ and $\bar{p}$ are the same functions as in Equation (6.2.6). The proof is based on the observation that the stress distribution for combined bending moment and axial force, shown in Figure 6.2.5(a), is the superposition of those shown in Figure 6.2.5(b) and (c). However, the stress distribution

(b) alone produces the same moment as (a), and the stress distribution (c) alone produces the same axial force as (a). If, now, (c) represents a block of shear stress rather than normal stress, then it produces a shear force, rather than an axial force, by exactly the same formula. When suitably nondimensionalized, therefore, the shear force is exactly the same function of the depth of the central zone as is the axial force.

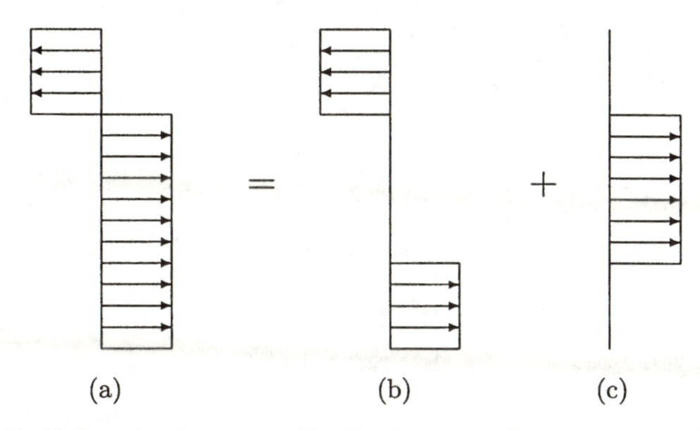

Figure 6.2.5. Fully plastic stress distribution at a beam cross-section under combined bending moment and axial force.

In a rectangular cantilever of length L carrying a concentrated transverse force F at its tip, collapse occurs when a plastic hinge forms at the built-in end. At this point $M = FL$ and $V = F$. Defining $f = FL/M_U$, we find that the lower bound based on Figure 6.2.5(b-c) predicts

$$f = 1 - \left(\frac{\sigma_Y}{2\tau_Y}\frac{h}{2L}\right)^2 f^2,$$

a quadratic equation for f that can be solved explicitly for f. Assuming the Tresca criterion ($\sigma_Y = 2\tau_Y$) and defining $\delta = h/2L$, we can write the solution as

$$f = \frac{1}{2\delta^2}\left(\sqrt{1+4\delta^2} - 1\right). \tag{6.2.12}$$

Upper Bound: Hodge Approach

The generalized strain rates conjugate to M and V are the curvature rate $\dot{\kappa}$ and the shear rate $\dot{\gamma}$, respectively. The strain-rate components conjugate to the stresses σ and τ are $\dot{\varepsilon} = -\dot{\kappa}y$ and $\dot{\gamma}$. Given $\dot{\kappa}$ and $\dot{\gamma}$, the associated flow rule (6.2.8) produces the stresses

$$\sigma = -\frac{\sigma_Y^2}{\dot{\lambda}}\dot{\kappa}y, \qquad \tau = \frac{\tau_Y^2}{\dot{\lambda}}\dot{\gamma}.$$

Satisfaction of the yield criterion (3.3.5) requires

$$\dot{\lambda} = \sqrt{(\tau_Y\dot{\gamma})^2 + (\sigma_y\dot{\kappa}y)^2} = \tau_Y\dot{\gamma}\sqrt{1+\nu y^2},$$

where $\nu = \sigma_Y \dot{\kappa}/\tau_Y \dot{\gamma}$. The stresses are therefore

$$\sigma = -\frac{\nu \sigma_Y y}{\sqrt{1+\nu^2 y^2}}, \qquad \tau = \frac{\tau_Y}{\sqrt{1+\nu^2 y^2}}. \tag{6.2.13}$$

The stress resultants are accordingly given in terms of the parameter ν as

$$M = \nu \sigma_Y \int_A \frac{y^2}{\sqrt{1+\nu^2 y^2}}\, dA,$$

$$V = \tau_Y \int_A \frac{1}{\sqrt{1+\nu^2 y^2}}\, dA.$$

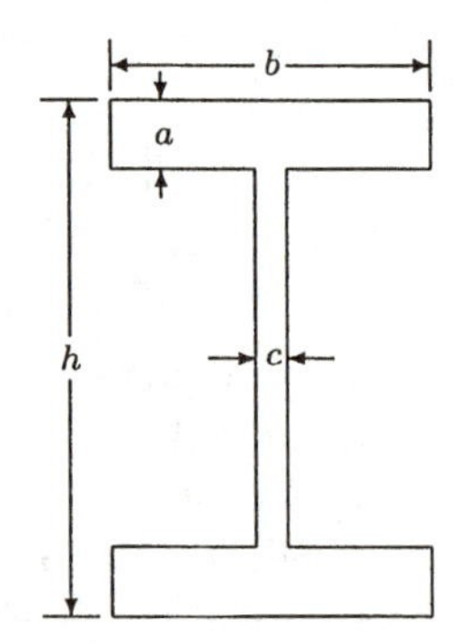

Figure 6.2.6. I-beam: geometry.

Hodge [1957b] evaluated the integrals in the preceding equations for the I-beam shown in Figure 6.2.6. If c/b and a/h are small, then the parametric form of the yield locus in dimensionless form is, after neglecting terms of order higher than the first in these quantities,

$$m = \frac{2\tanh\omega + j(\coth\omega - \omega \operatorname{csch}^2\omega)}{2+j},$$

$$v = \frac{\operatorname{sech}\omega + j\omega \operatorname{csch}\omega}{1+j},$$

where $\omega = \sinh^{-1}(\nu h/2)$, while $j = ch/2ab$ is a dimensionless shape parameter. The limiting cases $j = 0$ and $j = \infty$ correspond, respectively, to the ideal I-beam (with finite flange thickness but negligible web thickness) and the rectangular beam. The calculated interaction curves are shown in Figure 6.2.7.

Hodge also derived the same stress distribution by solving the constrained extremum problem, as in 6.2.3. On this basis Hodge regards the yield locus as also providing a lower bound. However, the shear stress given

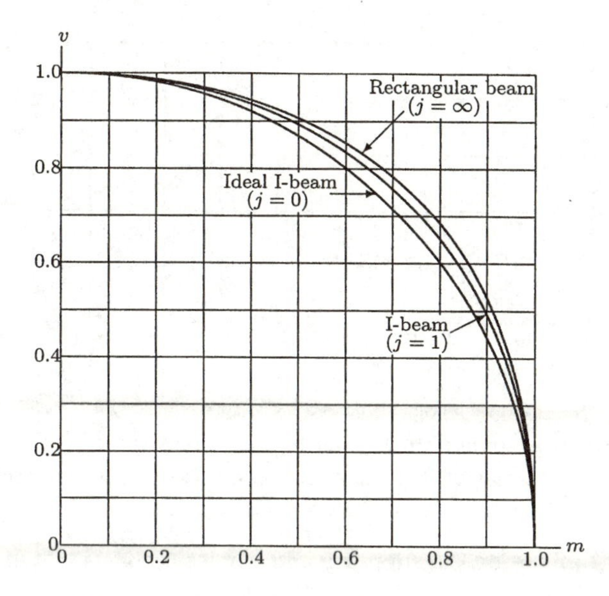

Figure 6.2.7. Interaction curves for I-beam (Hodge [1959]).

by $(6.2.13)_2$ does not vanish at the extreme fibers, and it therefore appears difficult to accept the stress distribution as statically admissible.

Plots of the lower-bound curve defined by Equation (6.2.12), the upper-bound curve due to Hodge [1957b], and the lower-bound and upper-bound curves due to Drucker [1956] (see 6.1.2) are shown in Figure 6.2.8. It can be seen that all the curves have the limit $f = 1$ as $\delta \to 0$, so that the effect of shear is negligible for beams that are not overly deep, and that all tend asymptotically to the hyperbola $f = 1/\delta$ for δ large. It must be remembered, however, that values of δ greater than about 0.5 cannot reasonably be regarded as describing beams.

Exercises: Section 6.2

1. Using the assumptions of elementary beam theory, derive an expression for the internal virtual work in a beam of arbitrary cross-section subject to a variable bending moment M and axial force P.

2. Find the ultimate yield locus for a beam having the idealized section of Figure 3.5.2.

3. For a beam whose cross-section is an isosceles triangle, find the ultimate yield loci for combined axial force and bending moment (a) perpendicular and (b) parallel to the axis of symmetry.

4. Analyze the collapse of a semicircular arch of ideal sandwich section

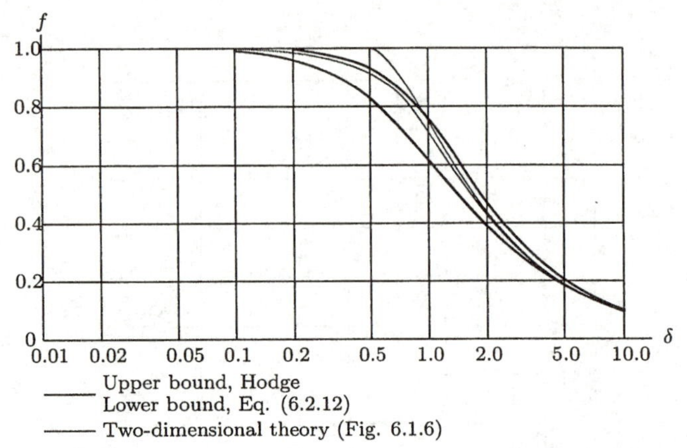

Figure 6.2.8. Lower-bound and upper-bound curves under combined bending and shear of a rectangular beam, load against depth-span ratio.

that is built in at the supports and carries a concentrated load $2F$ at the vertex.

5. Analyze the collapse of a symmetric simply supported arch of ideal sandwich section, forming a circular segment of angle 2α, and loaded by (a) a concentrated load $2F$ at the vertex, (b) a uniform vertical load of intensity q, and (c) a uniform radial load of intensity q.

6. Show that for combined torsion and axial force or bending moment of a circular bar, the unit circles are lower bounds to the yield loci in both the pt- and mt-planes.

7. Find a result analogous to (6.2.12) for an end-loaded cantilever of circular cross-section with radius a, letting $\delta = a/L$.

8. Evaluate the integrals following Equation (6.2.13) for a rectangular cross-section. Compare with the cited result of Hodge [1957b] for $j = \infty$.

Section 6.3 Limit Analysis of Trusses, Beams and Frames

6.3.1. Trusses

A truss is an assemblage of stiff bars that are more or less flexibly connected to one another at their ends. In an ideal truss, the connection is through frictionless pins, with the center of each pin coinciding with the intersection

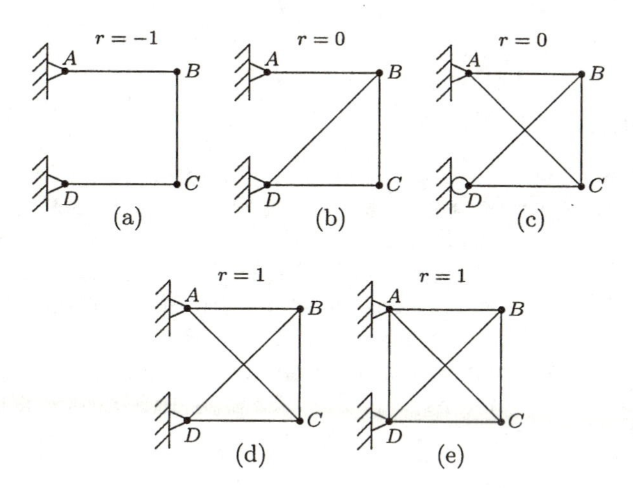

Figure 6.3.1. Some simple trusses.

of the centroidal lines of the bars meeting there. Moreover, all the loads are applied at the joints. Consequently, all bars carry axial forces only, and if n is the number of bars, then the bar forces $P_1, \ldots, P_n$ in effect constitute the stress field.

In order for each pin to be in equilibrium, the vector sum of all the bar forces in the members connected through that pin must be zero. If the number of joints is j, then the total number of joint equilibrium equations is $3j$, unless all the bars and loads lie in the same plane, in which case the truss is a *plane* (or *planar*) *truss* and the number of equations is $2j$. A nonplane truss is usually called a *space truss*.

The unknowns of the problem, in addition to the bar forces, are the reaction components, s in number. If $n + s = kj$ (where $k = 2$ or 3 in the plane truss or space truss, respectively), then the equilibrium equations are precisely enough to determine the unknowns, and the truss is *statically determinate* (or *isostatic*). If $r = n + s - kj$ is positive, then the truss is *statically indeterminate* (or *hyperstatic* or *redundant*), and r is called the *degree of static indeterminacy* (or, more simply, the *indeterminacy number* or *redundancy number*). If this number is negative, then the truss is unstable (or *hypostatic*) and is, in fact, a mechanism. In a stable truss, the number s of reaction components must be at least equal to the number of equilibrium equations for the truss as a whole, three for the plane truss and six for the space truss. If s is greater than this number, then the truss is *externally indeterminate*.

Some simple trusses, with r equal to 0, -1 and 1 are shown in Figure 6.3.1. In particular, the truss in (e) is statically indeterminate of degree one even though, apparently, $r = 6+4-2{\cdot}4 = 2$. In fact, bar AD cannot deform and therefore cannot carry any force. This bar may, as a result, be ignored in any analysis of the truss.

Limit Analysis of Trusses

A truss member will be said to fail if it can undergo significant lengthening or shortening with no significant change in the bar force. Failure in this sense can result from yielding, if the material is perfectly plastic or nearly so, or, in the case of a compression member, from buckling. Since the bar force in a failed member is no longer determined by equilibrium but by the failure criterion, it can be presumed as known if the properties of the bar are known, and the number n of unknown bar forces drops by one, as does the indeterminacy number r. The truss therefore becomes unstable if $r+1$ bars fail. In particular, a statically determinate truss collapses as soon as one bar fails.

Any choice of $r+1$ bars that fail provides a mechanism with one degree of freedom that can be used with the upper-bound theorem. Given a reference velocity v, the elongation rates $\dot{\Delta}_i$ of the ith failed bar may be determined by geometry. Let P_{Ui}^+ denote the ultimate bar force in tension, and P_{Ui}^- the magnitude of that in compression; the latter is the lesser of the yield force and the buckling force. The internal dissipation in a failed bar is $P_{Ui}^{\pm}|\dot{\Delta}_i|$, the superscript sign being that of $\dot{\Delta}_i$.

The truss of Figure 4.1.1 has three bars, six reaction components, and four joints, so that $r = 1$ and two bars must fail for collapse to occur. In 4.1.4 we read, however, that all three bars must fail for this truss to collapse. With P_U^+ the same in all three bars, the collapse load in this case is $F_U = P_U^+(1 + 2\cos\alpha)$.

If we were to assume a mechanism in which bars AD and CD fail, while BD remains rigid, the external work rate would be zero and the upper-bound theorem would fail to give a finite upper bound to F_U. Consider, now, an asymmetric mechanism with bars BD and CD only failing. Bar AD then rotates rigidly about its support, and if the downward component of the velocity of pin D is v, then its leftward component is $v\cot\alpha$. The velocity component in the direction of bar CD, equal to its elongation rate, is $v(\cos\alpha + \cot\alpha\sin\alpha) = 2v\cos\alpha$. Equating the external work rate to the total internal dissipation,

$$Fv = P_U^+ v(1 + 2\cos\alpha),$$

leads to the an upper bound equal to F_U. The equality can be explained as follows: suppose that the symmetry of the system is disturbed ever so slightly (e.g., by giving the load a small leftward component or by making bar CD just a little bit weaker than AD). The correct collapse mechanism would then, indeed, be the asymmetric one just considered. But a minimal asymmetry should not significantly affect the collapse load calculated on

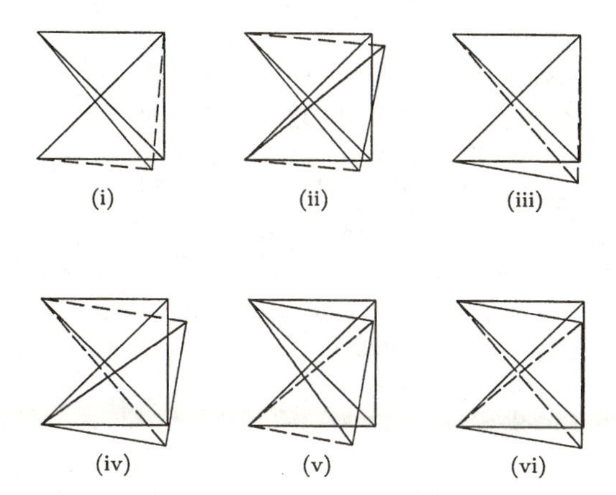

Figure 6.3.2. Collapse mechanisms for the truss of Figure 6.3.1(d).

the basis of assumed symmetry, and hence the collapse loads given by the symmetric and asymmetric mechanisms should be the same.

Let us now look at the possible collapse mechanisms of the truss of Figure 6.3.1(d) under a downward force F applied at joint C. First, it is apparent that any mechanism in which bars AC and CD do not deform does not permit any translation of C, so that the force F cannot do any work; such a mechanism is not admissible. A mechanism in which both AC and CD fail but the other bars remain rigid allows rotation of BC about B, leading to a zero external work rate for the given loading. The only mechanisms to be considered, therefore, are those in which either AC or CD fails, but not both; these mechanisms number six, and are shown in Figure 6.3.2. Dashed lines represent bars that fail.

We assume that the bars are of identical cross-section and sufficiently stiff so that $P_U^+ = P_U^- = P_U$ in every bar. The mechanism (i) in which CD and BC fail presents a rigid rotation of AC about A; if the downward velocity of C is v, then the angular velocity of rotation of AC is v/a, where a is the length of the nondiagonal bars, so that the shortening rate of CD and the lengthening rate of BC are also v. The total internal dissipation is thus $2P_U v$, and the resulting upper bound on F is $2P_U$. If (ii) CD and AB fail, then AC rotates about A, BD rotates about D, and BC rotates about its midpoint. The shortening rate of CD is again v, as is the lengthening rate of AB, and the upper bound is once more $2P_U$.

The mechanisms in which AC fails along with either (iii) BC or (iv) AB, and the one (v) in which CD and BD fail, all give the same upper bound of $(1 + 1/\sqrt{2})P_U$. Finally, in the shear mechanism (vi) in which AC and BD fail, these bars lengthen and shorten, respectively, at the rate $v/\sqrt{2}$, leading to the upper bound $\sqrt{2}P_U$. This upper bound, being the smallest, must equal the ultimate load.

To confirm this equality, we analyze the truss statically under this value of F, assuming that $P_{AC} = P_U$ and $P_{BD} = -P_U$. We find that all the other bar forces are of magnitude $P/\sqrt{2}$, so that the "stress field" (the distribution of bar forces) is plastically admissible.

It is clear that if the cross-sectional areas of all three nondiagonal bars were to be reduced by a factor of $\sqrt{2}$, their ultimate bar forces would fall to $1/\sqrt{2}$ times those of the diagonals, and the stress field would still be — just barely — plastically admissible under the just-calculated collapse load. Collapse would then occur with all bars yielding, indicating the most efficient use of material; such a truss would be an example of **minimum-weight design**.

A kinematic analysis of the minimum-weight truss shows that all six mechanism based on failure in two bars produce the same upper bound. The most general collapse mechanism, then, is a linear combination of the six. It is easy to see that this mechanism has four degrees of freedom, represented by the horizontal and vertical displacements of joints B and C.

Limit Design of Trusses

The design of structures on the basis of limit analysis is known as **limit design**. Limit design for minimum weight is particularly simple for trusses in which no bars buckle, because the weight of a bar is directly proportional to its area, and hence to its strength. Let the ultimate bar force of the ith bar be P_{Ui}, and its length L_i. In minimum-weight design, the object is then to minimize

$$\sum_{i=1}^{n} P_{Ui}L_i$$

subject to inequality constraints on the P_{Ui}, ensuring that the truss is strong enough to carry the prescribed load. The problem is one of **linear programming**; it is discussed in its general form after a study of the present example.

Suppose, for simplicity, that $P_{U(AB)} = P_{U(BC)} = P_{U(CD)} = P_{U1}$ and $P_{U(AC)} = P_{U(BD)} = P_{U2}$. We are thus led to minimize

$$3P_{U1} + 2\sqrt{2}P_{U2}$$

subject to the following inequalities:

$$2P_{U1} \geq F, \quad P_{U1} + \frac{P_{U2}}{\sqrt{2}} \geq F, \quad \sqrt{2}P_{U2} \geq F;$$

in this elementary case the six mechanisms furnish only three independent inequalities, and the second of the three is redundant, since it is satisfied whenever the other two are. The solution is simple, as we already know: all the inequalities are satisfied as equalities when $P_{U2} = \sqrt{2}P_{U1} = F/2$.

Linear-Programming Formulation of Limit Design

If a truss has n bars, then the number of different combinations of $r+1$ bars — that is, the number of different mechanisms — is

$$m = \frac{n!}{(n-r-1)!\,(r+1)!}.$$

Let v denote a positive reference velocity in the kth mechanism ($k = 1, \ldots, m$), such that the elongation rate of the ith bar in this mechanism is $\alpha_{ik}v$, and the velocity conjugate to the applied force F_j ($j = 1, \ldots, l$) is $\beta_{jk}v$. According to the upper-bound theorem, then,

$$\sum_{i=1}^{n} \alpha_{ik} P_{Ui} \geq \sum_{j=1}^{l} \beta_{jk} F_j, \quad k = 1, \ldots, m. \tag{6.3.1}$$

The problem of minimum-weight design of a truss is then one of minimizing the linear function

$$G(P_{U1}, \ldots, P_{Un}) = \sum_{i=1}^{n} L_i P_{Ui}$$

subject to the linear constraints (6.3.1); this problem is one of linear programming. The connection between limit design and linear programming was apparently first noted by Heyman [1951].

The standard linear-programming problem is one of maximizing, rather than minimizing, the *objective function* $\mathbf{c}^T\mathbf{x}$, where $\mathbf{c}$ and $\mathbf{x}$ are $1 \times n$ column matrices, subject to the constraints

$$\mathbf{a}_k^T \mathbf{x} \leq b_k, \quad k = 1, \ldots, m, \tag{6.3.2}$$

where the $\mathbf{a}_k$ are $1 \times n$ column matrices, and the b_k are real numbers. Often the additional constraint that the x_i be nonnegative is imposed. The problem of minimum-weight design of a truss becomes a standard problem if we identify x_i with P_{Ui}, c_i with $-L_i$, $\mathbf{a}_k$ with the kth column of the matrix $-[\alpha_{ik}]$, and $-b_k$ with the right-hand side of (6.3.1). By defining the $m \times n$ matrix $\mathbf{A}$ through $\mathbf{A}^T = [\mathbf{a}_1, \ldots, \mathbf{a}_m]$ and the $m \times 1$ matrix $\mathbf{b}$ through $\mathbf{b} = (b_1, \ldots, b_m)$, inequality (6.3.2) may be rewritten in the short form

$$\mathbf{A}\mathbf{x} \leq \mathbf{b}, \tag{6.3.3}$$

and the additional constraint as

$$\mathbf{x} \geq 0. \tag{6.3.4}$$

In an alternative formulation of the linear-programming problem, the constraint conditions (6.3.3)–(6.3.4) are replaced by

$$\mathbf{A}\mathbf{x} = \mathbf{b}, \quad \mathbf{x} \geq 0. \tag{6.3.5}$$

An efficient numerical solution method due to Dantzig, called the **simplex algorithm**, exists for the problem in this form (see Dantzig [1963]).

The constraints (6.3.3)–(6.3.4) can be converted to the form (6.3.5) by introducing the *slack variables* y_k ($k = 1, \ldots, m$), which form the $m \times 1$ matrix $\mathbf{y}$, and rewriting (6.3.3) as

$$\mathbf{A}\mathbf{x} + \mathbf{y} = \mathbf{b}, \quad \mathbf{y} \geq 0. \tag{6.3.6}$$

The $(n+m) \times 1$ matrix $\bar{\mathbf{x}}$ is now defined by $\bar{x}_i = x_i$ ($i = 1, \ldots, n$) and $\bar{x}_{n+k} = y_k$ ($k = 1, \ldots, m$). The $m \times (n+m)$ matrix $\bar{\mathbf{A}}$ is similarly defined by $\bar{A}_{ki} = A_{ki}$ ($i = 1, \ldots, n$) and $\bar{A}_{k,n+j} = \delta_{jk}$ ($j = 1, \ldots, m$). The constraint inequalities (6.3.4) and (6.3.6) together can now be written as

$$\bar{\mathbf{A}}\bar{\mathbf{x}} = \mathbf{b}, \quad \bar{\mathbf{x}} \geq 0,$$

a form identical with (6.3.5).

6.3.2. Beams

Any transversely loaded beam, except an ideal sandwich beam, is statically indeterminate in the sense of Section 4.1 — that is, the stress field cannot be deduced from the loading independently of unknown properties: at any section, an infinity of stress distributions can be found that give the same resultant moment M and shear force V. It is conventional, however, to call a beam statically determinate (or indeterminate) if it is externally determinate (or indeterminate) in the same sense as a truss, that is, if the number of independent reaction components is the same as (or greater than) the number of equilibrium equations available to determine them. In the absence of internal hinges, this number is three for plane bending. Any hinge, whether frictionless or a plastic hinge, provides an additional equilibrium equation: at a frictionless hinge, $M = 0$, since such a hinge cannot transmit moment, while at a plastic hinge $M = M_U^+$ or $M = -M_U^-$. The indeterminacy number of a beam is accordingly $r = s - h - 3$, where s is the number of reaction components[1] and h is the number of hinges. Like a plane truss, the beam collapses when r is reduced to -1, so that if h_0 hinges are present initially, the number of plastic hinges required for collapse is $s - h_0 - 2$, and specifically, one if the beam is statically determinate, and two or more if it is statically indeterminate. A plastic hinge may form at any point of the beam at which the condition $|M| = M_U$ is possible, that is, in the interior of a

[1]Provided that these components do not include three collinear forces.

span, at a built-in end, or at an intermediate support. A collapse mechanism is admissible if it does not violate any support condition (possibly relaxed by the formation of a plastic hinge) and if it produces a positive external work rate.

If the effect of shear on the formation of a plastic hinge can be neglected, as will be assumed, then a hinge that has rotated by an angle $\Delta\theta$ can be thought of as the limit of a small segment, of length, say, Δx, in which the curvature is $\Delta\theta/\Delta x$ and the plastic dissipation per unit length is $M_U|\Delta\dot{\theta}|/\Delta x$. The total internal dissipation in the hinge is therefore $M_U|\Delta\dot{\theta}|$.

A moment distribution is statically and plastically admissible if it is in equilibrium with the applied loads, is consistent with all force and moment end conditions and frictionless hinge conditions (if any), and is such that $|M| \leq M_U$ everywhere. In the moment distribution at collapse, the points where $M = \pm M_U$ are precisely the ones where plastic hinges form.

Example: Beam with Point Loads

If the beam carries point loads only, then the moment can vary only in straight-line fashion between points where concentrated forces (loads or reactions) act, and therefore the actual collapse mechanism must be one in which hinges form at load points, built-in ends, or intermediate supports. Consider the beam shown in Figure 6.3.3(a), indeterminate to the first degree, and requiring two plastic hinges for collapse. The possible mechanisms are (1) with hinges at A and B, (2) with hinges at A and C, and (3) with hinges at B and C. They are shown as Figure 6.3.3(b), (c), and (d), respectively.

Let us look first at mechanism 1. If the downward displacement of point B is Δ, then that of point C is $\frac{1}{2}\Delta$, so that the external work rate is $\alpha F\dot{\Delta} + \frac{1}{2}(1-\alpha)F\dot{\Delta} = \frac{1}{2}(1+\alpha)F\dot{\Delta}$. The angles of rotation of the hinges at A and B, assumed small, are respectively $3\Delta/L$ and $9\Delta/2L$; the total internal dissipation is therefore $15M_U\dot{\Delta}/2L$, and if this is equated to the external work rate, the resulting upper bound on FL/M_U is $15/(1+\alpha)$.

A similar analysis of mechanisms 2 and 3 leads to the respective upper bounds of $12/(2-\alpha)$ and $6/(1-\alpha)$; but the second of these is greater than the first for any α between 0 and 1, so that mechanism 3 may be discarded. It can easily be seen that mechanism 1 gives the lesser upper bound when $\alpha > 2/3$, and mechanism 2 when $\alpha < 2/3$. The ultimate load is therefore given by

$$\frac{F_U L}{M_U} = \begin{cases} \dfrac{12}{2-\alpha}, & \alpha \leq \dfrac{2}{3}, \\[2ex] \dfrac{15}{1+\alpha}, & \alpha \leq \dfrac{2}{3}. \end{cases}$$

The preceding result may also be cast in the form of two inequalities,

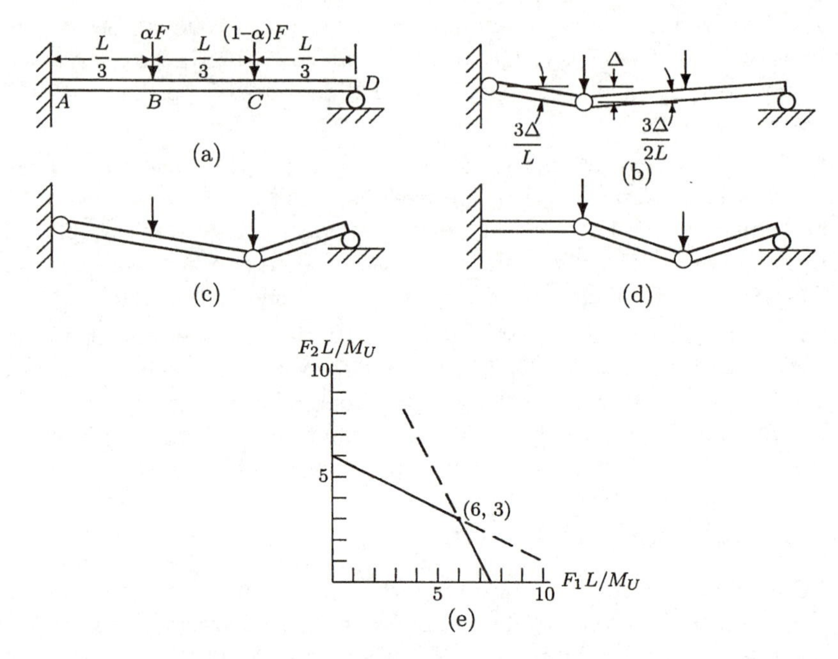

Figure 6.3.3. Single-span beam simply supported at one end and built in at the other, with concentrated loads at the third points: (a) geometry and loading; (b)–(d) collapse mechanisms; (e) interaction diagram.

parametrically dependent on α, that the total load F must obey:

$$\frac{(2-\alpha)FL}{M_U} \leq 12, \quad \frac{(1+\alpha)FL}{M_U} \leq 15.$$

Yet another way of presenting the result would be to regard the loads at B and C as two independent loads F_1 and F_2. The inequalities are accordingly rewritten as

$$\begin{aligned} F_1 + 2F_2 &\leq 12\frac{M_U}{L}, \\ 2F_1 + F_2 &\leq 15\frac{M_U}{L}, \end{aligned} \tag{6.3.7}$$

represented graphically by the interaction diagram shown in Figure 6.3.3(e).

The last pair of inequalities, Equations (6.3.7), can also be derived by means of an equilibrium analysis. Suppose that a plastic hinge has already formed at A, with $M_A = -M_U$; the beam is then statically determinate, and the bending moments at B and C can easily be calculated to be, respectively, $(2F_1 + F_2)L/9 - 2M_U/3$ and $(F_1 + 2F_2)L/9 - M_U/3$. The requirement that these moments not exceed M_U gives precisely the inequalities (6.3.7).

The inequalities are also design criteria: they give the minimum value of M_U that the section must have in order to carry a given set of loads F_1, F_2.

Example: Beam with Distributed Load

Suppose, now, that the beam just examined carries a uniformly distributed load of intensity F/L rather than the point loads. It can still be assumed with some certainty that one of the plastic hinges necessary for collapse will form at the built-in end, but the other hinge can be, in principle, anywhere along the span of the beam. In fact it will form, of course, at the section where the bending moment has a local extremum.

If a hinge has formed at the built-in end ($x = 0$), then the bending moment at any x, $0 \le x \le L$, is

$$M(x) = M_U \left[-\left(1 - \frac{x}{L}\right) + \frac{f}{2}\frac{x}{L}\left(1 - \frac{x}{L}\right) \right],$$

where $f = FL/M_U$. The maximum of the quantity in brackets occurs at $x/L = \frac{1}{2} + 1/f$, and equals $(f - 4 + 4/f)/8 \stackrel{\text{def}}{=} \phi(f)$. Any value of f for which $\phi(f)$ does not exceed unity is a lower bound for $f_U = F_U L/M_U$. For example, $\phi(10) = 0.8$, so that 10 is a lower bound. On the other hand, $\phi(12) = 1.04$, so that f_U must be somewhat less than 12.

The actual value of f_U is obtained by setting $\phi(f)$ equal to unity, giving the quadratic equation

$$f^2 - 12f + 4 = 0.$$

This equation has the two roots $6 \pm 4\sqrt{2}$. Clearly, since what is sought is the greatest lower bound, the larger root must be chosen. We thus obtain $f_U = 11.657$

In a kinematic solution, a mechanism with a plastic hinge at $x = 0$ and another at $x = \alpha L$, with α to be determined, is assumed, as in Figure 6.3.4. If the downward displacement of the hinge is Δ, then the average

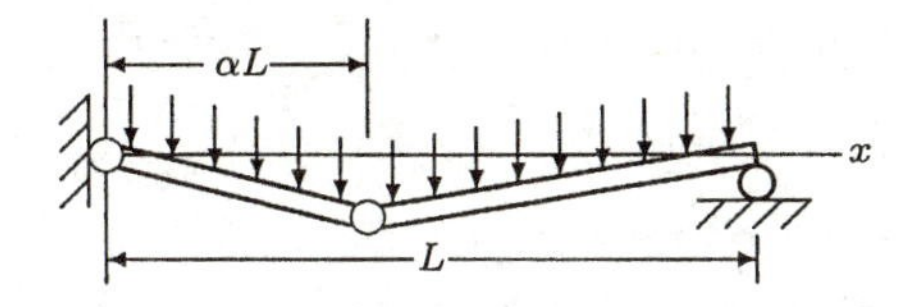

Figure 6.3.4. Uniformly loaded single-span beam, simply supported at one end and built in at the other: collapse mechanisms.

displacement of both rigid portions of the beam is $\Delta/2$, so that the external work rate is $F\dot{\Delta}/2$. The hinge at the built-in end rotates by an angle $\Delta/\alpha L$, and the other hinge by the angle $\Delta/\alpha L + \Delta/(1-\alpha)L$. The total internal dissipation is therefore

$$\frac{2-\alpha}{\alpha(1-\alpha)} M_U \frac{\dot{\Delta}}{L}.$$

It follows that an upper bound to f_U is

$$f = \frac{2(2-\alpha)}{\alpha(1-\alpha)}.$$

The least upper bound is found by minimizing f with respect to α:

$$\begin{aligned}\frac{1}{2}\frac{df}{d\alpha} &= \frac{2-\alpha}{\alpha(1-\alpha)^2} - \frac{2-\alpha}{\alpha^2(1-\alpha)} - \frac{1}{\alpha(1-\alpha)} \\ &= \frac{1}{[\alpha(1-\alpha)]^2}[\alpha(2-\alpha) - (1-\alpha)(2-\alpha) - \alpha(1-\alpha)] = 0,\end{aligned}$$

leading to the quadratic equation

$$\alpha^2 - 4\alpha + 2 = 0.$$

Since α must be less than 1, the only relevant root is $\alpha = 2 - \sqrt{2}$ and gives the least upper bound $f = 6 + 4\sqrt{2} = 11.657$, which, of course, coincides with the previously found greatest lower bound.

Without the analytical solution, assumed values of α give upper bounds that may be satisfactory; for example, $\alpha = 0.5$ leads to $f = 12$, and $\alpha = 0.6$ leads to $f = 11.667$. Moreover, assumed mechanisms can also be used to give lower bounds without resorting to an analytical solution. Consider, for example, the mechanism with $\alpha = 0.5$, corresponding to $f = 12$. The moment distribution corresponding to this mechanism is given by

$$M(x) = -M_U \left[1 - 7\frac{x}{L} + 6\left(\frac{x}{L}\right)^2\right],$$

and is not plastically admissible because $|M|_{max} = M(7L/12) = (25/24)M_U$. If, however, all the moments are multiplied by 24/25, then the distribution becomes plastically admissible, and in equilibrium with a load for which $f = (24/25)12 = 11.52$, which is thus a lower bound.

Example: Continuous Beam

A beam with two or more spans, separated by intermediate supports that exert transverse force reactions, is called a *continuous beam*. The rule governing the degree of static indeterminacy is the same for continuous as for simple beams. However, collapse of a continuous beam may occur in one span only, and does not in general require $r + 1$ plastic hinges. The collapse of a span between two intermediate supports, or between an intermediate support and a built-in end support, requires three hinges. A span between an intermediate support and a simple end support will collapse with only two hinges. It is thus possible for a continuous beam to remain statically indeterminate at collapse. For this reason the kinematic method is preferable by far for the limit analysis of continuous beams.

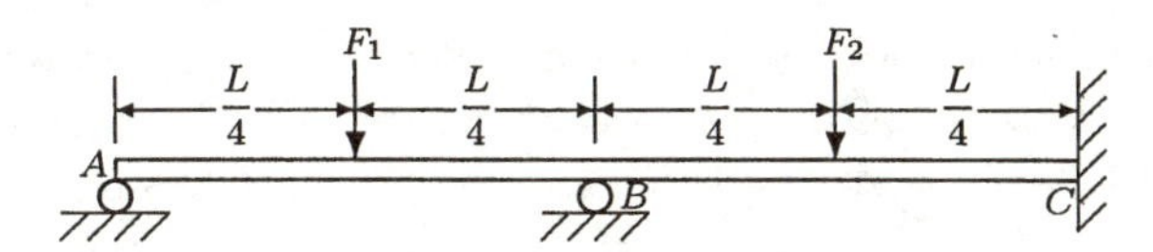

Figure 6.3.5. Continuous beam, simply supported at one end, built in at the other, and with an intermediate support, carrying a concentrated load at the midpoint of each span.

Looking at the beam shown in Figure 6.3.5, we see immediately that the feasible collapse mechanisms are (a) the one in which span AB collapses like the beam of Figure 6.3.4, and (b) the one in which span BC collapses like a beam with both ends built in. In mechanism (b) hinges develop at B (or rather, just to the right of B), at C, and at midspan. The loads F_1, F_2 are governed by the uncoupled inequalities,

$$F_1 \leq (6 + 2\sqrt{2})\frac{M_U}{L}, \qquad F_2 \leq 16\frac{M_U}{L}.$$

These inequalities also specify the minimum value of M_U.

6.3.3. Limit Analysis of Frames

A rigid frame (or simply a frame) is an assemblage of bars that are joined together rigidly, so that they cannot rotate with respect to one another. The joints transmit bending moment, and the members resist the applied loads primarily through bending; axial force and shear are considered secondary effects. Collapse is assumed to occur when sufficient plastic hinges have formed to produce a mechanism. In a multistory frame, collapse may be limited to a single story, and therefore the overall degree of static indeterminacy is not a relevant parameter for the determination of the necessary number of hinges.

Simple Frame

A one-story, one-bay frame such as shown in Figure 6.3.6 is statically indeterminate of degree three, and the collapse of the frame as a whole indeed requires four hinges, as shown in Figures 6.3.6(a) and (c). Consider, however, Figure 6.3.6(d), which illustrates the *beam mechanism*. This mechanism does not entail collapse in the sense of unlimited displacements; the deflection of the beam is limited by that of the columns. In practice, however, a structure may be said to collapse when its displacements can become significantly greater than those in the elastic range. If the axial elongation of the beam is neglected, a deflection Δ of the central hinge requires that the beam-column joints move laterally inward by a distance $\frac{1}{2}L - \sqrt{\frac{1}{4}L^2 - \Delta^2} \doteq \Delta^2/L$. This distance represents the elastic deflection of the columns, Δ_e, which is of the same order of magnitude as the beam deflection when the whole frame is elastic. Now $\Delta/\Delta_e \doteq \sqrt{L/\Delta_e}$ is a large number. We are therefore justified in regarding the beam mechanism as a collapse mechanism.

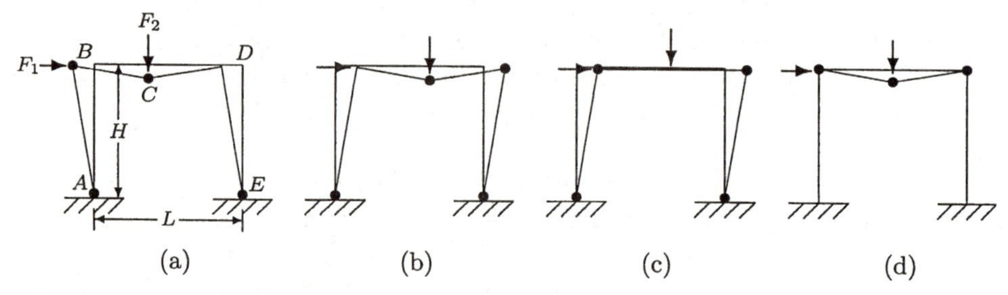

Figure 6.3.6. One-story, one-bay frame: (a)–(c) four-hinge mechanisms; (d) beam mechanism.

The only pertinent collapse mechanisms for the frame of Figure 6.3.6 are the beam mechanism (d), the panel or sidesway mechanism (c), and the composite mechanism (b), which is a superposition of (c) and (d) in which the hinge at B is eliminated. The composite mechanism (a) — the mirror image of (b) — in which joint D is rigid, entails negative work done by the horizontal force and therefore is viable only when this force is zero, in which case it is equivalent to (b). When a hinge is assumed to be at a joint such as B or D, it will actually be in the weaker of the two members meeting there — that is, it forms when the bending moment (which, for equilibrium, must be the same in both members as the joint is approached) reaches the smaller of the two values of M_U.

Let M_{U1}, M_{U2}, and M_{U3} denote the values of M_U in AB, BD, and DE, respectively. If it is assumed that the horizontal load F_1 is, for example, a wind load which is just as likely to act to the left at D as to the right at B (so that the loading of Figure 6.3.6 represents only one of two mirror-image cases), then the frame design should be symmetric, and $M_{U3} = M_{U1}$.

The upper-bound theorem applied to the three mechanisms (b)-(d) gives the following inequalities:

$$F_2L \leq 4M_{U2} + 4\min(M_{U1},\, M_{U2}),$$

$$F_1H \leq 2M_{U1} + 2\min(M_{U1},\, M_{U2}), \tag{6.3.8}$$

$$2F_1H + F_2L \leq 4M_{U1} + 4M_{U2} + 4\min(M_{U1},\, M_{U2}).$$

As before, these inequalities serve both analysis and design. For the purposes of analysis, let us assume equal values of M_U for all three members. The interaction diagram between F_1 and F_2 is then as shown in Figure 6.3.7(a). The design implications of inequalities (6.3.8) are discussed in the following subsection.

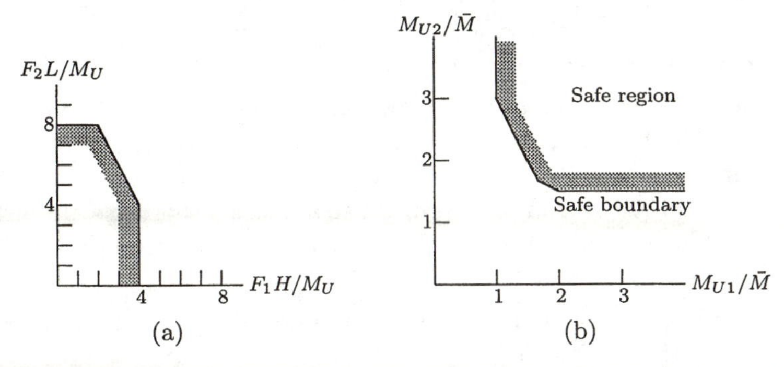

Figure 6.3.7. Interaction diagrams for the frame of Figure 6.3.6: (a) load plane; (b) design plane (see page 391).

Complex Frames

In a frame comprising several stories and bays, the number of possible collapse mechanisms can become quite large. Every transversely loaded member may form a beam mechanism, and each story may produce a panel mechanism. Furthermore, at any joint at which three or more members come together, a plastic hinge may form independently in each member near the joint (if only two members meet, the hinge can form only in the weaker member).

It is convenient to establish a basis of independent mechanisms, called *elementary mechanisms*, such that all mechanisms may be regarded as superpositions of the elementary ones. These elementary mechanisms, as first discussed by Neal and Symonds [1952], consist of all the beam and panel mechanisms, and in addition, of the *joint mechanisms* constituted by the formation of plastic hinges, at a joint, in every one of the members that come together there, resulting in a rotation of the joint [see Figure 6.3.8(e)]. The joint mechanisms are not in themselves collapse mechanisms, since the external work rate associated with them is zero (unless an external moment acts at the joint), but they are used in combination with beam and/or panel mechanisms in order to cancel superfluous hinges.

Let r denote, as before, the degree of redundancy of the frame. A simple method of determining r is to cut the frame at a sufficient number of sections so that it just becomes statically determinate, that is, equivalent to a set of simply supported beams and/or cantilevers; r is then the number of stress resultants (moments, axial forces and shear forces) that can arbitrarily be specified at the cuts. Equivalently, r is the number of sections at which the moment can be arbitrarily prescribed. Suppose, now, that the number of *critical sections* — that is, sections at which a plastic hinge can form — is n. It follows that there are $n-r$ independent relations among the n moments at the critical sections, and these relations are equilibrium equations. Each such equation can be associated, by means of the principle of virtual work, with a mechanism. Consequently, there are $n-r$ independent mechanisms.

As an example, consider the two-bay frame shown in Figure 6.3.8(a).[1] By means of two cuts, the frame can be transformed into three disconnected cantilevers, and therefore $r = 2\times 3 = 6$. The critical sections, as shown, number 10. Consequently the frame has four independent mechanisms. In terms of elementary mechanisms, these are (b)–(c) the two beam mechanisms, (d) the panel mechanism, and (e) the joint mechanism.

In the method of **superposition of mechanisms** due to Neal and Symonds [1952], the analysis begins by determining the upper bounds predicted by the elementary beam and panel mechanisms. Because of the symmetry of the structure, the two beam mechanisms give the same upper bounds. We thus obtain the two inequalities

$$(2F)L\dot{\theta} \le M_0\dot{\theta} + 2M_0(2\dot{\theta}) + 2M_0\dot{\theta} = 7M_0\dot{\theta}, \qquad \text{(b, c)}$$

$$F(2L)\dot{\theta} \le (4\cdot M_0 + 2\cdot 1.5M_0)\dot{\theta} = 7M_0\dot{\theta}, \qquad \text{(d)}$$

both of which yield the upper bound $F = 3.5M_0/L$.

In order to improve the upper bound, we proceed to study composite mechanisms. Mechanism (f) is a superposition of (c), (d), and (e) in which the hinges at sections 5 and 6 are eliminated, while a hinge is created at 4. The internal dissipation of the mechanisms can therefore be obtained by subracting from the sum of the right-hand members of the two preceding inequalities the quantity $(1.5M_0 + 2M_0 - 2M_0)\dot{\theta} = 1.5M_0\dot{\theta}$, or $(7+7-1.5)M_0\dot{\theta} = 12.5M_0\dot{\theta}$. The external work rate is just the sum of the left-hand sides, or $4FL\dot{\theta}$. Mechanism (f) therefore gives the upper bound $F = 3.125M_0/L$.

Mechanism (g) is a further superposition of (f) and (b) in which the hinge at 2 is eliminated. The internal dissipation is $(12.5+7-1)M_0\dot{\theta} = 18.5M_0\dot{\theta}$, while the external work rate is $6FL\dot{\theta}$. We thus obtain the even smaller upper bound of $F = 3.083M_0/L$.

In the present example, it appears that we have run out of reasonable mechanisms, and the result should give us the collapse load. In more complex cases, it may be quite difficult to make sure that all the possible collapse

[1]The circled numbers next to the members mean that the value of M_U for a member is the given number times a reference moment M_0.

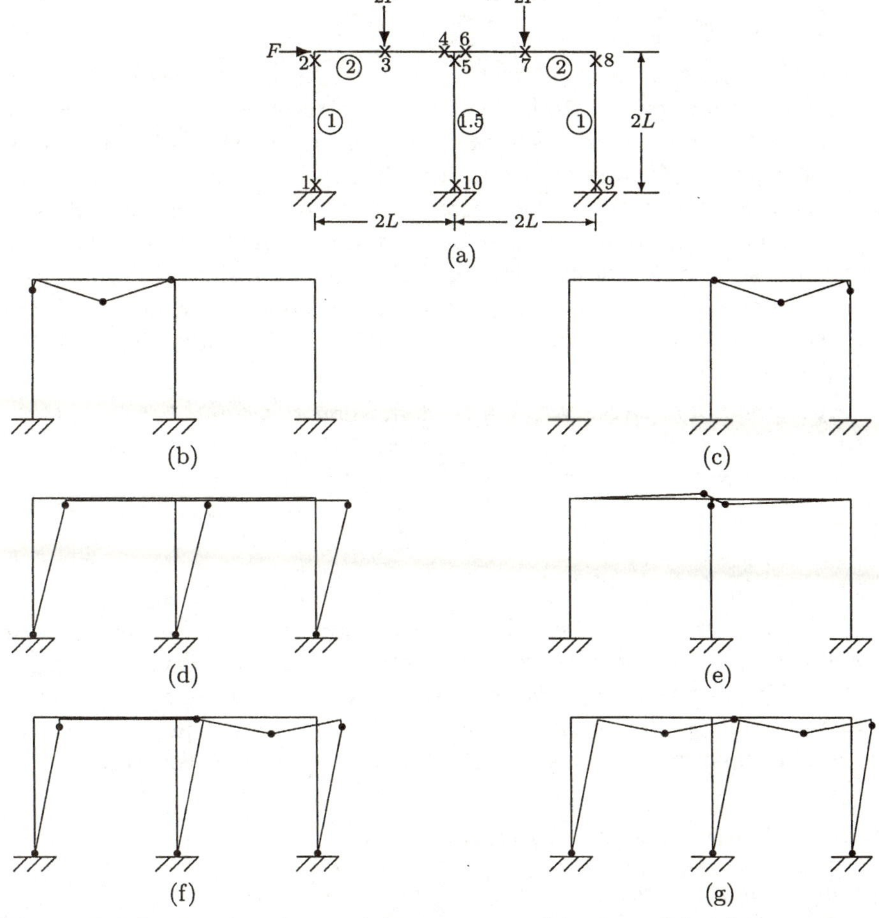

Figure 6.3.8. Two-bay frame: (a) geometry and loading; (b)–(c) beam mechanisms; (d) panel mechanism; (e) joint mechanism; (f)–(g) composite mechanisms.

mechanisms have been explored. The only way to check whether the best upper bound that has been found indeed gives the collapse load is to see if it is also a lower bound, that is, to find a statically admissible moment distribution such that $|M| = M_U$ at all sections corresponding to hinges in the mechanism, and $|M| \le M_U$ elsewhere. In the present example this is easy, since the optimal mechanism represents total collapse and thus involves $r+1 = 7$ hinges, leaving the structure statically determinate at collapse. The four independent equilibrium equations (which can be formed by applying the principle of virtual work to the elementary mechanisms) and the seven hinge conditions give eleven equations for the ten critical-section moments and the load F. It turns out that the moments at 2, 5, and 6 do not exceed the local values of M_U, and the load is in fact equal to the best upper-bound value derived above.

The situation is more difficult in multistory frames, in which the best mechanism found by the method of superposition of mechanisms often represents partial collapse, with fewer than $r+1$ plastic hinges and therefore not enough equations for a rigorous moment check. If the degree of redundancy remaining at failure is small (one or two), a trial-and-error procedure is usually applied: guesses are made for a sufficient number of critical-section moments so that the equilibrium equations can be solved. In more complicated cases, the moment-distribution method due to Horne [1954] and English [1954] may be used; for examples of application, see Hodge [1959], Chapter 3.

A method of analysis based on the lower-bound theorem is the **method of inequalities**, due to Neal and Symonds [1951]. The critical-section moments M_i ($i = 1, \ldots, n$) are governed by the $2n$ inequalities

$$-M_{Ui} \le M_i \le M_{Ui},$$

and by the $n - r$ equilibrium equations, at least one of which contains the load F. The equilibrium equations can be transformed by means of linear combinations so that only one of them contains F. The problem now is again one of linear programming: the equation containing F serves as the defining equation for F as the function of the M_i that is to be maximized, with the remaining equations, as well as the inequalities, serving as constraints.

6.3.4. Limit Design of Frames

Limit Design of a Simple Frame

The achievement of a minimum-weight design is not so simple for frames as it is for trusses, because there is no simple proportionality, or indeed any one-to-one relation, between weight and strength. In simple frames, such as the one analyzed in the preceding subsection, a trial-and-error approach is usually the easiest (see Heyman [1953]). In complex frames a more systematic approach, based on some simplifying assumptions, is necessary in order that the problem may be converted into one of mathematical (not necessarily linear) programming.

As a first example of limit design of a frame, we consider the fixed-base rectangular frame of Figure 6.3.6 (page 386), for which we derived inequalities (6.3.8). Let the design loads (working loads times appropriate safety factors) be such that $F_2L = 3F_1H \stackrel{\text{def}}{=} 12\bar{M}$, where $\bar{M}$ is a reference quantity having the dimensions of a moment. The sections chosen for the columns and the beam must then satisfy the following inequalities:

$$M_{U2} + \min(M_{U1}, M_{U2}) \ge 3\bar{M}, \tag{a}$$

Table 6.3.1. Plastic Moduli of Selected Wide-Flange Sections

Z (in.^{3})	Shape	Z (in.^{3})	Shape	Z (in.^{3})	Shape
287	W14×159	212	W14×120	150	W16×77
260	W14×145	198	W16×100	130	W16×67
234	W14×132	175	W16×89	105	W16×57

$$M_{U1} + \min(M_{U1},\, M_{U2}) \ge 2\bar{M}, \tag{b}$$

$$M_{U1} + M_{U2} + \min(M_{U1},\, M_{U2}) \ge 5\bar{M}. \tag{c}$$

The solution of these inequalities is shown in Figure 6.3.7(b) (page 387) as the "safe region" whose polygonal boundary will be called the *safe boundary*. Any point in the safe region or on the safe boundary will be said to represent a *safe design*.

We note that M_{U1} must be at least equal to $\bar{M}$ and M_{U2} to $1.5\bar{M}$; but if these criteria are met, there is no need for M_{U1} to be greater than $2\bar{M}$, or for M_{U2} to be greater than $3\bar{M}$. Thus the choice of sections can be made from a rather restricted range. Let the frame dimensions be L = 24 ft. and H = 12 ft., so that the total weight of the frame is proportional to $w_1 + w_2$, where w_1 and w_2 denote the weight per unit length of the column and beam, respectively. The design loads will be taken as $F_1 = 1.0 \times 10^5$ lb and $F_2 = 1.5 \times 10^5$ lb, corresponding to $\bar{M} = 3.6 \times 10^6$ lb-in. With the usual value of $\sigma_Y = 36 \times 10^3$ lb/in.^{2} for A36 structural steel, it follows that that $\bar{M}/\sigma_Y = 100$ in.^{3}. Then the ranges of the plastic modulus $Z = M_U/\sigma_Y$ for the columns and the beam are

$$100 \text{ in.}^3 \le Z_1 \le 200 \text{ in.}^3, \qquad 150 \text{ in.}^3 \le Z_2 \le 300 \text{ in.}^3,$$

and they must obey the inequalities

$$2Z_1 + Z_2 \ge 500 \text{ in.}^3, \quad Z_1 + 2Z_2 \ge 500 \text{ in.}^3.$$

It will be assumed that for architectural reasons, the section depth is to be limited to 16 in. The choice will be made from standard wide-flange sections, where the designation W$d \times w$ refers to a section whose depth is d (in inches) and whose weight per unit length is w (in pounds per foot). The strength of a wide-flange beam, for a given weight, increases sharply with the depth, and therefore the deepest available sections should be chosen for economy. A listing of section properties for selected wide-flange sections is shown in Table 6.3.1.

Four designs will be tried.

1. In a design with the lightest possible columns, a W16×57 section presents $Z_1 = 105$ in.^{3}, requiring $Z_{U2} \ge 290$ in.^{3}. Unfortunately, no standard

W16 section has a plastic modulus close to this value. We consequently choose, for the beam, a W14×159 section, giving $Z_2 = 287$ in.^{3}. If this is assumed to be close enough, we obtain a design with $w_1 + w_2 = 216$ lb/ft.

2. The lightest possible beam section is a W16×77, with $Z_2 = 150$ in.^{3}, requiring columns with $Z_1 \doteq 200$ in.^{3}, which is provided by a W16×100 ($Z = 198$ in.^{3}); this choice yields $w_1 + w_2 = 177$ lb/ft., a considerable improvement over the first trial.

3. The beam and column sections of design 2 can be reversed, yielding the same weight.

4. A beam section intermediate between those in designs 2 and 3 is a W16×89 and gives $Z_2 = 175$ in.^{3}; this requires $Z_1 \geq 162.5$ in.^{3}, and the lightest section satisfying this criterion is again W16×89; thus $w_1 + w_2 = 178$ lb/ft., virtually the same as designs 2 and 3, and superior to them by virtue of the greater ease of connections resulting from having the same section throughout. The minimum required value of $Z_1 = Z_2$ is 167 in.^{3}, so that this design carries an overdesign factor of 1.05. The additional margin of safety provides an allowance for axial force. If, for example, Equation (6.2.7) is used for the interaction, then each member can carry an axial force up to $(1 - 0.85/1.05)P_U = 0.19P_U$ with no loss in moment-carrying capacity.

On the basis of design 4, $M_{U1} = M_{U2} = 6.3 \times 10^6$ lb-in. With the overdesign factor included, the collapse forces are $F_1 = 1.05 \times 10^5$ lb and $F_2 = 1.575 \times 10^5$ lb The weight of the beam is about 2000 lb, and we are therefore justified, in retrospect, in having neglected it.

Since the structure is statically determinate at collapse, an equilibrium analysis can easily be performed. The moment at the only other critical section, namely B, is found to be of magnitude 3.78×10^6 lb-in., so that the yield criterion is nowhere violated, and the correct mechanism was chosen. While this last result is obvious in the present example, in frames in which a large number of possible mechanisms exists, the moment check is a necessity, as in analysis.

The same equilibrium analysis shows that the beam and column DE carry compressive axial forces of 0.875×10^5 lb, and that the maximum shear force (in CD and DE) has the same magnitude. We consider, first, the effect of axial force. The slenderness ratio of the columns, even if they are taken as doubly pinned, is about 20, so that they are not expected to buckle elastically. The cross-sectional area of a W16 × 89 section is 26.2 in.^{2}, so that $P_U = 9.432 \times 10^5$ lb, and $|P/P_U| = 0.093 < 0.19$.

We consider, finally, the effect of shear. The web area of the section is 7.9 in.^{2}, and therefore the average shear stress in the web is some 11×10^3 lb/in.^{2}, well below the shear yield stress of about 20×10^3 lb/in.^{2} for A36 steel. It follows that the frame can be safely designed on the basis of bending alone.

Foulkes' Theorems

For a complex frame, the time required for a trial-and-error method of limit design would be prohibitively long. Any systematic approach is based on the assumption of a functional relation between the weight per unit length and the flexural strength (as measured by M_U or Z) of a beam. Clearly, no such relation exists in general, but an approximate relation can be established for a limited range of sections that is used in the design of a frame. If a relation of the form

$$w = a + bM_U \tag{6.3.9}$$

can be found, then the problem of minimum-weight design of a frame can also be transformed into one of linear programming: the total weight is

$$W = a\sum_i L_i + b\sum_i L_i M_{Ui},$$

the summation being over all members, and therefore the objective function is

$$G(M_{U1}, \ldots) = \sum_{i=1}^{n} L_i M_{Ui},$$

independently of the parameters a and b in the approximate representation (6.3.9); this approach was presented by Foulkes [1953].

In the just-studied simple frame, for the three W16 sections considered, the unit weight (in lb/ft.) is very nearly given by

$$w \doteq 5 + 0.48Z,$$

with Z in cubic inches. The total weight of the frame, to within an additive constant, is therefore proportional to $Z_1 + Z_2$. The theoretical minimum-weight design is thus achieved by finding the point (Z_1, Z_2) on the safe boundary where the constant-weight line $Z_1 + Z_2 =$ const. is tangent to the boundary. This point is $Z_1 = Z_2 = 167$ in.^{3}, and would represent the actual minimum-weight design if a section with this value of Z could be found.

Note that the point of tangency between the constant-weight line and the safe boundary is a vertex of the boundary, that is, an intersection of two of the lines forming this boundary.[1] Since each such line represents a particular collapse mechanism, the theoretical minimum-weight frame can collapse in either of two mechanisms, or in a linear combination of the two. In particular, a linear combination of the two mechanisms, with nonnegative coefficients, can be found so that the inequality produced by the combined

[1]If the constant-weight lines are parallel to one of the boundary lines, then any point on the boundary segment of this line, including the two vertices, represents a minimum-weight design,

mechanism is represented by a line that is parallel to the constant-weight lines. Such a mechanism is known as a *Foulkes mechanism* or a *weight-compatible mechanism.*

If n independent values of M_U may be used in a design, then the *design space* is n-dimensional. The theoretical minimum-weight design is represented by the point of tangency of a constant-weight hyperplane with the safe boundary, which is made up of intersecting hyperplanes. If the point is unique, then it is a vertex of the boundary and is therefore an intersection of at least n of the hyperplanes making up the boundary, each representing a distinct collapse mechanism. A Foulkes mechanism can be formed by combining these mechanisms linearly so that the combined mechanism is represented by a hyperplane that is parallel to the constant-weight hyperplanes.

Consider a frame design for which a Foulkes mechanism can be assumed. Let L_i denote the combined length of all the members whose ultimate moment is M_{Ui}. In a Foulkes mechanism, the plastic hinges forming in these members have the property that the sum of the absolute values of their angular velocities is proportional to L_i. If the proportionality factor is c, then the total dissipation in the mechanism is

$$c \sum_{i=1}^{n} M_{Ui} L_i = cG.$$

If, moreover, a statically and plastically admissible distribution of bending moments compatible with the design can be found, then the design loads F_j are the collapse loads, and therefore, by virtual work

$$\sum_j F_j v_j = cG,$$

where the v_j are the velocities conjugate to the F_j in the Foulkes mechanism.

Consider, now, any other safe frame design, described by values M_{Ui}^* of the ultimate moments. Since the Foulkes mechanism is a kinematically admissible mechanism, it follows from upper-bound theorem that

$$\sum_j F_j v_j \le c \sum_{i=1}^{n} M_{Ui}^* L_i = cG^*,$$

and therefore $G^* \ge G$. This result, due to Foulkes [1954], may be stated in words as follows: *A frame design that admits a Foulkes mechanism and a compatible, statically and plastically admissible distribution of bending moments is a minimum-weight design.*

Upper-bound and lower-bound theorems for minimum-weight design were also established by Foulkes. The upper-bound theorem is obvious, since any safe design provides an upper bound to the minimum weight. Conversely, any design based on a Foulkes mechanism, without necessarily satisfying the condition of admissible moments, provides a lower bound to the minimum weight. A general method, based on Foulkes' theorems, for the mimimum-weight design of highly redundant frames was first developed by Heyman and Prager [1958].

The concept of a Foulkes mechanism can be extended to structures other than frames, including those modeled as continua. If, for example, a beam can be designed with an arbitrarily varying cross-section, then a minimum-weight design, based on the relation (6.3.9), is one in which the moment distribution at collapse is such that $|M| = M_U$ everywhere. Plastic flow, in this case, is not localized in hinges but occurs throughout the beam, with a curvature-rate distribution $\dot{\kappa}$. Since the plastic dissipation in a Foulkes mechanism is proportional to G, it follows that

$$\int_0^L M_U|\dot{\kappa}|\,dx = c\int_0^L M_U\,dx,$$

and the Foulkes mechanism is one in which $|\dot{\kappa}| = c$.

Similarly, a Foulkes mechanism for a plate obeying the Tresca criterion is one in which

$$|\dot{\kappa}_1| + |\dot{\kappa}_2| + |\dot{\kappa}_1 + \dot{\kappa}_2| = 2c,$$

where $\dot{\kappa}_1$ and $\dot{\kappa}_2$ are the principal curvature rates.

The general criterion for minimum-weight design of continua is due to Drucker and Shield [1957]. Applications to plate design were discussed by Hopkins and Prager [1955], Prager [1955b], and Freiberger and Tekinalp [1956], and to shells by Onat and Prager [1955], Freiberger [1956b] and Onat, Schumann, and Shield [1957]. The results for variable-section beams were applied to frame design by Heyman [1959, 1960] and by Save and Prager [1963]. Further contributions are due to Chan [1969], Maier, Srinivasam, and Save [1972], and Munro [1979]. Many of these results are reviewed in the books by Neal [1963], Massonnet and Save [1965], Heyman [1971], Save and Massonnet [1972], Rozvany [1976], Horne [1979], and Borkowski [1988].

Additional Remarks

1. In both the analysis and the design of frames, all loads were assumed to be concentrated, thus fixing in advance the locations of the critical sections. If any span carries a distributed load, then the critical section in that span must be assumed, and therefore any mechanism gives an upper bound to the collapse load; improvements to the upper bound can be achieved by changing the hinge locations. On the other hand, if any distributed load is replaced by a statically equivalent (equipollent) set of concentrated loads, then the collapse load calculated on the basis of the concentrated loads is

a lower bound on the collapse load for the distributed load. This result, derived by Symonds and Neal [1951], is known as the **load-replacement theorem**.

2. Only rectangular frames were studied as examples. Frames with inclined members, such as the gable frame of Figure 6.3.9, can be studied analogously. The frame shown has three independent mechanisms; these can

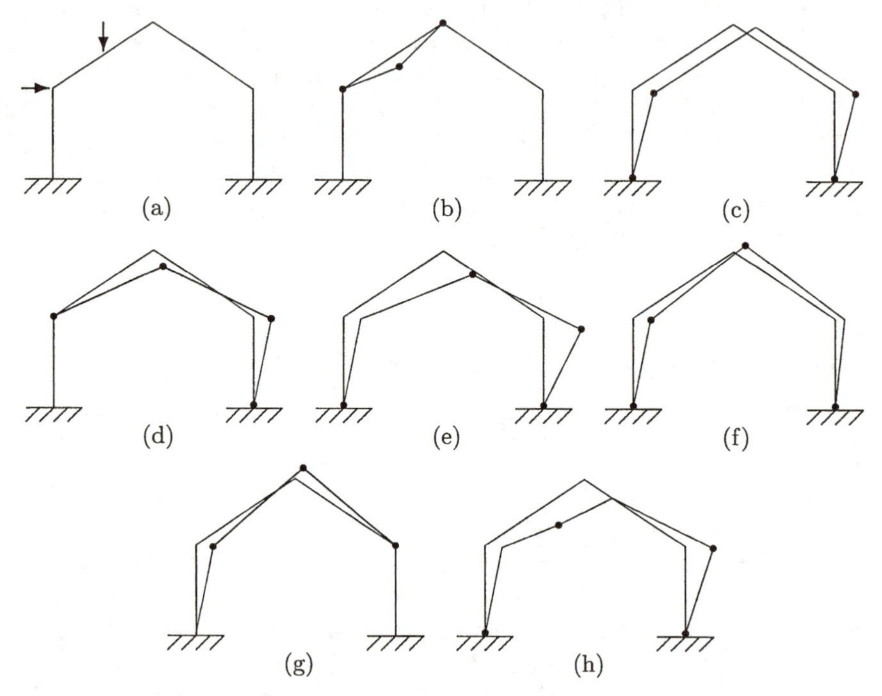

Figure 6.3.9. Gable frame: (a) geometry and loading; (b) beam mechanism; (c) panel mechanism; (d) gable mechanism; (e)–(f) other panel mechanisms; (h) composite mechanism.

be taken as the beam mechanism (b), the panel mechanism (c), and the *gable mechanism* (d), which may also be regarded as a kind of panel mechanism. Other panel mechanisms, each of which is a combination of (c) and (d) with one hinge eliminated, are shown as (e)–(g), and a composite mechanism as (h).

3. A relation between weight and strength given by (6.3.9) can be a reasonable approximation to the properties of actual sections if their number is small. For a larger range of sections, a better approximation is usually obtained with a nonlinear relation such as $w \propto M_U^\alpha$, where $\alpha = 2/3$ for geometrically similar sections, and $\alpha \doteq 0.6$ gives a good approximation for many standard I-beam sections. The problem of minimum-weight design then becomes one of *nonlinear programming*. The computational implemen-

tation of nonlinear programming is not yet at a point where it can be readily applied to the plastic design of complex structures.

Exercises: Section 6.3

1. The truss of Figure 6.3.1(d) is subject to equal horizontal forces F acting to the right at B and C, and a downward vertical force $2F$ acting at B. If the ultimate bar force P_U is the same in all bars and is equal in tension and compression, find the smallest upper bound for F_U by analyzing mechanisms, and verify that it is also a lower bound.

2. Find a minimum-weight design for the truss of the Exercise 1, assuming that $P_U = \pm\sigma_Y A$ in every bar.

3. A beam of span L, built in at the left end and simply supported at the right end, carries a load αF uniformly distributed over the left half and a concentrated load $(1-\alpha)F$ at a distance $L/4$ from the right end.

   (a) Find F_U as a function of α by means of both a kinematic and a static analysis.

   (b) Show how some reasonable upper and lower bounds can be found by means of assumed mechanisms.

   (c) Find the minimum value of M_U as a function of α.

4. In a continuous beam like that of Figure 6.3.5, the loads F_1 and F_2 are uniformly distributed over the respective spans rather than concentrated. Plot an interaction diagram between F_1L/M_U and F_2L/M_U.

5. If the rate of deflection of a beam is denoted $v(x)$ and the curvature rate is $\dot{\kappa} = v''(x)$, show how a Foulkes mechanism for a beam built in at both ends can be generated by means of a deflection curve made up of parabolic arcs. Find a minimum-weight design for an ideal sandwich beam of uniform depth but variable flange area.

6. An asymmetric frame is shown in the adjacent figure.

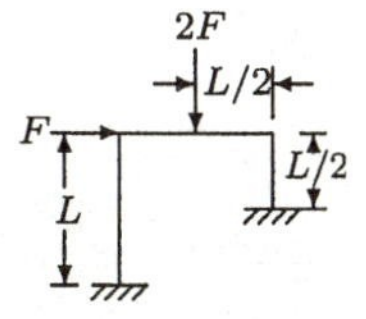

   (a) Assuming M_U to be the same in all members, find F_U.

   (b) Assuming only that the value of M_U in the short column is one-half of that in the long column, find a minimum-weight design for the frame.

7. A two-story one-bay frame has the geometry and loading shown in the adjacent figure, and is assumed to collapse under the loading.

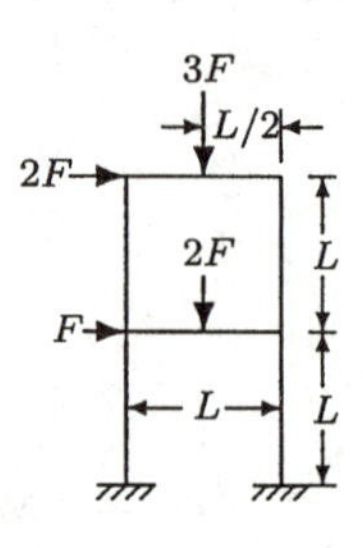

   (a) Find the value of M_U if this value is the same in all members.

   (b) Find the value of M_U if this is the value of the ultimate moment in the lower columns and beam, while in the upper columns and beam this value is $0.5M_U$.

   (c) Assuming only that M_U has one value (say M_{U1}) in the lower columns and beam and another value (say M_{U2}) in the upper columns and beam, find the minimum-weight design, and plot the design diagram relating M_{U1}/FL and M_{U2}/FL.

Section 6.4 Limit Analysis of Plates and Shells

The concept of plate yielding was already developed in Article 5.2.3, with the yield criterion given in terms of the moments $M_{\alpha\beta}$. As shown in 6.4.1, these moments serve as the generalized stresses in the limit analysis of plates.

In shells (as in arches), the equilibrium of moments is coupled with membrane forces (analogous to axial forces), and therefore both moments and membrane forces appear as generalized stresses. With the greater number of generalized stresses, the yield loci become more difficult to represent and approximations leading to piecewise linear yield criteria often become necessary. The theory is developed in 6.4.2, and examples are studied in 6.4.3. Thorough coverage of the material in this section can be found in the book by Save and Massonnet [1972].

6.4.1. Limit Analysis of Plates

Plastic Flow of Plates

With the stresses given by Equation (5.2.9), the plastic dissipation per unit volume at plastic collapse (or, equivalently, under the hypothesis of rigid–plastic behavior) is

$$D_p = \sigma_{\alpha\beta}\dot{\varepsilon}_{\alpha\beta} = -\frac{4}{h^2}M_{\alpha\beta}\operatorname{sgn} x_3(-x_3\dot{\kappa}_{\alpha\beta}) = \frac{4}{h^2}|x_3|M_{\alpha\beta}\dot{\kappa}_{\alpha\beta},$$

where h is the plate thickness. Integrating through the thickness, we obtain the plastic dissipation per unit area,

$$\bar{D}_p = M_{\alpha\beta}\dot{\kappa}_{\alpha\beta}. \tag{6.4.1}$$

Now consider any stress distribution $\sigma^*_{\alpha\beta}$ that does not violate the yield criterion and gives zero membrane forces. If the moments resulting from $\sigma^*_{\alpha\beta}$ are $M^*_{\alpha\beta}$, then the inequality (3.2.4) expressing the principle of maximum plastic dissipation can be integrated through the thickness, giving

$$(M_{\alpha\beta} - M^*_{\alpha\beta})\dot{\kappa}_{\alpha\beta} \geq 0. \tag{6.4.2}$$

The moments and the curvature rates are thus respectively the generalized stresses and generalized strain rates for the analysis of the bending collapse of plates. Since yield criteria in terms of moments have already been formulated (see 5.2.2), the associated flow rule can be deduced from (6.4.2) to have the form

$$\kappa_{\alpha\beta} = \dot{\lambda}\frac{\partial f}{\partial M_{\alpha\beta}}.$$

For the Mises, Tresca, and Johansen criteria, respectively, we obtain the plastic dissipations per unit area as

$$\bar{D}_p(\dot{\boldsymbol{\kappa}}) = \tfrac{2}{\sqrt{3}} M_U \sqrt{\dot{\kappa}_1^2 + \dot{\kappa}_1\dot{\kappa}_2 + \dot{\kappa}_2^2} \qquad \text{(Mises)},$$
$$\bar{D}_p(\dot{\boldsymbol{\kappa}}) = \tfrac{1}{2} M_U (|\dot{\kappa}_1| + |\dot{\kappa}_2| + |\dot{\kappa}_1 + \dot{\kappa}_2|) \qquad \text{(Tresca)},$$
$$\bar{D}_p(\dot{\boldsymbol{\kappa}}) = M_U (|\dot{\kappa}_1| + |\dot{\kappa}_2|) \qquad \text{(Johansen)}.$$

With moments replacing stresses and curvature rates replacing strain rates, the theorems of limit analysis can be applied to the estimation of collapse loads of plates as in any other problems. As with the plane problems studied in the preceding section, much more use can be made of the upper-bound theorem than of the lower-bound theorem.

Hinge Curves

A *hinge curve* or *yield curve* (hinge line or yield line if straight) is a curve in the region A occupied by the middle plane, across which the slope of the deflection w is discontinuous. Let $\Delta\theta$ denote the change in slope encountered along a line normal to the curve, and suppose that this change takes place uniformly over a narrow zone of width δ. Within this zone, the numerically larger principal curvature rate, say $\dot{\kappa}_1$, is given by $|\dot{\kappa}_1| = \Delta\dot{\theta}/\delta$, with $|\dot{\kappa}_1| \gg |\dot{\kappa}_2|$. If we integrate $\bar{D}_p(\dot{\boldsymbol{\kappa}})$ through the width of the zone, we obtain, for the plastic dissipation per unit length of the hinge curve, $(2M_U/\sqrt{3})|\Delta\dot{\theta}|$ for the Mises criterion and $M_U|\Delta\dot{\theta}|$ for the Tresca and Johansen criteria. The total internal dissipation needed in limit analysis,

$$\mathcal{D}_{int} = \int_R D_p(\dot{\varepsilon})dV,$$

is therefore

$$\mathcal{D}_{int} = \frac{2}{\sqrt{3}} M_U \left[\int_A \sqrt{\dot\kappa_1^2 + \dot\kappa_1\dot\kappa_2 + \dot\kappa_2^2}\, dA + \int_{HC} |\Delta\dot\theta|\, ds \right] \quad \text{(Mises)},$$

$$\mathcal{D}_{int} = M_U \left[\int_A \frac{1}{2}(|\dot\kappa_1| + |\dot\kappa_2| + |\dot\kappa_1 + \dot\kappa_2|) dA + \int_{HC} |\Delta\dot\theta|\, ds \right] \quad \text{(Tresca)},$$

$$\mathcal{D}_{int} = M_U \left[\int_A (|\dot\kappa_1| + |\dot\kappa_2|)\, dA + \int_{HC} |\Delta\dot\theta| ds \right] \quad \text{(Johansen)},$$

where HC denotes the hinge curve.

Comparing the Mises and Tresca results, we note that the expressions inside the square brackets coincide if

$$\frac{1}{2}(|\dot\kappa_1| + |\dot\kappa_2| + |\dot\kappa_1 + \dot\kappa_2|) = \sqrt{\dot\kappa_1^2 + \dot\kappa_1\dot\kappa_2 + \dot\kappa_2^2},$$

and this occurs if and only if one of three conditions $\dot\kappa_1 = 0$, $\dot\kappa_2 = 0$, or $\dot\kappa_1 + \dot\kappa_2 = 0$ is met. Consequently, an upper-bound load obtained for a Tresca plate on the basis of a velocity field obeying one of these conditions almost everywhere[1] serves for a Mises plate as well if multiplied by $2/\sqrt{3}$.

Yield-line theory was developed by Johansen [1932] for the ultimate-load design of reinforced-concrete slabs. In a polygonal plate, the yield curves are yield lines and divide the plate into portions that move as rigid bodies, so that all the dissipation takes place on the yield lines. If any edge is clamped, then either the portion of the plate adjacent to it does not move, or the edge itself becomes a yield line. In accordance with the upper-bound theorem, the yield-line pattern must be such as to minimize $\mathcal{D}_{int}/\mathcal{D}_{ext}$, where $\mathcal{D}_{ext} = \int_A q\dot w\, dA$ is the external work rate.

Applications of Yield-Line Theory

As an example, consider a simply supported rectangular plate of dimensions $2a \times 2b$ under a uniform load $F/4ab$. "Simply supported" is interpreted in the traditional sense, that is, the plate deflection vanishes at the edges; such a plate is called "position-fixed" by Johnson and Mellor [1973]. According to a weaker definition of simple support, used by Johnson and Mellor, only downward deflection is prevented, and the plate is free to lift off.

For the traditional definition, the yield-line pattern may be one of the two shown in Figure 6.4.1.

Actually, pattern (a) is a special case of pattern (b), with $c = a$, but because of its relative simplicity it is instructive to study it separately. It can easily be shown that, if v_0 is the center velocity, then the slope-discontinuity rate on the yield lines is $\sqrt{(v_0/a)^2 + (v_0/b)^2}$, so that, for a Tresca material,

$$\mathcal{D}_{int} = 4M_U v_0 \sqrt{a^2 + b^2} \sqrt{\frac{1}{a^2} + \frac{1}{b^2}} = 4M_U v_0 \left(\frac{a}{b} + \frac{b}{a} \right),$$

[1] "Almost everywhere" means everywhere except on a set of points whose total area is zero.

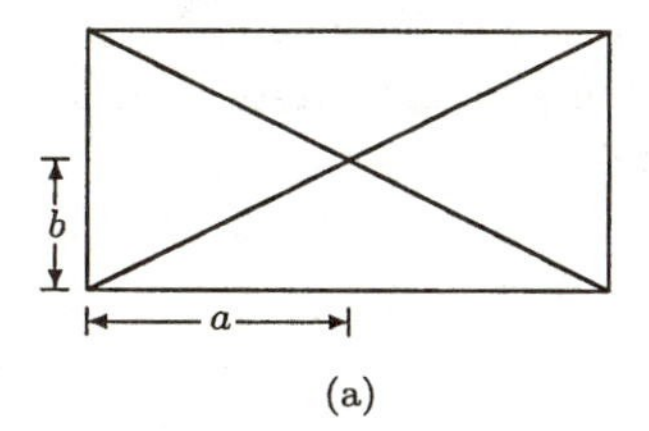

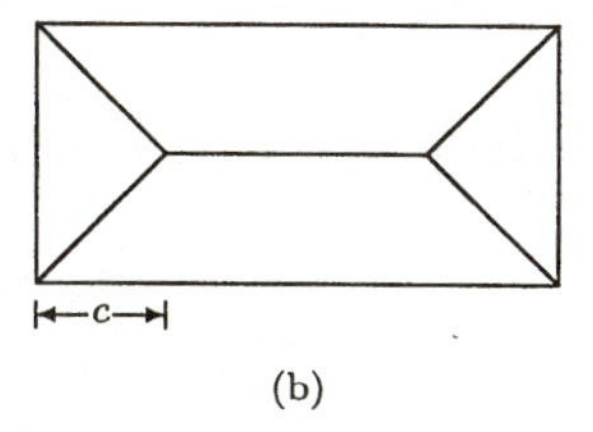

Figure 6.4.1. Yield-line patterns for a simply supported rectangular plate. Pattern (a) is a special case of pattern (b), with $c = a$.

while $\mathcal{D}_{ext} = 1thirdFv_0$, so that the upper bound on F is $12M_U(a/b + b/a)$. For pattern (b), we have

$$\mathcal{D}_{int} = 4M_U v_0 \left(\frac{c}{b} + \frac{b}{c}\right) + 2(a-c)M_U \cdot 2\frac{v_0}{b} = 4M_U v_0 \left(\frac{a}{b} + \frac{b}{c}\right)$$

and

$$\mathcal{D}_{ext} = \frac{F}{4ab} v_0 \left[\frac{1}{3} \cdot 4bc + \frac{1}{2} \cdot 4b(a-c)\right] = \frac{1}{6} F v_0 \left(3 - \frac{c}{a}\right).$$

The ratio $\mathcal{D}_{int}/\mathcal{D}_{ext}$ is a minimum at c given by

$$\frac{c}{a} = \frac{b}{a}\left[\sqrt{3 + \left(\frac{b}{a}\right)^2} - \frac{b}{a}\right],$$

which can be seen to be less than one for all $b/a < 1$. Consequently,

$$F \le \frac{8}{3} M_U \frac{a}{b} \left[\sqrt{3 + \left(\frac{b}{a}\right)^2} + \frac{b}{a}\right]^2$$

which is less than the upper bound given by pattern (a) except in the case of the square, when the two upper bounds are necessarily equal. The preceding results hold for a plate made of a Mises material if M_U is replaced by $2M_U/\sqrt{3}$.

If the plate is clamped, then the same yield-line pattern as above can be assumed, provided that *the edges become yield lines*. It can easily be verified that the total dissipation along the edges is just equal to that on the interior yield lines, so that the upper-bound load obtained by this method is twice what it is for the simply supported plate. It is shown later, however, that a better upper-bound load can be found for clamped plates in general.

A lower bound for the simply supported rectangular plate can be obtained by assuming the statically admissible moment field

$$m_{11} = 1 - \left(\frac{x_1}{a}\right)^2, \quad m_{22} = 1 - \left(\frac{x_2}{b}\right)^2, \quad m_{12} = -\lambda \frac{x_1 x_2}{ab},$$

where $m_{\alpha\beta} = M_{\alpha\beta}/M_U$, and λ is a constant to be determined. It can readily be verified that the center and the midpoints of the edges yield according to all three criteria. The equilibrium condition

$$m_{\alpha\beta,\alpha\beta} = -\frac{F}{4abM_U}$$

gives

$$F = 8M_U\left(\frac{a}{b} + \frac{b}{a} + \lambda\right).$$

The best lower bound is therefore obtained with the largest λ such that the yield criterion is not violated anywhere. This occurs, for each of the three yield criteria, if the criterion is satisfied at the corners, where $m_{11} = m_{22} = 0$ and $|m_{12}| = \lambda$. Thus the largest admissible value of λ turns out to be 1 for the Johansen, $\frac{1}{2}$ for the Tresca, and $1/\sqrt{3}$ for the Mises criterion. For a square plate ($a = b$), we obtain upper-to-lower-bound ratios of 1, 1.2, and 1.344, respectively, for the three criteria. For the square Johansen plate, we thus have the exact collapse load $F_U = 24M_U$. Upper and lower bounds for uniformly loaded rectangular Johansen plates with various combinations of simply supported and clamped edges have been calculated by Manolakos and Mamalis [1986].

A plate in the shape of a regular polygon will, according to yield-line theory, deform plastically into a pyramid, with yield lines on all diagonals. If we think of a circle as the limit as $n \to \infty$ of an n-sided polygon, then it becomes clear that for a circular plate to undergo plastic flow (collapse), the entire plate must become plastic — that is, the yield condition must be met everywhere, as was assumed in Section 5.3.

Applications of circular-plate results

The results for the clamped circular plate derived in 5.2.3 may be applied to finding an upper-bound load for a clamped plate of arbitrary shape. Consider the largest circle that can fit into the area A occupied by the plate; we can then assume a velocity field such that this circle is a hinge circle and the material inside it collapses like a clamped circular plate, while the material outside it remains rigid. If, for example, the plate is uniformly loaded and square, then the *total* load that would make the plate collapse in this mode is $4/\pi$ times the load carried by the largest inscribed circle, or $(4/\pi) \times 5.63 \times 2\pi M_U = 45.04M_U$; this is less, and therefore a better upper bound, than the one of $48M_U$ given by yield-line theory.

If a clamped plate is carrying a single concentrated load F, then the upper-bound collapse load of $2\pi M_U$ is obtained for any inscribed circle centered at the load. It was shown by Haythornthwaite and Shield [1958] that the moment field $M_r = -M_U$, $M_\theta = 0$ inside the circle can be extended outside it without violating equilibrium or the yield criterion, so that $2\pi M_U$ is in fact the collapse load.

The same hinge-circle mechanism may be used to obtain $2\pi M_U$ as the upper bound for a concentrated load carried by a simply supported plate of arbitrary shape. Since the moment field is $M_r = 0$, $M_\theta = M_U$ inside the circle, it can be continued outside it in a discontinuous but statically and plastically admissible manner as $M_r = M_\theta = 0$, so that $2\pi M_U$ is the collapse load in this case as well (see Zaid [1959]).

All the collapse loads for the axisymmetric Tresca plate may be used as bounds on the corresponding loads for the Mises plate, by virtue of the following reasoning: (1) Any moment field that is in equilibrium with the load and that obeys the Tresca criterion represented by the largest hexagon inscribed in the Mises ellipse (as in Figure 5.2.2, page 292), obviously does not violate the Mises criterion, and therefore the load is a lower bound for the Mises collapse load. (2) All the velocity fields associated with the sides of the Tresca hexagon satisfy one of the conditions $\dot{\kappa}_1 = 0$, $\dot{\kappa}_2 = 0$, $\dot{\kappa}_1 + \dot{\kappa}_2 = 0$; consequently, the Tresca collapse load multiplied by $2/\sqrt{3}$ is an upper bound for the Mises collapse load. Therefore, if f_M and f_T are the ultimate values of $f = F/2\pi M_U$ for the Mises and Tresca plates, respectively, then

$$f_T \le f_M \le \frac{2}{\sqrt{3}} f_T.$$

In order to obtain f_M exactly, we must solve the quadratic equation

$$M_r^2 - M_r M_\theta + M_\theta^2 - M_U^2 = 0$$

for M_θ, making sure that the correct root is chosen, and substitute it in the equilibrium equation. Let $\rho = r/a$, and let $\phi(\rho)$ be a function such that the distributed load (assumed to be acting downward) is given by $q(\rho) = -(fM_U/a^2)\phi'(\rho)/\rho$; note that $\phi(0) = 0$ and $\phi(1) = 1$. The differential equation for $m = M_r/M_U$,

$$\rho \frac{dm}{d\rho} + \frac{1}{2} m - \sqrt{1 - \frac{3}{4} m^2} = -f\phi(\rho),$$

must be solved subject to the initial condition $m(0) = 1$, and f_M is the value of f for which the solution satisfies $m(1) = 0$ for a simply supported plate and $m(1) = -1$ for a clamped plate.

If the plate is uniformly loaded, then $\phi(\rho) = \rho^2$. A numerical solution of the differential equation leads to $f_M = 3.26$ for the simply supported plate and $f_M = 5.92$ for the clamped plate. These values may be compared with the respective lower bounds of 3 and 5.63, and the upper bounds of 3.46 ($= 2\sqrt{3}$) and 6.50. Other results relating to plate collapse may be found in the books by Hodge [1959], Chapter 10; Hodge [1963]; Johnson and Mellor [1973], Chapter 15; and Save and Massonnet [1972].

6.4.2. Limit Analysis of Shells: Theory

A shell is to a plate essentially as an arch is to a beam. As we saw in 6.2.2, the use of a nonlinear P-M interaction locus makes the collapse analysis of arches difficult, and in practice a piecewise linear locus is needed. This simplification is all the more necessary for shells, in which more than one component of both membrane force and moment must in general be accounted for. As with one-dimensional members, a piecewise linear yield locus is generated by giving the shell an ideal sandwich structure. Alternatively, the piecewise linear locus may be viewed as an approximation to the "exact" nonlinear one for a solid (or nonideal sandwich) shell.

The geometry of a shell is usually described by its *middle surface*, analogous to the middle plane of a plate, so that the two free surfaces are at a distance $h/2$ from it. At any point of the middle surface, a local Cartesian basis $(\mathbf{e}_i)$ may be formed such that $\mathbf{e}_3$ is perpendicular to the middle surface, while $\mathbf{e}_1$ and $\mathbf{e}_2$ define its tangent plane. With respect to this basis, the stress resultants $N_{\alpha\beta}$, Q_α, and $M_{\alpha\beta}$ may be defined in the same way as for plates. The equilibrium equations they satisfy are, of course, different; they require a global curvilinear coordinate system. The general theory of shells is not presented here; only some special cases are studied.

The deformation of the middle surface can be described by means of the strain tensor with components $\bar{\varepsilon}_{\alpha\beta}$, describing stretching, and the curvature tensor with components $\kappa_{\alpha\beta}$, describing bending. With shearing deformation neglected, the internal virtual work per unit surface area is

$$M_{\alpha\beta}\,\delta\kappa_{\alpha\beta} + N_{\alpha\beta}\,\delta\bar{\varepsilon}_{\alpha\beta},$$

and consequently, the $M_{\alpha\beta}$ and $N_{\alpha\beta}$ constitute the generalized stresses for the most general shearless theory of limit analysis of shells; the shear forces Q_α are reactions and do not enter the yield locus.

If the mechanical behavior of the shell is isotropic in the tangent plane, then the yield locus is expressible in terms of the principal moments M_1, M_2, and the principal membrane forces N_1, N_2. If, in addition, symmetry or another constraint require one of the principal strains $\bar{\varepsilon}_1$, $\bar{\varepsilon}_2$ or the principal curvatures κ_1, κ_2 to be zero, then the conjugate force or moment ceases to be a generalized force and becomes an internal reaction instead; it can therefore be eliminated from the yield locus, further reducing the number of dimensions of the space in which the yield locus must be described.

Piecewise Linear Yield Locus

Following Hodge [1959], we derive the piecewise linear yield locus on the basis of the ideal sandwich shell, made up of two thin sheets of thickness t separated by a core of thickness h. The sheets are elastic–perfectly plastic and obey the Tresca yield criterion with a uniaxial yield stress σ_Y. The principal stresses in the two sheets will be denoted σ_1^+, σ_2^+ and σ_1^-, σ_2^-, respectively, the sign convention being chosen so that

$$M_\alpha = \frac{1}{2}(\sigma_\alpha^- - \sigma_\alpha^+)ht, \quad \alpha = 1, 2,$$

and

$$N_\alpha = (\sigma_\alpha^- + \sigma_\alpha^+)t, \quad \alpha = 1, 2.$$

Consequently,

$$\sigma_\alpha^\pm = N_{\frac{\alpha}{2t}} \mp M_{\frac{\alpha}{ht}}.$$

If the principal stresses $\sigma_\alpha^\pm$ are not to violate the Tresca yield criterion, they must satisfy the six inequalities

$$|\sigma_1^\pm| \le \sigma_Y, \quad |\sigma_2^\pm| \le \sigma_Y, \quad |\sigma_1^\pm - \sigma_2^\pm| \le \sigma_Y.$$

In order to express these inequalities in terms of the M_α and N_α, we define $M_U = \sigma_Y ht$ and $N_U = 2\sigma_Y t$, as well as the dimensionless quantities $m_\alpha = M_\alpha/M_U$, $n_\alpha = N_\alpha/N_U$. We thus obtain

$$\begin{aligned} |m_1 - n_1| \le 1, &\quad |m_1 + n_1| \le 1, \\ |m_2 - n_2| \le 1, &\quad |m_2 + n_2| \le 1, \\ |m_1 - m_2 + n_1 - n_2| \le 1, &\quad |m_1 - m_2 - n_1 + n_2| \le 1. \end{aligned} \tag{6.4.3}$$

The six absolute-value inequalities (6.4.3) are equivalent to twelve algebraic inequalities, so that the yield locus is bounded by twelve hyperplanes in the four-dimensional $m_1 m_2 n_1 n_2$-space.

If one of the quantities m_α, n_α represents an internal reaction rather than a generalized stress, then it can be eliminated from the yield locus. Suppose this quantity to be m_2 (i.e., suppose the shell to be constrained so that $\kappa_2 = 0$); then the inequalities involving m_2 may be rewritten as

$$\begin{aligned} -1 + n_2 \le m_2 \le 1 + n_2, &\qquad -1 - n_2 \le m_2 \le 1 - n_2, \\ -1 + m_1 + n_1 - n_2 \le m_2 &\le 1 + m_1 + n_1 - n_2, \\ -1 + m_1 - n_1 + n_2 \le m_2 &\le 1 + m_1 - n_1 + n_2. \end{aligned} \tag{6.4.4}$$

The actual values of m_2 do not matter, as long as some m_2 can be found so that all the inequalities (6.4.4) can be satisfied, and this is the case whenever the first member of each inequality is no greater than the third member of *every* inequality. The first two inequalities give

$$|n_2| \le 1, \tag{6.4.5a}$$

and the second two give

$$|n_1 - n_2| \le 1. \tag{6.4.5b}$$

Combining inequalities from the first and second pairs leads to the following additional nontrivial inequalities:

$$|2n_2 - n_1 + m_1| \le 2, \quad |2n_2 - n_1 - m_1| \le 2. \qquad (6.4.5c\text{-}d)$$

Together with $(6.4.3)_{1,2}$, we thus have six absolute-value inequalities involving m_1, n_1 and n_2, or a total of twelve algebraic inequalities. The yield locus is therefore a dodecahedron in the three-dimensional $m_1n_1n_2$-space, illustrated in Figure 6.4.2. The derivation of this locus is due to Hodge [1954],

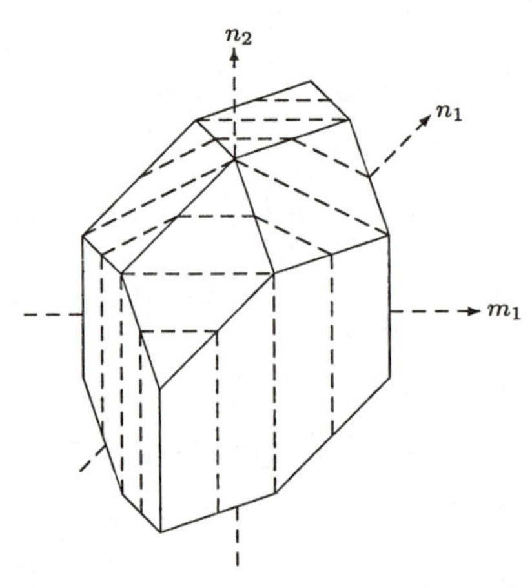

Figure 6.4.2. Piecewise linear yield locus for a cylindrical shell (from Prager [1959]).

who also derived the exact nonlinear yield locus for a solid shell made of uniform material. As discussed previously, the piecewise linear locus may be thought of as an approximation to the exact one when the appropriate values of M_U and N_U are used, namely $M_U = \sigma_Y h^2/4$ and $N_U = \sigma_Y h$, with σ_Y the uniaxial yield stress of the solid-shell material.

The plastic dissipation per unit area of the ideal sandwich shell, given rigid–plastic behavior, is

$$\bar{D}_p = \frac{\sigma_Y t}{2}\left(|\dot{\varepsilon}_1^+| + |\dot{\varepsilon}_2^+| + |\dot{\varepsilon}_1^+ + \dot{\varepsilon}_2^+| + |\dot{\varepsilon}_1^-| + |\dot{\varepsilon}_2^-| + |\dot{\varepsilon}_1^- + \dot{\varepsilon}_2^-|\right).$$

But

$$\dot{\varepsilon}_\alpha^\pm = \dot{\bar{\varepsilon}}_\alpha \mp \frac{h}{2}\dot{\kappa}_\alpha, \quad \alpha = 1, 2.$$

Consequently,

$$\begin{aligned}\bar{D}_p = \frac{1}{4}\,[&|N_U\dot{\bar{\varepsilon}}_1 + M_U\dot{\kappa}_1| + |N_U\dot{\bar{\varepsilon}}_1 - M_U\dot{\kappa}_1| \\ &+ |N_U\dot{\bar{\varepsilon}}_2 + M_U\dot{\kappa}_2| + |N_U\dot{\bar{\varepsilon}}_2 - M_U\dot{\kappa}_2| \\ &+ |N_U(\dot{\bar{\varepsilon}}_1 + \dot{\bar{\varepsilon}}_2) + M_U(\dot{\kappa}_1 + \dot{\kappa}_2)| + |N_U(\dot{\bar{\varepsilon}}_1 + \dot{\bar{\varepsilon}}_2) - M_U(\dot{\kappa}_1 + \dot{\kappa}_2)|]\,.\end{aligned} \qquad (6.4.6)$$

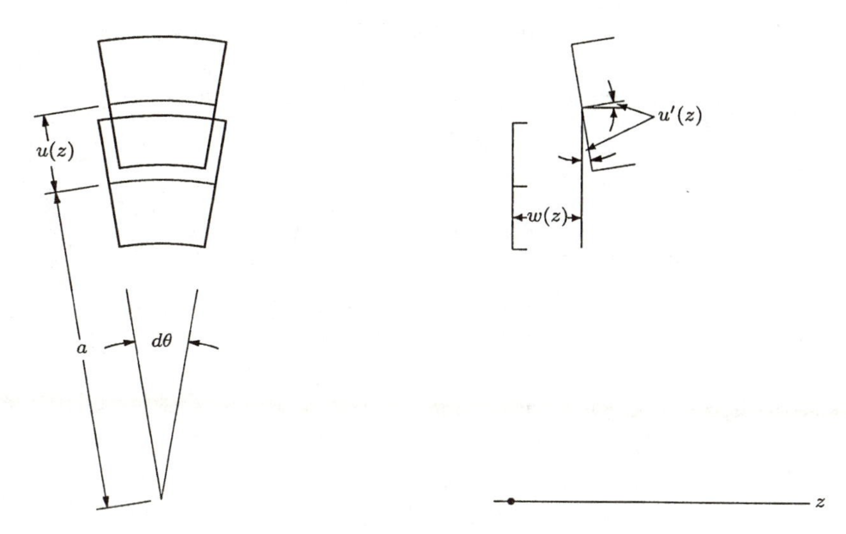

Figure 6.4.3. Displacements in a radially loaded cylindrical shell.

Yield criteria that are appropriate for the approximate treatment of reinforced-concrete shells are discussed by Save and Massonnet [1972], Chapter 9.

6.4.3. Limit Analysis of Shells: Examples

Radially Loaded Cylindrical Shell: Basic Equations

As a first example of the application of the theory discussed above, we consider a circular cylindrical shell of uniform mean radius a and thickness h, subject to an axisymmetric radial pressure distribution that may vary along the axial coordinate z. Because of axial symmetry the displacement of the middle surface has only the radial component $u(z)$ and the axial component $w(z)$. The displacements throughout the shell are assumed to be governed by the "plane sections remain plane" hypothesis applied to a longitudinal strip subtending a small angle $d\theta$ (see Figure 6.4.3):

$$u_r(r, z) = u(z), \qquad u_\theta(r, z) = 0,$$

$$u_z(r, z) = w(z) - (r - a)u'(z).$$

In accordance with Equations (1.2.1), the only nonvanishing strain components are

$$\varepsilon_\theta = \frac{u}{r} \doteq \frac{u}{a}$$

and

$$\varepsilon_z = w' - yu'',$$

where $y = r - a$. The internal virtual work per unit area is

$$\int_{-h/2}^{h/2} (\sigma_\theta \,\delta\varepsilon_\theta + \sigma_z \,\delta\varepsilon_z)\, dy = N_\theta \frac{\delta u}{a} + N_z \,\delta w' + M_z \,\delta u'',$$

where

$$N_\theta = \int_{-h/2}^{h/2} \sigma_\theta \, dy, \quad N_z = \int_{-h/2}^{h/2} \sigma_z \, dy, \quad M_z = -\int_{-h/2}^{h/2} y\sigma_z \, dy.$$

The generalized stresses are thus three in number. The conjugate generalized strains are $\bar{\varepsilon}_\theta = u/a$, $\bar{\varepsilon}_z = w'$, and $\kappa_z = u''$. The fact that $\kappa_\theta = 0$ (i.e., circles remain circles) removes M_θ from the rank of generalized stresses. The yield locus is therefore given by Equations $(6.4.3)_{1,2}$ and (6.4.5a)–(6.4.5d), with $M_1 = M_z$, $N_1 = N_z$, and $N_2 = N_\theta$.

The plastic dissipation per unit area is

$$\begin{aligned} \bar{D}_p = \frac{1}{4} \Bigg[& 2N_U \frac{|\dot{u}|}{a} + \left| N_U \frac{\dot{w}'}{a} + M_U \dot{u}'' \right| + \left| N_U \frac{\dot{w}'}{a} - M_U \dot{u}'' \right| \\ & + \left| N_U \left(\frac{\dot{u}}{a} + \dot{w}' \right) + M_U \dot{u}'' \right| + \left| N_U \left(\frac{\dot{u}}{a} + \dot{w}' \right) - M_U \dot{u}'' \right| \Bigg] . \end{aligned} \tag{6.4.7}$$

The equilibrium equations can be obtained by applying the principle of virtual work. If the shell axis occupies the interval $-L \le z \le L$, then the internal virtual work is

$$\overline{\delta W}_{int} = 2\pi a \int_{-L}^{L} (N_\theta \frac{\delta u}{a} + N_z \,\delta w' + M_z \,\delta u'')\, dz.$$

Through integration by parts this becomes

$$\begin{aligned} \overline{\delta W}_{int} = 2\pi a \Big\{ & (M_z \,\delta u' - M_z' \,\delta u + N_z \,\delta w)\big|_{-L}^{L} \\ & + \int_{-L}^{L} \left[\left(M_z'' + \frac{N_\theta}{a} \right) \delta u - N_z' \,\delta w \right] dz \Big\}. \end{aligned}$$

In addition to the radial pressure p (positive outward), let the shell be loaded by a bending moment M_z^+, an axial force N_z^+, and a shear force Q_r^+, all per unit length of circumference, at $z = L$, and similarly, M_z^-, N_z^-, and Q_r^- at $z = -L$. The external virtual work is therefore

$$\begin{aligned} \overline{\delta W}_{ext} = 2\pi a \Big[& \int_{-L}^{L} p \,\delta u \, dz + N_z^+ \delta w(L) + M_z^+ \delta u'(L) \\ & + Q_r^+ \delta u(L) - N_z^- \delta w(-L) - M_z^- \delta u'(-L) - Q_r^- \delta u(-L) \Big] . \end{aligned}$$

Equating the internal and external virtual work leads to the equilibrium equations

$$N_z' = 0, \quad M_z'' + \frac{N_\theta}{a} = p \tag{6.4.8}$$

and the boundary conditions

$$(M_z - M_z^{\pm})\,\delta u' = 0, \quad (M_z' + Q_r^{\pm})\,\delta u = 0, \quad (N_z - N_z^{\pm})\,\delta w = 0, \quad z = \pm L.$$

Note that the axial end forces must be equal and opposite for equilibrium. If these are zero, then $N_z = 0$ everywhere.

For the shell without end load, the piecewise linear yield locus reduces to a hexagon in the M_zN_θ-plane, formed by the three pairs of parallel lines described by

$$|m| = 1, \quad |2n + m| = 2, \quad |2n - m| = 2, \tag{6.4.9}$$

where $m = M_z/M_U$ and $n = N_\theta/N_U$ (see Figure 6.4.4, which also shows the "exact" nonlinear yield locus as well as the simplified square locus given by $|m| \leq 1$, $|n| \leq 1$).

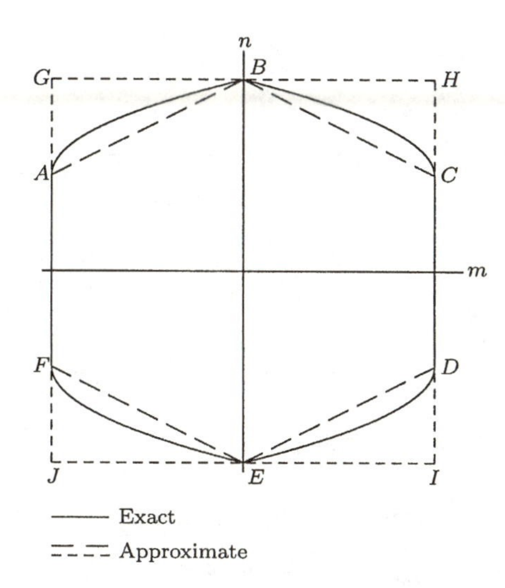

Figure 6.4.4. Yield loci for a cylindrical shell without end load.

The plastic dissipation for the yield criterion described by (6.4.9) may be obtained from (6.4.7) by substituting, as in uniaxial stress, $\dot{\bar{\varepsilon}}_1 = -\frac{1}{2}\dot{\bar{\varepsilon}}_2$, or $\dot{w}' = -\dot{u}/2a$. The result is

$$\bar{D}_p = \frac{1}{2}\left(N_U\frac{|\dot{u}|}{a} + \left|N_U\frac{\dot{u}}{2a} + M_U\dot{u}''\right| + \left|N_U\frac{\dot{u}}{2a} - M_U\dot{u}''\right|\right). \tag{6.4.10}$$

When the state of generalized stress can be assumed to be represented by points on one of the inclined lines AB, BC, DE, and EF of Figure 6.4.4, then N_θ can be expressed in terms of M_z, and the result substituted in the equilibrium equation $(6.4.8)_2$. The equation then becomes a linear differential equation for M_z, which can be solved subject to appropriate boundary conditions. Since the normality rule gives $2M_U\dot{u}'' = \pm N_U\dot{u}/a$ along the aforementioned lines, the dissipation is just $N_U|\dot{u}|/a$. The same result obtains at the vertices B and E. Along the lines AF and CD, normality requires $\dot{u}/\dot{u}'' = 0$. A nonvanishing velocity thus implies $\dot{u}'' = \pm\infty$, that is, the formation of a plastic hinge circle.

A simple case occurs when the pressure p is constant and the ends of the tube are free. Then $M_z = 0$, and the yield criterion reduces to $|N_\theta| = N_U$. The ultimate pressure is therefore given by $|p_U| = N_U/a$.

A solution for a pressurized tube with clamped ends was derived by Hodge [1954]. Here the central portion of the tube is in regime DE, and the outer portions in EF; the boundary is at $z = \pm\eta L$. The parameter η and the dimensionless ultimate pressure $\bar{p} = p_U a/N_U$ are given implicitly as functions of the dimensionless shell parameter ω, defined by

$$\omega^2 = \frac{N_U L^2}{2M_U a}.$$

The results are

$$\sinh\omega\eta = \frac{\sin\omega(1-\eta)}{\sqrt{2}\cos\omega(1-\eta)+1},$$

$$\bar{p} = \frac{2-\cos\omega(1-\eta)}{2[1-\cos\omega(1-\eta)]}$$

for $\omega \le 1.65$, and

$$\tan\omega(1-\eta) = \coth\omega\eta,$$

$$\bar{p} = 1 + \frac{1}{2(2\cosh\omega\eta - 1)}$$

for $\omega > 1.65$. For more details, see Hodge [1959], Section 11-2.

Cylindrical Shell with a Ring Load

As another example, suppose that a free-ended shell is subject to a radially inward ring loading at $z = 0$, its intensity being F per unit length of circumference; that is,

$$p(z) = -F\,\delta(z),$$

where $\delta(\cdot)$ is the Dirac delta function. Note that

$$\delta(z) = \frac{1}{2}\frac{d}{dz}\operatorname{sgn} z = \frac{1}{2}\frac{d^2}{dz^2}|z|.$$

It is reasonable to assume initially that the hoop stress is compressive, that is, $N_\theta \le 0$. If the shell is extremely short, then its collapse should not depend very much on whether the force is applied around a ring or uniformly distributed over the surface in a statically equivalent manner. For very short shells, then, we should expect $F_U \doteq 2N_U L/a$, with bending having little or no effect.

When bending is taken into account, it can be seen that at least in a central portion of the tube, $M_z \geq 0$. It will be assumed, to begin with, that the entire shell is plastic and in regime AB of Figure 6.4.4. Eliminating N_θ, we obtain the following dimensionless differential equation for $m(\zeta)$, using the dimensionless variable $\zeta = z/L$, and the parameters $f = Fa/(2N_U L)$ and ω as previously defined:

$$m''(\zeta) + \omega^2 m(\zeta) = 2\omega^2 - 4\omega^2 f\,\delta(\zeta).$$

The general solution of this equation that is even in ζ is

$$m(\zeta) = 2 - 2\omega f \sin(\omega|\zeta|) + C\cos\omega\zeta,$$

with C an arbitrary constant. The free-end condition $m(1) = m'(1) = 0$ yields $C = -2\cos\omega$ and

$$f = \frac{\sin\omega}{\omega}. \tag{6.4.11}$$

The solution may accordingly be written as

$$m(\zeta) = 2[1 - \cos\omega(1 - |\zeta|)].$$

The requirement that $0 \leq m \leq 1$ limits the validity of this solution to $\omega \leq \pi/3$. For sufficiently short shells, then, Equation (6.4.11) gives a lower bound to the collapse load; in particular, the limit as $\omega \to 0$ is $f = 1$, as previously discussed. An associated kinematically admissible velocity field, however, can easily be found, namely,

$$\dot{u}(\zeta) = -v_0 \cos\omega\zeta. \tag{6.4.12}$$

It can readily be checked that the generalized strain rates derived from this velocity field satisfy the normality condition for regime AB, and hence Equation (6.4.11) gives the collapse load for $\omega \leq \pi/3$.

Alternatively, the upper-bound theorem can be applied directly to the velocity field (6.4.12), the dissipation per unit length being $2\pi N_U|\dot{u}|$, and the external work rate $2\pi a F v_0$. The result is Equation (6.4.11) as an upper bound, as long as $\dot{u}(\zeta)$ does not change sign; this condition is met for $\omega \leq \pi/2$. Equation (6.4.11) is therefore an upper bound for $\pi/3 \leq \omega \leq \pi/2$. However, a better upper bound can be obtained for this range by assuming a velocity field with a plastic hinge circle at $\zeta = 0$, namely,

$$\dot{u}(\zeta) = -v_0(\cos\omega\zeta + \beta\sin\omega|\zeta|). \tag{6.4.13}$$

The additional dissipation at the hinge circle is $4\pi a M_U \beta v_0/L = 2\pi N_U L\beta v_0/\omega^2$. Equating dissipation and external work rate yields

$$f = \sin\omega - \frac{\beta}{2\omega^2}(1 - 2\cos\omega)$$

if $\dot{u}$ does not change sign, that is, if $\beta \leq \omega \cot \omega$. Choosing this limiting value for β gives the smallest value of f, namely,

$$f = \frac{2 - \cos\omega}{2\omega \sin\omega},$$

which is less than the right-hand side of (6.4.11) for $\pi/3 < \omega < \pi/2$.

A lower bound for $\omega > \pi/3$ can be obtained by assuming the solution

$$m(\zeta) = 2 - \cos\omega\zeta - 2\omega f \sin(\omega|\zeta|)$$

valid for $|\zeta| < \eta$, where η is such that $m(\eta) = m'(\eta) = 0$, and continuing it statically as $M_z = 0$, $N_\theta = 0$ for $|\zeta| > \eta$. The conditions at η lead to $\omega\eta = \pi/3$ and $f = \sqrt{3}/(2\omega)$, or, in dimensional form,

$$F = \sqrt{\frac{6M_U N_U}{a}},$$

a result that can be seen to be independent of the length. As can be seen, however, as the length increases, so does the discrepancy between the upper and lower bounds.

For longer shells the solution based on the hexagonal yield locus becomes difficult. For an infinitely long shell, Drucker [1954b] found the collapse load

$$F = 2\sqrt{\frac{3M_U N_U}{a}} \tag{6.4.14}$$

or $f = \sqrt{3/2}/\omega$. This load is based on the following moment distribution:

$$\begin{aligned}
m(\zeta) &= 2 - \cos\omega\zeta - \frac{\sqrt{6}}{\omega}\sin\omega|\zeta|, && 0 \leq |\zeta| \leq \zeta_1 \quad (AB),\\
&= -2 + \cosh\omega(|\zeta| - \zeta_2), && \zeta_1 \leq |\zeta| \leq \zeta_3 \quad (BC),\\
&= -2 + 2\cos\omega(|\zeta| - \zeta_4), && \zeta_3 \leq |\zeta| \leq \zeta_4 \quad (DE),\\
&= 0, && \zeta_4 \leq |\zeta| \quad \text{(rigid)},
\end{aligned}$$

where

$$\omega\zeta_1 = \cos^{-1}\frac{2 + 3\sqrt{2}}{7} = 0.469, \qquad \omega(\zeta_2 - \zeta_1) = \cosh^{-1} 2 = 1.317,$$

$$\omega(\zeta_3 - \zeta_2) = \frac{1}{2}\cosh^{-1} 4 = 1.032, \qquad \omega(\zeta_4 - \zeta_3) = \frac{1}{2}\cos^{-1}\frac{1}{4} = 0.659.$$

It follows that $\omega\zeta_4 = 3.477$, and therefore the result is valid for $\omega \geq 3.477$. Note that $m(0) = 1$ and $m(\zeta_2) = -1$, so that plastic hinges develop at those points. At ζ_3 the hoop stress N_θ changes abruptly from a negative to a positive value, and at ζ_4 back to zero.

As pointed out by Drucker [1954b], the mathematics is greatly simplified if the the yield locus is replaced by the rectangle $|M_z| \leq M_U$, $|N_\theta| \leq N_U$. With this yield criterion, Eason and Shield [1955] found complete solutions for shells of all lengths, and with the load not necessarily at the center. Since the rectangle circumscribes the hexagon, the collapse loads found by Eason and Shield are upper bounds on those that would be found for the hexagonal yield locus. Furthermore, a rectangle with vertices at $(\pm M'_U, \pm N'_U)$ may be inscribed in the hexagon; then the collapse load for the rectangle is a lower bound for the hexagon when M_U and N_U are respectively replaced by M'_U and N'_U. The values of M'_U and N'_U may be chosen to as to maximize the lower bound.

With the rectangular yield locus, both kinematic and static solutions are quite easy. The sides $|M_z| = M_U$ of the rectangle (like those of the hexagon) correspond to hinge circles, while the sides $|N_\theta| = N_U$ describe velocities $\dot{u}$ varying linearly with position. For each side, M_z and N_θ are polynomial functions of z.

For the symmetric problem under consideration, Eason and Shield's results are

$$f = \frac{1}{2}\left(\frac{1}{\omega^2} + 1\right), \quad \omega \leq 1 + \sqrt{2},$$
$$f = \frac{\sqrt{2}}{\omega}, \quad \omega \geq 1 + \sqrt{2}.$$

The latter result is equivalent to

$$F = 4\sqrt{\frac{M_U N_U}{a}}, \tag{6.4.15}$$

independent of the length. Equation (6.4.15) also gives the lower bound

$$F = 4\sqrt{\frac{M'_U N'_U}{a}},$$

where

$$\frac{M'_U}{M_U} + 2\frac{N'_U}{N_U} \leq 2, \quad M'_U \leq M_U.$$

The lower bound is maximized for $M'_U = M_U$, $N'_U = \frac{1}{2}N_U$, and equals $2\sqrt{2M_U N_U/a}$. The bounds thus bracket Drucker's collapse load (6.4.14) for the long shell.

Spherical Cap Under Pressure

In shells of revolution, as a rule, all four generalized stresses are active. An approximate theory, in which the effects of one of them are ignored, was proposed by Drucker and Shield [1959], but its applicability is limited.

It was shown by Hodge [1959] that for a shell of revolution, a point on the four-dimensional piecewise linear yield locus (6.4.3) corresponds to plastic deformation only if it lies on the intersection of two of the twelve hyperplanes, one of which represents yielding of the top sheet and the other represents yielding of the bottom sheet. The generalized stresses thus satisfy two yield equations and two equilibrium equations, making the problem "statically determinate." Furthermore, there are two normality conditions on the generalized strain rates, and since the latter are derived from two velocity components, the problem is "kinematically determinate" as well.

We consider here a spherical cap of radius a subtending a half-angle ϕ_0, clamped around its edge, and carrying a uniform radial pressure p (see Figure 6.4.5). The problem was treated by Onat and Prager [1954] on the basis of a nonlinear yield criterion and by Hodge [1959, Section 11-6] on the basis of the piecewise linear yield criterion (6.4.3).

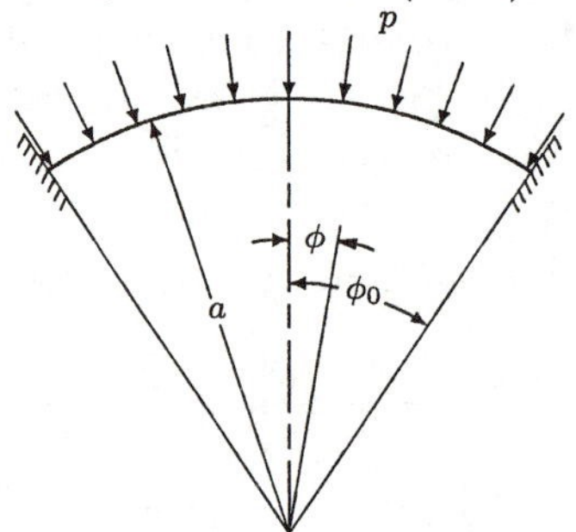

Figure 6.4.5. Spherical cap under uniform external pressure: geometry and loading.

If the radial displacement is $u_r = u$ and the meridional displacement is $u_\phi = v$, then the generalized strains are

$$\bar{\varepsilon}_\theta = \frac{u + v\cot\phi}{a}, \qquad \bar{\varepsilon}_\phi = \frac{u + v'}{a},$$

$$\kappa_\theta = \cot\phi\frac{u' - v}{a^2}, \qquad \kappa_\phi = \frac{u'' - v'}{a^2},$$

where $(\cdot)' = d(\cdot)/d\phi$.

To find an upper bound on the collapse pressure, we assume the velocity field

$$\dot{u} = -v_0(\cos\phi - \cos\phi_0), \qquad \dot{v} = 0.$$

The generalized strain rates are accordingly

$$\dot{\bar{\varepsilon}}_\theta = \dot{\bar{\varepsilon}}_\phi = -\frac{v_0}{a}(\cos\phi - \cos\phi_0), \qquad \dot{\kappa}_\theta = \dot{\kappa}_\phi = \frac{v_0}{a^2}\cos\phi.$$

The plastic dissipation per unit area is then, from Equation (6.4.6),

$$\bar{D}_p = N_U v_0[|\cos\phi_0 - (1-k)\cos\phi| + |(1+k)\cos\phi - \cos\phi_0|]$$

$$= \begin{cases} 2N_U v_0(\cos\phi - \cos\phi_0), & 0 \le \phi \le \phi^*, \\ 2kN_U v_0 \cos\phi, & \phi^* \le \phi \le \phi_0, \end{cases}$$

where

$$k = \frac{M_U}{N_U a}, \qquad \cos\phi^* = \frac{\cos\phi_0}{1-k}.$$

The expression for $\phi > \phi^*$ is necessary, of course, only if $\phi_0 > \phi^*$.

In addition, since $\dot{u}'$ does not vanish at the edge $\phi = \phi_0$, a plastic hinge circle forms there, necessitating the additional dissipation (per unit length) given by $M_U v_0 \sin\phi_0/a = kN_U v_0 \sin\phi_0$. The total internal dissipation is thus

$$\mathcal{D}_{int} = 2\pi a^2 \int_0^{\phi_0} \bar{D}_p \sin\phi\, d\phi + 2\pi a k N_U v_0 \sin^2\phi_0.$$

The external work rate is

$$\mathcal{D}_{ext} = 2\pi a^2 \int_0^{\phi_0} p\dot{u} \sin\phi\, d\phi = \pi a^2 p v_0 (1 - \cos\phi_0)^2.$$

Setting $\mathcal{D}_{ext} = \mathcal{D}_{int}$ yields the upper bound

$$p = 2\frac{N_U}{a}\left[1 + k\frac{1+\cos\phi_0}{1-\cos\phi_0} + \frac{k^2}{1-k}\left(\frac{\cos\phi_0}{1-\cos\phi_0}\right)^2\right], \quad \cos\phi_0 \le 1-k,$$
$$p = 4k\frac{N_U}{a}\frac{1+\cos\phi_0}{1-\cos\phi_0}, \quad \cos\phi_0 \ge 1-k.$$

A lower bound may be obtained by assuming that the stress field in the shell is one of simple membrane compression, $N_\theta = N_\phi = -N_U$, with $M_\theta = M_\phi = 0$. The corresponding lower-bound pressure is $p = 2N_U/a$. A lower bound that is better for sufficiently small cap angles was found by Hodge [1959] by assuming N_θ, N_ϕ, M_θ and M_ϕ to be sinusoidally varying functions of ϕ, substituting into the equilibrium equations, and choosing the free coefficients so as to maximize the pressure subject to the yield inequalities. The resulting best lower bound is given by

$$\frac{pa}{N_U} = 2 + \frac{1}{1-\cos\phi_0}\left[1 - \sqrt{\left(\frac{1-k}{1+k}\right)^2 + 4\left(\frac{1-\cos\phi_0}{1+\cos\phi_0}\right)^2}\,\right].$$

Clearly, this result is an improvement over the previous one if and only if the quantity under the square-root sign is less than unity.

Exercises: Section 6.4

1. Using yield-line theory, find upper bounds for a uniformly loaded, simply supported Tresca plate having the shape of (a) an equilateral triangle, (b) a right isosceles triangle, and (c) a regular hexagon.

2. Repeat Exercise 1 for clamped plates.

3. Find lower bounds for some of the plates in Exercises 1 and 2.

4. Using yield-line theory, find upper bounds for the plates of Exercises 1 and 2 when they carry a single concentrated load at the center or centroid. Compare with the result $F_U = 2\pi M_U$.

5. A square Tresca plate carrying a uniform downward load is simply supported along its edges against downward deflection but is free to lift off. Using yield-line theory, find an upper bound to the collapse load.

6. Using some of the methods of Section 6.2, find lower and upper bounds for the yield locus of an ideal sandwich shell obeying the Mises criterion and its associated flow rule (a) when M_2 is not a generalized stress and (b) when, in addition, $N_1 = 0$.

7. Find the complete solution for the clamped pressurized cylindrical shell with clamped ends obeying the "square" yield locus of Figure 6.4.4. Plot the nondimensional ultimate pressure $\bar{p} = p_U a/N_U$ against the shell parameter ω.

8. Repeat Exercise 7 for a shell that is (a) clamped at one end and free at the other and (b)simply supported at both ends.

9. Repeat Exercise 8 for the hexagonal yield locus.

10. For a short free-ended cylindrical shell subject to a noncentered ring load, find the collapse load based on the "square" yield locus.

Chapter 7

Dynamic Problems

Section 7.1 Dynamic Loading of Structures

7.1.1 General Concepts

All the problems studied so far have been static or quasi-static. In any complete solution of such a problem, the stress field satisfies the equilibrium equation with the prescribed body force and the static boundary conditions with the prescribed surface tractions. The effects of inertia are neglected.

In a body made of a standard perfectly plastic material, the ultimate loading is the greatest loading under which a solution to the *static* problem can be found such that the yield criterion is nowhere violated. If a loading in excess of the limit loading is applied, then, obviously, the static problem (in which, by definition, inertia is ignored) has no solution, and inertia effects must be taken into account. If the time of loading is short, enough of the external work may be transformed into kinetic energy so that excessive deformation is prevented; for example, when a nail is struck by a hammer, it may experience a force in excess of its static ultimate load without permanent deformation.

The problem of impact or impulsive loading of structural elements such as beams, plates and shells has been most often treated within the constitutive framework of limit analysis: rigid–perfectly plastic behavior, with the yield criterion in terms of generalized stresses. This approach, which has been reasonably successful with regard to the determination of permanent deformations, is presented in this section; it is generally regarded as justified when the energy imparted to the body greatly exceeds the elastic energy that can be stored. If local constitutive equations are used, waves propagate through the body. The propagation of one-dimensional and multidimensional waves in elastic-plastic bodies is studied in Sections 7.2 and 7.3, respectively.

It must be pointed out at the outset that the solution of dynamic problems on the basis of rate-independent plasticity is in conflict with the definition of rate-independent plastic behavior as the limiting behavior of viscoplastic bodies at very slow rates (see 3.1.2). The use of rate-independent plasticity for such problems is based on the assumption of a "dynamic" yield stress that is independent of rate *in the range of rates encountered in the dynamic problem*, but not, in general, identical with the static yield stress. The adequacy of this procedure, as opposed to the use of viscoplasticity theory, has long been the subject of debate and is discussed in 7.2.3. Until then, the rate-independent model will be tacitly assumed.

Dynamic Behavior of Rigid–Perfectly Plastic Bodies: General Results

Most of the structures for which static collapse loads have been determined on the basis of the rigid–perfectly plastic model have also been subject to dynamic analysis; the solutions have been surveyed by Krajcinovic [1973] and Jones [1975]. A common feature of the solutions is that if the body is restrained against rigid-body motion, then the velocity field eventually takes on a *mode form* in the sense that its spatial variation becomes independent of time. In other words, the velocity becomes the product of a time-dependent amplitude and a function of position:

$$\mathbf{v}(\mathbf{x}, t) = \dot{\Delta}(t)\mathbf{w}(\mathbf{x}).$$

A proof that all solutions must converge to mode form is due to Martin [1980]; it is based on an extremum principle, according to which the final mode is the one that minimizes $\mathcal{D}_{int}/\sqrt{K}$, where $\mathcal{D}_{int}$ is the total plastic dissipation and K is the kinetic energy.

The fact that at least the final phase of the motion is in mode form means that all points of the body come to rest at once. If the time at which this occurs is t_s, then the permanent displacement field $\mathbf{u}_p(\mathbf{x})$ can be found by integrating the velocity over time from $t = 0$ to $t = t_s$.

A theorem due to Martin [1975] gives a lower bound on t_s for a body on which all prescribed loads and prescribed surface velocities are zero for $t > t_0$. Let $\mathbf{v}^*$ be any kinematically admissible velocity field, with a strain-rate field $\dot{\boldsymbol{\varepsilon}}^*$ derived from it. When the body force, prescribed surface traction and prescribed surface velocity are all zero, the dynamic principle of virtual velocities reduces to

$$-\int_R \rho\dot{\mathbf{v}} \cdot \mathbf{v}^* \, dV = \int_R \sigma_{ij}\dot{\varepsilon}^*_{ij} \, dV,$$

where $\mathbf{v}$ is the actual velocity field and $\boldsymbol{\sigma}$ is the actual stress field. Combining this result with the maximum-plastic-dissipation postulate, we obtain

$$-\int_R \rho\dot{\mathbf{v}} \cdot \mathbf{v}^* \, dV \leq \int_R D_p(\dot{\boldsymbol{\varepsilon}}^*) \, dV.$$

The right-hand side of this inequality is independent of time, while on the left-hand side only $\dot{\mathbf{v}}$ depends on time. If $\mathbf{v}_0(\mathbf{x}) = \mathbf{v}(\mathbf{x}, t_0)$, then, since $\mathbf{v}(\mathbf{x}, t_s) = 0$, integration of both sides from t_0 to t_s leads to

$$\int_R \rho \mathbf{v}_0 \cdot \mathbf{v}^* \, dV \leq (t_s - t_0) \int_R D_p(\dot{\varepsilon}^*) \, dV.$$

Consequently,

$$t_s - t_0 \geq \frac{\int_R \mathbf{v}_0 \cdot \mathbf{v}^* \, dV}{\int_R D_p(\dot{\varepsilon}^*) \, dV}. \tag{7.1.1}$$

Another result due to Martin [1975] governs the permanent displacement field $\mathbf{u}_p(\mathbf{x}) = \mathbf{u}(\mathbf{x}, t_s)$. If $\boldsymbol{\sigma}^*$ is any statically and plastically admissible stress field, then the static principle of virtual work implies that

$$\int_{\partial R} \sigma_{ij}^* v_i n_j \, dS = \int_R \sigma_{ij}^* \dot{\varepsilon}_{ij} \, dV,$$

while the dynamic principle of virtual velocities gives

$$-\int_R \rho \dot{\mathbf{v}} \cdot \mathbf{v} \, dV = \int_R \sigma_{ij} \dot{\varepsilon}_{ij} \, dV.$$

Again using the maximum-plastic-dissipation postulate, we obtain

$$\int_{\partial R} \sigma_{ij}^* v_i n_j \, dS \leq -\int_R \rho \dot{\mathbf{v}} \cdot \mathbf{v} \, dV = -\frac{d}{dt} K,$$

where

$$K = \frac{1}{2} \int_R \rho \mathbf{v} \cdot \mathbf{v} \, dV.$$

Since $K(t_s) = 0$, integration of both sides of the inequality leads to

$$\int_{\partial R} \sigma_{ij}^* (u_{pi} - u_{0i}) n_j \, dS \leq K_0, \tag{7.1.2}$$

where $\mathbf{u}_0$ is the displacement field at t_0. In the case of impulsive loading lasting a very short time, $\mathbf{u}_0$ can be taken as zero.

The preceding results can easily be formulated in terms of generalized stresses and strains.

7.1.2 Dynamic Loading of Beams

Equations of Motion for Elastic–Plastic Beams

The equations of motion of a beam according to the elementary (Euler–Bernoulli) theory can be obtained by adding inertial forces to the distributed load. We consider only a beam of doubly symmetric cross-section, with the centroidal axis along the x-axis, and with bending confined to the xy-plane.

If the deflection in the positive y-direction is $u(x,\, t)$, then the inertial force per unit length is $-\bar{\rho}\partial^2 u/\partial t^2$, where $\bar{\rho} = \rho A$ is the beam mass per unit length, ρ being the mass density and A the cross-sectional area. With q denoting the distributed load per unit length, the equation of motion is

$$\frac{\partial^2 M}{\partial x^2} = q - \bar{\rho}\frac{\partial^2 u}{\partial t^2}. \tag{7.1.3}$$

When $q = 0$, and a functional relation is assumed between the moment M and the curvature $\kappa = \partial^2 u/\partial x^2$, it can be shown that the deflection can be expressed as

$$u(x,\, t) = tf(\eta),$$

where $\eta = x^2/t$, and that κ (and hence M) as well as the velocity $v = \partial u/\partial t$ are functions of the single variable η alone. If, say, $\kappa = \Psi(M)$, then the moment is governed by the differential equation

$$\frac{d^2}{d\eta^2}\left(\sqrt{\eta}\frac{dM}{d\eta}\right) + \frac{\bar{\rho}}{16}\Psi'(M)\sqrt{\eta}\frac{dM}{d\eta} = 0. \tag{7.1.4}$$

For a linearly elastic beam, $\Psi'(M) = 1/EI$, so that the equation is linear; a general solution is due to Boussinesq. For an elastic–plastic beam, Equation (7.1.4) is in general nonlinear, and must be solved numerically. It is, however, piecewise linear if a moment-curvature relation with linear work-hardening is assumed, and an analytic solution can then be obtained. The first such solution was found by Bohnenblust (see Duwez, Clark, and Bohnenblust [1950]), who treated an infinitely long beam in which one point is suddenly given a velocity that remains constant in time — equivalent to being struck by a very heavy concentrated mass traveling at that velocity (so heavy that the resistance of the beam is insufficient to decelerate it), or as in the experiments that were performed by Duwez et al., to the beam itself moving transversely as a rigid body before it impinges on a concentrated rigid obstacle. The deformation remains elastic if the impact velocity v_0 does not exceed $v_E = \kappa_E\sqrt{EI/\bar{\rho}}$, where κ_E is the elastic-limit curvature as defined in 4.4.1. When $v_0 > v_E$ and the hardening is linear, different deformation patterns develop for $v_E < v_0 < 2.087v_E$ and for $v_0 > 2.087v_E$.

The analogous problem for rigid–plastic beams, with various types of work-hardening and with extensions to variable velocity and to semi-infinite beams, was treated by Conroy [1952, 1955, 1956, 1963]. Equation (7.1.4) for such a beam is valid only in the plastic portions of the beam. In the rigid portions, the moment is governed by Equation (7.1.3) with $q = 0$ and with the deflection given by a straight-line function of x, say $u(x,\, t) = f(t) + g(t)x$. The boundary between the rigid and plastic portions is itself an unknown function of time. It was shown by Conroy [1952] how the plastic portions shrink to points, representing plastic hinges, in the limit as the work-hardening decreases to zero. For linear hardening, Conroy [1955] showed that certain problems can be solved by means of linear elastic analysis.

Impact Loading of a Perfectly Plastic Free-Ended Beam

The impact loading of a free-ended beam by a suddenly applied force that is a prescribed function of position and time was studied by Lee and Symonds [1952] for a rigid–perfectly plastic beam loaded at the center and by Bleich and Salvadori [1953, 1955] for an elastic–perfectly plastic beam under a loading that is symmetric about the midpoint but is otherwise arbitrary.

We consider, first, the problem treated by Lee and Symonds [1952]. A free-ended beam of length $2L$ is suddenly loaded at its midpoint by a concentrated force $F(t)$, which rises to a peak value F_m and declines to zero in a short time. Let $\Delta(t)$ denote the midpoint displacement. It can be assumed that at first the beam is accelerated as a rigid body, so that the displacement is Δ everywhere, and the load F is balanced by a uniformly distributed inertial force $\bar{\rho}\ddot{\Delta}$, with $2\bar{\rho}L\ddot{\Delta} = F$ [see Figure 7.1.1(a)]. The bending moment at a distance x from the midpoint may be obtained from the free-body diagram of Figure 7.1.1(b), and is

$$M(x, t) = \frac{F(t)}{4L}(L - x)^2;$$

the maximum moment is $M(0, t) = F(t)L/4$. The assumption of rigid-body motion is therefore valid as long as this moment is less than M_U, or $F(t) < 4M_U/L$.

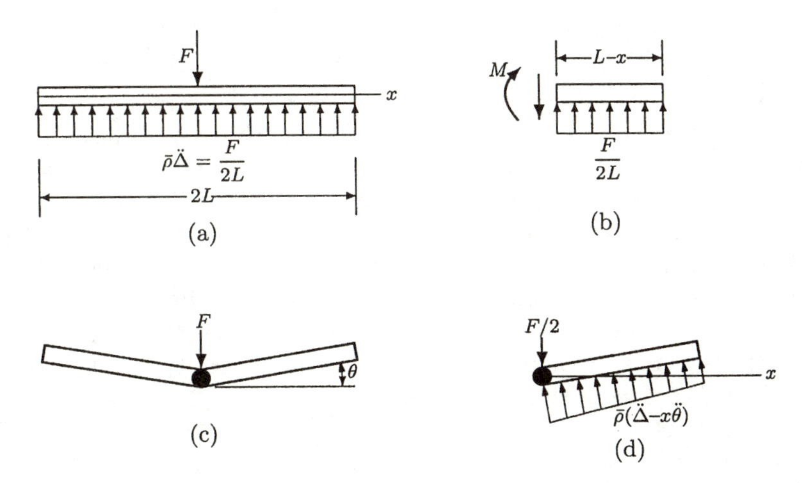

Figure 7.1.1. Impact loading of a perfectly plastic free-ended beam: (a) geometry and loading; (b) free-body diagram; (c) hinge rotation; (d) motion of one-half of the beam.

When $F(t) \geq 4M_U/L$, a plastic hinge must form at $x = 0$. Let the half-angle of rotation of the hinge be $\theta(t)$, as in Figure 7.1.1(c). Because of symmetry, it is sufficient to study the motion of half the beam, say $0 < x < L$

[see Figure 7.1.1(d)]. Since the deflection is

$$u(x,\, t) = \Delta(t) - x\theta(t)$$

as long as $\theta(t) \ll 1$, the global force equation of motion may be obtained by balancing the force $F/2$ acting on half the beam with the total inertial force:

$$\frac{F}{2} = \bar{\rho}\int_0^L (\ddot{\Delta} - x\ddot{\theta})dx = \bar{\rho}L\ddot{\Delta} - \bar{\rho}\frac{L^2}{2}\ddot{\theta}.$$

Similarly, the moment equation of motion is obtained by balancing the moment M_U with the moment about $x = 0$ of the inertial force:

$$M_U = \bar{\rho}\int_0^L (\ddot{\Delta} - x\ddot{\theta})x\, dx = \bar{\rho}\frac{L^2}{2}\ddot{\Delta} - \bar{\rho}\frac{L^3}{6}\ddot{\theta}.$$

The linear and angular accelerations are, respectively,

$$\ddot{\Delta} = \frac{2M_U}{\bar{\rho}L^2}(\phi - 3),$$

$$\ddot{\theta} = \frac{3M_U}{\bar{\rho}L^3}(\phi - 4),$$

where $\phi(t) = F(t)L/M_U$. The distribution of bending moment is now

$$M(x,\, t) = M_U\left(1 - \frac{x}{L}\right)^2\left\{1 - \frac{1}{2}[\phi(t) - 4]\frac{x}{L}\right\}.$$

It can easily be shown that this distribution has a local extremum in the interior of the beam when $\phi(t) > 6$, and that this extremum is located at $x/L = \phi(t)/3[\phi(t) - 4]$ and has the value $-2M_U[\phi(t) - 6]^3/27[\phi(t) - 4]^2$. This value equals $-M_U$ when ϕ reaches 22.9; an additional plastic hinge then forms in each half of the beam. For $F(t) \ge 22.9M_U/L$, therefore, the equations of motion must reflect the additional hinges, which move in toward the middle as the load increases. Further results relative to this problem were obtained by Symonds and Leth [1954] and by Cotter and Symonds [1955]. The effect of distributed rather than concentrated loads was studied by Salvadori and DiMaggio [1953] and by Seiler and Symonds [1954] (see also Seiler, Cotter and Symonds [1956]). An extension of the method to elastic–perfectly plastic beams is due to Alverson [1956].

Bleich and Salvadori [1953, 1955] examined the motion of an impulsively loaded elastic–perfectly plastic beam by using the natural modes of vibration of elastic beams — for the entire beam during the initial elastic phase, and for the portions of the beam separated by plastic hinges after the beam yields, provided that the hinges are stationary. Salvadori and DiMaggio [1953] studied the development of hinges for various degrees of concentration of the load, ranging from uniformly distributed to concentrated. For a comparison between the results of Lee and Symonds and those of Bleich and Salvadori, see the discussion by Symonds and by Bleich and Salvadori following the paper by Bleich and Salvadori [1955]. The impact loading of a rigid–perfectly plastic beam that is built in at both ends was studied by Symonds and Mentel [1958], who took into account the axial forces that develop when such a beam deflects.

Cantilever Struck by a Mass at Its Tip

The method of Lee and Symonds was applied by Parkes [1955] to study the impact on a cantilever beam of length L by an object of mass m traveling with a velocity v_0. The beam is again assumed to be rigid–perfectly plastic, so that the kinetic energy of the striker can be absorbed only in a plastic hinge. Initially the hinge is at the tip, but it moves in time toward the built-in end.

Let x denote distance along the beam measured from the tip and $\Delta(t)$ the tip deflection, and let $x = \bar{x}(t)$ give the position of the hinge at t. The portion $x > \bar{x}(t)$ remains undisturbed. In the portion $x < \bar{x}(t)$, the curvature rate $\dot{\kappa} = \partial^2 v/\partial x^2$ (where $v = \partial u/\partial t$) vanishes, so that the velocity is given by

$$v(x,\, t) = \dot{\Delta}(t)\left[1 - \frac{x}{\bar{x}(t)}\right].$$

Since the plastic hinge occurs at a maximum of the bending moment, the shear force there is zero, and therefore the net force on the moving portion of the beam is zero. The translational equation of motion for this portion of the beam is therefore

$$m\ddot{\Delta} + \frac{\bar{\rho}}{2}\frac{d}{dt}(\bar{x}\dot{\Delta}) = 0, \tag{7.1.5}$$

while the rotational motion can be obtained by taking moments about the tip, yielding

$$\frac{\bar{\rho}}{6}\frac{d}{dt}(\bar{x}^2\dot{\Delta}) = M_U. \tag{7.1.6}$$

Equations (7.1.5)–(7.1.6) can be integrated to give

$$\left[m + \frac{\bar{\rho}}{2}\bar{x}(t)\right]\dot{\Delta}(t) = A, \quad \frac{\bar{\rho}}{6}[\bar{x}(t)]^2\dot{\Delta}(t) - M_U t = B,$$

where A and B are constant. Since $\bar{x}(0) = 0$ and $\dot{\Delta}(0) = v_0$, it follows that $A = mv_0$ and $B = 0$. Hence, if we define $\beta = \bar{\rho}L/2m$ and $\xi = \bar{x}(t)/L$, then

$$\frac{\dot{\Delta}(t)}{v_0} = \frac{1}{1+\beta\xi}$$

and

$$t = \frac{mv_0 L}{3M_U}\frac{\beta\xi^2}{1+\beta\xi}, \tag{7.1.7}$$

provided that $\xi < 1$. The plastic hinge reaches the built-in end at time $t = t_1$, given by Equation (7.1.7) when $\xi = 1$. Note that $\dot{\Delta}(t_1) = v_0/(1+\beta)$.

For $t > t_1$, the motion is in mode form:

$$v(x, t) = \dot{\Delta}(t)\left(1 - \frac{x}{L}\right).$$

The equation for $\Delta(t)$ can be obtained by taking moments about $x = L$, namely,

$$\left(mL + \frac{\bar{\rho}L^2}{3}\right)\ddot{\Delta} = -M_U.$$

Integration, with continuity at $t = t_1$, leads to

$$\frac{\dot{\Delta}(t)}{v_0} = \frac{1}{1+\beta} - \frac{3M_U(t-t_1)}{mv_0L(3+2\beta)} = \frac{3M_U}{mv_0L(3+2\beta)}(t_s - t),$$

where

$$t_s = t_1 + \frac{mv_0L}{3M_U}\frac{3+2\beta}{1+\beta} = \frac{mv_0L}{M_U}$$

is the time at which the motion stops. Consequently, the permanent deflection u_p may be obtained by integrating the velocity over time up to t_s.

If the mass of the striker is much greater than that of the beam, that is, if $\beta \ll 1$, then the permanent deflection is approximately

$$u_p(x) \equiv \frac{mv_0^2L}{2M_U}\left(1 - \frac{x}{L}\right).$$

Thus the beam remains straight, and the deflection is that which is necessary so that all the kinetic energy is absorbed by the hinge at the built-in end.

If, on the other hand, the mass of the striker is much less than that of the beam, the final shape is a superposition of a rotation and a local deformation near the tip. The form is, approximately, given by

$$u_p(x) = \frac{2m^2v_0^2}{3M_U\bar{\rho}}\ln\frac{L}{x} \qquad \text{if } \beta x/L \gg 1,$$

with

$$u_p(0) = \frac{2m^2v_0^2}{3M_U\bar{\rho}}\ln\beta.$$

Parkes [1955] also carried out experiments in which mild-steel cantilever beams were subjected to impact by relatively heavy weights dropped on their tips, as well as by bullets. Reasonably good agreement with the theory was obtained when M_U was given a "dynamic" value based on the data of Manjoine [1944] for the rate sensitivity of the yield stress of mild steel.

A numerical study of the Parkes problem by Symonds and Fleming [1984] shows that when elasticity is taken into account, then the initial phase of the deflection is not even approximately represented by the traveling-hinge solution — that is, plastic deformation is not concentrated in a narrow zone in the interior. On the other hand, convergence to mode form is a prominent feature of the solution.

7.1.3 Dynamic Loading of Plates And Shells

Dynamic Loading of Rigid–Plastic Circular Plates

The dynamic problem of a simply supported rigid–plastic circular plate that is suddenly loaded by a uniformly distributed pressure — a blast loading — was studied by Hopkins and Prager [1954], as a sequel to the same authors' static solution (Hopkins and Prager [1953]) discussed in 5.2.3. The equilibrium equation (5.2.10) is replaced by the equation of motion,

$$(rM_r)' - M_\theta = -\int_0^r (p + \mu\ddot{w})r\,dr. \tag{7.1.8}$$

where we write u' for $\partial u/\partial r$ and $\dot{u}$ for $\partial u/\partial t$, while μ is the mass per unit area and p is the pressure (assumed acting in the negative z-direction). The Tresca yield criterion, as represented by the hexagon of Figure 5.2.2 (page 310), is assumed. As in the static case, the plate is assumed to be in regime BC, with point B corresponding to the center of the plate. A conical deformation can therefore be assumed, given by

$$w(r,\,t) = -\Delta(t)\left(1 - \frac{r}{a}\right).$$

Inserting this expression in the equation of motion (7.1.8) and substituting $M_\theta = M_U$, we can integrate the equation to obtain

$$M_r(r,\,t) = M_U - p\frac{r^2}{6} + \mu\ddot{\Delta}(t)\left(\frac{r^2}{6} - \frac{r^3}{12a}\right).$$

The boundary condition $M_r(a,\,t) = 0$ yields $\mu\ddot{\Delta} = 2(p - p_U)$, where $p_U = 6M_U/a^2$ is the static ultimate pressure. Eliminating $\mu\ddot{\Delta}$, we may write the following expression for M_r:

$$M_r = M_U - \frac{r^2}{6}\left[p - (p - p_U)\left(2 - \frac{r}{a}\right)\right].$$

This result is valid, however, only if the right-hand side does not exceed M_U, that is, if the quantity in brackets is nonnegative for all r. The necessary and sufficient condition for this is $p \le 2p_U$. Pressures satisfying this condition may be called moderate pressures, while pressures greater than $2p_U$ will be called high pressures.

At high pressures, the deflection may be assumed in the shape of a truncated cone, with a hinge circle of radius r_0 separating the outer conical region from the inner region, which moves rigidly with a velocity $\dot{w} = -\dot{\Delta}$. The latter region is at point B of the Tresca hexagon, so that $M_r = M_\theta = M_U$, and the equation of motion for $r < r_0$ reduces to

$$\mu\ddot{\Delta} = p.$$

For $r > r_0$, the velocity is

$$\dot{w}(r,\, t) = -\dot{\Delta}(t)\frac{a - r}{a - r_0},$$

and the acceleration is therefore given by

$$\mu\ddot{w} = -p\frac{a - r}{a - r_0}.$$

The equation of motion (7.1.8) now becomes

$$(rM_r)' = M_U - \frac{p}{6(a - r_0)}(2r^3 - 3r_0r^2 + r_0^3),$$

which when integrated subject to the initial condition $M_r = M_U$ at $r = r_0$, leads to

$$rM_r = rM_U - p\frac{(r - r_0)^3(r + r_0)}{12(a - r_0)}.$$

Finally, the condition $M_r = 0$ at $r = a$ gives r_0 as the solution of the cubic equation

$$\left(1 - \frac{r_0}{a}\right)^2\left(1 + \frac{r_0}{a}\right) = \frac{2p_U}{p}. \tag{7.1.9}$$

Suppose, now, that the pressure is suddenly removed at time $t = t_0$. For $p \le 2p_U$, the form of the preceding solution is still valid for $t > t_0$, with p replaced by zero. The midpoint acceleration is now given by $\mu\ddot{\Delta} = -2p_U$. The time history of the midpoint deflection is therefore

$$\begin{aligned}
\Delta(t) &= \frac{1}{\mu}[p(2t_0t - t_0^2) - p_Ut^2], \quad 0 < t < t_0,\\
&= \frac{1}{\mu}(p - p_U)t^2, \qquad\qquad\qquad t_0 < t < t_s,
\end{aligned}$$

where t_s is the time at which the motion stops, that is, $\dot{\Delta}(t_s) = 0$, so that $t_s = (p/p_U)t_0$. The permanent midpoint deflection is thus

$$\Delta_p = \Delta(t_s) = \frac{p(p - p_U)}{\mu p_U}t_0^2.$$

The result for t_s coincides with the lower bound given by Martin's theorem, Equation (7.1.1), when the conical velocity field is used, say $\dot{w}^* =$

$\dot{\Delta}^*(1 - r/a)$. The denominator of the right-hand side of (7.1.1) is then $2\pi M_U \dot{\Delta}^*$, while the numerator is

$$2\pi \int_0^a \mu\dot{\Delta}(t_0)\dot{\Delta}^* \left(1 - \frac{r}{a}\right)^2 r\, dr = 2\pi \frac{a^2}{12}\mu\dot{\Delta}(t_0)\dot{\Delta}^* = 2\pi\frac{a^2}{6}(p - p_U)t_0\dot{\Delta}^*.$$

Since $M_U = p_U a^2/6$, it follows that

$$t_s - t_0 \geq \left(\frac{p}{p_U} - 1\right) t_0,$$

and the right-hand side coincides with the exact value.

When $p > 2p_U$, the removal of the pressure means that in the central region the acceleration becomes zero, that is, $\ddot{\Delta} = 0$, and therefore

$$\dot{\Delta}(t) = \dot{\Delta}(t_0) = \frac{pt_0}{\mu}.$$

Since the velocity in the conical region is proportional to $\dot{\Delta}$, a nonvanishing acceleration there requires the radius of the hinge circle to be a function of time, say $\bar{r}(t)$, with $\bar{r}(t_0) = r_0$. For $r > \bar{r}(t)$, then, the velocity and the acceleration are, respectively,

$$\dot{w}(r, t) = -\frac{pt_0}{\mu}\frac{a - r}{a - \bar{r}(t)} \quad \text{and} \quad \ddot{w}(r, t) = -\frac{pt_0\dot{\bar{r}}(t)}{\mu}\frac{a - r}{[a - \bar{r}(t)]^2},$$

so that the equation of motion is

$$(rM_r)' = M_U + \frac{pt_0\dot{\bar{r}}}{6(a - \bar{r}^2)^2}[3a(r^2 - \bar{r}^2) - 2(r^3 - \bar{r}^3)].$$

Integrating, with $M_r = M_U$ at $r = \bar{r}$, we obtain

$$rM_r = rM_U + \frac{pt_0\dot{\bar{r}}(r - \bar{r})^2}{12(a - \bar{r})^2}[2a(r + 2\bar{r}) - (r^2 + 2r\bar{r} + 3\bar{r}^2)].$$

Setting $M_r = 0$ at $r = a$, we obtain for $\bar{r}$ the differential equation

$$(a^2 + 2a\bar{r} - 3\bar{r}^2)\frac{d\bar{r}}{dt} = -\frac{2p_U a^3}{pt_0}.$$

The factor on the left-hand side is positive, and therefore the radius of the hinge circle will decrease. The equation can be integrated to give

$$\frac{t}{t_0} = 1 + \frac{p}{2p_U a^3}[a^2(r_0 - \bar{r}) + a(r_0^2 - \bar{r}^2) - (r_0^3 - \bar{r}^3)].$$

Let t_1 denote the time at which the radius of the hinge circle goes to zero. Then

$$\frac{t_1}{t_0} = 1 + \frac{p}{2p_U}\left[\frac{r_0}{a} + \left(\frac{r_0}{a}\right)^2 - \left(\frac{r_0}{a}\right)^3\right],$$

where r_0 is given by Equation (7.1.9). For $t > t_1$, the velocity field (though not the deflection) is fully conical, and is consequently in mode form.

Perzyna [1958] studied the effect of different time profiles of the pressure pulse. The corresponding problem for a clamped plate was treated by Florence [1966]. The problem of impact loading of a circular plate, in which a uniform velocity is suddenly imparted to the plate (except the edge), was treated in a similar manner by Wang [1955] for a simply supported plate and by Wang and Hopkins [1954] for a clamped plate. Non-axisymmetric problems were treated by Hopkins [1957].

For simply supported plates, the results based on the Tresca criterion are immediately transferable to the Johansen criterion, since regime BC of Figure 5.2.2 is common to both. For clamped plates the criteria differ. The blast loading of square clamped plates obeying the Johansen criterion was studied by Cox and Morland [1959].

Pressure-Pulse Loading of a Cylindrical Shell

Relatively few dynamic problems have been solved for elastic–plastic or rigid–plastic shells. We consider here a solution due to Hodge [1955] for a circular cylindrical shell, made of a rigid–perfectly plastic material and clamped at both ends, that is suddenly loaded by a uniform radial pressure which does not decrease in time, the initial value of the pressure being greater than the static ultimate pressure p_U. For mathematical simplicity, the square yield criterion of Figure 6.4.4 (page 409) is adopted; an analysis by Hodge and Paul [1957] based on the hexagonal locus shows that the solution is not greatly affected by changes in shape of the yield locus.

The static ultimate pressure may be obtained by assuming that $N_\theta = N_U$ throughout. The equilibrium equation $(6.4.8)_2$ is then

$$M_z'' = p - \frac{N_U}{a}.$$

A collapse mechanism may be assumed to consist of hinge circles at the center ($z = 0$) and at the built-in ends ($z = \pm L$). For generalized-stress points on side GH of the square of Figure 6.4.4, the associated flow rule requires that $\dot{u}'' = 0$. The radial velocity field is accordingly

$$\dot{u}(z) = \dot{\Delta}\left(1 - \frac{|z|}{L}\right). \tag{7.1.10}$$

A moment distribution consistent with this velocity field has $M_z(0) = -M_U$ and $M_z(\pm L) = M_U$, and is given by

$$M_z = M_U\left(2\frac{z^2}{L^2} - 1\right),$$

so that

$$p_U = \frac{N_U}{a} + \frac{4M_U}{L^2} = \frac{N_U}{a}\left(1 + \frac{2}{\omega^2}\right),$$

where $\omega = \sqrt{N_U L^2/2M_U a}$ is the shell parameter defined in 6.4.3.

In the dynamic problem it can still be assumed that $N_\theta = N_U$, so that the velocity field is

$$\dot{u}(z, t) = \dot{\Delta}(t)\left(1 - \frac{|z|}{L}\right).$$

The equilibrium equation is replaced by the equation of motion,

$$M_z'' = p - \frac{N_U}{a} - \mu\ddot{u}, \tag{7.1.11}$$

where μ is again the mass per unit area. Integrating the equation of motion subject to the condition $M_z = -M_U$ at $z = 0$ leads to

$$M_z = -M_U + \left(p - \frac{N_U}{a}\right)\frac{z^2}{2} - \mu\ddot{\Delta}\left(\frac{z^2}{2} - \frac{|z|^3}{6L}\right).$$

The end condition $M_z = M_U$ at $z = \pm L$ then gives

$$\mu\ddot{\Delta} = \frac{3}{2}(p - p_U).$$

Eliminating $\mu\ddot{\Delta}$ from the expression for M_z, we obtain

$$M_z = M_U\left(2\frac{z^2}{L^2} - 1\right) - \frac{p - p_U}{4}\left(z^2 - \frac{|z|^3}{L}\right).$$

This solution is valid only if $|M_z| \le M_U$ everywhere, requiring

$$M_z''(0) \ge 0, \quad M_z'(L) \ge 0, \quad M_z'(-L) \le 0.$$

The first condition is satisfied if

$$p \le p_U + 8\frac{M_U}{L^2} = \frac{N_U}{a}\left(1 + \frac{6}{\omega^2}\right) \stackrel{\text{def}}{=} p_1.$$

The last two conditions are fulfilled if

$$p \ge p_U - 16\frac{M_U}{L^2} = \frac{N_U}{a}\left(1 - \frac{6}{\omega^2}\right). \tag{7.1.12}$$

If the pressure has a constant value p_0 for $0 < t < t_0$ and then suddenly drops to zero, then inequality (7.1.12) is obeyed after unloading only if $\omega^2 \le 6$; a shell satisfying this criterion may be called, following Hodge [1955], a *short* shell. For a short shell with $p_U < p_0 < p_1$ (a *moderate* pressure), the solution thus far derived is valid both for $0 < t < t_0$ and for $t > t_0$, with p given by p_0 and 0, respectively. The midpoint velocity is

$$\begin{aligned}\dot{\Delta}(t) &= \frac{3}{2\mu}(p_0 - p_U)t, \qquad 0 < t < t_0,\\ &= \frac{3}{2\mu}(p_0 t_0 - p_U t), \quad t_0 < t < t_s,\end{aligned}$$

where $t_s = (p_0/p_U)t_0$ is the time at which the motion stops. The permanent radial expansion at the midsection is

$$\Delta(t_s) = \frac{3p_0 t^2}{4\mu}\left(\frac{p_0}{p_U} - 1\right).$$

In a *long* shell ($\omega^2 > 6$) under moderate pressure, the solution is valid during the first phase but not after unloading. In order to obtain a valid solution, we must abandon the hypothesis that the entire shell is in regime GH of Figure 6.4.4. Instead, only a central portion of the shell, say $|z| < \bar{z}(t)$, will be assumed to be in this regime, with the sections $z = \pm\bar{z}(t)$ at point G, and the remainder of the shell in regime GJ. In the latter regime the flow rule gives $\dot{u} = 0$, and therefore the outer portions of the shell undergo no further motion. The velocity field is thus

$$\begin{aligned}\dot{u}(z, t) &= \dot{\Delta}(t)\left(1 - \frac{|z|}{\bar{z}(t)}\right), \quad |z| < \bar{z}(t),\\ &= 0, \qquad\qquad\qquad\qquad |z| > \bar{z}(t).\end{aligned}$$

The equation of motion for $|z| < \bar{z}$ is

$$M_z'' = -\left(\frac{N_U}{a} + \mu\ddot{\Delta}\right) + \mu\left(\frac{\ddot{\Delta}}{\bar{z}} - \frac{\dot{\Delta}\dot{\bar{z}}}{zbar^2}\right)|z|.$$

Since $M_z'(0, t) = 0$, a first integration gives

$$M_z' = -\left(\frac{N_U}{a} + \mu\ddot{\Delta}\right)z + \mu\left(\ddot{\Delta} - \frac{\dot{\Delta}\dot{\bar{z}}}{\bar{z}}\right)\frac{|z|z}{2\bar{z}}.$$

Furthermore, $M_z' = 0$ at $|z| = \bar{z}$, so that

$$\mu\left(\frac{\dot{\Delta}\dot{\bar{z}}}{\bar{z}} + \ddot{\Delta}\right) = -2\frac{N_U}{a}. \tag{7.1.13}$$

Integrating again yields, upon elimination of N_U by means of Equation (7.1.13) and with $M_z(0, t) = -M_U$,

$$M_z = -M_U + \mu\left(\frac{\dot{\Delta}\dot{\bar{z}}}{\bar{z}} - \ddot{\Delta}\right)\left(\frac{z^2}{4} - \frac{|z|^3}{6\bar{z}}\right).$$

Finally, since $M_z = M_U$ at $|z| = \bar{z}$, we obtain

$$\mu\left(\frac{\dot{\Delta}\dot{\bar{z}}}{\bar{z}} - \ddot{\Delta}\right) = 24\frac{M_U}{\bar{z}^2}. \tag{7.1.14}$$

Combining Equations (7.1.13) and (7.1.14) produces

$$\mu\ddot{\Delta} = -\frac{N_U}{a} - 12\frac{M_U}{\bar{z}^2}, \tag{7.1.15}$$

$$\mu\dot{\Delta}\dot{\bar{z}} = -\frac{N_U}{a}\bar{z} + 12\frac{M_U}{\bar{z}}. \tag{7.1.16}$$

Differentiating (7.1.16) with respect to time and substituting for $\mu\ddot{\Delta}$ from (7.1.15), we find that

$$\mu\dot{\Delta}\ddot{\bar{z}} = 0.$$

Since $\dot{\Delta}$ cannot be zero while the shell is in motion, it follows that $\ddot{\bar{z}} = 0$, so that $\dot{\bar{z}}$ is a constant. We obtain its value from Equation (7.1.16) at $t = t_0$, when $\bar{z} = L$ and $\dot{\Delta} = 3(p_0 - p_U)t_0/2\mu$. Hence

$$\dot{\bar{z}} = -\frac{2L}{3t_0}\frac{N_U}{a}\frac{1 - \frac{6}{\omega^2}}{p_0 - p_U},$$

which is always negative for a long shell ($\omega^2 > 6$); consequently, the hinge circles move toward the center, and their location is given by

$$\bar{z}(t) = L + (t - t_0)\dot{\bar{z}}. \tag{7.1.17}$$

The motion stops when the right-hand side of (7.1.16) vanishes, that is, when $\bar{z} = \sqrt{6}L/\omega$. The time when this occurs is

$$t_s = t_0 + \frac{L}{|\dot{\bar{z}}|}\left(1 - \frac{\sqrt{6}}{\omega}\right) = t_0\left[1 + \frac{3}{2}\frac{p_0 - p_U}{(1 + \sqrt{6}/\omega)(N_U/a)}\right].$$

The permanent radial expansion at the midsection can be obtained by integrating $\dot{\Delta}(t)$, as given by (7.1.16) and with $\bar{z}$ given by (7.1.17), from t_0 to t_s, and adding the result to $\Delta(t_0) = 3(p_0 - p_U)t_0^2/4\mu$.

Under high pressures, that is, $p_0 > p_1$, during the initial phase hinge circles may be assumed at $z = \pm z_0$ and at $z = \pm L$, with $M = -M_U$ and $N = N_U$ for $|z| \le z_0$, and the velocity given by

$$\begin{aligned} \dot{u}(z, t) &= \dot{\Delta}(t), & |z| < z_0, \\ &= \dot{\Delta}(t)\frac{L - z}{L - z_0}, & |z| > z_0. \end{aligned}$$

Equation (7.1.11) for $|z| < z_0$ gives

$$\mu\ddot{\Delta} = p - \frac{N_U}{a}.$$

For $|z| > z_0$, therefore, the equation of motion may therefore be written as

$$M_z'' = \left(p - \frac{N_U}{a}\right)\frac{|z| - z_0}{L - z_0},$$

and may be integrated twice, subject to the initial conditions $M_z' = 0$ and $M_z = -M_U$ at $|z| = z_0$, to give

$$M_z = \left(p - \frac{N_U}{a}\right)\frac{(|z| - z_0)^3}{6(L - z_0)} - M_U.$$

The conditions $M_z = M_U$ at the built-in ends $|z| = z_0$ lead to

$$z_0 = L - \sqrt{\frac{12M_U}{p_0 - (N_U/a)}},$$

and $z_0 > 0$ if $p_0 > p_1$.

After unloading, the inner hinge circles must be assumed to move, their location being, say, $|z| = \bar{z}(t)$. The equation of motion for $|z| < \bar{z}(t)$ reduces to $\mu\ddot{\Delta} = -N_U/a$. Continuity of the velocity at $t = t_0$ gives the midsection radial velocity for $t_0 < t < t_1$ as

$$\dot{\Delta}(t) = \frac{1}{\mu}\left(p_0 t_0 - \frac{N_U}{a}t\right),$$

where t_1 is the time at which the hinge circles coalesce at the midsection.

For $|z| > \bar{z}(t)$ the equation of motion is

$$M_z'' = -\frac{N_U}{a}\frac{|z| - \bar{z}}{L - \bar{z}} - \mu\dot{\Delta}\dot{\bar{z}}\frac{L - z}{(L - \bar{z})^2}.$$

Integrating subject to $M_z' = 0$ and $M_z = -M_U$ at $|z| = \bar{z}$ and substituting $M_z = M_U$ at $|z| = L$ gives

$$\mu\dot{\Delta}\dot{\bar{z}} = -\frac{N_U}{a}(L - \bar{z}) - 12\frac{M_U}{L - \bar{z}}.$$

Substituting for $\dot{\Delta}$ as a function of time, we find the the differential equation for $\bar{z}(t)$ is separable and can be written as

$$\frac{dt}{\bar{t} - t} = -\frac{(L - \bar{z})d\bar{z}}{(L - \bar{z})^2 + 3L^2/\omega^2},$$

where $\bar{t} = (p_0 a/N_U)t_0$. We can integrate the equation subject to $\bar{z}(t_0) = z_0$, obtaining

$$\left(\frac{\bar{t} - t_0}{\bar{t} - t}\right)^2 = \frac{(L - \bar{z})^2 + 3L^2/\omega^2}{(L - z_0)^2 + 3L^2/\omega^2}.$$

We finally find t_1 as

$$t_1 = t_0 \left[1 + \left(\frac{p_0 a}{N_U} - 1 \right) \left(1 - \sqrt{\frac{3 + \omega^2 (1 - z_0/L)^2}{3 + \omega^2}} \right) \right].$$

For $t > t_1$, the velocity field may again be assumed to be of the form (7.1.10) (mode form). Calculation of the permanent deformation may be carried out by analogy with the preceding examples. Details may be found in the paper by Hodge [1955] (see also Hodge [1956]).

As the reader can see, there are a great many similarities between the dynamic shell and plate problems formulated according to the rigid–perfectly plastic model, and these in turn resemble the beam problems. The comments expressed previously concerning the validity of the traveling-hinge solution may therefore be expected to be applicable to structural impact problems in general.

Exercises: Section 7.1

1. Formulate Martin's inequality (7.1.2) in terms of generalized stresses to give a bound on the permanent deflection for (a) beams and (b) plates.

2. Determine the motion due to impact loading of a free-ended rigid–perfectly plastic beam by a concentrated force at the center if (a) the force has the constant value F_m for $0 < t < t_0$ and is removed at $t = t_0$, and (b) the force rises linearly from zero to F_m at $t = t_0/2$ and then declines linearly to zero at $t = t_0$.

3. Analyze the motion of a rigid–perfectly plastic cantilever struck by a mass m that is uniformly distributed over a length a at the free end of the beam. In particular, find the stopping time t_s and compare with the bound (7.1.1). Show that the solution reduces to that of Parkes as $a/L \to 0$.

4. Analyze the motion of a rigid–perfectly plastic beam struck by a concentrated mass m at midspan when the beam is (a) simply supported and (b) clamped.

5. Analyze the motion of a rigid–perfectly plastic circular Tresca plate of radius a that is suddenly loaded by a pressure p uniformly distributed over the circle $r < b$ (where $b < a$) when the plate is (a) simply supported and (b) clamped.

6. Analyze the motion of a simply supported rigid–perfectly plastic circular Tresca plate that is struck by a concentrated mass m at the center.

7. Compare the stopping time t_s found in the text for a long clamped cylindrical shell under a moderate impulsive pressure with the bound (7.1.1).

8. Analyze the motion of a long clamped cylindrical shell under a high pressure for $t > t_1$. Find the stopping time t_s and compare with (7.1.1).

9. Analyze the motion of a clamped cylindrical shell under a pressure-pulse loading. Study various cases.

10. Analyze the motion of an infinitely long cylindrical shell suddenly loaded by a ring load.

Section 7.2 One-Dimensional Plastic Waves

7.2.1 Theory of One-Dimensional Waves

In studying the propagation of longitudinal stress and strain waves in a thin bar, it is common to represent the problem by a "one-dimensional" approximation in which the only nonvanishing stress component is assumed to be the longitudinal one, and the contribution of the transverse displacement to the deformation and the inertia is ignored. It is known from elastic bar theory that this approximation yields good results at points of the bar whose distance from the bar ends is more than a few diameters; near the ends, three-dimensional corrections are necessary.

We let x denote the Lagrangian coordinate along the bar axis and $u(x, t)$ the corresponding displacement. The conventional ("engineering") strain $\varepsilon(x, t)$ and velocity $v(x, t)$ are then given by

$$\varepsilon = \frac{\partial u}{\partial x}, \quad v = \frac{\partial u}{\partial t},$$

and satisfy the *kinematic compatibility relation*

$$\frac{\partial \varepsilon}{\partial t} = \frac{\partial v}{\partial x}. \tag{7.2.1}$$

The equations of motion, in the absence of body force, reduce to

$$\frac{\partial \sigma}{\partial x} = \rho \frac{\partial v}{\partial t}, \tag{7.2.2}$$

where σ is the nominal ("engineering") stress and ρ is the mass density in the undeformed state.

The equations governing the problem of torsional motion of a thin-walled circular tube are exactly the same as the preceding ones if u and v are interpreted as the circumferential displacement and velocity (u_θ and v_θ), respectively, and ε and σ as the conventional shear strain and shear stress ($\gamma_{z\theta}$ and $\tau_{z\theta}$). Except for some comments concerning finite strain, the following theory applies to the torsional problem as well.

Shock Fronts

A *shock front* is said to occur at a point $x = \alpha(t)$ of the bar if the velocity v is discontinuous there. We suppose that the shock front is moving at a finite speed c in the positive x-direction (say from left to right), that is, $c = \alpha'(t) > 0$, and we designate the values of v just to the right (in front) of the shock and just to the left of (behind) the shock by v^+ and v^-, respectively. The *jump* in v is defined as

$$[\![v]\!] = v^- - v^+.$$

Similar definitions apply to ε and σ. It is readily apparent that with v a discontinuous function of x and of t, its partial derivatives with respect to both variables are in effect infinite at the shock front. By Equation (7.2.2), the same is true of ε and σ.

Relations among the jumps in velocity, strain, and stress may be derived by treating the shock front as a thin zone in which these quantities change very rapidly, their partial derivatives having large constant values. If the shock thickness is h, then, approximately,

$$[\![v]\!] \doteq \pm h\frac{\partial v}{\partial x},$$

for a front moving to the left and to the right, respectively, with similar relations for ε and σ.

Another approximation for the jump may be obtained by following it along the time axis. Since v^- is the earlier and v^+ the later value, and since the duration of the shock passage at a given point is h/c, it follows that

$$[\![v]\!] \doteq \frac{h}{c}\frac{\partial v}{\partial t}.$$

Applying these approximations to Equations (7.2.1)–(7.2.2), we obtain the *shock relations*

$$[\![\varepsilon]\!] = \pm\frac{1}{c}[\![v]\!], \quad [\![\sigma]\!] = \pm\rho c[\![v]\!], \tag{7.2.3}$$

where the + and − signs apply to fronts moving to the left and to the right, respectively.

Eliminating $[\![v]\!]$, we are left with the *shock-speed equation*

$$\rho c^2 = \frac{[\![\sigma]\!]}{[\![\varepsilon]\!]}. \tag{7.2.4}$$

Impact of a Rigid–Plastic Bar

As an application of the shock relations (7.2.3), we study the problem of longitudinal impact of a bar treated as rigid–plastic. Suppose that the bar, of length L, is moving rigidly at a speed v_0 in the direction of its axis until, at time $t = 0$, it squarely strikes a rigid target. The situation at $t > 0$ is shown in Figure 7.2.1; it is assumed that plastic deformation takes place only in an infinitely narrow zone at $x = \xi(t)$, where x is measured from the free end [so that $\xi(0) = L$], with the remaining portions of the bar rigid. In

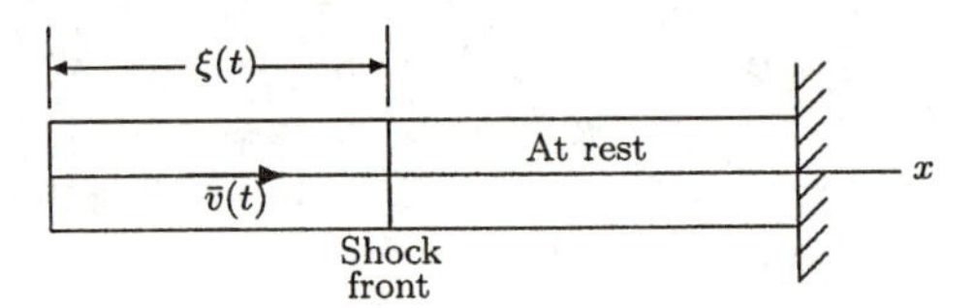

Figure 7.2.1. Impact of a rigid-plastic bar on a rigid wall.

particular, the portion near the target, $x > \xi(t)$, is assumed to have come to rest, while the portion near the free end is moving with a uniform velocity:

$$\begin{aligned} v(x,\,t) &= \bar{v}(t), \quad & x < \xi(t), \\ v(x,\,t) &= 0, & x > \xi(t). \end{aligned}$$

The section $x = \xi(t)$, since the velocity is discontinuous there, is the location of a shock front moving in the negative x-direction with a speed $c = -\dot{\xi}$ with respect to the undeformed material. For convenience, stress and strain will be taken as positive in compression, so that the signs in Equations (7.2.3) are reversed.

We begin by assuming the bar as rigid–perfectly plastic, so that it is at yield at the shock front. The true stress, then, equals σ_Y at $x = \xi(t)$ and is continuous there. The nominal stress is also σ_Y ahead of the front, but it equals $\sigma_Y/(1-\bar{\varepsilon})$ behind the front if $\bar{\varepsilon}$ is the strain there. The discontinuities at the front are therefore

$$[\![\sigma]\!] = \frac{\sigma_Y}{1-\bar{\varepsilon}} - \sigma_Y = \frac{\sigma_Y \bar{\varepsilon}}{1-\bar{\varepsilon}}, \quad [\![\varepsilon]\!] = \bar{\varepsilon}, \quad [\![v]\!] = -\bar{v},$$

and the shock relations are

$$\bar{\varepsilon} = -\frac{\bar{v}}{\dot{\xi}}, \quad \frac{\sigma_Y \bar{\varepsilon}}{1-\bar{\varepsilon}} = -\rho\dot{\xi}\bar{v},$$

from which, on the one hand,

$$\rho\dot{\xi}^2 = \frac{\sigma_Y}{1-\bar{\varepsilon}} \tag{7.2.5}$$

while, on the other hand,

$$\rho \bar{v}^2 = \frac{\sigma_Y \bar{\varepsilon}^2}{1-\bar{\varepsilon}}, \tag{7.2.6}$$

and upon differentiating,

$$\rho \bar{v}\, d\bar{v} = \frac{\sigma_Y}{2} d\left(\frac{\bar{\varepsilon}^2}{1-\bar{\varepsilon}}\right). \tag{7.2.7}$$

In addition, the equation of motion of the undeformed portion is

$$\rho \xi(t)\dot{\bar{v}}(t) = -\sigma_Y,$$

so that

$$\rho \bar{v}\dot{\bar{v}} = -\frac{\sigma_Y \bar{v}}{\xi} = \sigma_Y \bar{\varepsilon}\frac{\dot{\xi}}{\xi}. \tag{7.2.8}$$

Combining (7.2.7) and (7.2.8) gives

$$2\frac{d\xi}{\xi} = \frac{1}{\bar{\varepsilon}} d\left(\frac{\bar{\varepsilon}^2}{1-\bar{\varepsilon}}\right). \tag{7.2.9}$$

If the initial impact velocity is $\bar{v}(0) = v_0$, then the initial strain at the impact end is obtained in terms of v_0 through (7.2.6), namely,

$$\frac{\rho v_0^2}{\sigma_Y} = \frac{\varepsilon_0^2}{1-\varepsilon_0}.$$

Equation (7.2.9) may now be integrated, giving

$$\ln\left(\frac{\xi}{L}\right)^2 = \frac{1}{1-\bar{\varepsilon}} - \frac{1}{1-\varepsilon_0} - \ln\frac{1-\bar{\varepsilon}}{1-\varepsilon_0}.$$

If ξ_0 is the value of ξ where $\bar{\varepsilon} = 0$, then

$$\xi_0 = L\sqrt{1-\varepsilon_0}\exp\left(-\frac{\varepsilon_0}{2(1-\varepsilon_0)}\right).$$

and the shock front stops when $\xi(t) = \xi_0$, so that the portion $0 < x < \xi_0$ remains undeformed.

The preceding analysis is equivalent to that of G. I. Taylor [1948], who based it on a different approach. By means of some approximating assumptions Taylor derived a formula for the dynamic yield stress in terms of the impact speed and the specimen dimensions before and after impact.

For a bar made of work-hardening material, the problem was treated by Lee and Tupper [1954]. The geometry and notation are the same as in the preceding treatment. If the conventional stress-strain relation is given by $\sigma = F(\varepsilon)$ and the initial yield stress is σ_E, then the material just ahead of the shock front may be assumed to be about to yield, so that $\sigma = \sigma_E$ there, while immediately behind the front the stress is $\bar{\sigma} = F(\bar{\varepsilon})$. The stress jump is therefore $[\![\sigma]\!] = \bar{\sigma} - \sigma_E$. From the shock relations we obtain, first,

$$\rho\dot{\xi}^2 = \frac{\bar{\sigma} - \sigma_E}{\bar{\varepsilon}},$$

which may be contrasted with (7.2.5), and second,

$$\rho\bar{v}^2 = (\bar{\sigma} - \sigma_E)\bar{\varepsilon}.$$

Proceeding as before, we obtain

$$2\frac{d\xi}{\xi} = \frac{d[(\bar{\sigma} - \sigma_E)\bar{\varepsilon}]}{\sigma_E\bar{\varepsilon}}.$$

The initial strain at the impact end is related to the impact speed v_0 by means of

$$\rho v_0^2 = (\sigma_0 - \sigma_E)\varepsilon_0,$$

where $\sigma_0 = F(\varepsilon_0)$. The relation between $\bar{\varepsilon}$ and ξ, which provides the distribution of permanent strain along the bar, is obtained as

$$\ln\left(\frac{\xi}{L}\right)^2 = \int_{\varepsilon_0}^{\bar{\varepsilon}} \frac{d\{[F(\varepsilon) - \sigma_E]\varepsilon\}}{\sigma_E\varepsilon},$$

provided that $\bar{\varepsilon}$ is positive; setting $\bar{\varepsilon} = 0$ gives ξ_0, which defines, as before, the extent of the undeformed portion. The result was used by Lee and Tupper [1954] in order to predict the permanent deformation of steel specimens, and experiments performed by them showed fairly good agreement.

7.2.2 Waves in Elastic-Plastic Bars

Taking the elasticity of the bar into account changes the nature of the problem drastically. In an elastic solid, disturbances cannot be propagated at a speed faster than the elastic wave speed. It would therefore be impossible for the entire portion of the bar ahead of the plastic shock front to move as a rigid body from the moment of impact; an elastic front has to intervene. Moreover, when an elastic wave of compression reaches the free end, it is reflected as a tension wave and will bring about unloading when it reaches the plastically deforming material.

At a section of the bar where the material is elastic, with Young's modulus E, we have $[\![\sigma]\!]/[\![\varepsilon]\!] = E$, so that the shock speed is just the elastic bar-wave speed $c_e = \sqrt{E/\rho}$, independent of the local state variables. The same is true when the behavior is elastic-plastic, provided that the stress change across the shock represents unloading.

Donnell Theory

To understand the nature of a front of loading into a plastic state, it is simplest to begin, following an approach due to Donnell [1930], by considering a material with a "bilinear" stress-strain relation (Figure 7.2.2). We

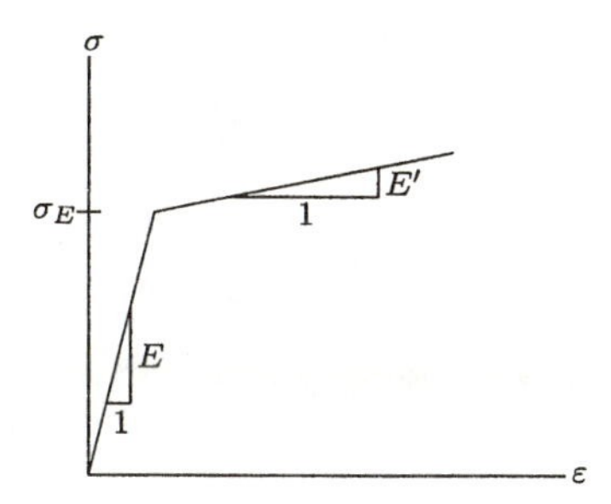

Figure 7.2.2. Bilinear stress-strain diagram.

return to the convention of stress and strain as positive in tension. If the initial state is stress-free, then any disturbance involving a stress increase up to the yield stress σ_E will be propagated at the aforementioned elastic speed c_e. On the other hand, stress increases above σ_E will travel at the slower "plastic" speed $c_p = \sqrt{E'/\rho}$. Thus in a semi-infinite bar occupying the half-line $x > 0$ in which at $t = 0$ a stress $\sigma_0 > \sigma_E$ is applied at the left end, we distinguish three regions, whose time dependence is shown in Figure 7.2.3:

$$\begin{array}{lll} 0: & c_e t < x & \sigma = 0, \\ I: & c_p t < x < c_e t & \sigma = \sigma_E, \\ II: & 0 < x < c_p t & \sigma = \sigma_0. \end{array}$$

The values of strain and velocity in each of the regions can be obtained from the shock relations (7.2.3); they are shown in Figure 7.2.3.

Finite Bars

If the bar is of finite length, say L, and is supposed to occupy the interval $0 < x < L$, then a front propagating to the right from $x = 0$ will, upon reaching the end $x = L$, be *reflected* as one propagating to the left. The nature of the reflected front depends on the end condition at $x = L$, and if the bar is elastic-plastic, on the intensity of the incident front. End conditions are assumed as being of two kinds: (a) *free*, characterized by $\sigma = 0$, and (b) *fixed*, characterized by $v = 0$ if the frame of reference is stationary with respect to the fixed end.

Consider, now, an elastic shock front, with a stress $-\sigma_0$ and zero velocity behind it, moving into a region of zero stress and velocity $-v_0$ (so that from the shock relations, $\sigma_0 = \rho c_e v_0$) until it reaches a *free* right end. This is the problem of an elastic bar traveling freely to the left, at a speed v_0, until it

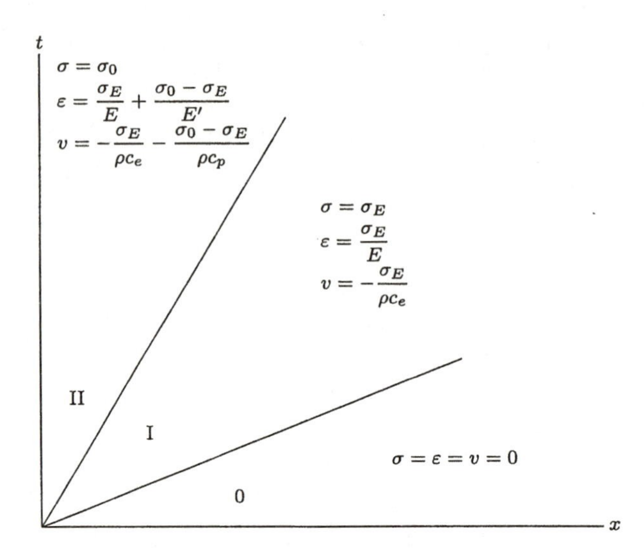

Figure 7.2.3. Semi-infinite bar of bilinear material: regions in xt-plane.

impinges upon a fixed rigid obstacle.[1] Behind the reflected front the stress must be zero, so that it is a front of *unloading*, and therefore, by the shock relations, the velocity behind it will be $+v_0$.

When the reflected front reaches $x = 0$, at the time $t = 2L/c_e$ after the initial impact, it will in turn be reflected to the right. If this end is regarded as a fixed end, then, in order to have $v = 0$ behind the second reflected front, the stress there must be $+\sigma_0$, that is, a tensile stress. Unless the bar somehow fuses with the obstacle, there is nothing to transmit a tensile stress between them, and therefore the assumption that the left end is fixed for $t > 2L/c_e$ must be abandoned in favor of the one that the end is free. In this case, the state behind the second reflected front has $\sigma = 0$ and $v = +v_0$; that is, the bar *rebounds* from the obstacle. The *rebound time* is consequently $2L/c_e$.

If the bar is elastic-plastic, with a compressive yield stress σ_E, then the preceding results are valid as long as $\sigma_0 \le \sigma_E$, that is, if $v_0 \le \sigma_E/\rho c_e \stackrel{\text{def}}{=} v_E$. If $v_0 > v_E$, and if the material is modeled by a bilinear stress-strain relation as on the previous page, then an elastic and a plastic shock front propagate from the impact end at the respective speeds c_e and c_p, and the results for the semi-infinite bar hold until the initial elastic front, with $\sigma = -\sigma_E$ and $v = v_E - v_0$ behind it, reaches the free end. The reflected front, being one of unloading, is elastic, the stress behind it being zero and the velocity $2v_E - v_0$.

Let $\alpha \stackrel{\text{def}}{=} c_p/c_e$. The reflected elastic and initial plastic fronts meet at point A (Figure 7.2.4), with $x_A = 2\alpha l/(1+\alpha)$ and $t_A = 2L/(1+\alpha)c_e$.

[1] This problem was studied by Lenskii [1948] and De Juhasz [1949].

From this point, fronts may propagate both to the left and to the right. We assume, first, that only elastic fronts propagate, leaving behind them a region with $\sigma = -\sigma_1$ and $v = v_1$. The *strain* cannot be uniform in this region, since plastic deformation has taken place to the left of A but not to the right of A. Thus a stationary discontinuity front, at which only the strain is discontinuous, forms at A.

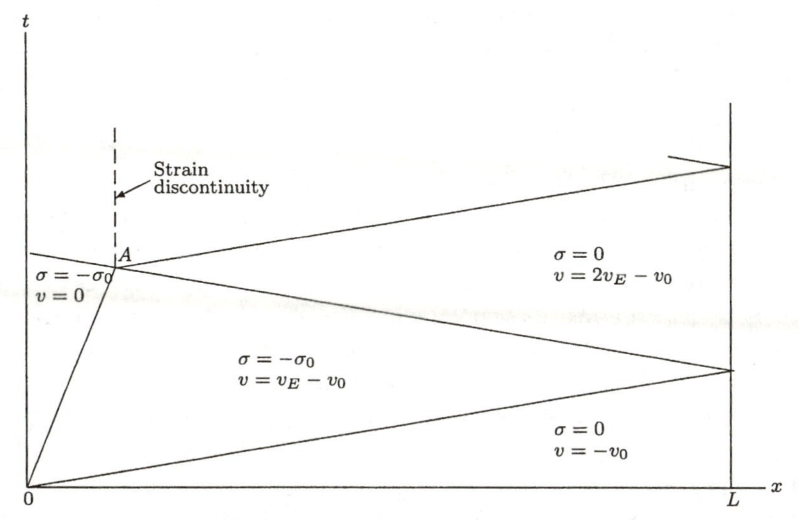

Figure 7.2.4. Finite bar of bilinear material: wave reflection and intersection.

The shock relations along the rightward and leftward fronts, respectively, are

$$\sigma_1 = \rho c_e(v_1 - 2v_E + v_0), \qquad \sigma_1 - \sigma_0 = -\rho c_e v_1.$$

Noting that $\sigma_0 = \sigma_E + \rho c_p(v_0 - v_E)$, we obtain from these relations the solutions $\sigma_1 = \frac{1}{2}\rho c_e(1+\alpha)(v_0 - v_E)$ and $v_1 = \frac{1}{2}[(3-\alpha)v_E - (1-\alpha)v_0]$. In order to justify the assumption that only elastic fronts propagate from A (in other words, that the plastic front is *absorbed* at A), we must have $\sigma_1 < \sigma_E$, the condition for which is $v_0 < [(3+\alpha)/(1+\alpha)]v_E$.

Let us assume that this last condition is met, and that the leftward elastic front is reflected from the impact end as though rebound occurred, as with the elastic bar, at $t = 2L/c$. The rebound velocity is found to be $2v_E - v_0$; but this is positive only if $v_0 < 2v_E$. This, then, is the condition for rebound at $t = 2L/c_e$. It can be ascertained that no further plastic deformation occurs, and therefore the bar is left with a permanent deformation of $\varepsilon_0 - \sigma_0/E$ to the left of point A.

If $v_0 > 2v_E$, then the bar remains in contact with the obstacle at least until the next front reaches the left end. If $2v_E < v_0 < [(3+\alpha)/(1+\alpha)]v_E$, then the compressive stress holding them together is $\rho c_e(v_0 - 2v_E)$.

If $v_0 > [(3+\alpha)/(1+\alpha)]v_E$, then the plastic front continues to the right beyond A. Ahead of it, we have $\sigma = -\sigma_E$ and $v = 3v_E - v_0$. If the stress and velocity behind it are again denoted $-\sigma_1$, v_1, then the shock relations are

$$\sigma_1 - \sigma_E = \rho c_p(v_1 - 3v_E + v_0), \qquad \sigma_1 - \sigma_0 = -\rho c_e v_1,$$

yielding $v_1 = 2\alpha v_E/(1+\alpha)$.

It is clear that the numerical problem of determining the rebound time becomes more complex as the impact velocity increases. Some solutions are shown in Figure 7.2.5.

Kármán–Taylor–Rakhmatulin Theory

It is not difficult to extend the Donnell theory to materials whose stress-strain curve consists of more than two line segments, provided that the overall curve is convex downward. Such a curve may in turn be regarded as an approximation to one that is smooth beyond the yield stress (Figure 7.2.6). It is clear that as the segments become shorter, the jumps in the field quantities at the plastic shock fronts become smaller. If the smooth curve is seen as the limit of the segmented one, then it follows that in a bar made of a material described by the smooth curve, no plastic shock fronts can be propagated.

The theory that was developed independently by Taylor [1946, 1958], von Kármán (see Kármán and Duwez [1950]) and Rakhmatulin [1945] during World War II is based on complementing the basic field equations (7.2.1)–(7.2.2) with the constitutive equation

$$\sigma = F(\varepsilon). \tag{7.2.10}$$

While its form is that of a nonlinear elastic stress-strain relation, Equation (7.2.10) is assumed in this theory to hold only during an initial loading process, unloading being linearly elastic. An initial yield stress σ_E may be incorporated if $F(\varepsilon) = E\varepsilon$ for $|\varepsilon| < \varepsilon_0$, where $\varepsilon_0 = \sigma_E/E$ (here it is assumed that the initial loading curves are the same in tension and compression). The function $F(\cdot)$ is assumed to be continuously differentiable, with $F'(\varepsilon) > 0$, and convex in the sense that $\varepsilon F''(\varepsilon) \le 0$.

A solution to the system of equations (7.2.1), (7.2.2), and (7.2.10) that may be valid for some problems, at least in a certain region of the xt-plane, can be obtained by assuming that the three variables σ, v, ε depend on x and t only through the combination $c = x/t$, that is, that they are constant along lines of slope c emanating from (0, 0). (Such a solution is sometimes called a "similarity" solution.) Since $\partial c/\partial x = 1/t$ and $\partial c/\partial t = -c/t$, Equations (7.2.1)–(7.2.2) become

$$dv = -c\, d\varepsilon, \qquad d\sigma = -\rho c\, dv. \tag{7.2.11}$$

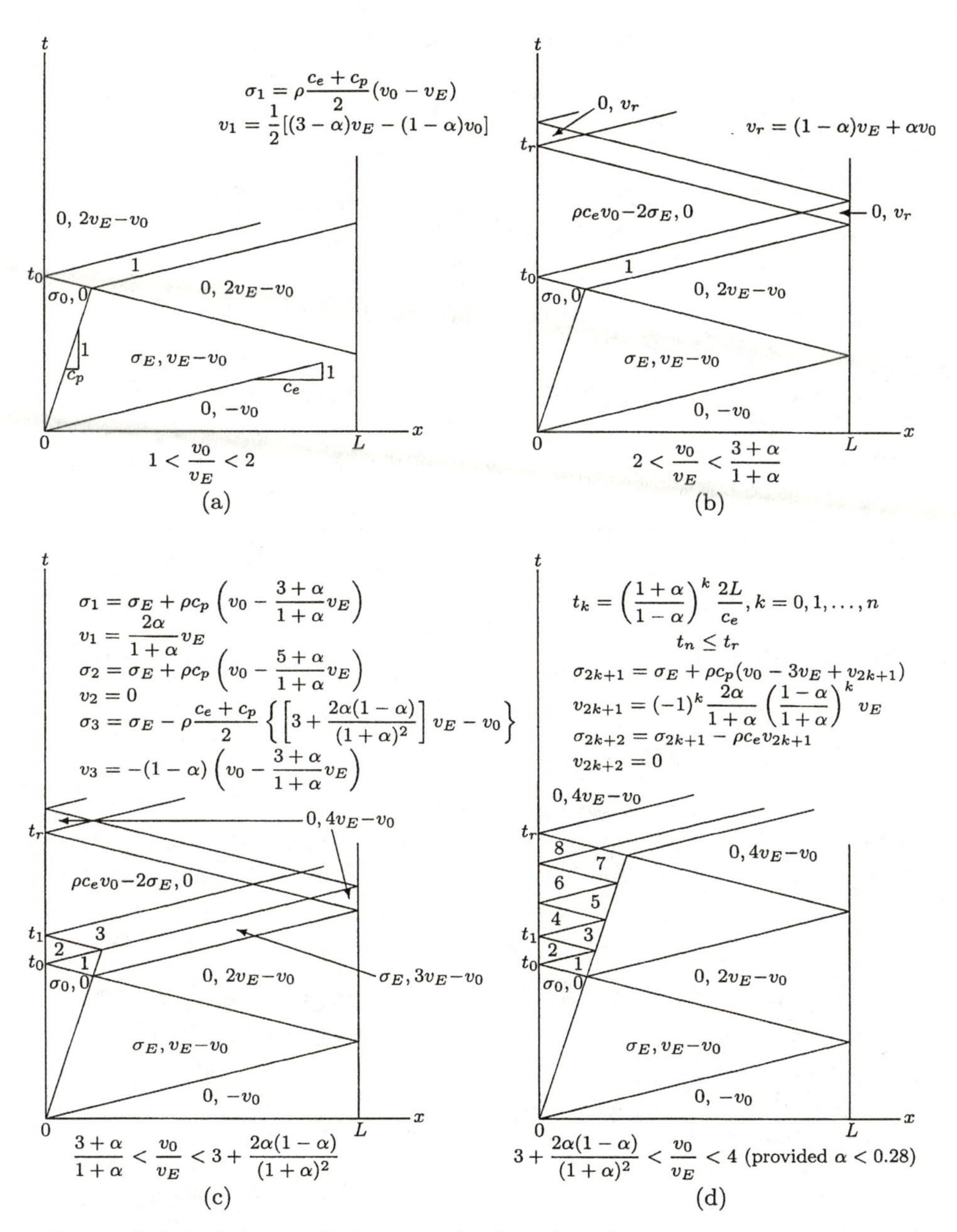

Figure 7.2.5. Solutions for impact of a finite bar of bilinear material at various impact velocities.

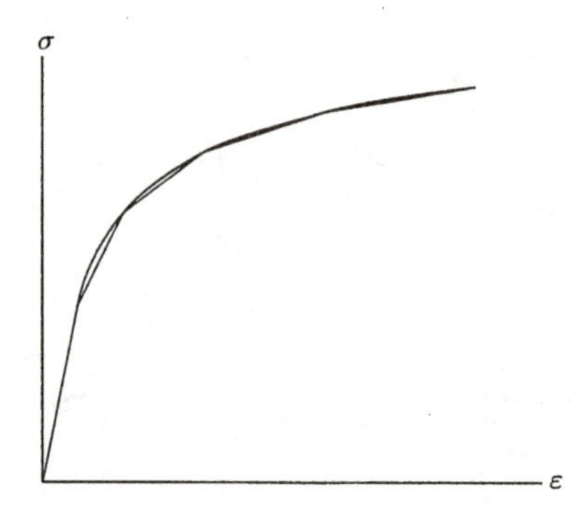

Figure 7.2.6. Piecewise linear and smooth stress-strain curves.

These equations are clearly of the same form as the shock relations (7.2.3), with the jumps replaced by differentials. Indeed, the limit of the shock relations for "infinitesimal shocks" propagating to the right takes precisely the form of Equations (7.2.11).

Eliminating dv between the two equations (7.2.11) yields

$$\rho c^2 = \frac{d\sigma}{d\varepsilon}.$$

Combining this result with (7.2.10) leads to a relation between c and ε:

$$c(\varepsilon) = \sqrt{\frac{F'(\varepsilon)}{\rho}}. \tag{7.2.12}$$

Provided that the sign of ε is given, this relation will be assumed invertible if $|\varepsilon| > \varepsilon_E$. Thus to any given $c > c_e$ there corresponds a value of ε and, by (7.2.10), of σ. From the first of Equations (7.2.11) we deduce, in addition, a relation for the velocity:

$$v = -\int c(\varepsilon)\, d\varepsilon, \tag{7.2.13}$$

the lower limit of the integral depending on the initial conditions.

The preceding solution can be applied to the problem of impact on a semi-infinite bar, treated previously according to the Donnell model. The notation will be simplified by treating stress and strain as positive in compression, and consequently the minus signs in Equations (7.2.11) and (7.2.13) will be eliminated. The relations between the impact velocity v_0, the maximum stress σ_0 and the maximum strain ε_0 are accordingly given by

$$v_0 = \int_0^{\varepsilon_0} c(\varepsilon)\, d\varepsilon, \qquad \sigma_0 = F(\varepsilon_0). \tag{7.2.14}$$

As in the Donnell theory, an elastic shock front propagates into the undeformed region, with $\sigma = \sigma_E$ *immediately* behind it. In the region

$c(\varepsilon_0)t < x < c_e t$, however, the state varies continuously, the solution being the "similarity" solution described previously. For $x < c(\varepsilon_0)t$, the state is again uniform, as in the Donnell theory.

In a finite bar, with the right end $x = L$ free, the incident elastic front is again reflected as an unloading front, with a speed that is, at least at first, equal to c_e.[2] However, this front propagates, from the outset, into a region of plastically deformed material, with strains that are not necessarily infinitesimal. At finite strain, the elastic relation $\Delta\sigma = E\Delta\varepsilon$ between stress and strain changes, with E the Young's modulus measured at small strains, is not valid if σ and ε are the nominal stress and conventional strain, respectively. Experiments show, on the other hand, that the slope of the unloading line is very nearly E if the stress and strain are taken as the true (Cauchy) stress and the logarithmic strain, respectively. As we noted before, when the strain is positive in compression then, with incompressibility assumed, the true stress σ_t is related to the nominal stress σ by $\sigma = \sigma_t/(1-\varepsilon)$. Since the change of strain in an elastic process is infinitesimal, we accordingly have $\Delta\sigma \doteq \Delta\sigma_t/(1-\varepsilon)$, and $\Delta \ln(1-\varepsilon) = \Delta\varepsilon/(1-\varepsilon)$. The *unloading* relation is, consequently,

$$\frac{\partial\sigma}{\partial\varepsilon} = \frac{E}{(1-\varepsilon)^2}. \tag{7.2.15}$$

Assuming, first, that ε_0 is sufficiently small for the right-hand side of (7.2.15) to be closely approximated by E, we can show that as in the Donnell theory, rebound occurs at $t = 2L/c_e$ if $v_0 < 2v_E$. After rebound, the entire bar is permanently deformed. As before, a *plateau of permanent strain*, of value $\varepsilon_0 - \sigma_0/E$, extends to the left of point A, while the permanent strain decreases from this value to zero between A and the right end. The distribution of permanent compressive strain is shown as a thickening of the bar, since volume remains essentially constant. Thus the final shape of the bar is as shown in Figure 7.2.7, and this shape has frequently been found in tests.

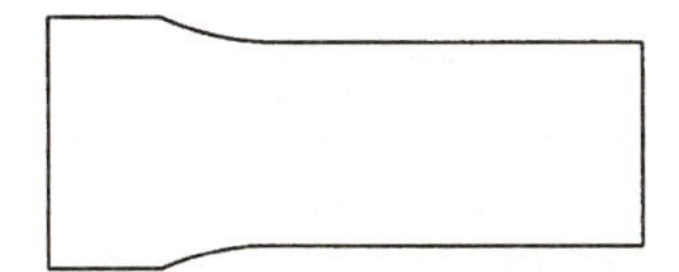

Figure 7.2.7. Final shape of a bar with a smooth stress-strain curve.

If the strains are too large to be considered infinitesimal, then it follows from Equation (7.2.15) that the elastic unloading front does not propagate at the constant speed c_e. Instead, its trace in the xt-plane is curved in the

[2]The problem of the finite bar was studied by Lee [1953] and others (see Cristescu [1967], Section II.9).

region $c(\varepsilon_0)t < x < c_e t$, the slope being $-c_e/(1-\varepsilon^+)$ where $x/t = c(\varepsilon^+)$. The determination of the state behind this front must be accomplished numerically, for example, by the method of characteristics discussed in Section 5.1.

7.2.3 Rate Dependence

Dynamic Stress-Strain Relations

It was already pointed out by Kármán and Duwez [1950] that an assumed stress-strain relation such as $\sigma = F(\varepsilon)$ used in solving dynamic problems is not necessarily the same as the stress-strain relation obtained from static or quasi-static tests. Most metals are rate-dependent, or viscoplastic, and rate-independent plasticity, as was pointed out in Chapter 3, is an approximation valid for sufficiently slow processes. The work of Kármán, Taylor, Rakhmatulin, and others who have followed the "rate-independent" approach is based on the assumption of a "dynamic" stress-strain relation valid in the range of strain rates that is encountered in dynamic tests, but not necessarily identical with the static one. Kolsky and Douch [1962], among other experimenters, performed tests in which they deduced dynamic stress-strain relations for several metals from wave-propagation experiments, and compared them with the static relations. The results showed that the relations were markedly different, with the dynamic curve well above the static one, for commercially pure copper and aluminum, but that they were identical for a precipitation-hardened aluminum alloy. Many other experimenters have obtained results, reviewed by Clifton [1983], in which the dynamic stresses are well above the static ones. Bell [1968] is relatively alone among major investigators in maintaining that the static stress-strain relation governs the dynamic problem.

The appearance of a plateau of permanent strain, predicted by the KTR theory, has been seen by many as justifying the "rate-independent" approach. Many others, however, have criticized this approach. In particular, it has been pointed out that in most tests the plateau occurs in a region that is too close to the impact end for the one-dimensional approximation, which neglects shear deformation and transverse inertia, to be valid.

More generally, the very assumption of a dynamic stress-strain relation that is more or less rate-independent in the dynamic range of rates has been questioned. For example, it was found by Sternglass and Stuart [1953], Bianchi [1964], and other investigators that in wires or bars that have been statically prestressed into the plastic range (and not unloaded), additional pulses travel at the elastic wave speed, in clear contradiction with the rate-independent model (see Clifton [1983] for a review of the relevant evidence).

Viscoplastic Theory

An alternative theory, treating the material as viscoplastic, was proposed by Sokolovskii [1948] and with somewhat greater generality by Malvern [1951a,b]; the viscoplasticity model of Perzyna [1963], discussed in Section 3.1, is essentially a three-dimensional generalization of the model underlying the Sokolovskii–Malvern theory. The constitutive equation used by Malvern is

$$\frac{\partial \varepsilon}{\partial t} = \frac{1}{E}\frac{\partial \sigma}{\partial t} + g(\sigma, \varepsilon), \tag{7.2.16}$$

where the function $g(\sigma, \varepsilon)$ vanishes for stresses whose magnitudes are below those given by the static stress-strain curve. For example, if the static relation is given by $\sigma = F(\varepsilon)$, then a possible form of g is

$$g(\sigma, \varepsilon) = G(|\sigma| - |F(\varepsilon)|)\operatorname{sgn}\sigma,$$

where G is a function that is positive for positive argument and zero otherwise. In the Sokolovskii model, $|F(\varepsilon)| = \sigma_E$ for $|\varepsilon| \geq \varepsilon_E$, that is, the static behavior of the material is taken as perfectly plastic. With G given by $G(x) = A < x >$, A being constant, the resulting equation for $g(\sigma, \varepsilon)$ is just the one-dimensional version of the Hohenemser–Prager model [Equation (3.1.3)].

The use of constitutive equations of the form (7.2.16) is not limited to materials that are viscoplastic in the classical sense, that is, those that are characterized by a static yield stress below which the behavior is elastic. It is used in creep models such as the Bailey–Norton–Nadai law [see Equation (2.1.6)], with $g(\sigma, \varepsilon) = B|\sigma|^m|\varepsilon - \sigma/E|^{-n}\operatorname{sgn}\sigma$.

A generalized constitutive equation that includes both the rate-independent and the Malvern models as special cases was proposed in the early 1960s by Simmons, Hauser, and Dorn [1962], Cristescu [1964], and Lubliner [1964]. This equation has the form

$$\frac{\partial \varepsilon}{\partial t} = f(\sigma, \varepsilon)\frac{\partial \sigma}{\partial t} + g(\sigma, \varepsilon) \tag{7.2.17}$$

for loading ($\sigma\partial\sigma/\partial t > 0$); for unloading, $f(\sigma, \varepsilon)$ is replaced by $1/E$, or if finite strains must be taken into account, $(1 - \varepsilon)^2/E$. The Malvern theory corresponds to the special case $f(\sigma, \varepsilon) \equiv 1/E$, and the KTR theory is recovered if $f(\sigma, \varepsilon) = 1/F'(\varepsilon)$ and $g(\sigma, \varepsilon) \equiv 0$. Explicit calculations by Lubliner [1965] and by Lubliner and Valathur [1969] show that solutions of impact problems according to the generalized theory tend to those of the KTR theory if the bar is short, the impact is strong, and if the viscoplastic relaxation time is long, while under the converse conditions the results of the Malvern theory are approached.

7.2.4 Application of the Method of Characteristics

Characteristic Relations

The three equations (7.2.1), (7.2.2), and (7.2.17) form a system of partial differential equations for the unknown variables σ, v, and ε. (Note that in using these equations we are returning to treating stress and strain as positive in tension.) As in Section 5.1, it is simpler to consider first a single first-order partial differential equation of the form

$$A\frac{\partial v}{\partial x} + B\frac{\partial v}{\partial t} = C,$$

where A, B, and C are functions of x, t, and v. Rewriting the equation as

$$B\left(\frac{\partial v}{\partial x}\frac{A}{B}\,dt + \frac{\partial v}{\partial t}\,dt\right) = C\,dt,$$

we find that the left-hand side is proportional to $dv = (\partial v/\partial x)\,dx + (\partial v/\partial t)\,dt$ if $dx = c\,dt$, where $c = A/B$; the characteristic direction is thus defined by $dx/dt = c$. Along a characteristic, then,

$$\frac{dx}{A} = \frac{dt}{B} = \frac{dv}{C}.$$

If a discontinuity in v, resulting from the intersection of two characteristics of the same family, should propagate, then it forms a shock front, as discussed in 7.2.1. Any discontinuity in $\partial v/\partial t$ or $\partial v/\partial x$ that is propagated through the xt-plane must (as discussed in 5.1.1) be across a characteristic, and $dx/dt = c$ is just the speed with which the front carrying the jump is propagated. If v is the velocity, then a front across which the acceleration $\partial v/\partial t$ is discontinuous is called an *acceleration wave*.

The system formed by Equations (7.2.1)–(7.2.2) and (7.2.17) is equivalent to

$$\lambda_1\left(\rho\frac{\partial v}{\partial t} - \frac{\partial \sigma}{\partial x}\right) + \lambda_2\left(\frac{\partial \varepsilon}{\partial t} - \frac{\partial v}{\partial x}\right) + \lambda_3\left(\frac{\partial \varepsilon}{\partial t} - f\frac{\partial \sigma}{\partial t} - g\right) = 0,$$

where λ_1, λ_2, λ_3 are arbitrary multipliers. Let us regroup the terms of this equation so that it takes the form

$$\left[\lambda_1\frac{\partial \sigma}{\partial x} + \lambda_3 f\frac{\partial \sigma}{\partial t}\right] + \left[\lambda_2\frac{\partial v}{\partial x} - \lambda_1\rho\frac{\partial v}{\partial t}\right] - \left[(\lambda_2 + \lambda_3)\frac{\partial \varepsilon}{\partial t}\right] + \lambda_3 g = 0.$$

The determination of the characteristics of the system is tantamount to finding the values of dx/dt for which the bracketed terms in the last equation are proportional, respectively, to $d\sigma$, dv, and $d\varepsilon$, where

$$d\sigma = \frac{\partial \sigma}{\partial x}\,dx + \frac{\partial \sigma}{\partial t}\,dt$$

and likewise for dv and $d\varepsilon$. The possible values of dx/dt must satisfy

$$\frac{dx}{dt} = \frac{\lambda_1}{\lambda_3 f} = -\frac{\lambda_2}{\lambda_1 \rho} = \frac{0}{\lambda_2 + \lambda_3}.$$

One way that all these equations can be obeyed is if

(a) $\lambda_1 = \lambda_2 = 0$, λ_3 arbitrary (let $\lambda_3 = 1$).

Two other possibilities are given by

(b) $\lambda_2 = -\lambda_3$

and

(c) $\dfrac{\lambda_1}{\lambda_2} = \pm\sqrt{\dfrac{f}{\rho}}.$

In case (a) we have

$$\frac{dx}{dt} = 0.$$

In cases (b) and (c),

$$\frac{dx}{dt} = -\frac{\lambda_1}{\lambda_2 f} = \pm c,$$

where $c = 1/\sqrt{\rho f}$.

If f is positive, then at each point there exist three distinct characteristic directions, which accordingly define three families of characteristic curves in the xt-plane. Since the number of characteristic directions equals the number of unknown variables, the system of equations is hyperbolic, as in 5.1.1.

Along an infinitesimal characteristic segment of family (a) we have $dx = 0$, and therefore changes in the unknown variables are given by Equation (7.2.17), that is,

$$f\,d\sigma - d\varepsilon + g\,dt = 0 \quad \text{along } dx = 0. \tag{7.2.18a}$$

Along characteristic segments of families (b) and (c), respectively, we obtain the relations

$$d\sigma = \rho c\,dv - \frac{g}{f}\,dt \quad \text{along } dx = c\,dt, \tag{7.2.18b}$$

$$d\sigma = -\rho c\,dv - \frac{g}{f}\,dt \quad \text{along } dx = -c\,dt. \tag{7.2.18c}$$

Viscoplastic Bars

For a bar made of an elastic-viscoplastic material governed by (7.2.16) (with infinitesimal strains), the characteristic relations (7.2.18) become

$$\frac{1}{E}d\sigma - d\varepsilon + g\,dt = 0 \quad \text{along } dx = 0,$$

$$d\sigma = \rho c_e\,dv - Eg\,dt \quad \text{along } dx = c_e\,dt,$$

$$d\sigma = -\rho c_e\,dv - Eg\,dt \quad \text{along } dx = -c_e\,dt.$$

Since the characteristics are straight, the numerical integration of these equations is quite straightforward. For certain simple forms of the function g, the equations may be solved analytically. Examples include solutions by Sokolovskii [1948] and by Kaliski and Wlodarczyk [1967] for $g(\sigma, \varepsilon) =< |\sigma| - \sigma_Y > (\operatorname{sgn}\sigma)/\eta$ (the Hohenemser–Prager model), and by Cristescu and Predeleanu [1965] for g given by

$$g(\sigma, \varepsilon) = \frac{1}{\eta}\left[\left(1 + \frac{E'}{E}\right)\sigma - E'\varepsilon - \sigma_E \operatorname{sgn}\sigma\right], \quad |\sigma| > \sigma_E,$$

with $g = 0$ for $|\sigma| < \sigma_E$.

Rate-Independent Bars; Unloading Wave

Since the KTR theory corresponds to $g == 0$, it is clear that Equations (7.2.11) representing the "similarity" solution are just the characteristic relations along characteristics of family (c), these characteristics being curved. The lines $x = ct$ along which the state variables are constant are the characteristics of family (b). The characteristic network for the problem of constant-velocity impact on a semi-infinite bar made of a rate-independent material is shown in Figure 7.2.8(a).

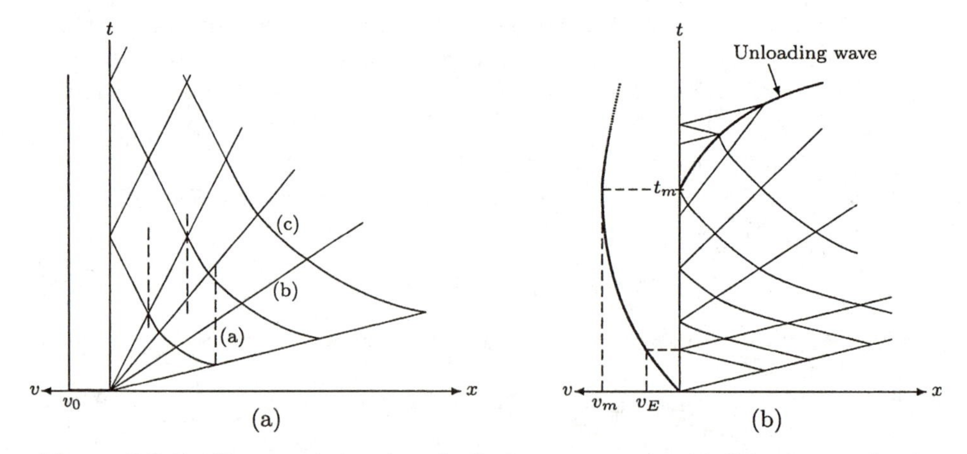

Figure 7.2.8. Characteristic networks for impact on a semi-infinite bar made of a rate-independent material: (a) constant-velocity impact; (b) impact velocity rising to a peak and declining.

Suppose, now, that the impact is not sudden, but that the boundary condition at $x = 0$ has the form $v(0, t) = v_0(t)$, where $v_0(0) = 0$ and, for $t > 0$, v_0 is a continuous function that increases with t until it reaches a maximum value v_m at $t = t_m$, after which it decreases.

As before, we may suppose the characteristics of family (a) to be straight lines along which σ, v, and ε are constant. Rather than all of them emanating from (0, 0), however, each one originates from a point on the t-axis such that $v = v_0(t)$. We may integrate the characteristic relations along families (b) and (c) to obtain

$$v = \int_0^\varepsilon c(\varepsilon)d\varepsilon, \quad \sigma = F(\varepsilon).$$

The second of these equations, however, is valid only in a loading process, that is, only as long as $\partial\sigma/\partial t > 0$. Since σ and v are monotonically related, at $x = 0$ this condition prevails only for $t < t_m$.

What happens then? As first demonstrated by Rakhmatulin [1945], a front behind which $\partial\sigma/\partial t \le 0$, called an *unloading wave* [see Figure 7.2.8(b)], begins to propagate into the plastically deforming region of the bar. Since in this region $\partial\sigma/\partial t > 0$, across this front the stress rate $\partial\sigma/\partial t$, and therefore also the strain rate $\partial\varepsilon/\partial t$ and the acceleration $\partial v/\partial t$, are discontinuous. An unloading wave is therefore an acceleration wave.

The calculation of the unloading wave in the xt-plane must be performed step by step. Its speed, say c_u, is initially $c(\varepsilon_m)$, where ε_m is the strain at $(0, t_m)$. An example of a graphic construction, due to Rakhmatulin and Shapiro [1948], is shown in Figure 7.2.9 (taken from Cristescu [1967]) for a bar in which the stress at the end of the bar first increases monotonically, then remains constant for a while, and finally decreases monotonically to zero. As can be seen, the speed of the unloading wave increases as the wave

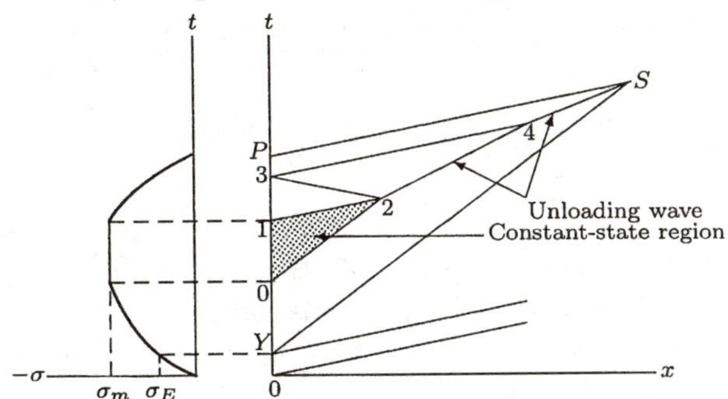

Figure 7.2.9. Approximate construction of unloading wave (Cristescu [1967]).

propagates. It was shown by Lee [1953] that the unloading-wave speed is bounded by

$$c(\varepsilon) \le c_u \le c_e.$$

A proof of this result is given in Section 7.3, in which the general theory of elastic-plastic acceleration waves is studied. In his first paper, Rakhmatulin [1945] erroneously showed the unloading wave as attaining speeds faster than that of the elastic wave (see the discussion by Bodner and Clifton [1966]). More information on wave propagation in elastic-plastic and elastic-viscoplastic bars can be found in the books by Cristescu [1967], Chapters II and III, and Nowacki [1978], Chapter III.

Exercises: Section 7.2

1. Discuss how a shock front might form in an elastic-plastic bar whose stress-strain curve beyond the initial yield stress σ_E is given by $\sigma = \sigma_E\{1 + [(\varepsilon - \sigma_E/E)/\alpha]^2\}$. Consider the limit of a rigid–plastic bar ($E = \infty$).

2. For a given impact velocity v_0, find the distribution of permanent strain following impact in a rigid–plastic bar of length L when the stress-strain relation beyond the initial yield stress is given by $F(\varepsilon) = \sigma_E[1+(\varepsilon/\alpha)^{1/m}]$. Letting $\alpha = 1$ and $m = 5$, plot the strain distributions for various values of $v_0/\sqrt{\sigma_E/\rho}$.

3. Repeat Exercise 2 for the stress-strain relation of Exercise 1 with $\alpha = 1$.

4. An elastic-plastic bar with linear work-hardening, of initial length L, strikes a rigid obstacle at $t = 0$. Find the rebound time in terms of L/c_e and $\alpha = c_p/c_e$ when v_0/v_E is 3, 4, and 5.

5. A semi-infinite elastic-plastic bar with linear work-hardening (with tangent modulus E' in the plastic range) is subjected at time $t = 0$ to a stress $\sigma_0 > \sigma_E$ which remains constant until $t = t_0$, whereupon it is removed. Discuss the motion of the bar for various ranges of σ_0.

6. Repeat Exercise 5 for bars with the stress-strain relations of Exercises 1 and 2.

7. A semi-infinite viscoplastic bar with the constitutive equation (7.2.16) is subjected to a suddenly applied stress or velocity. Find the equation governing the decay of the stress discontinuity along the initial shock front. Show that the decay is exponential if $g(\sigma, \varepsilon)$ is a linear function of its arguments for $\sigma > \sigma_E$.

Section 7.3 Three-Dimensional Waves

Three-dimensional waves in rate-independent elastic-plastic solids appear to have been studied first by Craggs [1957], who found relations for the speeds of plastic waves on the basis of the characteristics of the governing differential equations. As in the one-dimensional case, the characteristic speeds coincide with those of acceleration waves, and in this section only the acceleration-wave approach is followed. Furthermore, only rate-independent elastic-plastic solids are discussed. In elastic-viscoplastic solids the wave speeds are those of elastic waves, while the study of specific dynamic boundary-value problems must, as a rule, be pursued numerically, except when the constitutive model is so simple that analytical methods, analogous to those used for linearly viscoelastic solids, may be used; examples of both approaches are given in the book by Nowacki [1978]. Acceleration waves in elastic-plastic solids were first investigated by Thomas [1958, 1961] for perfect plasticity, and with greater generality by Hill [1962] and Mandel [1962].

7.3.1 Theory of Acceleration Waves

General Theory

The general theory of acceleration waves is due to Hadamard [1903]; a modern exposition may be found in the monograph by Truesdell and Toupin [1960]. It is presented here in a simplified form suitable for infinitesimal deformations. We consider an acceleration wave moving in $x_1x_2x_3$-space at the speed c. The unit normal vector to the wave surface, pointing in the direction of wave motion, will be denoted $\mathbf{n}$. The jump operator $[\![\cdot]\!]$ will be used as previously defined, and the superposed dot will denote $\partial/\partial t$. A time-dependent field ϕ is assumed to be continuous across the wave but to have possibly discontinuous derivatives. At time t, ϕ has the value $\phi(x_1, x_2, x_3, t)$ at a point of the wave front. At an infinitesimally later time $t+dt$, the given point will have moved to $x_i + n_i c\, dt$, $i = 1, 2, 3$, and ϕ will accordingly have the value $\phi(x_1, x_2, x_3, t) + d\phi$, where

$$d\phi = \phi_{,i}\, n_i c\, dt + \dot{\phi} dt.$$

The continuity of ϕ requires $[\![d\phi]\!] = 0$, or

$$cn_i[\![\phi_{,i}]\!] = -[\![\dot{\phi}]\!].$$

Since, however, it is only the component of *del*ϕ along $\mathbf{n}$ that undergoes a jump, $[\![\phi_{,i}]\!]$ must be proportional to n_i, and therefore, since $n_j n_j = 1$,

$$c[\![\phi_{,j}]\!] = -[\![\dot{\phi}]\!]\, n_j.$$

The last result is known as the **Hadamard compatibility condition.**

When applied to the velocity field $\mathbf{v}$, the condition yields

$$c[\![v_{i,j}]\!] = -[\![\dot{v}_i]\!]\, n_j.$$

When applied to the stress field $\boldsymbol{\sigma}$ and combined with the momentum-balance equation (1.3.1), it gives

$$[\![\dot{\sigma}_{ij}]\!]\, n_j = -\rho c[\![\dot{v}_i]\!].$$

If, finally, the material is characterized by an effective tangent modulus tensor $\bar{\mathsf{C}}$ such that

$$[\![\dot{\sigma}_{ij}]\!] = \bar{C}_{ijkl}[\![\dot{\varepsilon}_{ij}]\!] ,$$

then combining all the results of this paragraph yields the eigenvalue problem

$$(Q_{ik} - \rho c^2 \delta_{ik})[\![\dot{v}_k]\!] = 0,$$

where

$$Q_{ik} \stackrel{\text{def}}{=} \bar{C}_{ijkl} n_j n_l \tag{7.3.1}$$

defines the *acoustic tensor* $\mathbf{Q}$. The eigenvalues of $\mathbf{Q}$, if positive, equal ρc^2 with c the wave speed, while the eigenvectors are the directions of $[\![\dot{\mathbf{v}}]\!]$ or *polarization directions*.

Elastic Waves

If the material is linearly elastic, then the tangent modulus tensor $\bar{\mathsf{C}}$ is just the elastic modulus tensor C, and the corresponding acoustic tensor will be denoted $\mathbf{Q}^e$:

$$Q^e_{ik} = C_{ijkl} n_j n_l. \tag{7.3.2}$$

If the material is also isotropic, then C is given by Equation (1.4.10), and $\mathbf{Q}^e$ is given by

$$Q^e_{ik} = (\lambda + \mu) n_i n_k + \mu \delta_{ik}.$$

One eigenvector is obviously $\mathbf{n}$, with the corresponding eigenvalue being $\lambda + 2\mu$, so that the wave speed is

$$c^e_1 = \sqrt{\frac{\lambda + 2\mu}{\rho}} = \sqrt{\frac{1 - \nu}{(1 + \nu)(1 - 2\nu)}} c_e,$$

where $c_e = \sqrt{E/\rho}$ is the elastic bar-wave speed defined in the preceding section. Since the polarization direction coincides with that of propagation, a wave of this nature is called *longitudinal*; it is also known a *dilatational wave* or a *P-wave*.

Any vector perpendicular to $\mathbf{n}$ is also an eigenvector of $\mathbf{Q}$ and corresponds to the degenerate eigenvalue μ. The polarization being perpendicular to the wave normal, the wave is called *transverse* (also a *shear wave* or an *S-wave*), and its speed is

$$c^e_2 = \sqrt{\frac{\mu}{\rho}}.$$

As in the bar studied previously, the elastic acceleration-wave speeds in three-dimensional solids are independent of the local field variables, so that all disturbances, including sufficiently weak shocks, travel at these speeds. A shock is "sufficiently weak" if it does not entail temperature and/or density changes large enough to affect the elastic moduli.

Elastic-Plastic Waves

We now consider an acceleration wave propagating in a standard elastic-plastic solid as described in Section 3.2, with a yield function f. If $f < 0$ at the wave front, then the solid is behaving elastically, and the wave is an elastic one. It will therefore be assumed that $f = 0$ at the wave front. Then the relation (3.2.10) between stress rate and strain rate, with the normality rule (3.2.7) incorporated, may be written, following Hill [1959], in the form

$$\dot{\sigma}_{ij} = C_{ijkl}\dot{\varepsilon}_{kl} - \frac{1}{l}C_{ijmn}m_{mn}\langle g\rangle, \tag{7.3.3}$$

where

$$m_{ij} = \frac{\dfrac{\partial f}{\partial \sigma_{ij}}}{\sqrt{\dfrac{\partial f}{\partial \sigma_{kl}}\dfrac{\partial f}{\partial \sigma_{kl}}}}, \qquad g = C_{pqkl}m_{pq}\dot{\varepsilon}_{kl}, \qquad l = h + C_{klmn}m_{kl}m_{mn},$$

h being a plastic modulus, not exactly the same as in Section 3.2, but to be specified more precisely later. Note that $\mathbf{m}$ is a unit tensor, that is, $m_{ij}m_{ij} = 1$.

Applying the jump operator to Equation (7.3.3) yields

$$\llbracket\dot{\sigma}_{ij}\rrbracket = C_{ijkl}\llbracket\dot{\varepsilon}_{kl}\rrbracket - \frac{1}{l}C_{ijmn}m_{mn}\llbracket\langle g\rangle\rrbracket .$$

We may now write

$$\llbracket\langle g\rangle\rrbracket = \eta\llbracket g\rrbracket ,$$

where η is a real number that depends on the ranges of g^+ and g^-.

Several cases may be distinguished:

$$\begin{array}{ll} \text{(a)} & g^+ > 0,\ g^- > 0:\ \ \eta = 1, \\ \text{(b)} & g^+ > 0,\ g^- \le 0:\ \ 0 < \eta < 1, \\ \text{(c)} & g^+ \le 0,\ g^- > 0:\ \ 0 < \eta < 1, \\ \text{(d)} & g^+ \le 0,\ g^- \le 0:\ \ \eta = 0. \end{array}$$

In addition, we may consider the situation where the material ahead of the wave is about to yield (i.e., $g^+ > 0$), but is still elastic (i.e., $\dot{\boldsymbol{\sigma}}^+ = \mathsf{C}\cdot\dot{\boldsymbol{\varepsilon}}^+$). Since $g^- > 0$ also, we have

$$\eta = \frac{g^-}{g^- - g^+},$$

and thus obtain two additional cases:

$$\begin{array}{ll} \text{(e)} & g^- > g^+:\ \ \eta > 1, \\ \text{(f)} & g^- < g^+:\ \ \eta < 0. \end{array}$$

The waves represented by these cases can be categorized as follows, according to Mandel [1962]: case (a) represents a *plastic* wave, with the material deforming plastically on both sides; case (b) an *unloading* wave; and case (d) an *elastic* wave. Case (c), which was not considered by Mandel, can be called a *reloading* wave. Cases (e) and (f) represent *loading* waves; the former will here be called *strong* and the latter *weak*.

The effective tangent modulus $\bar{\mathsf{C}}$ may now be expressed as

$$\bar{C}_{ijkl} = C_{ijkl} - \eta N_{ij} N_{kl},$$

where $N_{ij} = C_{ijkl} m_{kl}/\sqrt{l}$. The acoustic tensor $\mathbf{Q}$ accordingly may be written in the form

$$Q_{ij} = Q^e_{ij} - \eta d_i d_j,$$

where $\mathbf{Q}^e$ is given by (7.3.2), and $d_i \stackrel{\text{def}}{=} N_{ik} n_k$ defines a vector $\mathbf{d}$ that Hill [1962] calls the "traction" exerted by $\mathbf{N}$ on the wave front.

Analysis of Wave Speeds

The analysis to be presented is based on that of Mandel [1962]. Since $\mathbf{Q}$ is symmetric, it has real eigenvalues and mutually perpendicular eigenvectors. Whenever such an eigenvalue — say A — is positive, it represents a a wave propagated at the speed $c = \sqrt{A/\rho}$. The symmetry of $\mathbf{Q}$ is due to the assumption of an associated flow rule; the propagation of acceleration waves in nonstandard elastic-plastic materials was also studied by Mandel [1964], whose conclusions are summarized later.

In what follows, the eigenvalues of $\mathbf{Q}$ and $\mathbf{Q}^e$ will be denoted A_α and A^e_α ($\alpha = 1, 2, 3$), respectively, with the ordering $A_1 \geq A_2 \geq A_3$ and $A^e_1 \geq A^e_2 \geq A^e_3$. If a Cartesian basis is chosen along the eigenvectors of $\mathbf{Q}^e$, then the characteristic equation of $\mathbf{Q}$ is

$$F(A) = 0 \tag{7.3.4}$$

where

$$\begin{aligned} F(A) &= \det(\mathbf{Q} - \mathbf{I}A) \\ &= \begin{vmatrix} A^e_1 - \eta d_1^2 - A & -\eta d_1 d_2 & -\eta d_1 d_3 \\ -\eta d_1 d_2 & A^e_2 - \eta d_2^2 - A & -\eta d_2 d_3 \\ -\eta d_1 d_3 & -\eta d_2 d_3 & A^e_1 - \eta d_3^2 - A \end{vmatrix} \\ &= (A^e_1 - A)(A^e_2 - A)(A^e_3 - A) - \eta[d_1^2(A^e_2 - A)(A^e_3 - A) \\ &\quad + d_2^2(A^e_1 - A)(A^e_3 - A) + d_3^2(A^e_1 - A)(A^e_2 - A)]. \end{aligned}$$

The character of the roots of Equation (7.3.4) depends on η. The simplest case is $\eta = 0$, corresponding to an elastic wave. In this case, of course, $A_\alpha = A^e_\alpha$ ($\alpha = 1, 2, 3$).

If $\eta > 0$, as in cases (a)–(c) and in case (e), then $F(A_1^e) = -\eta d_1^2(A_1^e - A_2^e)(A_1^e - A_3^e) \leq 0$. Similarly, $F(A_2^e) \geq 0$, $F(A_3^e) \leq 0$, and $F(-\infty) = +\infty$. Consequently,[1]

$$A_3 \leq A_3^e \leq A_2 \leq A_2^e \leq A_1 \leq A_1^e.$$

Furthermore, the A_α decrease monotonically with η. If A_α^p denotes the value of A_α for $\eta = 1$ (plastic wave), and A_α^u a value of A_α for $0 < \eta < 1$ (unloading and reloading), then we have the inclusion

$$A_\alpha^p \leq A_\alpha^u \leq A_\alpha^e, \quad \alpha = 1, 2, 3.$$

Here we see the bounding of the wave speed of an unloading wave, previously obtained by Lee [1953] for waves in a thin bar.[2]

The speeds of strong loading waves, case (e), are even slower than those of plastic waves. The inclusions are

$$A_3^{sl} \leq A_3^p, \quad A_3^e \leq A_2^{sl} \leq A_2^p, \quad A_2^e \leq A_1^{sl} \leq A_1^p.$$

In fact, A_3^{sl} may be zero, in which case the surface does not propagate.

In a weak loading wave, case (f), η is negative and therefore the nature of the roots changes drastically. The ordering of the wave speeds is

$$A_3^e \leq A_3^{wl} \leq A_2^p, \quad A_2^e \leq A_2^{wl} \leq A_1^p, \quad A_1^e \leq A_1^{wl}.$$

The existence of waves propagating faster than elastic waves (in a bar) was apparently first noted by Lee [1953]. Mandel [1962] gave, as an example of such a wave moving at an *infinite* speed, the loading wave in a finite bar following reflection from a fixed end.

An analysis of plastic waves in a nonstandard material, also due to Mandel [1964], shows that (a) the eigenvectors of $\mathbf{Q}$ are not mutually perpendicular, (b) A_3^p may be negative (representing a slip or rupture surface rather than a wave), and (c) A_1^p may be greater than A_1^e. While soils are usually modeled as nonstandard in static problems, theories of waves in soils have typically been based on standard elastic-plastic models, for example by Grigorian [1960] and by Chadwick, Cox, and Hopkins [1964].

Isotropically Elastic Materials

If the material is elastically isotropic, with the modulus tensor given by (1.4.10), and if in addition it is plastically incompressible (tr $\dot{\varepsilon}^p = 0$), so that $m_{kk} = 0$, then

$$C_{ijkl}m_{ij}m_{kl} = 2\mu m_{ij}m_{ij} = 2\mu,$$

---

[1] It was also shown by Mandel [1962] that $A_3 > 0$ for $\eta \leq 1$, except that A_3^p may be zero if the material is perfectly plastic.

[2] As pointed out in the preceding section, Rakhmatulin's [1945] finding of an unloading wave propagating faster than the elastic wave is erroneous.

$$l = h + 2\mu, \qquad g = 2\mu m_{kl}\dot{\varepsilon}_{kl},$$

and

$$N_{ij} = \frac{2\mu}{\sqrt{h+2\mu}} m_{ij}.$$

If plastic isotropy is also assumed, then the hardening modulus h can be related very simply to the uniaxial tangent modulus as well as to the tangent modulus in shear. Recalling the definition of g and multiplying every term of Equation (7.3.3) by m_{ij}, with contraction implied, we obtain

$$m_{ij}\dot{\sigma}_{ij} = \left(1 - \frac{1}{h+2\mu} C_{ijkl} m_{ij} m_{kl}\right) g = \frac{2\mu h}{h+2\mu} m_{kl}\dot{\varepsilon}_{kl}.$$

In a quasi-static uniaxial test, the left-hand side is just $m_{11}\dot{\sigma}_{11}$. Moreover, $m_{ij} = 0$ for $i! = j$ and $m_{22} = m_{33} = -\frac{1}{2} m_{11}$, so that $m_{11} = \sqrt{2/3}$, and

$$m_{kl}\dot{\varepsilon}_{kl} = \sqrt{\frac{2}{3}} (\dot{\varepsilon}_{11} - \dot{\varepsilon}_{22}).$$

Consequently,

$$\dot{\sigma}_{11} = \frac{2\mu h}{2\mu + h} (\dot{\varepsilon}_{11} - \dot{\varepsilon}_{22}).$$

Plastic incompressibility further implies that

$$\dot{\sigma}_{11} = (3\lambda + 2\mu)(\dot{\varepsilon}_{11} + 2\dot{\varepsilon}_{22}).$$

Eliminating $\dot{\varepsilon}_{22}$, we obtain the tangent compliance

$$\frac{\dot{\varepsilon}_{11}}{\dot{\sigma}_{11}} = \frac{\lambda + \mu + 2\mu(3\lambda + 2\mu)/3h}{(3\lambda + 2\mu)\mu} = \frac{1}{E} + \frac{2}{3h},$$

so that $h = \frac{2}{3} H$, where $H = d\sigma_Y / d\varepsilon^p$ is the uniaxial tangent plastic modulus. It can also be easily shown that h is twice the shearing tangent plastic modulus $d\tau_Y / d\gamma^p$.

The eigenvector of $\mathbf{Q}^e$ corresponding to the fastest elastic wave (the longitudinal wave) is of course $\mathbf{n}$. If this is identified with $\mathbf{e}_1$, then $\mathbf{d}$ is given by

$$d_i = \delta_{ik} N_{k1} = \frac{2\mu}{\sqrt{h+2\mu}} m_{i1}.$$

Since any vector perpendicular to $\mathbf{n}$ is also an eigenvector of $\mathbf{Q}^e$, we are free to choose $\mathbf{e}_2$ and $\mathbf{e}_3$ in the tangent plane of the wave; let us choose them so that $d_3 = 0$. The eigenvalue problem now reads

$$0 = (\mathbf{Q} - A\mathbf{I})[\dot{\mathbf{v}}] = \begin{bmatrix} \lambda + 2\mu - \eta d_1^2 - A & -\eta d_1 d_2 & 0 \\ -\eta d_1 d_2 & \mu - \eta d_2^2 - A & 0 \\ 0 & 0 & \mu - A \end{bmatrix} [\dot{\mathbf{v}}]. \tag{7.3.5}$$

Obviously, one eigenvector is $\mathbf{e}_3$, with the corresponding eigenvalue $A_2 = \mu = A_2^e = A_3^e$ (necessary because $A_3^e \le A_2 \le A_2^e$). Hence, at any η, an elastic transverse wave can propagate. If $\eta > 0$, then we also have $A_3 \le A_2^e \le A_1 \le A_1^e$, a result first derived by Craggs [1957]. The waves corresponding to A_1 and A_3 are in general neither longitudinal nor transverse, unless $d_1 d_2 = 0$. When neither d_1 nor d_2 is zero, the waves are called *longitudinal–transverse.*

7.3.2 Special Cases

Purely Longitudinal and Transverse Waves

(a) If $d_2 = 0$, then $A_3 = \mu$ and the polarization is along $\mathbf{e}_3$. In this case both transverse waves are elastic, while for the longitudinal wave we obtain

$$A_1 = \lambda + 2\mu - \eta d_1^2 = \lambda + 2\mu - \eta \frac{4\mu^2}{h + 2\mu} m_{11}^2.$$

The tensor $\mathbf{m}$ can be shown to have the same form as in the quasi-static uniaxial test: $m_{13} = 0$ because $d_3 = 0$ by choice of basis, $m_{12} = 0$ because $d_2 = 0$ by hypothesis, $m_{23} = 0$ because $[\dot{\varepsilon}_{23}] = 0$, and $m_{22} = m_{33} = -\frac{1}{2} m_{11}$ from the resulting symmetry; hence $m_{11} = \sqrt{2/3}$, and

$$A_1 = \lambda + 2\mu - \frac{4\eta\mu}{3(1 + h/2\mu)}. \tag{7.3.6}$$

Note that for $h = 0$ (perfect plasticity) and $\eta = 1$ (plastic wave) we have $A_1 = \lambda + \frac{2}{3}\mu = K$, the elastic bulk modulus. In this case the longitudinal wave may be said to propagate as though the material were a fluid.

(b) If $d_1 = 0$, then, by contrast, it is the longitudinal wave that is elastic ($A_1 = \lambda + 2\mu$), and in addition to the elastic transverse wave with $A_2 = \mu$, there is a transverse wave with

$$A_3 = \mu - \eta d_2^2 = \mu - \eta \frac{4\mu^2}{h + 2\mu} m_{12}^2.$$

It can easily be shown that this case is equivalent to quasi-static simple shear, so that $m_{12} = 1/\sqrt{2}$, and

$$A_3 = \mu - \frac{\eta\mu}{1 + h/2\mu}. \tag{7.3.7}$$

For $\eta = 1$ (plastic wave), A_3 reduces to the tangent modulus in shear, $\mu h/(2\mu + h)$. If $h = 0$ (perfect plasticity), then $A_3 = (1 - \eta)\mu$, so that a transverse plastic wave ($\eta = 1$) or a transverse strong loading wave ($\eta > 1$) cannot be propagated.

Longitudinal–Transverse Waves

In an isotropic and homogeneous half-space, say $x_1 > 0$, whose boundary $x_1 = 0$ is subject to time-varying tractions (or displacements) that are uniform over the plane (i.e., independent of x_2 and x_3), all the field variables may be assumed to be functions of x_1 and t only, and the wave is said to be *plane*. A plane wave is longitudinal if the traction or displacement is purely normal, and transverse if the traction or displacement is purely tangential. If the boundary tractions or displacements include both normal and tangential components, then the wave is longitudinal–transverse.

The propagation of longitudinal–transverse plane waves in an elastic-perfectly plastic half-space was studied by Bleich and Nelson [1966], and in an elastic-plastic half-space with work-hardening by Ting and Nan [1969]; for further references, see Nowacki [1978], Section 22. Here, we limit ourselves to finding expressions for the wave speeds through a direct application of Equation (7.3.5).

If the tangential traction is in the x_2-direction, then the half-space can be assumed to be in a state of plane strain, so that, if the material is isotropic, $\sigma_{13} = \sigma_{23} = 0$. Furthermore, all field quantities, including the velocity, can be assumed to be independent of x_2, so that $\dot{\varepsilon}_{22} = 0$. The Lévy–Mises flow rule requires that $\sigma_{22} = \sigma_{33}$, and the stress-deviator tensor is given by

$$\mathbf{s} = \begin{bmatrix} s & \tau & 0 \\ \tau & -\frac{1}{2}s & 0 \\ 0 & 0 & -\frac{1}{2}s \end{bmatrix},$$

where $s = \frac{2}{3}(\sigma_{11} - \sigma_{22})$ and $\tau = \sigma_{12}$. The Mises yield criterion may be written as

$$\frac{3}{4}s^2 + \tau^2 = k^2. \tag{7.3.8}$$

The "traction" vector $\mathbf{d}$ has the components

$$d_1 = \sqrt{\frac{2}{h+2\mu}}\frac{\mu}{k}s, \quad d_2 = \sqrt{\frac{2}{h+2\mu}}\frac{\mu}{k}\tau, \quad d_3 = 0.$$

The characteristic equation is, accordingly,

$$\left(\lambda + 2\mu - A - \eta\frac{2\mu^2}{h+2\mu}\frac{s^2}{k^2}\right)\left(\mu - A - \eta\frac{2\mu^2}{h+2\mu}\frac{\tau^2}{k^2}\right) - \eta^2\frac{4\mu^4}{(h+2\mu)^2}\frac{s^2\tau^2}{k^4} = 0.$$

Upon multiplying the equation by $2k(h+2\mu)/\mu$, we may rewrite it as

$$3s^2(A-\mu)\{A - [K + \tfrac{4}{3}(1-\eta)\mu]\}$$
$$+4\tau^2(A - \lambda - 2\mu)[A - (1-\eta)\mu] + 2k^2\frac{h}{\mu}(A - \lambda - 2\mu)(A - \mu) = 0.$$

It can easily be ascertained that the special cases $\tau = 0$ and $s = 0$ lead to Equations (7.3.6) and (7.3.7), respectively. For a plastic wave ($\eta = 1$) in an elastic-perfectly plastic medium ($h = 0$) the equation reduces to the result of Bleich and Nelson [1966]:

$$3s^2(A-\mu)(A-K) + 4\tau^2 A(A-\lambda-2\mu) = 0.$$

A problem closely related to that of longitudinal–transverse waves plane waves in a half-space is that of longitudinal–torsional waves in a long thin-walled cylindrical tube. This problem was studied by Clifton [1966, 1968] and T. C. T. Ting [1969].

Spherical Radial Wave

Purely longitudinal waves arise when the loading is spherically or cylindrically radial, such as that produced by the explosion of a point charge or a line charge in an infinite medium. The cylindrical problem is relatively complicated, since the condition of plane strain must be enforced, and the problem depends on the yield criterion chosen. The spherical problem is fairly simple; it was first treated by Lunts [1949].

In an isotropic and plastically incompressible elastic-plastic solid, a spherically symmetric state of deformation requires $\varepsilon_\theta^p = -\frac{1}{2}\varepsilon_r^p$, and is equivalent to a state of uniaxial stress equal to $\sigma_r - \sigma_\theta$. The yield criterion may be written as

$$|\sigma_r - \sigma_\theta| = \sigma_Y(\varepsilon_r^p),$$

where $\sigma = \sigma_Y(\varepsilon^p)$ is the uniaxial relation between yield stress and plastic strain. If, now, H is again the uniaxial plastic modulus, then

$$\dot{\varepsilon}_r^p = \frac{1}{H} < \operatorname{sgn}(\sigma_r - \sigma_\theta)(\dot{\sigma}_r - \dot{\sigma}_\theta) >$$

when the yield criterion is obeyed. It will be assumed that the loading is compressive, that is, that $\sigma_r - \sigma_\theta = -\sigma_Y$. If v denotes the radial velocity, and if $(\cdot)' = \partial(\cdot)/\partial r$, then the radial and tangential strain rates are given by

$$\begin{aligned} v' &= \frac{\dot{\sigma}_r - 2\nu\dot{\sigma}_\theta}{E} - \frac{<\dot{\sigma}_\theta - \dot{\sigma}_r>}{H}, \\ \frac{v}{r} &= \frac{(1-\nu)\dot{\sigma}_\theta - \nu\dot{\sigma}_r}{E} + \frac{<\dot{\sigma}_\theta - \dot{\sigma}_r>}{2H}. \end{aligned} \tag{7.3.9}$$

The equation of motion is

$$\sigma_r' + 2\frac{\sigma_r - \sigma_\theta}{r} = \rho\dot{v}.$$

Applying the jump operator to this equation leads to $[\![\sigma_r']\!] = \rho[\![\dot{v}]\!]$. If, as before, $A = \rho c^2$, where c is the wave speed, then the Hadamard compatibility conditions give $[\![v']\!] = \rho[\![\dot{\sigma}_r]\!]/A$. Since the velocity itself is continuous at an acceleration wave front, application of the jump operator to Equations (7.3.9) produces the two homogeneous linear equations for $\lfloor\dot{\sigma}_r\rfloor$, $\lfloor\dot{\sigma}_\theta\rfloor$:

$$\left(\frac{1}{H}+\frac{1}{E}-\frac{1}{A}\right)\lfloor\dot{\sigma}_r\rfloor - \left(\frac{1}{H}+\frac{2\nu}{E}\right)\lfloor\dot{\sigma}_\theta\rfloor = 0,$$

$$-\left(\frac{1}{2H}+\frac{\nu}{E}\right)\lfloor\dot{\sigma}_r\rfloor + \left(\frac{1}{2H}+\frac{1-\nu}{E}\right)\lfloor\dot{\sigma}_\theta\rfloor = 0,$$

if it is assumed that the wave is a compressive plastic wave, that is, $\sigma_\theta - \sigma_r = \sigma_Y$ and $\dot{\sigma}_\theta - \dot{\sigma}_r > 0$. Setting the determinant of the two equations equal to zero gives

$$A = \frac{E + 2(1-\nu)H}{3E + 2(1+\nu)H}\frac{E}{1-2\nu}.$$

Since $h = \frac{2}{3}H$, it can easily be seen that this equation is equivalent to (7.3.6) with $\eta = 1$.

Cylindrical Shear Wave

A purely transverse wave will be propagated cylindrically if, for example, a rigid axle, embedded in an infinite medium, suddenly has a torque applied to it. If the medium is isotropic, then it may be assumed that the only velocity component is the circumferential velocity $v_\theta = v(r, t)$ and the only stress component is $\sigma_{r\theta} = \sigma_{\theta r} = \tau(r, t)$. The equation of motion is

$$\tau' + 2\frac{\tau}{r} = \rho\dot{v},$$

from which $\lfloor v'\rfloor = \lfloor\dot{\tau}\rfloor / A$. If the shear strain is γ, then

$$\dot{\gamma} = v' - \frac{v}{r} = \frac{\dot{\tau}}{\mu} + \frac{3\langle\dot{\tau}\,\mathrm{sgn}\,\tau\rangle}{H},$$

so that

$$\lfloor v'\rfloor = \left(\frac{1}{\mu}+\frac{3}{H}\right)\lfloor\dot{\tau}\rfloor ,$$

and thus

$$A = \frac{\mu H}{H + 3\mu},$$

a result equivalent to (7.3.7) with $\eta = 1$. The propagation of the wave was studied by Rakhmatulin [1948].

Exercises: Section 7.3

1. Show that in an elastically isotropic material, the plastic modulus h is twice the tangent plastic modulus in shear.

2. Find the eigenvalues A_1 and A_3 of Equation (7.3.5) in the general case.

3. Show that the state of stress in a purely transverse wave is that of simple shear.

4. Using suitable assumptions, establish the governing equations of combined longitudinal–torsional wave propagation in a long thin-walled elastic-plastic cylindrical tube.

5. For a Mises material with linear work-hardening, find explicit expressions for the wave speeds of longitudinal–transverse plastic waves in terms of the ratio of normal stress to shear stress at the wave.

6. Find the speed of a spherically radial elastic-plastic wave as a function of η.

Chapter 8

Large-Deformation Plasticity

Section 8.1 Large-Deformation Continuum Mechanics

8.1.1 Continuum Deformation

As a model of mechanical behavior, plasticity theory is applicable primarily to those solids that can experience inelastic deformations considerably greater than elastic ones. But the resulting total deformations, and the rotations accompanying them, may still be small enough so that many problems can be solved with small-displacement kinematics, and this is the situation that has, with a few exceptions, thus far been addressed in this book. However, when strains or rotations become so large that they cannot be neglected next to unity, the mechanician must resort to the theory of large or finite deformations.

Since the reference and displaced configurations of a body, as discussed in Section 1.2, may be quite different, it is appropriate to use a notation that makes the difference apparent, a notation based on that introduced by Noll [1955] and made current by the monograph of Truesdell and Toupin [1960]. A material point is denoted simply $\mathbf{X}$, and its Lagrangian coordinates are denoted X_I ($I = 1, 2, 3$); a point in the current or displaced configuration is denoted $\mathbf{x}$, and its Cartesian coordinates (sometimes called *Eulerian* coordinates) are x_i ($i = 1, 2, 3$). Different indices are used for the two coordinate systems because the corresponding bases, the Lagrangian basis ($\mathbf{e}_I$) and the Eulerian basis ($\mathbf{e}_i$), are, in principle, independent of each other.

Whenever direct notation is used for tensors in this chapter, vectors are treated as column matrices and second-rank tensors as square matrices; thus $\mathbf{u}^T\mathbf{v}$ is used for the scalar product $\mathbf{u} \cdot \mathbf{v}$, while $\mathbf{AB}$ denotes the second-rank tensor with components $A_{ik}B_{kj}$. Furthermore, the scalar product of two second-rank tensors is $\mathbf{A} : \mathbf{B} = \operatorname{tr}(\mathbf{A}^T\mathbf{B})$. If $\mathbf{\Gamma}$ is a fourth-rank tensor, then $\mathbf{\Gamma} : \mathbf{A}$ denotes the second-rank tensor with components $\Gamma_{ijkl}A_{kl}$. In particular, a fourth-rank tensor with components $\partial U_{ij}/\partial V_{kl}$ will be denoted $\partial \mathbf{U}/\partial \mathbf{V}$, and the second-rank tensor with components $\partial U_{ij}/\partial V_{kl}A_{kl}$ will be denoted $(\partial \mathbf{U}/\partial \mathbf{V}) : \mathbf{A}$.

Deformation Gradient

The *motion* of the body is described by the functional relation

$$\mathbf{x} = \chi(\mathbf{X}, t).$$

When χ is continuously differentiable with respect to $\mathbf{X}$, then the *deformation gradient* at $\mathbf{X}$ is the second-rank tensor $\mathbf{F}(\mathbf{X}, t)$ whose Cartesian components are

$$F_{iI} = \frac{\partial \chi_i}{\partial X_I}.$$

Note that these components are evaluated with respect to both bases simultaneously; in the terminology of Truesdell and Toupin [1960], the deformation gradient is a *two-point tensor*.

If, in a neighborhood of the material point $\mathbf{X}$, the function $\chi(\mathbf{X}, t)$ is invertible — in other words, if the material points in the neighborhood are in one-to-one correspondence with their displaced positions — then, by the **implicit function theorem** of advanced calculus, the matrix of components of $\mathbf{F}(\mathbf{X}, t)$ (the *Jacobian* matrix) must be nonsingular, that is, $J(\mathbf{X}, t) \neq 0$, where $J(\mathbf{X}, t) \stackrel{\text{def}}{=} \det \mathbf{F}(\mathbf{X}, t)$ is the *Jacobian determinant*. If we consider only displaced configurations that can evolve continuously from one another, then since $J = 1$ when the displaced and reference configurations coincide, we obtain the stronger condition $J(\mathbf{X}, t) > 0$.

The inverse of $\mathbf{F}(\mathbf{X}, t)$, denoted $\mathbf{F}^{-1}(\mathbf{X}, t)$, has the components $F_{Ii}^{-1}(\mathbf{X}, t) = \partial X_I/\partial x_i|_{\mathbf{x}=\chi(\mathbf{X}, t)}$. Note that $\mathbf{F}^{-1}$ is also a two-point tensor, but of a different kind from $\mathbf{F}$: while the components of the latter (F_{iI}) are such that the first index refers to an Eulerian and the second to a Lagrangian basis, in the former it is the reverse (F_{Ii}^{-1}).

Local Deformation

Consider two neighboring material points $\mathbf{X}$ and $\mathbf{X}' = \mathbf{X} + \mathbf{u}$, where $\mathbf{u}$ is a "small" Lagrangian vector emanating from $\mathbf{X}$. If the displaced positions of $\mathbf{X}$ and $\mathbf{X}'$ are respectively given by $\mathbf{x}$ and $\mathbf{x}'$, then, $\mathbf{F}$ being continuous, we have (in the matrix-based direct notation that is used throughout this chapter)

$$\mathbf{x}' = \mathbf{x} + \mathbf{F}(\mathbf{X}, t)\mathbf{u} + o(|\mathbf{u}|) \quad \text{as} \quad |\mathbf{u}| \to 0.$$

Since a rigid-body displacement does not change distances between points, let us compare the distance between $\mathbf{x}$ and $\mathbf{x}'$ with that between $\mathbf{X}$ and $\mathbf{X}'$:

$$\begin{aligned}|\mathbf{x}' - \mathbf{x}| &= \sqrt{(\mathbf{x}' - \mathbf{x})^T(\mathbf{x}' - \mathbf{x})} \\ &= \sqrt{\mathbf{u}^T\mathbf{F}^T(\mathbf{X}, t)\mathbf{F}(\mathbf{X}, t)\mathbf{u} + o(|\mathbf{u}|^2)} \\ &= \sqrt{\mathbf{u}^T\mathbf{C}(\mathbf{X}, t)\mathbf{u} + o(|\mathbf{u}|)},\end{aligned}$$

where $\mathbf{C} \stackrel{\text{def}}{=} \mathbf{F}^T\mathbf{F}$ is known, in the Noll–Truesdell terminology, as the *right Cauchy–Green tensor*. The notation will henceforth be simplified by writing $\mathbf{F}$ for $\mathbf{F}(\mathbf{X}, t)$ and so on. The components of $\mathbf{C}$ are given by

$$C_{IJ} = \chi_{i,I}\,\chi_{i,J}\,,$$

where $(\cdot)_{,I} = \partial(\cdot)/\partial X_I$. $\mathbf{C}$ is a *Lagrangian* tensor field which, moreover, is symmetric ($\mathbf{C} = \mathbf{C}^T$) and positive definite ($\mathbf{u}^T\mathbf{C}\mathbf{u} > 0$ for any $\mathbf{u} \neq 0$).

Stretch and Strain

The *stretch* at $(\mathbf{X}, t)$ along a direction $\mathbf{u}$ is defined by

$$\lambda_{\mathbf{u}} = \lim_{h\to 0} \frac{|\chi(\mathbf{X} + h\mathbf{u}, t) - \chi(\mathbf{X}, t)|}{h|\mathbf{u}|} = \sqrt{\frac{\mathbf{u}^T\mathbf{C}\mathbf{u}}{\mathbf{u}^T\mathbf{u}}}.$$

We have $\lambda_{\mathbf{u}} = 1$ for every $\mathbf{u}$ if and only if $\mathbf{C} = \mathbf{I}$; then the displacement is locally a rigid-body displacement. If $\mathbf{u}^{(\alpha)}$ ($\alpha = 1, 2, 3$) are the eigenvectors of $\mathbf{C}$ and if $\lambda_\alpha \stackrel{\text{def}}{=} \lambda_{\mathbf{u}^{(\alpha)}}$, then the λ_α^2 are the eigenvalues of $\mathbf{C}$, and the λ_α (the *principal stretches*) are the eigenvalues of $\mathbf{U} \stackrel{\text{def}}{=} \mathbf{C}^{\frac{1}{2}}$.

A strain is a measure of how much a given displacement differs locally from a rigid-body displacement. In particular, a *Lagrangian* strain is a measure of how much $\mathbf{C}$ differs from $\mathbf{I}$. The following Lagrangian strain tensors are commonly used: (a) the *Green–Saint-Venant* strain tensor[1] already mentioned in Section 1.2, $\mathbf{E} = \frac{1}{2}(\mathbf{C} - \mathbf{I})$, with eigenvalues $\frac{1}{2}(\lambda_\alpha^2 - 1)$; (b) the conventional strain tensor $\mathbf{E}_e = \mathbf{U} - \mathbf{I}$, with eigenvalues $\lambda_\alpha - 1$ (the principal conventional strains); and (c) the logarithmic strain tensor $\mathbf{E}_l = \ln \mathbf{U}$, with eigenvalues $\ln \lambda_\alpha$ (the principal logarithmic strains). Note that all three strains may be regarded as special cases of $(1/k)(\mathbf{U}^k - \mathbf{I})$, with (a), (b), and (c) corresponding respectively to $k = 2$, $k = 1$ and the limit as $k \to 0$. Furthermore,

$$\frac{1}{k}(\lambda^k - 1) = \frac{1}{k}[(1 + \lambda - 1)^k - 1] = \lambda - 1 + o(|\lambda - 1|),$$

so that

$$\left.\begin{array}{c}\mathbf{E}_e \\ \mathbf{E}_l\end{array}\right\} = \mathbf{E} + o(||\mathbf{E}||),$$

[1] Often called simply *the* Lagrangian strain tensor.

where $||\mathbf{E}||$ denotes some measure of the magnitude (a *norm*) of $\mathbf{E}$. In other words, the different Lagrangian strain tensors are approximately equal when they are sufficiently small. For large strains the Green–Saint-Venant strain tensor is analytically the most convenient, except in cases where the principal directions $\mathbf{u}^{(\alpha)}$ are known a priori.

Since $\mathbf{U}$ is symmetric and positive definite, we can form the two-point tensor $\mathbf{R} = \mathbf{F}\mathbf{U}^{-1}$. Note that

$$\mathbf{R}^T\mathbf{R} = \mathbf{U}^{-1}\mathbf{F}^T\mathbf{F}\mathbf{U}^{-1} = \mathbf{U}^{-1}\mathbf{U}^2\mathbf{U}^{-1} = \mathbf{I},$$

that is, $\mathbf{R}$ is orthogonal. Also, $\det \mathbf{U} = J$, so that $\det \mathbf{R} = J/J = 1$, and consequently, $\mathbf{R}$ is a *proper* orthogonal tensor, or a *rotation*. The decomposition $\mathbf{F} = \mathbf{R}\mathbf{U}$ is the **right polar decomposition** of $\mathbf{F}$, and $\mathbf{U}$ is called the *right stretch tensor*; this is why $\mathbf{C}$ is called the *right* Cauchy–Green tensor. $\mathbf{R}$ is usually called simply the *rotation tensor*.

If the displacement is locally a rigid-body one, then we simply have $\mathbf{F} = \mathbf{R}$. To study the general case, let us consider points near $\mathbf{X}$ given by

$$\mathbf{X}^{(\alpha)} = \mathbf{X} + h\mathbf{u}^{(\alpha)},$$

where $\mathbf{u}^{(\alpha)}$ is, as above, an eigenvector of $\mathbf{C}$. The displaced images of these points are

$$\mathbf{x}^{(\alpha)} = \mathbf{x} + h\mathbf{v}^{(\alpha)} + o(h),$$

where

$$\mathbf{v}^{(\alpha)} = \mathbf{F}\mathbf{u}^{(\alpha)} = \mathbf{R}\mathbf{U}\mathbf{u}^{(\alpha)}.$$

Since, however, $\mathbf{u}^{(\alpha)}$ is also an eigenvector of $\mathbf{U}$, it follows that $\mathbf{U}\mathbf{u}^{(\alpha)} = \lambda_\alpha \mathbf{u}^{(\alpha)}$, and therefore

$$\mathbf{v}^{(\alpha)} = \lambda_\alpha \mathbf{R}\mathbf{u}^{(\alpha)},$$

so that $\mathbf{R}$ represents the rotation of the eigenvectors of $\mathbf{C}$.

For an arbitrary vector $\mathbf{u}$, $\mathbf{U}\mathbf{u}$ is not in general parallel to $\mathbf{u}$. In fact, if we consider the ellipsoid centered at $\mathbf{X}$, given by

$$\mathbf{u}^T\mathbf{C}\mathbf{u} = r^2,$$

with principal semiaxes r/λ_α, we see that its displaced image is approximately the sphere of radius r centered at $\mathbf{x}$. The ellipsoid given by the preceding equation is called the *reciprocal strain ellipsoid*.

We may also ask what is the effect of the displacement on the sphere of radius r, centered at $\mathbf{X}$, in the reference configuration. If $\mathbf{x} + \mathbf{v}$ is the displaced image of $\mathbf{X} + \mathbf{u}$, then $\mathbf{v} = dot\mathbf{F}\mathbf{u}$, so that $\mathbf{u}^T\mathbf{u} = dot\mathbf{v}^T\mathbf{B}^{-1}\mathbf{v}$ (where $\mathbf{B} = \mathbf{F}\mathbf{F}^T$), and $\mathbf{v}^T\mathbf{B}^{-1}\mathbf{v} = r^2$ defines an ellipsoid centered at $\mathbf{x}$ with principal semiaxes $\lambda_\alpha r$; this ellipsoid is called simply the *strain ellipsoid*. The tensor $\mathbf{B}^{-1}$ is called the *Finger deformation tensor*, while $\mathbf{B}$ is called (again in the terminology of Truesdell et al.) the *left Cauchy–Green tensor*. It is an "Eulerian" tensor, and its eigenvalues are also λ_α^2. Its name derives, as may be surmised, from the left polar decomposition of $\mathbf{F}$ (see Exercise 1). It can easily be shown that $\mathbf{B} = \mathbf{RCR}^T$.

The most commonly used "Eulerian" strain tensor is the *Almansi strain tensor* $\mathbf{E}_a = \frac{1}{2}(\mathbf{I} - \mathbf{B}^{-1})$. It is easy to show that $\mathbf{E} = \mathbf{F}^T\mathbf{E}_a\mathbf{F}$, so that $\mathbf{E}_a = \mathbf{RU}^{-1}\mathbf{EU}^{-1}\mathbf{R}^T$. Hence, in order for $\mathbf{E}_a$ to be approximately equal to $\mathbf{E}$ (i.e. to have $\mathbf{E}_a = \mathbf{E} + o(||\mathbf{E}||)$), it is necessary not only for $||\mathbf{E}||$ but also for $||\mathbf{R} - \mathbf{I}||$ to be small.

In many treatments the right Cauchy–Green tensor $\mathbf{C}$ is given another definition, namely, as a *metric tensor*. If the Lagrangian coordinates X_I are used to describe points in the displaced body, then they are no longer Cartesian coordinates, since the surfaces X_I = constant are not necessarily planes. An infinitesimal vector $d\mathbf{x}$ in the displaced body has the square of its magnitude given by

$$d\mathbf{x} \cdot d\mathbf{x} = C_{IJ} dX_I dX_J.$$

Hence $\mathbf{C}$ is often called the *metric tensor in the displaced (strained) body*. The identity $\mathbf{I}$ is accordingly regarded as the metric tensor in the unstrained body. The analysis of continuum deformation based on metric tensors is usually carried out by using curvilinear coordinates to begin with (see, e.g., Green and Zerna [1968], Eringen [1962], Marsden and Hughes [1983]). For our purposes, its main usefulness is in the derivation of **compatibility conditions**, that is, the finite-deformation analogue of Equation (1.2.4), and to that end Cartesian coordinates are adequate. The result is usually given as the vanishing of a fourth-rank tensor called the *Riemann–Christoffel tensor* or the *curvature tensor*, being the necessary (*locally* also sufficient) condition for a symmetric positive definite tensor field $\mathbf{C}$ to be derivable from a configuration χ. It can be shown that the result is equivalent to the symmetric second-rank tensor equation

$$e_{IKM}e_{JLN}[E_{KL,MN} - \tfrac{1}{2}C_{PQ}^{-1}(E_{LP,M} + E_{MP,L} - E_{LM,P})(E_{KQ,N} + E_{NQ,K} - E_{KN,Q})] = 0, \tag{8.1.1}$$

which can be seen to reduce to (1.2.4) for sufficiently small strains.

A fundamental difference between the Lagrangian and the Eulerian deformation and strain tensors must be emphasized. The Lagrangian tensors transform according to the usual rules (Section 1.1) under a rotation of the Lagrangian basis, while they are invariant under a rotation of the Eulerian basis. To see the invariance, let $(\mathbf{e}_i^*)$ denote the rotated Eulerian basis, with $\mathbf{e}_i^* = Q_{ij}\mathbf{e}_j$, $\mathbf{Q}$ being orthogonal. Since $x_i^* = Q_{ij}x_j$, it follows that $\chi_i^*(\mathbf{X}, t) = Q_{ij}\chi_j(\mathbf{X}, t)$, and hence $F_{iI}^* = Q_{ij}F_{jI}$. These relations may be written as $\mathbf{x}^* = \mathbf{Qx}$, $\chi^*(\mathbf{X}, t) = \mathbf{Q}\chi(\mathbf{X}, t)$, and $\mathbf{F}^* = \mathbf{QF}$. Consequently, $\mathbf{C}^* =$

$\mathbf{F}^{starT}\mathbf{F}^* = \mathbf{F}^T\mathbf{Q}^T\mathbf{Q}\mathbf{F} = \mathbf{C}$. Since the Eulerian basis is, in a sense, fixed in space, invariance under its rotation is related to the isotropy of space itself, and is invoked in the formulation of fundamental physical principles.

Conversely, Eulerian tensors are invariant under a rotation of the Lagrangian basis; such a rotation may represent, for example, a rotation of the given body in space. This kind of invariance may, accordingly, be used to establish material isotropy — the fact that the mechanical response of a body does not depend on its orientation. Eulerian tensors transform according to the usual rules under a time-independent rotation of the Eulerian basis; but only the ones called *objective* do so when the rotation is time-dependent. The objectivity of tensors is discussed in 8.1.2.

Area and Volume Deformation

Consider again a material point $\mathbf{X}$ and two neighboring points $\mathbf{X}' = \mathbf{X} + h\mathbf{u}'$, $\mathbf{X}'' = \mathbf{X} + h\mathbf{u}''$, defined by two vectors $\mathbf{u}'$, $\mathbf{u}''$ and a small real number h. The vector area of the triangle formed by the three points is

$$\mathbf{A}_0 = \frac{1}{2}h^2\mathbf{u}' \times \mathbf{u}'',$$

or, in components

$$A_{0I} = \frac{1}{2}h^2 e_{IJK}u'_J u''_K.$$

The displaced positions of $\mathbf{X}$, $\mathbf{X}'$ and $\mathbf{X}''$ are $\mathbf{x}$, $\mathbf{x}' = \mathbf{x} + h\mathbf{v}' + o(h)$ and $\mathbf{x}'' = \mathbf{x} + h\mathbf{v}'' + o(h)$, respectively, where $\mathbf{v}' = \mathbf{F}\mathbf{u}'$ and $\mathbf{v}'' = \mathbf{F}\mathbf{u}''$. The displaced area is

$$\mathbf{A} = \frac{1}{2}h^2\mathbf{v}' \times \mathbf{v}'' + o(h^2)$$

or

$$\begin{aligned}
A_i &\doteq \tfrac{1}{2}h^2 e_{ijk}v'_j v''_k = \tfrac{1}{2}h^2 e_{ijk}F_{jJ}F_{kK}u'_J u''_K \\
&= \tfrac{1}{4}h^2 e_{ijk}(F_{jJ}F_{kK} - F_{jK}F_{kJ})u'_J u''_K \\
&= \tfrac{1}{4}h^2 e_{ijk}F_{jJ}F_{kK}(u'_J u''_K - u'_K u''_J) \\
&= \tfrac{1}{4}h^2 e_{ijk}F_{jJ}F_{kK}e_{IJK}e_{ILM}u'_L u''_M \\
&= \left(\tfrac{1}{2}e_{ijk}e_{IJK}F_{jJ}F_{kK}\right)\left(\tfrac{1}{2}h^2 e_{ILM}u'_L u''_M\right) = (\det\mathbf{F})F_{Ii}^{-1}A_{0I}.
\end{aligned}$$

In direct notation,

$$\mathbf{A} = J\mathbf{F}^{T-1}\mathbf{A}_0 + o(|\mathbf{A}_0|).$$

If, now, $\mathbf{X}''' = \mathbf{X} + h\mathbf{u}'''$ is a fourth point, then the volume of the tetrahedron in the reference configuration is $V_0 = \frac{1}{3}h\mathbf{A}_0 \cdot \mathbf{u}''' = \frac{1}{3}h\mathbf{A}_0^T\mathbf{u}'''$. Since the displaced position of $\mathbf{X}'''$ is $\mathbf{x}''' = \mathbf{X} + h\mathbf{F}\mathbf{u}''' + o(h)$, the displaced volume is

$$V = \frac{1}{3}h\mathbf{A}^T\mathbf{F}\mathbf{u}''' + o(h^3) = JV_0 + o(V_0).$$

The Jacobian determinant is thus seen as having the important property of measuring the ratio between displaced and referential infinitesimal volumes, that is,

$$dV = J\,dV_0. \tag{8.1.2}$$

If ρ and ρ_0 denote the mass density in the displaced and reference configurations, respectively, then conservation of mass in an infinitesimal volume requires that $\rho\,dV = \rho_0\,dV_0$. Therefore,

$$J = \frac{\rho_0}{\rho}.$$

Deformation-Rate Measures

The velocity field at time t is defined by

$$\dot{\chi}(\mathbf{X},\,t) \doteq \frac{\partial\chi(\mathbf{X},\,t)}{\partial t}.$$

In the Eulerian description, the velocity field is given by $\mathbf{v}(\mathbf{x},\,t)$, and the *velocity gradient* $\mathbf{L}$, an Eulerian tensor, is defined by $L_{ij} = \partial v_i/\partial x_j$. In view of the definition of $\mathbf{F}$, $\dot{\mathbf{F}}$ is given by

$$\dot{F}_{iJ} = \frac{\partial}{\partial t}\left(\frac{\partial\chi_i}{\partial X_J}\right) = \frac{\partial\dot{\chi}_i}{\partial X_J} = \frac{\partial v_i}{\partial x_j}\frac{\partial\chi_j}{\partial X_J} = L_{ij}F_{jJ},$$

or

$$\dot{\mathbf{F}} = \mathbf{LF}. \tag{8.1.3}$$

A rigid-body motion is characterized by the right Cauchy-Green tensor field $\mathbf{C}$ being constant in time (not necessarily equal to $\mathbf{I}$, since the initial configuration of the motion may be deformed with respect to the reference configuration), or, equivalently, $\dot{\mathbf{C}} = 0$. Using differential calculus, we obtain

$$\dot{\mathbf{C}} = \mathbf{F}^T\dot{\mathbf{F}} + \dot{\mathbf{F}}^T\mathbf{F} = \mathbf{F}^T(\mathbf{L}+\mathbf{L}^T)\mathbf{F} = 2\mathbf{F}^T\mathbf{DF},$$

where $\mathbf{D} \stackrel{\text{def}}{=} \frac{1}{2}(\mathbf{L}+\mathbf{L}^T)$ is the (Eulerian) *deformation-rate tensor* or (in Truesdell's terminology) *stretching* tensor; it was denoted $\mathbf{d}$ in Chapter 3. An immediate corollary of the last result is

$$\dot{\mathbf{E}} = \mathbf{F}^T\mathbf{DF}. \tag{8.1.4}$$

If $\mathbf{W} \stackrel{\text{def}}{=} \frac{1}{2}(\mathbf{L}-\mathbf{L}^T)$, then obviously $\mathbf{L} = \mathbf{D}+\mathbf{W}$. The antisymmetric tensor $\mathbf{W}$ is called the *spin, vorticity* or (Truesdell) *spinning* tensor.

The deformation-rate tensor $\mathbf{D}$ is of fundamental significance in the study of deformational motions of continua. Let us consider two neighboring points[1] $\mathbf{X}$, $\mathbf{X}+d\mathbf{X}$, with their displaced positions at time t designated

[1]The "differentials" $d\mathbf{X}$ have exactly the same meaning as the small vectors $h\mathbf{u}$ used in the preceding discussion, but their use makes terms of the type $o(h)$ unnecessary.

respectively by $\mathbf{x}$ and $\mathbf{x} + d\mathbf{x}$, so that $d\mathbf{x} = \mathbf{F}(\mathbf{X}, t)d\mathbf{X}$. Because of this last relation, it is easy to obtain the material (Eulerian) time derivative, as defined in Section 1.3, of $d\mathbf{x}$:

$$\frac{D}{Dt}d\mathbf{x} = \dot{\mathbf{F}}d\mathbf{X} = \mathbf{L}d\mathbf{x}.$$

The preceding equation shows us that $\mathbf{L}$ is the operator that transforms $d\mathbf{x}$ into its material time derivative. Now if the magnitude of $d\mathbf{x}$ is $ds = \sqrt{d\mathbf{x}^T d\mathbf{x}}$, then

$$\frac{D}{Dt}ds = \frac{1}{2ds}\frac{D}{Dt}(d\mathbf{x}^T d\mathbf{x}) = \frac{1}{ds}d\mathbf{x}^T\mathbf{D}\,d\mathbf{x}.$$

If $\mathbf{n}$ is a unit vector, so that $d\mathbf{x} = \mathbf{n}\,ds$, then

$$\frac{D}{Dt}\ln ds = \mathbf{n}^T\mathbf{D}\mathbf{n},$$

so that with respect to a Cartesian basis $(\mathbf{e}_i)$, the diagonal components D_{11}, D_{22}, and D_{33} are just the *logarithmic strain rates* (see Section 2.1) along $\mathbf{e}_1$, $\mathbf{e}_2$ and $\mathbf{e}_3$, respectively.

What is the meaning of the off-diagonal components? We may answer the question by considering two small Lagrangian vectors $d\mathbf{X}^{(1)}$, $d\mathbf{X}^{(2)}$, such that $\boldsymbol{\chi}(\mathbf{X} + d\mathbf{X}^{(\alpha)}, t) = \mathbf{x} + d\mathbf{x}^{(\alpha)} + o(|d\mathbf{x}^{(\alpha)}|)$, $\alpha = 1, 2$, with $d\mathbf{x}^\alpha = \mathbf{F}d\mathbf{X}^\alpha$. With, furthermore, $d\mathbf{x}^\alpha = \mathbf{n}^\alpha\, ds_\alpha$, we have $d\mathbf{x}^{(1)}\cdot d\mathbf{x}^{(2)} = ds_1\, ds_2 \cos\theta$, where $\cos\theta = \mathbf{n}^{(1)}\cdot\mathbf{n}^{(2)}$. But we also have $d\mathbf{x}^{(1)}\cdot d\mathbf{x}^{(2)} = d\mathbf{X}^{(1)T}\mathbf{C}d\mathbf{X}^{(2)}$, so that

$$\begin{aligned}\frac{D}{Dt}(d\mathbf{x}^{(1)}\cdot d\mathbf{x}^{(2)}) &= d\mathbf{X}^{(1)T}\dot{\mathbf{C}}d\mathbf{X}^{(2)} = 2d\mathbf{X}^{(1)T}\mathbf{F}^T\mathbf{D}\mathbf{F}d\mathbf{X}^{(2)}\\ &= 2d\mathbf{x}^{(1)T}\mathbf{D}\,d\mathbf{x}^{(2)} = 2ds_1\, ds_2\mathbf{n}^{(1)T}\mathbf{D}\mathbf{n}^{(2)}.\end{aligned}$$

On the other hand,

$$\frac{D}{Dt}(d\mathbf{x}^{(1)}\cdot d\mathbf{x}^{(2)}) = \cos\theta\frac{D}{Dt}(ds_1 ds_2) - ds_1\, ds_2 \sin\theta\frac{D}{Dt}\theta.$$

Now, if $\mathbf{n}^{(1)}$ and $\mathbf{n}^{(2)}$ are instantaneously perpendicular, with $\sin\theta = 1$, then

$$\mathbf{n}^{(1)T}\mathbf{D}\mathbf{n}^{(2)} = -\frac{1}{2}\frac{D}{Dt}\theta.$$

But $-(D/Dt)\theta$ is the rate at which an instantaneously right angle becomes acute — precisely what we know as the *shearing rate.*

We can easily determine the rate of change of infinitesimal *volumes* by performing the material time differentiation of Equation (8.1.2):

$$\frac{D}{Dt}dV = \left(\frac{D}{Dt}J\right)dV_0 = \dot{J}\,dV_0 = (\partial J/\partial\mathbf{F}) : \dot{\mathbf{F}}\,dV_0;$$

but $\partial J/\partial \mathbf{F} = J\mathbf{F}^{T-1}$, so that

$$\frac{1}{dV}\frac{D}{Dt}dV = \operatorname{tr}\mathbf{L} = \operatorname{tr}\mathbf{D} = \nabla\cdot\mathbf{v} = \frac{D}{Dt}\ln dV.$$

A motion in which the volume remains constant (called an *isochoric* motion) is therefore characterized by $\nabla\cdot\mathbf{v} = 0$ throughout. The velocity field $\mathbf{v}$ is then called *solenoidal.*

8.1.2 Continuum Mechanics and Objectivity

Stress

The stress tensor $\boldsymbol{\sigma}$ was introduced in Section 1.3 in connection with infinitesimal-displacement theory. However, the procedure used in defining it is valid under finite deformation, provided that the oriented surface element $\mathbf{n}\,dS$ is in the current configuration. The stress tensor is then called the *Cauchy stress*, and following Truesdell et al. it will be denoted $\mathbf{T}$. The equations of motion (1.3.3) are exact if all quantities occurring in them are Eulerian, that is, if ρ denotes mass per current volume and $(\cdot)_{,j} = \partial(\cdot)/\partial x_j$, the x_j being Eulerian coordinates. It can accordingly be shown that the deformation power P_d defined in Section 1.4 is given exactly by

$$P_d = \int_R \mathbf{T} : \mathbf{D}\,dV.$$

Note that as $\mathbf{T}$ is symmetric, $\mathbf{T} : \mathbf{D} = \mathbf{T} : \mathbf{L}$, since $\mathbf{T} : \mathbf{W} = 0$. Now, from Equation (8.1.3), $\mathbf{L} = \dot{\mathbf{F}}\mathbf{F}^{-1}$, and therefore

$$\mathbf{T} : \mathbf{L} = \operatorname{tr}(\mathbf{T}\dot{\mathbf{F}}\mathbf{F}^{-1}) = \operatorname{tr}(\mathbf{F}^{-1}\mathbf{T}\dot{\mathbf{F}}) = (\mathbf{T}\mathbf{F}^{T-1}) : \dot{\mathbf{F}},$$

so that an alternative expression for P_d is

$$P_d = \int_{R_0} \bar{\mathbf{T}} : \dot{\mathbf{F}}\,dV_0,$$

where R_0 is the region occupied by the body in the reference configuration, and $\bar{\mathbf{T}} \stackrel{\text{def}}{=} J\mathbf{T}\mathbf{F}^{T-1}$ is a two-point tensor known as the *first Piola-Kirchhoff stress* or, more simply, as the *Piola stress*. In indicial notation, $\bar{T}_{iJ} = JT_{ik}F_{Jk}^{-1}$.

Since the area deformation discussed in the preceding subsection can be written, for infinitesimal areas, as

$$d\mathbf{A} = J\mathbf{F}^{T-1}\,d\mathbf{A}_0,$$

it follows that

$$\bar{\mathbf{T}}\,d\mathbf{A}_0 = \mathbf{T}\,d\mathbf{A}.$$

Consequently, the Piola stress is the multiaxial generalization of the uniaxial nominal (engineering) stress.

It can be shown in several ways that the Lagrangian equations of motion can be written as

$$\bar{T}_{iJ},_J + \rho_0 b_i = \rho_0 a_i, \tag{8.1.5}$$

where **a** and **b** denote, as before, the acceleration and the body force per unit mass, respectively. The derivation of Equation (8.1.5) is left to an exercise.

Yet another Lagrangian form for the deformation power can be derived by using Equation (8.1.4). Since $\mathbf{D} = \mathbf{F}^{T-1}\dot{\mathbf{E}}\mathbf{F}^{-1}$, it follows that

$$\mathbf{T} : \mathbf{D} = \mathrm{tr}\,(\mathbf{T}\mathbf{F}^{T-1}\dot{\mathbf{E}}\mathbf{F}^{-1}) = \mathrm{tr}\,(\mathbf{F}^{-1}\mathbf{T}\mathbf{F}^{T-1}\dot{\mathbf{E}}),$$

and therefore

$$P_d = \int_{R_0} \mathbf{S} : \dot{\mathbf{E}}\, dV_0,$$

where $\mathbf{S} = \mathbf{F}^{-1}\bar{\mathbf{T}} = J\mathbf{F}^{-1}\mathbf{T}\mathbf{F}^{T-1}$ is a symmetric Lagrangian tensor known as the *second Piola–Kirchhoff stress* or the *Kirchhoff–Trefftz stress*, with components $S_{IJ} = F_{Ik}^{-1}\bar{T}_{kJ} = JF_{Ik}^{-1}F_{Jl}^{-1}T_{kl}$.

Virtual Work

The principle of virtual work introduced in 1.3.5 can readily be extended to finite deformations (there is, however, no large-deformation counterpart to the principle of virtual stresses). Consider two possible configurations χ and $\chi + \delta\mathbf{u}$, where $\delta\mathbf{u}$ is a virtual displacement field as before, and suppose that the acceleration field vanishes identically. Multiplying the left-hand side of Equation (8.1.5) by δu_i, with summation over i implied as usual, integrating over the reference volume R_0, and applying the divergence theorem leads to

$$\int_{R_0} \bar{T}_{iJ}\delta u_{i},_J\, dV_0 = \int_{R_0} \rho_0 b_i \delta u_i\, dV_0 + \int_{\partial R_{0t}} \bar{t}_i^a \delta u_i\, dS_0, \tag{8.1.6}$$

where $\bar{\mathbf{t}}^a$ is the prescribed surface traction per unit reference area, with $\bar{t}_i = n_{0J}\bar{T}_{iJ}$, $\mathbf{n}_0$ being the unit normal vector in the reference configuration.

The left-hand side of the above equation can be cast in another form by introducing the virtual strain field $\delta\mathbf{E}$, which, from the definition of $\mathbf{E}$, can easily be shown to be given by

$$\delta E_{IJ} = \frac{1}{2}(F_{kI}\delta u_{k},_J + F_{kJ}\delta u_{k},_I).$$

The principle of virtual work can now be written as

$$\int_{R_0} \mathbf{S} : \delta\mathbf{E}\, dV_0 = \int_{R_0} \rho_0 \mathbf{b}\cdot\delta\mathbf{u}\, dV_0 + \int_{\partial R_{0t}} \bar{\mathbf{t}}^a \cdot \delta\mathbf{u}\, dS_0,$$

A rate form of the principle of virtual work can be obtained by differentiating the left-hand side of (8.1.5) with respect to time (the right-hand side is again assumed to be zero) and then multiplying it by δv_i, where δv is a virtual velocity field. The result is

$$\int_{R_0} \dot{\bar{T}}_{iJ}\delta v_{i,J}\, dV_0 = \int_{R_0} \rho_0 \dot{b}_i \delta v_i\, dV_0 + \int_{\partial R_{0t}} \dot{\bar{t}}_i^a \delta v_i\, dS_0, \tag{8.1.7}$$

Energy Balance

Like the local equations of motion, the local energy-balance equation (1.4.1) is exact as an Eulerian equation if the term $\sigma_{ij}\dot{\varepsilon}_{ij}$ is replaced by $\mathbf{T} : \mathbf{D}$. It can easily be recast into the Lagrangian form

$$\rho_0 \dot{u} = \mathbf{S} : \dot{\mathbf{E}} + \rho_0 r - \text{Div}\bar{\mathbf{h}}, \tag{8.1.8}$$

where Div denotes the divergence operator with respect to Lagrangian coordinates, and $\bar{\mathbf{h}} = J\mathbf{F}^{-1}\mathbf{h}$ is the Lagrangian heat-flux vector. The term $\mathbf{S} : \dot{\mathbf{E}}$ may, of course, be replaced by $\bar{\mathbf{T}} : \dot{\mathbf{F}}$.

Objective Rates

Consider a possible motion $\chi(\mathbf{X}, t)$ of a body, and another possible motion $\chi^*(\mathbf{X}, t)$ of the same body that differs from the first motion by a *superposed rigid-body motion.* The relation between the two motions must have the form

$$\chi^*(\mathbf{X}, t) = \mathbf{Q}(t)\chi(\mathbf{X}, t) + \mathbf{c}(t),$$

where $\mathbf{Q}(t)$ is a proper orthogonal tensor and $\mathbf{c}(t)$ a vector, both continuous functions of time that are also, for simplicity, assumed to be continuously differentiable. It is clear that according to a basic principle of relativity, the relation between the two motions is equivalent to that between the descriptions of one and the same motion as seen by two observers who are moving relative to each other, with $\mathbf{Q}(t)$ describing the relative rotation of their respective Eulerian bases. If $\mathbf{g}$ is an Eulerian vector generated by the motion χ, then it is called *objective* if its counterpart $\mathbf{g}^*$, generated by χ^*, is related to $\mathbf{g}$ by $\mathbf{g}^* = \mathbf{Q}\mathbf{g}$. Similarly, if $\mathbf{G}$ is an Eulerian second-rank tensor generated by χ, then it is objective if $\mathbf{G}^*$ is related to it by $\mathbf{G}^* = \mathbf{Q}\mathbf{G}\mathbf{Q}^T$. Since $\mathbf{F}^* = \mathbf{Q}\mathbf{F}$, it follows that $d\mathbf{x} = \mathbf{F}\, d\mathbf{X}$ is an objective vector and $\mathbf{B}$ is an objective tensor. The area element $d\mathbf{A}$ is also an objective vector (since $dV = d\mathbf{x}^T d\mathbf{A}$). The objectivity of the surface tractions $\mathbf{t}$, and therefore of the Cauchy stress $\mathbf{T}$, is argued on physical grounds, in that the tractions represent contact forces within the body. The assertion that *if a motion* χ *generates a stress* $\mathbf{T}$ *then the transformed motion* χ^* *generates the stress* $\mathbf{T}^* = \mathbf{Q}\mathbf{T}\mathbf{Q}^T$ is known as the **principle of objectivity** or **principle of material frame indifference.**[1]

[1]The principle is usually stated as an axiom. For a critique, see Woods [1981].

On the other hand, vectors and tensors derived by means of material time differentiation are not, in general, objective. Thus, if $\mathbf{g}^* = \mathbf{Q}\mathbf{g}$, then $(D/Dt)\mathbf{g}^* = \mathbf{Q}(D\mathbf{g}/Dt + \boldsymbol{\Omega}\mathbf{g})$, where $\boldsymbol{\Omega} = \mathbf{Q}^T\dot{\mathbf{Q}}$. It can be readily shown that $\boldsymbol{\Omega}$ is an antisymmetric tensor: since $\mathbf{Q}^T\mathbf{Q} = \mathbf{I}$, it follows that

$$\frac{d}{dt}(\mathbf{Q}^T\mathbf{Q}) = \dot{\mathbf{Q}}^T\mathbf{Q} + \mathbf{Q}^T\dot{\mathbf{Q}} = \boldsymbol{\Omega}^T + \boldsymbol{\Omega} = 0.$$

It can also be shown that the vector $\boldsymbol{\omega}$ derived from $\boldsymbol{\Omega}$ by $\omega_j = \frac{1}{2}e_{ijk}\Omega_{ik}$ is the angular velocity of the rotation described by $\mathbf{Q}$.

Similarly, if $\mathbf{G}^* = \mathbf{Q}\mathbf{G}\mathbf{Q}^T$, then

$$\frac{D}{Dt}\mathbf{G}^* = \mathbf{Q}\left(\frac{D}{Dt}\mathbf{G} + \boldsymbol{\Omega}\mathbf{G} - \mathbf{G}\boldsymbol{\Omega}\right)\mathbf{Q}^T.$$

Since $\dot{\mathbf{F}}^* = \mathbf{Q}(\dot{\mathbf{F}} + \boldsymbol{\Omega}\mathbf{F})$, the velocity gradient $\mathbf{L} = \dot{\mathbf{F}}\mathbf{F}^{-1}$ transforms as

$$\mathbf{L}^* = \mathbf{Q}(\mathbf{L} + \boldsymbol{\Omega})\mathbf{Q}^T = \mathbf{Q}(\mathbf{D} + \mathbf{W} + \boldsymbol{\Omega})\mathbf{Q}^T.$$

Decomposing $\mathbf{L}^*$ into its symmetric and antisymmetric parts, we find that

$$\mathbf{D}^* = \mathbf{Q}\mathbf{D}\mathbf{Q}^T, \quad \mathbf{W}^* = \mathbf{Q}(\mathbf{W} + \boldsymbol{\Omega})\mathbf{Q}^T.$$

Thus the Eulerian deformation-rate tensor $\mathbf{D}$ is an objective tensor, a result that could also have been found from Equation (8.1.4), since the Lagrangian tensor $\mathbf{E}$, and hence also $\dot{\mathbf{E}}$, is invariant under the transformation.

Note that for an objective tensor $\mathbf{G}$, combinations such as

$$\frac{D}{Dt}\mathbf{G} - \mathbf{W}\mathbf{G} + \mathbf{G}\mathbf{W}, \quad \frac{D}{Dt}\mathbf{G} - \mathbf{L}\mathbf{G} - \mathbf{G}\mathbf{L}^T, \quad \frac{D}{Dt}\mathbf{G} + \mathbf{L}^T\mathbf{G} + \mathbf{G}\mathbf{L}$$

are objective tensors known as *objective rates* of $\mathbf{G}$; they are called, respectively, the *Jaumann, Oldroyd*, and *Cotter–Rivlin* rates. Note further that they differ from one another by terms that are bilinear in $\mathbf{D}$ and $\mathbf{G}$ and therefore themselves objective. Clearly, an infinity of objective rates can be constructed. Adding the term $(\operatorname{tr}\mathbf{D})\mathbf{G}$ to the Oldroyd rate yields the *Truesdell rate*, related to the material time derivative of the second Piola–Kirchhoff stress:

$$\dot{\mathbf{S}} = J\mathbf{F}^{-1}\left[\frac{D}{Dt}\mathbf{T} + (\operatorname{tr}\mathbf{D})\mathbf{T} - \mathbf{L}\mathbf{T} - \mathbf{T}\mathbf{L}^T\right]\mathbf{F}^{T-1},$$

and the quantity in brackets is just the Truesdell rate of the Cauchy stress.

Let $\mathbf{G}$ be a symmetric tensor, and let its Jaumann rate be denoted $\overset{*}{\mathbf{G}}$. An important property of this rate, not shared by the other rates given above, is the following:

$$\frac{D}{Dt}(\mathbf{G} : \mathbf{G}) = 2\mathbf{G} : \overset{*}{\mathbf{G}}, \tag{8.1.9}$$

since

$$\mathbf{G} : (\mathbf{GW}) - \mathbf{G} : (\mathbf{WG}) = \operatorname{tr}(\mathbf{GGW} - \mathbf{GWG}) = 0.$$

Another important property is that if $\mathbf{G}$ is a deviatoric tensor (i.e., if $\operatorname{tr}\mathbf{G} = 0$), then $\overset{*}{\mathbf{G}}$ is also deviatoric. Both properties, however, are shared by objective rates having the Jaumann form but with $\mathbf{W}$ replaced by some other antisymmetric tensor representing a spin.

Exercises: Section 8.1

1. Show that there exists a symmetric positive definite tensor $\mathbf{V}$ (the *leftstretch tensor*) such that $\mathbf{F} = \mathbf{VR}$, where $\mathbf{R}$ is the same as in the right polar decomposition of $\mathbf{F}$, and $\mathbf{B} = \mathbf{V}^2$.

2. Show that $\mathbf{E} = \mathbf{F}^T\mathbf{E}_a\mathbf{F}$, where $\mathbf{E}$ is the Green–Saint-Venant strain and $\mathbf{E}_a$ is the Almansi strain.

3. A state of deformation known as *simple shear* occurs when $\mathbf{F}$ is given by the component matrix (with respect to coincident Eulerian and Lagrangian bases)
$$\begin{bmatrix} 1 & \gamma & 0 \\ 0 & 1 & 0 \\ 0 & 0 & 1 \end{bmatrix}.$$
Find $\mathbf{C}$, $\mathbf{B}$, $\mathbf{R}$, the principal stretches, and the eigenvectors of $\mathbf{C}$ and $\mathbf{B}$.

4. With respect to the finite-strain compatibility condition (8.1.1), consult one of the references given in the text and show that the vanishing of the Riemann–Christoffel tensor is equivalent to Equation (8.1.1).

5. Find the rate of change $(D/Dt)\, d\mathbf{A}$ of an infinitesimal Eulerian area element $d\mathbf{A}$.

6. Show that the invariants of $\mathbf{C}$ are the same as those of $\mathbf{B}$, and then show that if ψ is a function of these invariants, then $\mathbf{F}(\partial\psi/\partial\mathbf{C})\mathbf{F}^T = \mathbf{B}(\partial\psi/\partial\mathbf{B})$. (*Hint*: Show that $\mathbf{FC}^k\mathbf{F}^T = \mathbf{B}^{k+1}$.)

7. Derive the Lagrangian equations of motion (8.1.5) by at least two methods.

8. Show that the Jaumann, Oldroyd, Cotter–Rivlin, and Truesdell rates of an objective Eulerian second-rank tensor field $\mathbf{G}$ are objective.

9. Show that if $\bar{\mathbf{G}}$ is a Lagrangian tensor field and $\mathbf{G}$ is an Eulerian tensor field related to it by $\mathbf{G} = \mathbf{F}^{T-1}\bar{\mathbf{G}}\mathbf{F}^{-1}$, then the Cotter–Rivlin rate of $\mathbf{G}$ is linearly related to $\dot{\bar{\mathbf{G}}}$, while if $\mathbf{G} = (1/J)\mathbf{F}\bar{\mathbf{G}}\mathbf{F}^T$, then the Truesdell rate is linearly related to $\dot{\bar{\mathbf{G}}}$.

10. Show that $(D/Dt)f + f\nabla \cdot \mathbf{v}$ is an objective rate if f is an objective scalar field.

Section 8.2 Large-Deformation Constitutive Theory

8.2.1 Thermoelasticity

A thermoelastic body was defined in 1.4.1, under small deformations, as one in which the strain is determined by the stress and the internal-energy density, and in the absence of internal constraints, "strain" and "stress" in the definition may be interchanged. For large deformations, and assuming no internal constraints, we may define a thermoelastic body as one in which the Cauchy stress $\mathbf{T}$ depends only on the deformation gradient $\mathbf{F}$ and the internal-energy density u. The second law of thermodynamics assures the existence of an entropy density $\eta = \bar{\eta}(u, \mathbf{F})$ and of an absolute temperature T such that $T^{-1} = \partial\bar{\eta}/\partial u$, as in Section 1.4.

The "isotropy of space" mentioned previously requires, however, that the scalar η be invariant under a rotation of the Eulerian basis. As we know, under such a rotation (by an orthogonal matrix $\mathbf{Q}$) $\mathbf{F}$ becomes $\mathbf{QF}$; its polar decomposition factors $\mathbf{R}$ and $\mathbf{U}$ become, respectively, $\mathbf{QR}$ and $\mathbf{U}$. Consequently, η (and hence T) must be independent of $\mathbf{R}$ and can depend on $\mathbf{F}$ only through $\mathbf{U}$ or, equivalently, through $\mathbf{C}$ or $\mathbf{E}$. The reduced local form of the second law becomes

$$(\mathbf{S} + T\rho_0\partial\bar{\eta}/\partial\mathbf{E}) : \dot{\mathbf{E}} = 0,$$

from which follows, in the absence of internal constraints (i.e., with the several components of $\dot{\mathbf{E}}$ independent),

$$\mathbf{S} = -T\rho_0\partial\bar{\eta}/\partial\mathbf{E}.$$

With the absolute temperature used as an independent variable, the reduced local form of the second law can be expressed in terms of the Helmholtz free-energy density $\psi(T, \mathbf{E})$ in the form

$$(\mathbf{S} - \rho_0\partial\psi/\partial\mathbf{E}) : \dot{\mathbf{E}} = 0, \tag{8.2.1}$$

and yields the Lagrangian finite-deformation analogue of Equation (1.4.3), namely

$$\mathbf{S} = \rho_0\partial\psi/\partial\mathbf{E}. \tag{8.2.2}$$

From this we can in turn obtain the relation for the Cauchy stress:

$$\mathbf{T} = \rho\mathbf{F}(\partial\psi/\partial\mathbf{E})\mathbf{F}^T. \tag{8.2.3}$$

Incompressible Elastic Continuum

If the continuum is subject to an internal constraint given by $\mathbf{G} : \dot{\mathbf{E}} = 0$, which may or may not be derivable from a holonomic constraint of the form $h(T, \mathbf{E}) = 0$, then Equation (8.2.1) may be changed to

$$(\mathbf{S} - \rho_0 \partial\psi/\partial\mathbf{E} + p\mathbf{G}) : \dot{\mathbf{E}} = 0,$$

where p is an undetermined Lagrangian multiplier. Equation (8.2.2) is accordingly replaced by

$$\mathbf{S} = -p\mathbf{G} + \rho_0 \partial\psi/\partial\mathbf{E}.$$

In particular, the incompressibility constraint $J = 1$ is equivalent to

$$\dot{J} = (\partial J/\partial\mathbf{E}) : \dot{\mathbf{E}} = \mathbf{C}^{-1} : \dot{\mathbf{E}} = 0.$$

The stress-deformation relations now become

$$\mathbf{S} = -p\mathbf{C}^{-1} + \rho\partial\psi/\partial\mathbf{E},$$

$$\mathbf{T} = -p\mathbf{I} + \rho\mathbf{F}(\partial\psi/\partial\mathbf{E})\mathbf{F}^T,$$

so that p is an indeterminate pressure (note that $\rho = \rho_0$ in this case).

The constitutive relations take a special form when the continuum is isotropic in the reference configuration, that is, when $\psi(T, \mathbf{E})$ is invariant under a rotation of the Lagrangian basis. In such a case ψ can depend on $\mathbf{E}$ only through a set of invariants, for example, the principal invariants of $\mathbf{C}$, defined in the same way as for stress in Section 1.3, namely $I_1 = \operatorname{tr}\mathbf{C}$, $I_2 = \frac{1}{2}(\mathbf{C} : \mathbf{C} - I_1^2)$, $I_3 = \det\mathbf{C}$. In the incompressible case the last of these is unity, and therefore I_1 and I_2 are the only independent deformation variables; moreover, it can be shown that in this case, $I_2 = -\operatorname{tr}\mathbf{C}^{-1}$. Consequently,

$$\partial\psi/\partial\mathbf{E} = 2\partial\psi/\partial\mathbf{C} = 2\frac{\partial\psi}{\partial I_1}\mathbf{I} + 2\frac{\partial\psi}{\partial I_2}\mathbf{C}^{-2}.$$

Since $\mathbf{F}\mathbf{I}\mathbf{F}^T = \mathbf{F}\mathbf{F}^T = \mathbf{B}$ and $\mathbf{F}\mathbf{C}^{-2}\mathbf{F}^T = \mathbf{B}^{-1}$, we obtain

$$\mathbf{T} = -p\mathbf{I} + 2\rho\left(\frac{\partial\psi}{\partial I_1}\mathbf{B} + \frac{\partial\psi}{\partial I_2}\mathbf{B}^{-1}\right).$$

Finally, noting that the invariants of $\mathbf{C}$ are the same as those of $\mathbf{B}$, the last relation may be written more concisely as

$$\mathbf{T} = -p\mathbf{I} + 2\rho(\partial\psi/\partial\mathbf{B})\mathbf{B}.$$

Compressible Elastic Continuum

For the compressible continuum, rather than dealing directly with the tensors $\mathbf{E}$ or $\mathbf{C}$, it is more convenient to separate volume deformation from distortion. Let us recall that the infinitesimal strain tensor $\boldsymbol{\varepsilon}$ can be decomposed additively as $\boldsymbol{\varepsilon} = \varepsilon_v \mathbf{I} + \mathbf{e}$, where $\varepsilon_v = \frac{1}{3}\mathrm{tr}\,\boldsymbol{\varepsilon}$ is the volume strain and $\mathbf{e}$ is the strain deviator (see, for example, Equation (1.4.16) for the strain-energy function of an isotropic, linearly elastic continuum). The corresponding decomposition for finite deformation is not additive but multiplicative, since volume deformation is defined by $J = \sqrt{\det \mathbf{C}}$. In particular, let us define

$$\bar{\mathbf{F}} = J^{-1/3}\mathbf{F}, \quad \bar{\mathbf{B}} = J^{-2/3}\mathbf{B}, \quad \bar{\mathbf{C}} = J^{-2/3}\mathbf{C}, \quad \bar{\mathbf{E}} = \frac{1}{2}(\bar{\mathbf{C}} - \mathbf{I});$$

note that $\det \bar{\mathbf{C}} = 1$. If we now set the Helmholtz free energy as $\psi = \psi(T, J, \bar{\mathbf{E}})$, then we may write

$$\mathbf{S} = \rho_0 \left[\frac{\partial \psi}{\partial v} J\mathbf{C}^{-1} + J^{-2/3}\mathrm{Dev}\,(\partial\psi/\partial\bar{\mathbf{E}}) \right],$$

$$\mathbf{T} = \rho \left[J\frac{\partial \psi}{\partial J}\mathbf{I} + \mathrm{dev}\,(\bar{\mathbf{F}}\partial\psi/\partial\bar{\mathbf{E}}\bar{\mathbf{F}}^T) \right];$$

here dev is the ordinary deviator operator, that is, $\mathrm{dev}\,\mathbf{H} = \mathbf{H} - \frac{1}{3}\mathrm{tr}\,\mathbf{H}$, while Dev is a "Lagrangian" deviator operator defined by $\mathrm{Dev}\,\mathbf{H} = \mathbf{H} - \frac{1}{3}(\mathbf{H} : \mathbf{C})\mathbf{C}^{-1}$. Note that for the Cauchy stress, the deviator is determined by the dependence on the distortion while the pressure is determined by the dependence on the volume deformation. If the free energy depends only on J then the continuum is a fluid.

Finally, for the isotropic continuum, with $\psi = \psi(T, J, \bar{\mathbf{B}})$, the stress is given by

$$\mathbf{T} = \rho \left[J\frac{\partial \psi}{\partial J}\mathbf{I} + 2\,\mathrm{dev}\,(\bar{\mathbf{B}}\partial\psi/\partial\bar{\mathbf{B}}) \right],$$

and if the invariants $\bar{I}_1$ and $\bar{I}_2$ are introduced, by

$$\mathbf{T} = \rho \left[J\frac{\partial \psi}{\partial v}\mathbf{I} + 2\,\mathrm{dev}\left(\frac{\partial \psi}{\partial \bar{I}_1}\mathbf{B} + \frac{\partial \psi}{\partial \bar{I}_2}\mathbf{B}^{-1} \right) \right].$$

8.2.2 Inelasticity: Kinematics

Multiplicative Decomposition

In Section 1.5 we introduced the additive decomposition of the infinitesimal strain tensor into elastic and inelastic (later renamed plastic) parts, which is basic to virtually all the subsequently developed theory. For strains that are too large to be treated as infinitesimal, we found in 2.1.2 that the decomposition still works, at least for longitudinal strain, when this is taken as the logarithmic strain. In order to formulate a theory of plasticity for large deformation, we have to establish the appropriate decomposition for arbitrary states of deformation.

An additive decomposition of the logarithmic strain is equivalent to a *multiplicative* decomposition of the stretch λ, that is, $\lambda = \lambda^e \lambda^p$. Let us consider, first, problems of three-dimensional deformation in which the principal directions of strain are known and remain constant and in which, in addition, the principal directions of elastic and plastic strain may be assumed to coincide. In such problems we may, with no loss of generality, let $\mathbf{R} = \mathbf{I}$, where $\mathbf{R}$ is the rotation tensor defined in 8.1.1, and apply the multiplicative decomposition to each of the principal stretches. The result can be written as $\mathbf{U} = \mathbf{U}_e\mathbf{U}_p = \mathbf{U}_p\mathbf{U}_e$ or as $\mathbf{V} = \mathbf{V}_e\mathbf{V}_p = \mathbf{V}_p\mathbf{V}_e$; here $\mathbf{U}$ and $\mathbf{V}$ are respectively the right and left stretch tensors defined in 8.1.1 and Exercise 1 of Section 8.1, and in the class of problems under discussion they coincide.

These formulas, of course, cannot be valid in the general case: if $\mathbf{U}_e$ and $\mathbf{U}_p$ are somehow defined but are not coaxial, then $\mathbf{U}_e\mathbf{U}_p \neq \mathbf{U}_p\mathbf{U}_e$. Furthermore, neither product is in general symmetric and therefore cannot equal $\mathbf{U}$. The same reasoning applies to $\mathbf{V}$. What form, then, should the multiplicative decomposition take in general?

An answer to this question may be based on a theory first explicitly formulated by Kröner [1960], and further developed by Lee and Liu [1967], Fox [1968], Lee [1969], Mandel [1972], and others on the basis of considerations of the behavior of crystals. The reasoning behind it goes like this: if an infinitesimal neighborhood $(\mathbf{x}, \mathbf{x}+, d\mathbf{x})$ (by which we mean the points contained between x_i and $x_i + dx_i$, $i = 1, 2, 3$) in a plastically deformed crystal could be cut out and instantaneously relieved of all stresses, it would be mapped into $(\hat{\mathbf{x}}, \hat{\mathbf{x}} + d\hat{\mathbf{x}})$, the transformation being composed of a rigid-body displacement and a purely elastic deformation ("unloading").[1] The position $\hat{\mathbf{x}}$ is arbitrary, but we may assume a linear relation between $d\mathbf{x}$ and $d\hat{\mathbf{x}}$, that is, $d\hat{\mathbf{x}} = \mathbf{F}_e^{-1}\, d\mathbf{x}$. Here $\mathbf{F}_e$ (or $\mathbf{F}_e^{-1}$) is not to be interpreted as a deformation gradient, since the transformation is not a displacement of the whole body; that is, it is *not* assumed that there exists a differentiable function $\hat{\mathbf{x}}(\mathbf{x})$ such that $\mathbf{F}_e^{-1} = \nabla\hat{\mathbf{x}}$. Nevertheless, since $\mathbf{F}_e$ is nonsingular, it may be subjected to the polar decompositions $\mathbf{F}_e = \mathbf{R}_e\mathbf{U}_e = \mathbf{V}_e\mathbf{R}_e$. The tensors $\mathbf{U}_e$ and $\mathbf{V}_e$ represent the elastic deformation that is removed along with the removal of the stresses. We may accordingly define the elastic right Cauchy–Green tensor $\mathbf{C}_e = \mathbf{U}_e^2 = \mathbf{F}_e^T\mathbf{F}_e$ and the elastic Green–Saint-Venant strain $\mathbf{E}_e = \frac{1}{2}(\mathbf{C}_e - \mathbf{I})$.

Now consider a stress-free reference configuration of the crystal, such that $d\mathbf{x} = \mathbf{F}\, d\mathbf{X}$; then $d\hat{\mathbf{x}} = \mathbf{F}_e^{-1}\mathbf{F}\, d\mathbf{X}$. We define $\mathbf{F}_p \stackrel{\text{def}}{=} \mathbf{F}_e^{-1}\mathbf{F}$, and accordingly write

$$\mathbf{F} = \mathbf{F}_e\mathbf{F}_p. \tag{8.2.4}$$

[1] As pointed out by Mandel [1972], the elastic unloading operation has to be regarded as a fictitious one in a material with a strong Bauschinger effect, in which real unloading produces reverse plastic deformation,

The right Cauchy–Green tensor $\mathbf{C}$ is then given by

$$\mathbf{C} = \mathbf{F}_p^T \mathbf{C}_e \mathbf{F}_p. \tag{8.2.5}$$

The transformations involved in the decomposition are shown schematically in Figure 8.2.1; the unloaded element $(\hat{\mathbf{x}}, \hat{\mathbf{x}} + d\hat{\mathbf{x}})$ is usually said to be in an "intermediate configuration."

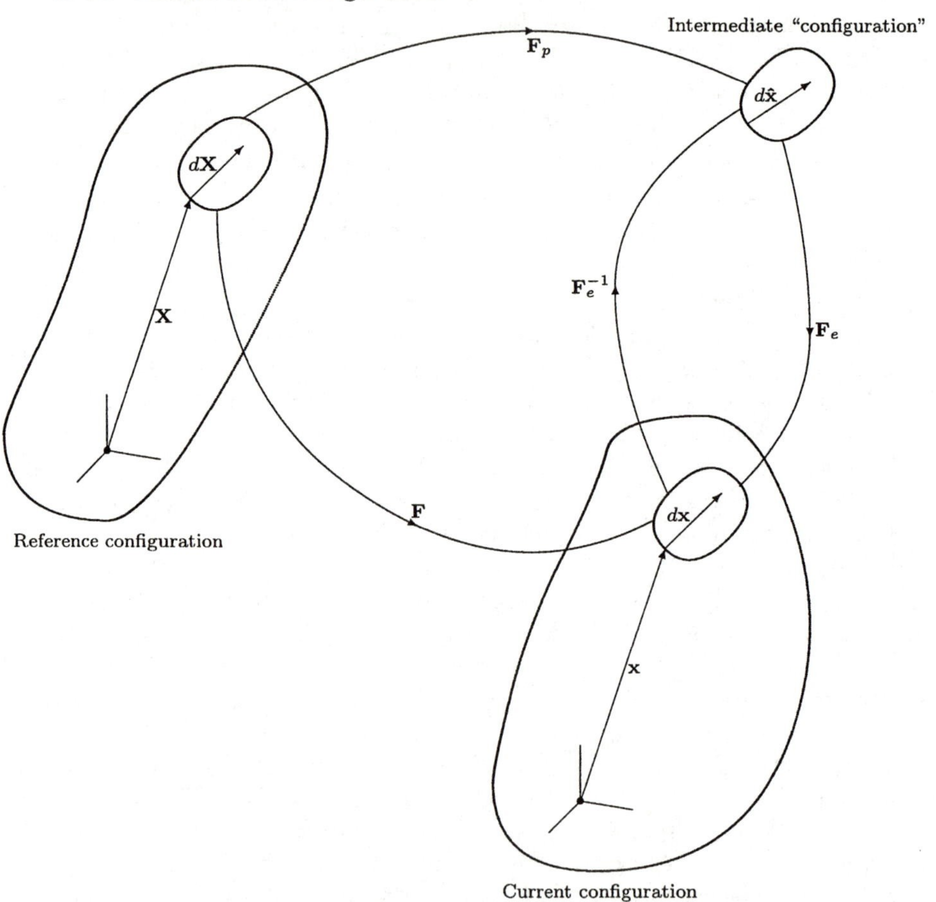

Figure 8.2.1. Multiplicative decomposition of deformation.

If we define $J_e = \det \mathbf{F}_e$ and $J_p = \det \mathbf{F}_p$, then

$$J = J_e J_p$$

and

$$J_e = \rho_p/\rho, \quad J_p = \rho_0/\rho_p,$$

where ρ_p is the density in the intermediate configuration. In metals, plastic deformation is in general virtually isochoric, so that J_p may be taken as equal to unity.

The Invariance Question

The process that moves the body from the reference configuration to the current one is responsible, in general, for both elastic and plastic deformations. By hypothesis, the elastic deformation is removed in the fictitious local unloading process producing the intermediate configuration. Consequently, the intermediate configuration may be said to be only plastically deformed, with the effects of the internal processes leading to plastic deformation embodied in $\mathbf{F}_p$. However, $\mathbf{F}_p$ represents more than just plastic deformation: if the polar decomposition $\mathbf{F}_p = \mathbf{R}_p\mathbf{U}_p$ is performed, then the deformation is represented by $\mathbf{U}_p$, or equivalently by $\mathbf{C}_p = \mathbf{U}_p^2 = \mathbf{F}_p^T\mathbf{F}_p$ or $\mathbf{E}_p = \frac{1}{2}(\mathbf{C}_p - \mathbf{I})$, while $\mathbf{R}_p$ represents *plastic rotation*.

Since the intermediate configuration is a fictitious one, its orientation is in principle arbitrary. Consequently it may be argued that if the current state is described by tensors based in the intermediate configuration, then the description should be invariant under an arbitrary rotation of this configuration, $d\hat{\mathbf{x}}^* = \mathbf{Q}\, d\hat{\mathbf{x}}$. If this is so, then $\mathbf{F}_p$ can influence the state only through $\mathbf{U}_p$ (or, equivalently, $\mathbf{C}_p$ or $\mathbf{E}_p$). Note that $\mathbf{E}_e$ is not invariant under such a rotation (the proof is left to an exercise), and therefore, under the present hypothesis, it cannot serve as a Lagrangian measure of elastic strain. The description of the state by the total strain $\mathbf{E}$, a plastic strain $\mathbf{E}_p$ (not necessarily defined by means of the multiplicative decomposition), and possibly other (scalar) internal variables, in addition to the temperature, has its origin in the work of Green and Naghdi [1965, 1971], and has more recently been discussed by Casey and Naghdi [1980, 1983], Simo and Ortiz [1985], Casey [1985], and Rubin [1986].

On the other hand, it has been argued by many other workers[1] that it is precisely the tensor $\mathbf{E}_e$ which, along with the hardening variables and the temperature, determines the current thermomechanical state. The argument may be based on the physics of crystal plasticity, with $\mathbf{E}_e$ taken as representing the distortion of the lattice, while the hardening variables represent dislocation densities and such. A consequence of this argument is that $\mathbf{F}_p$ cannot be a state variable if $\mathbf{E}_e$ is one (if $\mathbf{E}$ or $\mathbf{C}$ is used as a state variable, then $\mathbf{F}_p$ enters through the definition of $\mathbf{E}_e$). The physical aspect of the argument is as follows: a perfectly plastic crystal specimen that is loaded until it yields, plastically deformed, and finally unloaded is mechanically undistinguishable from its original state — in other words, plastic deformation alone (in the absence of hardening) does not produce a change of state.

[1]For example Lee [1969], Mandel [1972], Kratochvíl [1973], Dashner [1986], Stickforth [1986].

If $\mathbf{E}_e$ is an essential Lagrangian state variable, then the orientation of the intermediate configuration cannot be arbitrary (Mandel [1972]), but must be related to a "director frame" attached to the lattice. An intermediate configuration that rotates along with a director frame is called *isoclinic* (Mandel [1973]). If $\mathbf{F}_p$ is defined with respect to an isoclinic intermediate configuration, then it is a Lagrangian tensor in the sense of being invariant under a rotation of the Eulerian frame.

The two arguments can be reconciled under the following circumstances:

1. If the dependence of the thermomechanical state functions on $\mathbf{E}_e$ is isotropic, then it is equivalent to an isotropic dependence on $\mathbf{B}_e$, since $\mathbf{B}_e = \mathbf{R}_e\mathbf{C}_e\mathbf{R}_e^T$. But it is also equivalent to an isotropic dependence on $\mathbf{U}\mathbf{C}_p^{-1}\mathbf{U}$, since

$$\mathbf{B}_e = \mathbf{F}\mathbf{F}_p^{-1}\mathbf{F}_p^{T-1}\mathbf{F}^T = \mathbf{R}\mathbf{U}\mathbf{C}_p^{-1}\mathbf{U}\mathbf{R}^T,$$

and this is a special case of a dependence on $\mathbf{C}$ and $\mathbf{C}_p$, and therefore on $(\mathbf{E}, \mathbf{E}_p)$.

2. If it is assumed that $\mathbf{R}_p = \mathbf{I}$ (Haupt [1985]), then $\mathbf{C}_e = \mathbf{U}_p^{-1}\mathbf{C}\mathbf{U}_p^{-1}$, another special case.

Plastic Deformation Rates

The velocity gradient $\mathbf{L}$ is given by

$$\mathbf{L} = \mathbf{L}_e + \mathbf{F}_e\mathbf{L}_p\mathbf{F}_e^{-1},$$

where

$$\mathbf{L}_e = \dot{\mathbf{F}}_e\mathbf{F}_e^{-1}, \quad \mathbf{L}_p = \dot{\mathbf{F}}_p\mathbf{F}_p^{-1}.$$

Consequently,

$$\dot{\mathbf{C}} = \mathbf{F}_p^T[\dot{\mathbf{C}}_e + 2(\mathbf{C}_e\mathbf{L}_p)^S]\mathbf{F}_p, \tag{8.2.6}$$

with $\dot{\mathbf{C}}_e = 2\mathbf{F}_e^T\mathbf{D}_e\mathbf{F}_e$, $\mathbf{D}_e = \mathbf{L}_e^S$. We see that $\dot{\mathbf{C}}$ depends on $\mathbf{L}_e$ only through its symmetric part, and on $\mathbf{L}_p$ through the symmetric part of $\mathbf{C}_e\mathbf{L}_p$. The Eulerian deformation-rate tensor $\mathbf{D}$ is

$$\mathbf{D} = \mathbf{D}_e + (\mathbf{F}_e\mathbf{L}_p\mathbf{F}_e^{-1})^S.$$

The tensor $\mathbf{L}_p$ is referred to by Mandel [1972] as the "plastic distortion rate" and is related to the dislocation motion as follows: if the ith slip system consists of dislocations moving in the glide plane whose unit normal is $\mathbf{n}_i$, in the direction of the unit vector $\mathbf{m}_i$, and producing the shear rate $\dot{\gamma}_i$, then

$$\mathbf{L}_p = \sum_i \dot{\gamma}_i\mathbf{m}_i \otimes \mathbf{n}_i, \tag{8.2.7}$$

the summation being over all slip systems.

As regards a "plastic deformation rate" $\mathbf{D}_p$, however, there is no unequivocal definition; among the ones that have been proposed are $\mathbf{D} - \mathbf{D}_e =$

$(\mathbf{F}_e\mathbf{L}_p\mathbf{F}_e^{-1})^S$, $\mathbf{L}_p^S$, and $(\mathbf{C}_e\mathbf{L}_p)^S$. The first of these tensors is Eulerian. The other two are defined in the intermediate configuration, and are therefore Lagrangian tensors if this configuration is isoclinic.

8.2.3 Inelasticity: Thermomechanics

Internal Variables

Let $\boldsymbol{\alpha}$ denote the internal-variable vector describing dislocation densities, point-defect densities and other structural properties. In accordance with the argument of Mandel and others, the local thermomechanical state is then determined by $(\mathbf{E}_e, \boldsymbol{\alpha}, T)$, while in the hypothesis of Green and Naghdi it is determined by $(\mathbf{E}, \mathbf{E}_p, \boldsymbol{\alpha}, T)$. Mathematically, both postulates are special cases of of $(\mathbf{E}, \mathbf{F}_p, \boldsymbol{\alpha}, T)$, or of $(\mathbf{E}, \boldsymbol{\xi}, T)$ if we define $\boldsymbol{\xi} = (\mathbf{F}_p, \boldsymbol{\alpha})$ as the *apparent* internal-variable vector. Furthermore, if equations for the shear rates $\dot{\gamma}_i$ are given, then Equation (8.2.7) constitutes a rate equation for $\mathbf{F}_p$. The description of the state by $(\mathbf{E}, \boldsymbol{\xi}, T)$ is the basic one used in the remainder of this section, with the special cases invoked as appropriate.

Stress

The deformation power per unit mass is, as a result of Equation (8.2.6),

$$\frac{1}{\rho_0}\mathbf{S} : \dot{\mathbf{E}} = \frac{1}{\rho_0}(\mathbf{S}_e : \dot{\mathbf{E}}_e + \boldsymbol{\Sigma} : \mathbf{L}_p), \tag{8.2.8}$$

where

$$\mathbf{S}_e = \mathbf{F}_p\mathbf{S}\mathbf{F}_p^T, \quad \boldsymbol{\Sigma} = \mathbf{C}_e\mathbf{S}_e = J\mathbf{F}_e^T\mathbf{T}\mathbf{F}_e^{T-1}$$

are stress tensors defined in the intermediate configuration. Both $\mathbf{S}_e$ and $\boldsymbol{\Sigma}$ differ from the *Kirchhoff stress* (weighted Cauchy stress) tensor $\mathbf{P} = J\mathbf{T}$ by quantities of order $|\mathbf{F}_e - \mathbf{I}|$. Furthermore, $\mathbf{S}_e$ is a symmetric tensor that is conjugate to the elastic strain tensor $\mathbf{E}_e$, and $J_p^{-1}\mathbf{S}_e$ may be regarded as the *second Piola–Kirchhoff stress with respect to the intermediate configuration.* The tensor $\boldsymbol{\Sigma}$, which was introduced by Mandel [1972], is not in general symmetric.

Free-Energy Density

In accordance with the two arguments cited above, the free-energy density is given either by the form favored by Mandel and others, namely,

$$\psi = \bar{\psi}(\mathbf{E}_e, \boldsymbol{\alpha}, T), \tag{8.2.9}$$

or by the Green–Naghdi form,[1]

$$\psi = \tilde{\psi}(\mathbf{E}, \mathbf{E}_p, \boldsymbol{\alpha}, T), \tag{8.2.10}$$

[1]In the work of Green and Naghdi and their followers, $\boldsymbol{\alpha}$ is usually taken to consist of a single scalar variable κ.

both forms being special cases of $\psi(\mathbf{E}, \mathbf{F}_p, \boldsymbol{\alpha}, T)$ and therefore of $\psi(\mathbf{E}, \boldsymbol{\xi}, T)$.

If Equation (8.2.2) is assumed to be valid for a rate-independent plastic material, then, with the free-energy density given by (8.2.9), it can be shown to be equivalent to

$$\mathbf{S}_e = \rho_0 \partial\bar{\psi}/\partial\mathbf{E}_e; \tag{8.2.11}$$

the proof is left to an exercise. By analogy with the uncoupled form (1.5.4) for the free-energy density under infinitesimal deformation, the function $\bar{\psi}$ in Equation (8.2.9) is often restricted even further to be given by

$$\bar{\psi}(\mathbf{E}_e, \boldsymbol{\alpha}, T) = \psi_e(\mathbf{E}_e, T) - T\eta_p(\boldsymbol{\alpha}) + u_p(\boldsymbol{\alpha}). \tag{8.2.12}$$

Here ψ_e is the thermoelastic free energy, usually identified with the lattice energy, while u_p and η_p are respectively the stored energy and the configurational entropy due to dislocations and other relevant lattice defects. It follows that $\mathbf{S}_e = \rho_0 \partial\psi_e/\partial\mathbf{E}_e$; this means that at a given temperature, there is a one-to-one correspondence between $\mathbf{S}_e$ and $\mathbf{E}_e$, as well as between $\boldsymbol{\Sigma}$ and $\mathbf{E}_e$ (or $\mathbf{C}_e$) in view of the relation $\boldsymbol{\Sigma} = \mathbf{C}_e\mathbf{S}_e$. As was remarked above, $\boldsymbol{\Sigma}$ is not in general a symmetric tensor; however, $\mathbf{C}_e^{-1}\boldsymbol{\Sigma} = \mathbf{S}_e$ is symmetric, as is $\boldsymbol{\Sigma}\mathbf{C}_e = \mathbf{C}_e\mathbf{S}_e\mathbf{C}_e$. Consequently $\boldsymbol{\Sigma}$ must obey a constraint, for example

$$(\boldsymbol{\Sigma}\mathbf{C}_e)^T = \boldsymbol{\Sigma}\mathbf{C}_e.$$

With $\mathbf{C}_e$ a function of $\boldsymbol{\Sigma}$, such a constraint is equivalent to three nonlinear equations involving the components of $\boldsymbol{\Sigma}$, limiting $\boldsymbol{\Sigma}$ to a six-dimensional manifold in the nine-dimensional space of second-rank tensors. If $\mathbf{C}_e = \mathbf{I}$, then the equations are linear, and the manifold becomes "flat" — namely, the space of symmetric second-rank tensors. If $\mathbf{E}_e$ is small, then the manifold is "slightly curved" or "almost flat."

The Green–Naghdi form (8.2.10) for the free-energy density is often rewritten, with no loss in generality, as

$$\psi = \hat{\psi}(\mathbf{E} - \mathbf{E}_p, \mathbf{E}_p, \kappa, T), \tag{8.2.13}$$

with further specializations for "special materials" (Casey and Naghdi [1981]). While $\mathbf{E} - \mathbf{E}_p$ was identified with elastic strain in the original work of Green and Naghdi [1965], in later work (Green and Naghdi [1971]) this identification was dropped. Nevertheless, the decomposition

$$\mathbf{E} = \mathbf{E}_e + \mathbf{E}_p$$

is not infrequently used, for example by Simo and Ortiz [1985]. Casey [1985] justifies it as an approximation that is valid when (i) small plastic deformations are accompanied by moderate elastic strains, (ii) small elastic strains are accompanied by moderate plastic deformations, or (iii) small strains are accompanied by moderate rotations.

Elastic Tangent Stiffness

Let $\mathbf{\Gamma} = \partial\mathbf{S}/\partial\mathbf{E}$ denote the usual "Lagrangian" elastic tangent stiffness tensor. With the free-energy density given by (8.2.9), the elastic tangent stiffness tensor in the intermediate configuration may be defined as $\mathbf{\Gamma}^e = \partial\mathbf{S}_e/\partial\mathbf{E}_e = \rho_0\partial^2\bar{\psi}/\partial\mathbf{E}_e\partial\mathbf{E}_e$. The relation between $\mathbf{\Gamma}$ and $\mathbf{\Gamma}^e$ is

$$\mathbf{\Gamma}^e : \mathbf{A} = \mathbf{F}_p[\mathbf{\Gamma} : (\mathbf{F}_p^T\mathbf{A}\mathbf{F}_p)]\mathbf{F}_p^T, \tag{8.2.14}$$

where $\mathbf{A}$ is any symmetric second-rank tensor defined in the intermediate configuration, and $\mathbf{\Gamma}^e : \mathbf{A}$ denotes the tensor with components $\Gamma^e_{IJKL}A_{KL}$; the derivation of the relation (8.2.14) is left to an exercise. With the free-energy density given by Equation (8.2.12), we have $\mathbf{\Gamma}^e = \rho_0\partial^2\psi_e/\partial\mathbf{E}_e\partial\mathbf{E}_e$.

Plastic Dissipation

With the identification $\boldsymbol{\xi} = (\mathbf{F}_p, \boldsymbol{\alpha})$, the Kelvin inequality (1.5.3) becomes

$$D = -\rho_0[(\partial\psi/\partial\mathbf{F}_p) : \dot{\mathbf{F}}_p + (\partial\psi/\partial\boldsymbol{\alpha}) \cdot \dot{\boldsymbol{\alpha}}] \geq 0.$$

With ψ given by Equation (8.2.9), this becomes

$$D = -\rho_0[(\partial\bar{\psi}/\partial\mathbf{C}_e) : (\partial\mathbf{C}_e/\partial\mathbf{F}_p) : \dot{\mathbf{F}}_p + (\partial\bar{\psi}/\partial\boldsymbol{\alpha}) \cdot \dot{\boldsymbol{\alpha}}] \geq 0.$$

But $(\partial\mathbf{C}_e/\partial\mathbf{F}_p) : \dot{\mathbf{F}}_p = -2(\mathbf{C}_e\mathbf{L}_p)^S$ and $\partial\bar{\psi}/\partial\mathbf{C}_e = \mathbf{F}_p(\partial\psi/\partial\mathbf{C})\mathbf{F}_p^T = (1/2\rho_0)\mathbf{S}_e$. Consequently,

$$D = D_p - \rho_0(\partial\bar{\psi}/\partial\boldsymbol{\alpha}) \cdot \dot{\boldsymbol{\alpha}} \geq 0,$$

where D_p is the *plastic dissipation* per unit mass, variously given by

$$D_p = \mathbf{\Sigma} : \mathbf{L}_p = \mathbf{S}_e : (\mathbf{C}_e\mathbf{L}_p)^S = \mathbf{P} : (\mathbf{F}_e\mathbf{L}_p\mathbf{F}_e^{-1})^S.$$

With $\bar{\psi}$ given by Equation (8.2.12), the total dissipation per unit mass is

$$D = D_p + \rho_0 T\dot{\eta}_p - \rho_0\dot{u}_p \geq 0.$$

When the free-energy density is given by Equation (8.2.13), the dissipation is

$$D = \mathbf{S} : \dot{\mathbf{E}}_p - \rho_0\left[(\partial\hat{\psi}/\partial\mathbf{E}_p) : \dot{\mathbf{E}}_p + \frac{\partial\hat{\psi}}{\partial\kappa}\dot{\kappa}\right].$$

8.2.4 Yield Condition and Flow Rule

In the theory of Green and Naghdi, the yield function is given as $f(\mathbf{S}, T, \mathbf{E}_p, \kappa)$. The flow rule is given as a rate equation for $\mathbf{E}_p$, and the development of the theory is very much like that of infinitesimal-deformation plasticity.

Mandel [1972] assumed the yield function in stress space as $\tilde{f}(\boldsymbol{\Sigma}, \boldsymbol{\alpha}, T)$. He further established the following version of the maximum-plastic-dissipation postulate (3.2.4):

$$(\boldsymbol{\Sigma} - \boldsymbol{\Sigma}^*) : \mathbf{L}_p \geq 0 \text{ for any } \boldsymbol{\Sigma}^* \text{ such that } \tilde{f}(\boldsymbol{\Sigma}^*, \boldsymbol{\alpha}, T) \leq 0, \tag{8.2.15}$$

from which he derived the *nine-dimensional* associated flow rule

$$\mathbf{L}_p = \phi \partial \tilde{f}/\partial \boldsymbol{\Sigma}, \quad \phi \geq 0.$$

This result implies that not only the plastic deformation but also the plastic rotation $\mathbf{R}_p$ are determined by the same flow rule.

When it is remembered that $\boldsymbol{\Sigma}$, although not in general symmetric, is nonetheless limited to a six-dimensional manifold, then the validity of this result may be questioned (Lubliner [1986]). Let the yield function in strain space be defined as

$$\bar{f}(\mathbf{E}, \boldsymbol{\xi}, T) = \tilde{f}(\boldsymbol{\Sigma}(\mathbf{E}, \boldsymbol{\xi}, T), \boldsymbol{\alpha}, T),$$

and the total specific dissipation as

$$D(\mathbf{E}, \boldsymbol{\xi}, \dot{\boldsymbol{\xi}}, T) = -(\partial \psi/\partial \boldsymbol{\xi}) \cdot \dot{\boldsymbol{\xi}}.$$

With the free-energy density given by (8.2.8), it can easily be shown that inequality (8.2.15) is equivalent to the large-deformation analogue of (3.2.12), namely

$$D(\mathbf{E}, \boldsymbol{\xi}, \dot{\boldsymbol{\xi}}, T) - D(\mathbf{E}^*, \boldsymbol{\xi}, \dot{\boldsymbol{\xi}}, T) \geq 0 \tag{8.2.16}$$

for any $\mathbf{E}^*$ such that $\bar{f}(\mathbf{E}^*, \boldsymbol{\xi}, T) \leq 0$. Suppose, in particular, that the process is isothermal, that it is elastic from $(\mathbf{E}^*, \boldsymbol{\xi}, T)$ to $(\mathbf{E}, \boldsymbol{\xi}, T)$, and that its continuation from $(\mathbf{E}, \boldsymbol{\xi}, T)$ is plastic; the time derivatives of $(\mathbf{E}, \boldsymbol{\xi}, T)$ at $(\mathbf{E}, \boldsymbol{\xi}, T)$ are $(\dot{\mathbf{E}}, \dot{\boldsymbol{\xi}}, 0)$. If $\mathbf{E}^*$ is close to $\mathbf{E}$, then $\mathbf{E}^* = \mathbf{E} - h\dot{\mathbf{E}} + o(h)$ for some small positive h. For the left-hand side of (8.2.16) we therefore have

$$D(\mathbf{E}, \boldsymbol{\xi}, \dot{\boldsymbol{\xi}}, T) - D(\mathbf{E}^*, \boldsymbol{\xi}, \dot{\boldsymbol{\xi}}, T) = h(\partial D/\partial \mathbf{E}) : \dot{\mathbf{E}} + o(h),$$

where $\partial D/\partial \mathbf{E}$ is evaluated at $(\mathbf{E}, \boldsymbol{\xi}, \dot{\boldsymbol{\xi}}, T)$. Hence

$$(\partial D/\partial \mathbf{E}) : \dot{\mathbf{E}} \geq 0. \tag{8.2.17}$$

Furthermore,

$$\rho_0 \partial D/\partial \mathbf{E} = -\rho_0(\partial^2 \psi/\partial \mathbf{E} \partial \boldsymbol{\xi}) \cdot \dot{\boldsymbol{\xi}} = -(\partial \mathbf{S}/\partial \boldsymbol{\xi}) \cdot \dot{\boldsymbol{\xi}} = (\partial \mathbf{S}/\partial \mathbf{E}) : \dot{\mathbf{E}} - \dot{\mathbf{S}} = \boldsymbol{\Gamma} : \dot{\mathbf{E}}^{(p)}.$$

The last equality defines the *apparent Lagrangian plastic strain rate*

$$\dot{\mathbf{E}}^{(p)} \stackrel{\text{def}}{=} \dot{\mathbf{E}}|_{\dot{\mathbf{S}}=0, \dot{T}=0} = \dot{\mathbf{E}} - \boldsymbol{\Gamma}^{-1} : \dot{\mathbf{S}}.$$

The rate $\dot{\mathbf{E}}^{(p)}$, not to be confused with Green and Naghdi's $\dot{\mathbf{E}}_p$, is related to $\mathbf{L}_p$ by

$$\dot{\mathbf{E}}^{(p)} = \mathbf{F}_p^T[(\mathbf{C}_e\mathbf{L}_p)^S + 2\mathbf{\Gamma}^{e-1} : (\mathbf{L}_p\mathbf{S}_e)^S]\mathbf{F}_p \tag{8.2.18}$$

(Lubliner [1986]). Inequality (8.2.17) may consequently be written as

$$\dot{\mathbf{E}} : \mathbf{\Gamma} : \dot{\mathbf{E}}^{(p)} \geq 0, \tag{8.2.19}$$

the large-deformation counterpart to (3.2.15).

From the isothermal consistency condition in strain space,

$$\dot{\bar{f}} = (\partial\bar{f}/\partial\mathbf{E}) : \dot{\mathbf{E}} + (\partial\bar{f}/\partial\boldsymbol{\xi}) \cdot \dot{\boldsymbol{\xi}} = 0,$$

it follows, by analogy with the corresponding arguments in Section 3.2, that $\dot{\boldsymbol{\xi}} \neq 0$ only if $(\partial\bar{f}/\partial\mathbf{E}) : \dot{\mathbf{E}} > 0$, and this inequality is compatible with (8.2.17) only if

$$\dot{\mathbf{E}}^{(p)} = \phi\mathbf{\Gamma}^{-1} : (\partial\bar{f}/\partial\mathbf{E}), \quad \phi \geq 0. \tag{8.2.20}$$

The last result may be regarded as an associated flow rule in strain space. Let us now return to stress space. Suppose, first, that the yield surface is given in $\mathbf{S}$-space by $\hat{f}(\mathbf{S}, \boldsymbol{\xi}, T) = 0$. With $\mathbf{S} = \mathbf{S}(\mathbf{E}, \boldsymbol{\xi}, T)$, we have

$$\bar{f}(\mathbf{E}, \boldsymbol{\xi}, T) = \hat{f}(\mathbf{S}(\mathbf{E}, \boldsymbol{\xi}, T), \boldsymbol{\xi}, T),$$

so that $\partial f/\partial\mathbf{E} = \partial\hat{f}/\partial\mathbf{S} : \partial\mathbf{S}/\partial\mathbf{E} = \mathbf{\Gamma} : \partial\hat{f}/\partial\mathbf{S}$, since $\mathbf{\Gamma}$ is symmetric. Consequently, (8.2.20) is equivalent to

$$\dot{\mathbf{E}}^{(p)} = \phi\partial\hat{f}/\partial\mathbf{S}, \quad \phi \geq 0. \tag{8.2.21}$$

Next, suppose the yield surface to be given in $\mathbf{S}_e$ space by $f(\mathbf{S}_e, \boldsymbol{\xi}, T) = 0$; then $\hat{f}(\mathbf{S}, \boldsymbol{\xi}, T) = f(\mathbf{F}_p\mathbf{S}\mathbf{F}_p^T, \boldsymbol{\xi}, T)$, and therefore

$$\partial\hat{f}/\partial\mathbf{S} = \mathbf{F}_p^T(\partial f/\partial\mathbf{S}_e)\mathbf{F}_p. \tag{8.2.22}$$

Combining Equations (8.2.18), (8.2.21) and (8.2.22), we obtain

$$(\mathbf{C}_e\mathbf{L}_p)^S + 2\mathbf{\Gamma}^{e-1} : (\mathbf{L}_p\mathbf{S}_e)^S = \phi\partial f/\partial\mathbf{S}_e. \tag{8.2.23}$$

Finally, let $\mathbf{\Sigma}$ be the stress variable, as in Mandel's theory, with the previously considered yield function $\tilde{f}$. Since $f(\mathbf{S}_e, \boldsymbol{\xi}, T) = \tilde{f}(\mathbf{C}_e\mathbf{S}_e, \boldsymbol{\xi}, T)$, it can easily be shown by means of the chain rule that

$$\partial f/\partial\mathbf{S}_e = (\mathbf{C}_e\partial\tilde{f}/\partial\mathbf{\Sigma})^S + 2\mathbf{\Gamma}^{e-1} : [(\partial\tilde{f}/\partial\mathbf{\Sigma})\mathbf{S}_e]^S. \tag{8.2.24}$$

Now if the left-hand side of (8.2.23) is written as $\mathbf{\Delta} : \mathbf{L}_p$, $\mathbf{\Delta}$ being a fourth-rank tensor, then the right-hand side of (8.2.24) is just $\mathbf{\Delta} : (\partial\tilde{f}/\partial\mathbf{\Sigma})$. Combining the two equations we obtain $\mathbf{\Delta} : (\mathbf{L}_p - \phi\partial\tilde{f}/\partial\mathbf{\Sigma}) = 0$, or

$$\mathbf{L}_p = \phi\partial\tilde{f}/\partial\mathbf{\Sigma} + \mathbf{L}_p'', \tag{8.2.25}$$

where $\mathbf{L}_p''$ may be any tensor satisfying $\mathbf{\Delta} : \mathbf{L}_p'' = 0$. Within the nine-dimensional space of second-rank tensors, the tensors obeying this equation form a three-dimensional subspace. Therefore, the flow rule (8.2.25) determines only the projection of $\mathbf{L}_p$ into the complement of this subspace, itself a six-dimensional subspace. This last subspace is the so-called *cotangent space* of the aforementioned manifold containing $\mathbf{\Sigma}$; it is just the space that contains all possible values of $\partial\phi/\partial\mathbf{\Sigma}$ for continuously differentiable scalar-valued functions $\phi(\mathbf{\Sigma})$.

Negligible Elastic Strains

If the elastic strains are negligible next to unity, then the preceding theory can be considerably simplified. $(\mathbf{C}_e\mathbf{L}_p)^S$ is approximately equal to $\mathbf{L}_p^S$, and since $\mathbf{E}_e \doteq \mathbf{\Gamma}^{e-1} : \mathbf{S}_e$, the term $2\mathbf{\Gamma}^{e-1} : (\mathbf{L}_p\mathbf{S}_e)^S$ in (8.2.18) may be neglected. Consequently,

$$\dot{\mathbf{E}}^{(p)} \doteq \mathbf{F}_p^T\mathbf{L}_p^S\mathbf{F}_p = \dot{\mathbf{E}}_p,$$

and the "Lagrangian" flow rule (8.2.21) may be replaced by

$$\dot{\mathbf{E}}_p = \phi\,\partial\hat{f}/\partial\mathbf{S},$$

equivalent to the associated flow rule of the Green–Naghdi theory.

To formulate an approximate form in the intermediate configuration, we define the *rotated Kirchhoff stress* as $\tilde{\mathbf{P}} = J\mathbf{R}_e^T\mathbf{T}\mathbf{R}_e$; then $\mathbf{S}_e$ and $\mathbf{\Sigma}$ are both approximately equal to $\tilde{\mathbf{P}}$, and either (8.2.22) or (8.2.24) may be replaced by

$$\mathbf{L}_p^S = \phi\,\partial f/\partial\tilde{\mathbf{P}}.$$

An Eulerian version can be written if the dependence of f on $\tilde{\mathbf{P}}$ is isotropic, since this is then equivalent to an isotropic dependence on the Kirchhoff stress $\mathbf{P}$, and $\partial f/\partial\tilde{\mathbf{P}} = \mathbf{R}_e^T(\partial f/\partial\mathbf{P})\mathbf{R}_e$. Moreover, $\mathbf{D} - \mathbf{D}_e = (\mathbf{F}_e\mathbf{L}_p\mathbf{F}_e^{-1})^S \doteq \mathbf{R}_e\mathbf{L}_p^S\mathbf{R}_e^T$, and therefore the approximate associated flow rule reads

$$\mathbf{D} - \mathbf{D}_e = \phi\,\partial f/\partial\mathbf{P}. \qquad (8.2.26)$$

In addition, $J_e \doteq 1$, so that $J \doteq J_p$. If, as usual, plastic volume change is neglected, then $J \doteq 1$, and $\mathbf{P}$ can be replaced by $\mathbf{T}$. The resulting form of the associated flow rule is then the same as in small-deformation theory.

A similar result is obtained if the principal axes of strain do not rotate (i.e., if they are the same in the reference, intermediate, and current configurations). In this case $\mathbf{R} = \mathbf{R}_e = \mathbf{I}$, so that $\tilde{\mathbf{P}} = \mathbf{P} \doteq \mathbf{T}$, and the result is valid even if the yield function does not depend isotropically on $\tilde{\mathbf{P}}$. It is on this basis that large-deformation problems have been treated in previous chapters of this book.

While this subsection has been concerned with the possible forms taken by the associated flow rule, analogous forms may be used for a nonassociated flow rule if the various forms of the yield function f ($\tilde{f}$, $\bar{f}$, $\hat{f}$) are replaced by the corresponding forms of a plastic potential, say g. Finally, the results may be applied to viscoplastic flow laws if ϕ is replaced by a function of the state variables, g (or any of its variants) is a viscoplastic potential, and $\mathbf{F}_p$ is written as $\mathbf{F}_i$, $\mathbf{L}_p$ as $\mathbf{L}_i$, and so forth.

Exercises: Section 8.2

1. Show that if both sides of Equation (8.2.2) are differentiated with respect to time, then the result is equivalent to

$$\overset{\star}{\mathbf{P}} = \mathbf{\Lambda} : \mathbf{D},$$

where the left-hand side is the Jaumann rate of the Kirchhoff stress $\mathbf{P} = J\mathbf{T}$ and the Eulerian tangent elastic modulus tensor $\mathbf{\Lambda}$ is defined by

$$\Lambda_{ijkl} = \rho_0 F_{iI} F_{jJ} F_{kK} F_{lL} \frac{\partial^2 \psi}{\partial E_{IJ} \partial E_{KL}} + \tfrac{1}{2} J(\delta_{ik} T_{jl} + \delta_{il} T_{jk} + \delta_{jk} T_{il} + \delta_{jl} T_{ik}).$$

2. Derive Equation (8.2.6).

3. Show that if $\mathbf{E}_e$ is defined by means of the multiplicative decomposition, then it is not invariant under a rotation of the intermediate configuration.

4. Derive Equation (8.2.11).

5. Derive the relation (8.2.14) between $\mathbf{\Gamma}$ and $\mathbf{\Gamma}_e$.

6. Show that with the given assumptions on the free energy, inequality (8.2.16) is equivalent to (8.2.15).

7. Derive (8.2.18).

8. Show that if $|\mathbf{E}_e| \ll 1$, then $|\mathbf{\Delta} : \mathbf{L}_p - \mathbf{L}_p^S| \ll |\mathbf{L}_p|$, where $\mathbf{\Delta} : \mathbf{L}_p$ is the left-hand side of (8.2.23).

Section 8.3 Numerical Methods

As we saw in Section 4.5, numerical methods for rate-independent elastic–plastic solids are based on the rate form of the constitutive equations, Equations (4.5.5)–(4.5.7). It was natural that the extension of such methods to large-deformation problems, first undertaken in the late 1960s, would be based on some analogous form. While these *rate-based* formulations have been extensively used, they have been shown to possess some severe flaws, which are discussed in 8.3.1. An alternative "hyperelastic" formulation manages to avoid the difficulties posed by the rate-based formulations, though the versions available as of this writing rely on some simplifying assumptions on the constitutive equations. It is likely, however, that in future versions such assumptions will become unnecessary. The "hyperelastic" formulation is discussed in 8.3.2.

8.3.1 Rate-Based Formulations

Introduction

The earliest extensions to large deformations of the rate-based numerical methods for elastic–plastic solids were in Lagrangian form;[1] among the first was one by Hibbit, Marcal, and Rice [1970], who, on the basis of the large-deformation principle of virtual work, Equation (8.1.6), derived an incremental stiffness containing three parts in addition to the "small-strain" stiffness, namely the initial-load, initial-stress, and initial-strain stiffnesses.

Since, however, the problems to be solved have typically been ones of large-scale plastic flow, it was felt by many that an Eulerian form of the equations would be preferable. Most solution schemes have followed a formulation similar to that of McMeeking and Rice [1975], with the analogue of (4.5.5), in particular, usually taken as

$$\overset{\star}{\mathbf{P}} = \mathsf{C} : (\mathbf{D} - \mathbf{D}_p), \tag{8.3.1}$$

where the superposed asterisk denotes the Jaumann rate; this rate is generally preferred, in view of the properties discussed at the end of Section 8.1, though it was pointed out by Truesdell and Noll [1965, p. 404] that "despite claims and whole papers to the contrary, any advantage claimed for one such rate over another is pure illusion."

The Kirchhoff stress appears in Equation (8.3.1) because, unlike the Cauchy stress, its use leads to a symmetric global tangent stiffness matrix upon discretization. With isotropic elasticity assumed, as is usual, C is taken as given by (1.4.10), with the Lamé coefficients λ and μ constant. The plastic deformation rate $\mathbf{D}_p$ is almost invariably assumed to be governed by a flow rule that is associated (in Kirchoff-stress space) with a generalized Mises yield criterion that may incorporate kinematic hardening. Such a criterion takes the form

$$f(\mathbf{P}, \mathbf{H}, \kappa) = \frac{1}{2}(\mathbf{P} - \mathbf{H})' : (\mathbf{P} - \mathbf{H})' - [k(\kappa)]^2 = 0, \tag{8.3.2}$$

[1] Formulations based on convected coordinates, such as that due to Needleman [1972], are equivalent to Lagrangian ones.

where $(\cdot)' = \text{dev}\,(\cdot)$, $\mathbf{H}$ (replacing $\boldsymbol{\rho}$ in the small-deformation theory) is the tensor-valued internal variable denoting the center of the elastic region (the "back stress"), and κ is, as before, an internal variable representing isotropic hardening. The associated flow rule is then

$$\mathbf{D}_p = \phi(\mathbf{P} - \mathbf{H})', \tag{8.3.3}$$

where ϕ is determined by means of the consistency condition $\dot{f} = 0$. With appropriate rate equations for $\mathbf{H}$ and κ (unnecessary, of course, in the case of perfect plasticity), ϕ is found as a linear form in $\mathbf{D}$, and the subsequent substitution of (8.3.3) in (8.3.1) leads to

$$\overset{\star}{\mathbf{P}} = \mathbf{\Lambda} : \mathbf{D}, \tag{8.3.4}$$

where $\mathbf{\Lambda}$ is the Eulerian tangent modulus tensor analogous to C_{ep}.

The finite-element scheme is based on the rate form of the principle of virtual work, Equation (8.1.7), translated into Eulerian form. With the same notation as in Section 4.5, the discretization is

$$bv^h(\mathbf{x})\Big|_{\Omega^e} = \mathbf{N}^e(\mathbf{x})\dot{\mathbf{q}}^e, \qquad \mathbf{D}^h(\mathbf{x})\Big|_{\Omega^e} = \mathbf{B}^e(\mathbf{x})\dot{\mathbf{q}}^e.$$

The global rate equation then takes the form (4.5.28), with the load rate $\dot{\mathbf{F}}$ (not to be confused with the derivative of the deformation gradient appearing elsewhere in this chapter) defined by

$$\dot{\mathbf{F}} = \sum_e \mathbf{A}^{eT} \left[\int_{\Omega^e} \mathbf{N}^{eT} \dot{\mathbf{f}}\, d\Omega + \int_{\partial\Omega_t^e} \mathbf{N}^{eT} \dot{\mathbf{t}}^a \, d\Gamma \right],$$

while the tangent stiffness $\mathbf{K}_t$ is given by

$$\mathbf{K}_t = \sum_e \mathbf{A}^{eT} \left(\int_{\Omega^e} \mathbf{B}^{eT} \mathbf{\Lambda} \mathbf{B}^e \, d\Omega + \mathbf{K}_s^e \right) \mathbf{A}^e.$$

Here $\mathbf{K}_s^e$ is the additional element stiffness (the initial-stress stiffness) due to the various terms resulting from the conversion of the equations to Eulerian form.

The formulation given above runs into a number of difficulties. When the rate problem has been solved, the imposition of a load increment requires an integration in time that maintains objectivity, necessitating *incrementally objective* integration algorithms (Rubinstein and Atluri [1983], Pinsky, Ortiz, and Pister [1983]) that may be computationally expensive. The rate-based formulation itself, moreover, has flaws that makes its results suspect. As we saw in Section 8.2, there is no uniquely defined plastic strain rate, and a flow rule of the form (8.3.3) can be given only as an approximation under restricted circumstances. Other difficulties arise from the use of Equation (8.3.1) and from the choice of the Jaumann rate.

Hypoelasticity

In the absence of plastic deformation, Equation (8.3.1), with isotropic elasticity assumed, becomes

$$\overset{\star}{\mathbf{P}} = \lambda(\operatorname{tr}\mathbf{D})\mathbf{I} + 2\mu\mathbf{D}.$$

This equation represents a model of behavior known in the literature as *hypoelasticity* (Truesdell and Noll [1965]). However, as shown by Simo and Pister [1984], such an equation with *constant* coefficients[1] cannot be derived from an energy-based elastic (also called *hyperelastic*) stress-strain relation such as (8.2.2) (see also Exercise 1 of Section 8.2), and the result holds when the Jaumann rate is replaced by any other objective rate. Elasticity without a strain energy is difficult to motivate physically, since it may entail nonvanishing dissipation in a closed cycle of deformation.

Kinematic Hardening

Consider, next, the generalization to large deformation of the Melan–Prager kinematic hardening model, Equation (3.3.8), using the Jaumann rate:

$$\dot{\mathbf{H}} - \mathbf{WH} + \mathbf{HW} = c\mathbf{D}_p. \tag{8.3.5}$$

Note that this equation implies $\operatorname{tr}\dot{\mathbf{H}} = 0$, so that if $\mathbf{H} = 0$ initially then $\mathbf{H} = \mathbf{H}'$, that is, $\mathbf{H}$ is purely deviatoric.

Let the model described by Equations (8.3.2)–(8.3.4) now be applied to a problem in which the displacement is equivalent to simple shear. For simplicity, incompressible rigid–plastic behavior may be assumed, so that $\mathbf{D}_p = \mathbf{D}$, $J = 1$, and $\mathbf{P} = \mathbf{T}$. The deformation gradient is

$$\mathbf{F} = \begin{bmatrix} 1 & \gamma & 0 \\ 0 & 1 & 0 \\ 0 & 0 & 1 \end{bmatrix},$$

yielding

$$\mathbf{L} = \begin{bmatrix} 0 & \dot{\gamma} & 0 \\ 0 & 0 & 0 \\ 0 & 0 & 0 \end{bmatrix}, \quad \mathbf{D} = \frac{1}{2}\begin{bmatrix} 0 & \dot{\gamma} & 0 \\ \dot{\gamma} & 0 & 0 \\ 0 & 0 & 0 \end{bmatrix}, \quad \mathbf{W} = \frac{1}{2}\begin{bmatrix} 0 & \dot{\gamma} & 0 \\ -\dot{\gamma} & 0 & 0 \\ 0 & 0 & 0 \end{bmatrix}.$$

The stress-deviator and back-stress tensors can be assumed to be given, respectively, by

$$\mathbf{T}' = \begin{bmatrix} \sigma & \tau & 0 \\ \tau & -\sigma & 0 \\ 0 & 0 & 0 \end{bmatrix}, \quad \mathbf{H} = \mathbf{H}' = \begin{bmatrix} \alpha & \beta & 0 \\ \beta & -\alpha & 0 \\ 0 & 0 & 0 \end{bmatrix}.$$

[1] In the general theory of hypoelasticity the coefficients are assumed to depend on stress; they must do so isotropically.

For compatibility with the yield criterion (8.3.2) and the flow rule (8.3.3), it is necessary that

$$\sigma = \alpha, \qquad \tau = k + \beta.$$

The rate equation (8.3.5) now reduces to the two simultaneous equations

$$\dot{\alpha} - \beta\dot{\gamma} = 0, \qquad \dot{\beta} + \alpha\dot{\gamma} = \frac{1}{2}c\dot{\gamma}.$$

Assuming $\gamma = \gamma(t)$ with $\dot{\gamma} > 0$, these equations can be solved as

$$\alpha = \frac{1}{2}c(1 - \cos\gamma), \quad \beta = \frac{1}{2}c\sin\gamma.$$

This solution, first discussed by Lehmann [1972], implies that the shear stress is an oscillating function of the shear strain, clearly in contradiction with the notion of hardening. Numerous proposals have been put forward to deal with this contradiction. Some (e.g., Dafalias [1985a,b], Loret [1983]) involve a generalization of the flow rule into the nine-dimensional one envisaged by Mandel. Others simply suggest replacing the Jaumann rate in Equation (8.3.5) by some other objective rate. For example, Haupt and Tsakmakis [1986] propose the Truesdell rate, which, with the incompressibility condition $\operatorname{tr}\mathbf{L} = 0$ taken into account, becomes

$$\dot{\mathbf{H}} - \mathbf{L}\mathbf{H} - \mathbf{H}\mathbf{L}^T = \mathbf{F}\frac{D}{Dt}(\mathbf{F}^{-1}\mathbf{H}\mathbf{F}^{T-1})\mathbf{F}^T.$$

Since

$$\mathbf{D} = -\frac{1}{2}\mathbf{F}\left(\frac{D}{Dt}\mathbf{C}^{-1}\right)\mathbf{F}^T,$$

the rate equation can be integrated to yield

$$\mathbf{H} = \frac{1}{2}c\mathbf{F}(\mathbf{I} - \mathbf{C}^{-1})\mathbf{F}^T = \frac{1}{2}c(\mathbf{B} - \mathbf{I});$$

in other words, this proposal corresponds to the integrated form $\boldsymbol{\rho} = c\boldsymbol{\varepsilon}^p$ of the small-deformation kinematic hardening model. Note, however, that $\mathbf{H}$ now is not purely deviatoric.

With $\mathbf{F}$ as given above, $\mathbf{B}$ is

$$\mathbf{B} = \begin{bmatrix} 1+\gamma^2 & \gamma & 0 \\ \gamma & 1 & 0 \\ 0 & 0 & 1 \end{bmatrix},$$

and therefore

$$\mathbf{H} = \frac{1}{2}c\begin{bmatrix} \gamma^2 & \gamma & 0 \\ \gamma & 0 & 0 \\ 0 & 0 & 0 \end{bmatrix}, \qquad \mathbf{H}' = \frac{1}{2}c\begin{bmatrix} \frac{1}{2}\gamma^2 & \gamma & 0 \\ \gamma & -\frac{1}{2}\gamma^2 & 0 \\ 0 & 0 & 0 \end{bmatrix},$$

that is,

$$\alpha = \frac{1}{4}c\gamma^2, \quad \beta = \frac{1}{2}c\gamma,$$

reproducing straight-line hardening as for infinitesimal deformations.

Non-oscillatory behavior is also found when the rate used is of the Jaumann form — that is, a *corotational* rate — but with $\mathbf{W}$ replaced by $\dot{\mathbf{R}}\mathbf{R}^T$, where $\mathbf{R}$ is the rotation tensor defined in 8.1.1; see Dafalias [1983].

However, as pointed by Atluri [1984] (see also Reed and Atluri [1985]), non-oscillatory behavior per se does not necessarily represent a solution that agrees with experiment; for example, the just-discussed solution produces a normal stress $\sigma = \alpha = \frac{1}{2}(\tau - k)\gamma$ that is considerably larger than experimentally observed in metals. Instead, Atluri suggests that, in view of the fact that any objective rate of $\mathbf{H}$ differs from its Jaumann rate by terms depending on $\mathbf{H}$ and $\mathbf{D}$, these terms be determined on the basis of experimental data.

8.3.2 "Hyperelastic" Numerical Methods

An alternative methodology, which avoids the difficulties posed by the rate formulation, was proposed by Simo and Ortiz [1985]. It is based on the multiplicative decomposition of the deformation gradient, as discussed in 8.2.2, and the stress-deformation relations given by Equation (8.2.2), with the free-energy density, however, assumed to be given in general by (8.2.10) rather than by (8.2.9) (which is generally regarded as more consistent with the multiplicative decomposition); the circumstances under which the two representations coincide were discussed in 8.2.2. The formulation may, in principle, be given equivalently with respect to the reference, current, or intermediate configuration. What follows is a summary of the presentation given in Simo and Hughes [1988] (see also Simo [1988a,b]).

With the stress-deformation relations in direct form rather than in rate form, objectivity is enforced automatically when these relations are, as is usual, invariant under a rigid-body rotation of the current configuration — that is, when the free-energy density ψ is determined by tensors that are invariant under such a rotation. *Plastic incompressibility* may be conveniently incorporated by assuming, as in the infinitesimal-deformation theory, that the mean stress is determined only by the volume deformation, which is elastic. With the help of the multiplicative decomposition of the deformation gradient as in 8.2.1, namely, $\mathbf{F} = J^{1/3}\bar{\mathbf{F}}$ (see 8.2.1), with $J = J_e$ and $\bar{\mathbf{F}} = \bar{\mathbf{F}}_e\mathbf{F}_p$, the free-energy function may be assumed as a special (uncoupled) case of (8.2.12), with

$$\psi_e(\mathbf{E}_e, T) = \psi_{eV}(J_e, T) + \psi_{eD}(\bar{\mathbf{E}}_e, T),$$

where $\bar{\mathbf{E}}_e = \frac{1}{2}(\bar{\mathbf{F}}_e^T\bar{\mathbf{F}}_e - \mathbf{I})$; then $\frac{1}{3}\mathrm{tr}\,\mathbf{T} = \rho\partial\psi_V/\partial J_e$.

An efficient method for solving the governing equations turns out to be the elastic predictor—plastic corrector algorithm, similar to the one presented for infinitesimal deformation in Section 4.5. The problem is assumed to be displacement-driven, so that $\mathbf{F}$ is assumed to be known as a function of time. During the predictor phase, $\mathbf{F}_p$ is assumed to remain constant. If the trial elastic state does not violate the yield condition, then it is the correct state at the next time point. Otherwise, the plastic corrector must be activated.

In the corrector phase it is the total deformation gradient $\mathbf{F}$ that is assumed to remain constant,[1] and the problem is governed by the equations of evolution, that is, the flow equation [which is assumed, in Lagrangian form, as a rate equation for $\mathbf{C}_p$ (see 8.2.4 for a discussion of the limitations on this assumption)] and the rate equations for the 986]. that it is precisely the tensor $\mathbf{E}_e$ which, along with the hardening variables, if any.

An integration algorithm that automatically incorporates the equivalent of incremental objectivity is based on performing the time discretization of the rate equations (that is, their transformation into incremental equations by means of an appropriate integration scheme) in their Lagrangian form. The time-discretized equations are then transformed ("pushed forward") into Eulerian form for an implementation of the return-mapping algorithm, analogous to the one given in Section 4.5. As in the infinitesimal-deformation case, the use of consistent (algorithmic) tangent moduli is essential in order to achieve a quadratic rate of asymptotic convergence with Newton's method.

The finite-element discretization may be performed with respect to the last computed configuration as a reference configuration — the so-called *updated Lagrangian* approach — but with a mesh in which the nodes are fixed in the initial configuration, permitting the plotting of the deformed mesh at each computed configuration.

Several examples of the application of the method discussed here are given by Simo [1988b]; the one shown here describes the necking of a circular bar made of a nonlinearly work-hardening material. Isoparametric constant-volume $Q4$ elements are used. Figures 8.3.1(a)–(c) show three different meshes in the initial configuration — of 50, 200, and 400 elements, respectively — and Figures 8.3.1(d)–(f) show the corresponding deformed meshes in the deformed configuration; it is seen that there are no significant differences among the three. Figure 8.3.2 shows the ratio of the neck radius to the initial radius plotted against the relative elongation of the bar; the centered symbols represent experimental results, the solid curve represents results calculated by the "hyperelastic" method in 15 time steps, and the

---

[1]Recall that in the infinitesimal-deformation case only the plastic strain ε^p and the total strain ε, respectively, remain constant in the two phases, while the rotation may be ignored.

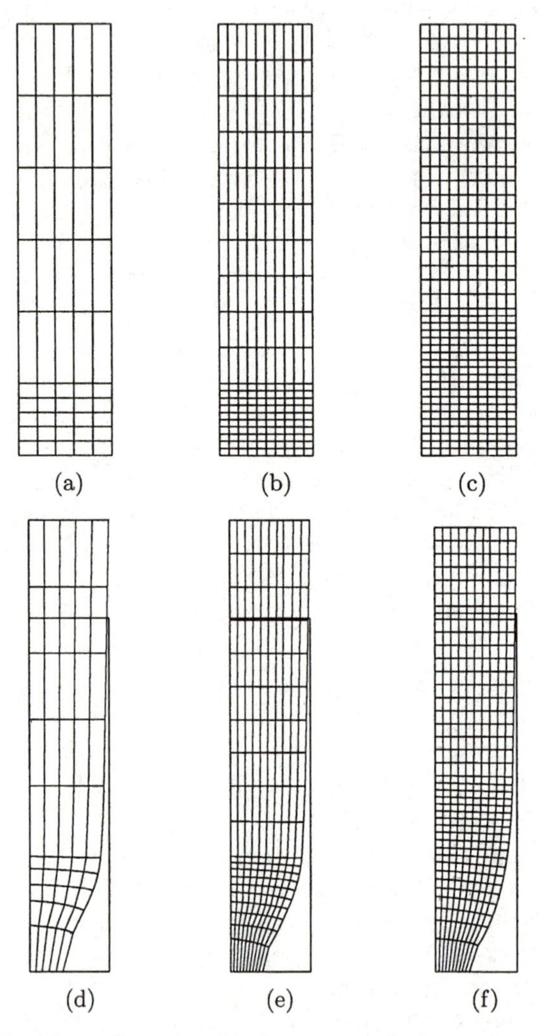

Figure 8.3.1. Necking of a circular bar, numerical results: (a)–(c) initial meshes; (d)–(f) deformed meshes after 14% elongation (from Simo [1988b]).

other two curves represent results calculated by means of other numerical methods at considerably greater computational cost: one (the dashed curve) in 100 load steps and the other (the dotted curve) in 29,000 (!) load steps. Moreover, the hyperelastic method is found to be fairly insensitive to both mesh refinement and step size.

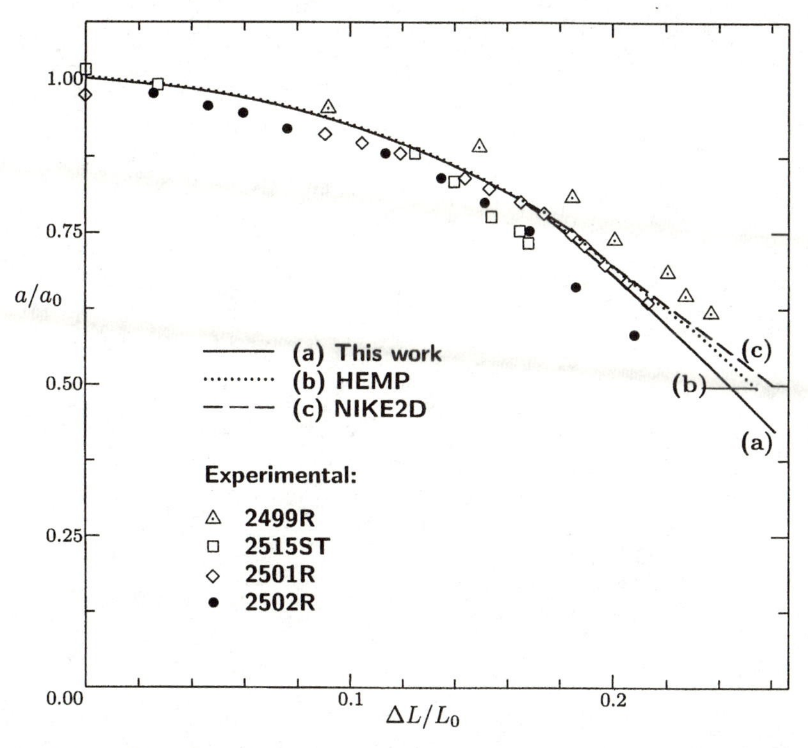

Figure 8.3.2. Necking of a circular bar, numerical and experimental results: ratio of neck radius to initial bar radius against relative elongation (from Simo [1988b]).

References

Alexander, J. M. [1961], *Q. J. Appl. Math.* **19**, 31.

Alverson, R. C. [1956], *J. Appl. Mech.* **23**, 411.

Atluri, S. N. [1984], *Computer Methods Appl. Mech. Engrg.* **43**, 137.

Baba, S., and T. Kajita [1982], *Int. J. Num. Methods Engrg.* **18**, 927.

Backhaus, G. [1968], *Z. angew. math. Mech.* **48**, 99.

Baltov, A., and A. Sawczuk [1965], *Acta Mech.* **1**, 81.

Bammann, D., and R. D. Krieg [1987], in *Unified Constitutive Equations for Creep and Plasticity* (ed. A. K. Miller), Elsevier Applied Science, London, p. 303.

Basinski, S. J., and Z. S. Basinski [1979], in *Dislocations in Solids* (ed. F. R. N. Nabarro), Vol. 4, North-Holland, Amsterdam, p. 261.

Bataille, J., and J. Kestin [1975], *J. Méc.* **14**, 365.

Bazant, Z. P. [1978], *Int. J. Solids Struct.* **14**, 691.

Bell, J. F. [1968], *The Physics of Large Deformation of Crystalline Solids*, Springer Tracts in Natural Philosophy, Vol. 14, Springer-Verlag, Berlin.

Bell, J. F. [1973], *The Experimental Foundations of Solid Mechanics,* in *Handbuch der Physik* (ed. S. Flügge), Vol. VIa/1, Springer-Verlag, Berlin.

Bergander, H. [1980], *Z. angew. Math. Mech.* **60**, 509.

Bianchi, G. [1964], in *Stress Waves in Anelastic Solids*(IUTAM Symp., Providence, 1963, ed. H. Kolsky and W. Prager), Springer-Verlag, Berlin, p. 101.

Bishop, J. F. W. [1953], *J. Mech. Phys. Solids* **2**, 44.

Bishop, J. F. W., and R. Hill [1951], *Philos. Mag.* **42**, 414.

Bishop, J. F. W., R. Hill, and N. F. Mott [1945], *Proc. Phys. Soc.* **57**, 147.

Bland, D. R. [1956], *J. Mech. Phys. Solids* **4**, 209.

Bleich, H. H., and I. Nelson [1966], *J. Appl. Mech.* **33**, 149.

Bleich, H. H., and M. G. Salvadori, [1953], *Proc. ASCE* **79**, Sep. 287.

Bleich, H. H., and M. G. Salvadori [1955], *Trans. ASCE* **120**, 499.

Bodner, S. R. [1968], in *Mechanical Behavior of Materials Under Dynamic*

Loads (Symp. San Antonio, 1967, ed. U. S. Lindholm), Springer-Verlag, New York, p. 176.

Bodner, S. R., and R. J. Clifton [1966], *J. Appl. Mech.* **34**, 91.

Bodner, S. R., and Y. Partom [1972], *J. Appl. Mech.* **39**, 751.

Bodner, S. R., and Y. Partom [1975], *J. Appl. Mech.* **42**, 385.

Bolton, M. [1979], *A Guide to Soil Mechanics*, Macmillan, London.

Boresi, A. P., and K. P. Chong [1987], *Elasticity in Engineering Mechanics*, Elsevier, New York.

Borkowski, A. [1988], *Analysis of Skeletal Structural Systems in the Elastic and Elastic-Plastic Range*, Elsevier, Amsterdam.

Bridgman, P. W. [1923], *Proc. Am. Acad. Arts Sci.* **58**, 163.

Bridgman, P. W. [1952], *Studies in Large Plastic Flow and Fracture*, McGraw-Hill, New York.

Budiansky, B. [1959], *J. Appl. Mech.* **26**, 259.

Calladine, C. R. [1973], *Int. J. Mech. Sci.* **15**, 593.

Casey, J. [1985], *Int. J. Solids Struct.* **21**, 671.

Casey, J., and P. M. Naghdi [1980], *J. Appl. Mech.* **47**, 672.

Casey, J., and P. M. Naghdi [1981], *J. Appl. Mech.* **48**, 285.

Casey, J., and P. M. Naghdi [1983], *Q. J. Mech. Appl. Math.* **37**, 231.

Ceradini, G. [1966], *Meccanica* **4**, 77.

Cernocky, E. P., and E. Krempl [1979], *Int. J. Non-Linear Mech.* **14**, 183.

Chaboche, J. L. [1977], *Bull. Acad. Polon. Sci. Sér. Sci. Tech.* **25**, 33.

Chadwick, P., A. D. Cox, and H. G. Hopkins [1964], *Philos. Trans. Roy. Soc. London* **A256**, 235.

Chakrabarty, J. [1987], *Theory of Plasticity*, McGraw-Hill, New York.

Chan, H. S. Y. [1969], *J. Appl. Mech.* **36**, 73.

Chan, K. S., S. R. Bodner, K. P. Walker, and U. S. Lindholm [1984], in *Proc. 2nd Symp. Nonlinear Constitutive Equations High-Temp. Appl.* (Cleveland, 1984).

Chen, W. F. [1975], *Limit Analysis and Soil Mechanics*, Elsevier, Amsterdam.

Clifton, R. J. [1966], in *Proc. 5th U.S. Natl. Cong. Appl. Mech.* (Minneapolis, 1966), American Society of Mechanical Engineers, New York, p. 465.

Clifton, R. J. [1968], *J. Appl. Mech.* **35**, 782.

Clifton, R. J. [1983], *J. Appl. Mech.* **50**, 941.

Collins, I. F. [1969], *J. Mech. Phys. Solids* **17**, 323.

Conroy, M. F. [1952], *J. Appl. Mech.* **19**, 465.

Conroy, M. F. [1955], *J. Appl. Mech.* **22**, 48.

Conroy, M. F. [1956], *J. Appl. Mech.* **23**, 239.

Conroy, M. F. [1963], *J. Méc.* **2**, 455.

Considère, F. [1891], *Cong. Int. Procédés de Constr.* (Paris, 1889), Vol. 3, p. 371.

Cotter, B. A., and P. S. Symonds [1955], *Proc. ASCE* **81**, 21.

Cottrell, A. H., and B. A. Bilby [1949], *Proc. Phys. Soc.* **A62**, 49.

Cowper, G. R. [1960], *J. Appl. Mech.* **47**, 496.

Cox, A. D., and L. W. Morland [1959], *J. Mech. Phys. Solids* **7**, 229.

Craggs, J. W. [1957], *J. Mech. Phys. Solids* **5**, 115.

Cristescu, N. [1964], in *Stress Waves in Anelastic Solids* (IUTAM Symp., Providence, 1963, ed. H. Kolsky and W. Prager), Springer-Verlag, Berlin, p. 118.

Cristescu, N. [1967], *Dynamic Plasticity*, North-Holland, Amsterdam.

Cristescu, N., and M. Predeleanu [1965], in *Proc. 4th Int. Cong. Rheol.* (Providence, 1963, ed. E. H. Lee), Vol. 3, Wiley, New York, p. 79.

Dafalias, Y. F. [1975], *On Cyclic and Anisotropic Plasticity*, Ph.D. thesis, Department of Civil Engineering, University of California, Berkeley.

Dafalias, Y. F. [1983], *J. Appl. Mech.* **50**, 561.

Dafalias, Y. F. [1985a], in *Plasticity Today* (IUTAM Symp., Udine, 1983, ed. A. Sawczuk and G. Bianchi), Elsevier Applied Science, London, p. 135.

Dafalias, Y. F. [1985b], *J. Appl. Mech.* **52**, 865.

Dafalias, Y. F., and E. P. Popov [1975], *Acta Mech.* **21**, 173.

Dantzig, G. B. [1963], *Linear Programming and Extensions*, Princeton University Press, Princeton, N.J.

Dashner, P. A. [1986], *J. Appl. Mech.* **53**, 55.

De Juhasz, K. J. [1949], *J. Franklin Inst.* **248**, 15, 113.

Derrington, M. G., and W. Johnson [1958], *Appl. Sci. Res.* **A7**, 408.

Donnell, L. H. [1930], *Trans. ASME* **52**, 153.

Drucker, D. C. [1950], *Q. Appl. Math.* **7**, 411.

Drucker, D. C. [1951], in *Proc. 1st U.S. Natl. Cong. Appl. Mech.*, American Society of Mechanical Engineers, New York, p. 487.

Drucker, D. C. [1953], *J. Mech. Phys. Solids* **1**, 217.

Drucker, D. C. [1954a], *J. Appl. Mech.* **21**, 71.

Drucker, D. C. [1954b], in *Proc. 1st Midwestern Conf. Solid Mech.* (Urbana, 1953), p. 158.

Drucker, D. C. [1956], *J. Appl. Mech.* **23**, 509.

Drucker, D. C. [1959], *J. Appl. Mech.* **26**, 101.

Drucker, D. C., and W. F. Chen [1968], in *Engineering Plasticity*(ed. J. Heyman and F. A. Leckie), Cambridge University Press, Cambridge, England, p. 119.

Drucker, D. C., H. J. Greenberg, and W. Prager [1951], *J. Appl. Mech.* **18**, 371.

Drucker, D. C., and H. G. Hopkins [1955], in *Proc. 2nd U.S. Natl. Cong. Appl. Mech.* (Ann Arbor, 1954), American Society of Mechanical Engineers, New York, p. 517.

Drucker, D. C., and W. Prager [1952], *Q. Appl. Math.* **10**, 157.

Drucker, D. C., W. Prager, and H. J. Greenberg [1952], *Q. Appl. Math.* **9**, 381.

Drucker, D. C., and R. T. Shield [1957], *Q. Appl. Math.* **15**, 269.

Drucker, D. C., and R. T. Shield [1959], *J. Appl. Mech.* **26**, 61.

Duwez, P. E., D. S. Clark, and H. F. Bohnenblust [1950], *J. Appl. Mech.* **17**, 27.

Eason, G. [1960a], *Appl. Sci. Res.* **A8**, 53.

Eason, G. [1960b], *Q. J. Mech. Appl. Math.* **12**, 334.

Eason, G., and R. T. Shield [1955], *J. Mech. Phys. Solids* **4**, 53.

Engesser, F. [1889], *Z. Archit. Ing.wesen* **35**, 455.

English, J. M. [1954], *Trans. ASCE* **119**, 1143.

Eringen, A. C. [1962], *Nonlinear Theory of Continuous Media*, McGraw-Hill, New York.

Florence, A. L. [1966], *J. Appl. Mech.* **33**, 256.

Foulkes, J. [1953], *Q. Appl. Math.* **10**, 347.

Foulkes, J. [1954], *Proc. Roy. Soc. London* **A223**, 482.

Fox, N. [1968], *Q. J. Mech. Appl. Math.* **21**, 67.

Frank, F. C. [1951], *Philos. Mag.* **42**, 809.

Freiberger, W. [1949], *Austral. J. Sci. Res.* **A2**, 354.

Freiberger, W. [1956a], *Q. Appl. Math.* **14**, 259.

Freiberger, W. [1956b], *J. Appl. Mech.* **23**, 576.

Freiberger, W., and B. Tekinalp [1956], *J. Mech. Phys. Solids* **4**, 294.

Frémond, M., and A. Friaâ [1982], *J. Méc. Théor. Appl.* **1**, 881.

Galin, L. A. [1949], *Prikl. Mat. Mekh.* **13**, 285.

Gaydon, F. A., and H. Nuttall [1957], *J. Mech. Phys. Solids* **6**, 17.

Geiringer, H. [1931], in *Proc. 3rd Int. Cong. Appl. Mech.* (Stockholm, 1930), Vol. 2, p. 185.

Geiringer, H. [1951], *Proc. Natl. Acad. Sci. U.S.A.* **37**, 214.

Gerard, G., and H. Becker [1957], *Handbook of Structural Stability, Part I,*

NACA Tech. Note 3781.

Germain, P., Q. S. Nguyen, and P. Suquet [1983], *J. Appl. Mech.* **50**, 1010.

Gilman, J. J. [1969], *Micromechanics of Flow in Solids*, McGraw-Hill, New York.

Green, A. E., and P. M. Naghdi [1965], *Arch. Ration. Mech. Anal.* **18**, 251.

Green, A. E., and P. M. Naghdi [1971], *Int. J. Engrg. Sci.* **9**, 1219.

Green, A. E., and W. Zerna [1968], *Theoretical Elasticity*, Oxford University Press, London.

Green, A. P. [1951], *Philos. Mag.* **42**, 900.

Green, A. P. [1953], *Q. J. Mech. Appl. Math.* **6**, 223.

Green, A. P. [1954a], *J. Mech. Phys. Solids* **3**, 1.

Green, A. P. [1954b], *J. Mech. Phys. Solids* **3**, 143.

Greenberg, H. J. [1949], *Q. Appl. Math.* **7**, 85.

Greenberg, H. J., and W. Prager [1951], *Proc. ASCE* **77**, Sep. 59.

Griggs, D. T. [1936], *J. Geol.* **44**, 541.

Grigorian, S. S. [1960], *Prikl. Mat. Mekh.* **24**, 1057.

Gvozdev, A. A. [1938], in *Proc. Conf. Plast. Def.*(in Russian), Akademiia Nauk SSSR, Moscow, p. 19.

Hadamard, J. [1903], *Leçons sur la propagation des ondes et les équations de l'hydrodynamique*, Hermann, Paris.

Halphen, B., and Q. S. Nguyen [1975], *J. Méc.* **14**, 39.

Handelman, G. H. [1944], *Q. Appl. Math.* **1**, 351.

Hart, E. W. [1976], *J. Engrg. Matls. Technol.* **98**, 193.

Haupt, P. [1985], *Int. J. Plast.* **1**, 303.

Haupt, P., and C. Tsakmakis [1986], *Int. J. Plast.* **2**, 279.

Haythornthwaite, R. M., and R. T. Shield [1958], *J. Mech. Phys. Solids* **6**, 127.

Helling, D. E., and A. K. Miller [1987], *Acta Mech.* **69**, 9.

Hencky, H. [1923], *Z. angew. Math. Mech.* **3**, 241.

Hencky, H. [1924], *Z. angew. Math. Phys.* **4**, 323. 353.

Heyman, J. [1951], *Q. Appl. Math.* **8**, 373.

Heyman, J. [1953], *Struct. Engr.* **31**, 125.

Heyman, J. [1959], *Proc. Instn. Civ. Engrs.* **12**, 39; *Q. J. Mech. Appl. Math.* **12**, 314.

Heyman, J. [1960], *Int. J. Mech. Sci.* **1**, 121.

Heyman, J. [1971], *Plastic Design of Frames*, Cambridge University Press, Cambridge, England.

Heyman, J. [1973], *Int. J. Mech. Sci.* **15**, 845.

Heyman, J., and W. Prager [1958], *J. Franklin Inst.* **266**, 339.

Hibbit, H. D., P. V. Marcal, and J. R. Rice [1970], *Int. J. Solids Struct.* **6**, 1069.

Hill, R. [1948a], *Q. J. Mech. Appl. Math.* **1**, 18.

Hill, R. [1948b], *Proc. Roy. Soc. London* **A193**, 281.

Hill, R. [1948c], *J. Iron Steel Inst.* **158**, 177.

Hill, R. [1950], *The Mathematical Theory of Plasticity*, Oxford University Press, London.

Hill, R. [1951], *Philos. Mag.* **42**, 868.

Hill, R. [1952], *Philos. Mag.* **43**, 353.

Hill, R. [1953], *J. Mech. Phys. Solids* **1**, 265.

Hill, R. [1958], *J. Mech. Phys. Solids* **6**, 236.

Hill, R. [1959], *J. Mech. Phys. Solids* **7**, 371.

Hill, R. [1962], *J. Mech. Phys. Solids* **10**, 1.

Hill, R., E. H. Lee, and S. J. Tupper [1947], *Proc. Roy. Soc. London*A, **188**, 273.

Hill, R., and M. P. L. Siebel [1951], *Philos. Mag.* **42**, 721.

Hill, R., and M. P. L. Siebel [1953], *J. Mech. Phys. Solids* **1**, 207.

Hill, R., and S. J. Tupper [1948], *J. Iron Steel Inst.* **159**, 353.

Hirsch, P. B. [1975], *The Physics of Metals*, Vol. 2: *Defects*(ed. P. B. Hirsch), Cambridge University Press, Cambridge, England, p. 189.

Hodge, P. G., Jr. [1953], *Div. Appl. Math. Report* B11-22, Brown University, Providence, R.I.

Hodge, P. G., Jr. [1954], *J. Appl. Mech.* **21**, 336.

Hodge, P. G., Jr. [1955], *J. Mech. Phys. Solids* **3**, 176.

Hodge, P. G., Jr. [1956], in *Proc. 2nd Midwestern Conf. Solid Mech.* (Lafayette, 1955), p. 150.

Hodge, P. G., Jr. [1957a], in *Proc. 9th Int. Cong. Appl. Mech.* (Brussels, 1956), Vol. 8, p. 65.

Hodge, P. G., Jr. [1957b], *J. Appl. Mech.* **24**, 453.

Hodge, P. G., Jr. [1959], *Plastic Analysis of Structures*, McGraw-Hill, New York.

Hodge, P. G., Jr. [1963], *Limit Analysis of Rotationally Symmetric Plates and Shells*, Prentice-Hall, Englewood Cliffs, N.J.

Hodge, P. G., Jr., and B. Paul [1957], in *Proc. 3rd Midwestern Conf. Solid Mech.* (Ann Arbor, 1957), p. 29.

Hodge, P. G. Jr., and G. N. White, Jr. [1950], *J. Appl. Mech.* **17**, 180.

Hohenemser, K., and W. Prager [1932], *Z. angew. Math. Mech.* **12**, 1.

Hopkins, H. G. [1957], *Proc. Roy. Soc. London* **A241**, 153.

Hopkins, H. G., and W. Prager [1953], *J. Mech. Phys. Solids* **2**, 1.

Hopkins, H. G., and W. Prager [1954], *Z. angew. Math. Phys.* **5**, 317.

Hopkins, H. G., and W. Prager [1955], *J. Appl. Mech.* **22**, 372.

Hopkins, H. G., and A. J. Wang [1954], *J. Mech. Phys. Solids* **3**, 117.

Horne, M. R. [1950], *J. Instn. Civ. Engrs.* **34**, 174.

Horne, M. R. [1954], *Proc. Instn. Civ. Engrs.* **3**, 51.

Horne, M. R. [1979], *Plastic Theory of Structures*, 2nd ed., Pergamon Press, Oxford.

Hudson, J. A., E. T. Brown, and C. Fairhurst [1971], in *Proc. 13th Symp. Rock Mech.* (Urbana, 1970), p. 180.

Hull, D., and D. J. Bacon [1984], *Introduction to Dislocations*, 3rd ed., Pergamon Press, Oxford.

Hutchinson, J. W. [1972], *J. Appl. Mech.* **39**, 155.

Hutchinson, J. W. [1974], in *Advances in Applied Mechanics*(ed. C.-S. Yih), Vol. 14, Academic Press, New York, p. 67.

Il'iushin A. A., [1954], *Prikl. Mat. Mekh.* **18**, 641.

Il'iushin A. A., [1961], *Prikl. Mat. Mekh.* **25**, 503.

Imegwu, E. O. [1960], *J. Mech. Phys. Solids* **8**, 141.

Ishlinskii, I. U. [1954], *Ukr. Mat. Zh.* **6**, 314.

Iwan, W. D. [1967], *J. Appl. Mech.* **34**, 612.

Johansen, K. W. [1932], *Proc. Int. Assn. Bridge and Struc. Engrg.* **1**, 277.

Johnson, D. [1988], *Int. J. Solids Struct.* **24**, 321.

Johnson, W., and P. B. Mellor [1973], *Engineering Plasticity*, Van Nostrand Reinhold, London.

Johnson, W., R. Sowerby, and R. Venter [1982], *Plane Strain Slip-Line Theory and Bibilography*, Pergamon Press, Oxford.

Jones, N. [1975], *Shock Vib. Digest* **8**, 89.

Josselin de Jong, G. [1965], in *Proc. IUTAM Symp. on Rheology and Soil Mechanics* (Grenoble, 1964), p. 69.

Josselin de Jong, G. [1974], in *Proc. Symp. on the Role of Plasticity in Soil Mechanics* (Cambridge, England, 1973), p. 12.

Kachanov, L. M. [1954], *Prikl. Mat. Mekh.* **19**,

Kachanov, L. M. [1971], *Foundations of Theory of Plasticity*, North-Holland, Amsterdam.

Kadashevich, Yu. I., and V. V. Novozhilov [1958], *Prikl. Mat. Mekh.* **22**, 104.

Kaliski, S., and E. Wlodarczyk [1967], *Arch. Mech. Stos.* **19**, 433.

Kármán, T. v. [1910], *VDI (Ver. deut. Ing.) Forschungsh.* **81**.

Kármán, T. v. [1911], *Z. Ver. deut. Ing.* **55**, 1749.

Kármán, T. v., and P. Duwez [1950], *J. Appl. Phys.* **21**, 987.

Kazinczy, G. [1914], *Betonszemle* **2**, 68.

Kestin, J., and Rice, J. R. [1970], in *A Critical Review of Thermodynamics* (ed. E. B. Stuart, B. Gal-Or, and A. J. Brainard), Mono Book Corp., Baltimore, p. 275.

Kliushnikov, V. D. [1959], *Prikl. Mat. Mekh.* **23**, 405.

Kobayashi, S., S.-I. Oh, and T. Altan [1989], *Metal Forming and the Finite-Element Method*, Oxford University Press, London.

Kochendörfer, A. [1938], *Z. Phys.* **108**, 244.

Kocks, U. F. [1987], in *Unified Constitutive Equations for Creep and Plasticity* (ed. A. K. Miller), Elsevier Applied Science, London, p. 1.

Koiter, W. T. [1953a], *Q. Appl. Math.* **11**, 350.

Koiter, W. [1953b], *Anniversary Volume on Applied Mechanics Dedicated to B. Biezeno*, H. Stam, Haarlem, p. 232.

Koiter, W. T. [1956], *Proc. Kon. Ned. Akad. Wet.* **B59**, 24.

Koiter, W. T. [1960], *Progress in Solid Mechanics* (ed. I. N. Sneddon and R. Hill), Vol. 1, North-Holland, Amsterdam, p. 206.

Kolsky, H., and L. S. Douch [1962], *J. Mech. Phys. Solids* **10**195.

König, J. A. [1987], *Shakedown of Elastic-Plastic Structures*, Elsevier, Amsterdam.

Kotsovos, M. D., and H. K. Cheong [1984], *ACI J.* **81**, 358.

Krajcinovic, D. [1973], *Shock Vibr. Digest* **5**, 1.

Kratochvíl, J. [1973], *Acta Mech.* **16**, 127.

Krempl, E. [1987], *Acta Mech.* **69**, 25.

Krempl, E., J. J. McMahon, and D. Yao [1986], *Mech. Matls.* **5**, 35.

Krieg, R. D. [1975], *J. Appl. Mech.* **42**, 641.

Krieg, R. D., J. C. Swearengen, and W. B. Jones [1987], in *Unified Constitutive Equations for Creep and Plasticity* (ed. A. K. Miller), Elsevier Applied Science, London, p. 245.

Krieg, R. D., J. C. Swearengen and R. W. Rohde [1978], in *Inelastic Behavior of Pressure Vessel and Piping Components* (ed. T. Y. Chang and E. Krempl), American Society of Mechanical Engineers, New York, p. 15.

Kröner, E. [1960], *Arch. Ration. Mech. Anal.* **4**, 273.

Lee, E. H. [1950], *Proc. 3rd Symp. Appl. Math.*(Ann Arbor, 1949), McGraw-Hill, New York, p. 213.

Lee, E. H. [1952], *Philos. Mag.* **43**, 549.

Lee, E. H. [1953], *Q. Appl. Math.* **10**, 335.

Lee, E. H. [1969], *J. Appl. Mech.* **36**, 1.

Lee, E. H., and D. T. Liu [1967], *J. Appl. Phys.* **38**, 19.

Lee, E. H., and P. S. Symonds [1952], *J. Appl. Mech.* **19**, 308.

Lee, E. H., and S. J. Tupper [1954], *J. Appl. Mech.* **21**, 63.

Lehmann, Th. [1972], *Ing.-Arch.* **41**, 297.

Lenskii, V. S. [1948], *Prikl. Mat. Mekh.* **12**, 165.

Leth, C.-F. A. [1954], *Div. Appl. Math. Report* A11-107, Brown University, Providence, R.I.

Lode, W. [1925], *Z. angew. Math. Mech.* **5**, 142.

Loret, B. [1983], *Mech. Matls.* **2**, 287.

Love, A. E. H. [1927], *Mathematical Theory of Elasticity*, 4th ed., Cambrdige University Press, Cambridge, England; reprinted by Dover, New York.

Lubahn, J. D., and R. P. Felgar [1961], *Plasticity and Creep of Metals*, Wiley, New York.

Lubliner, J. [1964], *J. Mech. Phys. Solids* **12**, 59.

Lubliner, J. [1965], *J. Méc.* **4**, 111.

Lubliner, J. [1972], *Int. J. Non-Linear Mech.* **7**, 237.

Lubliner, J. [1974], *Int. J. Solids Struct.* **10**, 310.

Lubliner, J. [1986], *Mech. Matls.* **5**, 29.

Lubliner, J., and M. Valathur [1969], *Int. J. Solids Struct.* **5**, 1275.

Ludwik, P. [1909], *Elemente der Technologischen Mechanik*, Springer-Verlag, Berlin.

Lunts, Ya. L. [1949], *Prikl. Mat. Mekh.* **13**, 55.

Maier, G. [1969], *Int. J. Solids Struct.* **5**, 261.

Maier, G. [1970], *J. Mech. Phys. Solids* **18**, 319.

Maier, G. [1983], *Mech. Res. Comm.* **10**, 45.

Maier, G., R. Srinivasam, and M. A. Save [1972], in *Proc. Int. Symp. Computer-Aided Structural Design* (Coventry, 1972), Vol. 1, p. 32.

Malvern, L. E. [1951a], *Q. Appl. Math.* **8**, 405.

Malvern, L. E. [1951b], *J. Appl. Mech.* **18**, 203.

Mandel, J. [1962], *J. Méc.* **1**, 30.

Mandel, J. [1964], in *Stress Waves in Anelastic Solids*(IUTAM Symp., Providence, 1963, ed. H. Kolsky and W. Prager), Springer-Verlag, Berlin, p. 331.

Mandel, J. [1967], *C. R. Acad. Sci. Paris* **264A**, 133.

Mandel, J. [1972], *Plasticité Classique et Viscoplasticité* (CISM, Udine, 1971), Springer-Verlag, Vienna–New York.

Mandel, J. [1973], *Int. J. Solids Struct.* **9**, 725.

Mandel, J. [1976], *Mech. Res. Comm.* **3**, 251, 483.

Mandel, J., J. Zarka, and B. Halphen [1977], *Mech. Res. Comm.* **4**, 309.

Manjoine, M. J. [1944], *J. Appl. Mech.* **11**, 211.

Manolakos, D. E., and A. G. Mamalis [1986], *Int J. Mech. Sci.* **28**, 815.

Marcal, P. V. [1965], *Int. J. Mech. Sci.* **7**, 229, 841.

Markov, A. A. [1947], *Prikl. Mat. Mekh.* **11**, 339.

Marsden, J. E., and T. J. R. Hughes [1983], *Mathematical Foundations of Elasticity*, Prentice-Hall, Englewood Cliffs, N.J.

Martin, J. B. [1975], *J. Mech. Phys. Solids* **23**, 123.

Martin, J. B. [1980], in *Variational Methods in the Mechanics of Solids* (ed. S. Nemat-Nasser), Pergamon Press, Oxford, p. 278.

Massonnet, C. E., and M. A. Save [1965], *Plastic Analysis and Design*, Blaisdell, Waltham, Mass.

McMeeking, R. M., and J. R. Rice [1975], *Int. J. Solids Struct.* **11**, 601.

Melan, E. [1938], *Ing.-Arch.* **9**, 116.

Mendelson, A. [1968], *Plasticity: Theory and Applications*, Macmillan, New York.

Michell, J. H. [1900], *Proc. London Math. Soc.* **31**, 141.

Miller, A. K. [1976], *J. Engrg. Matls. Technol.* **98**, 97.

Mises, R. v. [1913], *Göttinger Nachr. Math.-Phys. Kl.*, 582.

Mises, R. v. [1926], *Z. angew. Math. Mech.* **6**, 199 (footnote to paper by F. Schleicher).

Mises, R. v. [1928], *Z. angew. Math. Mech.* **8**, 161.

Moreau, J. J. [1963], *C. R. Acad. Sci. Paris* **257A**, 4117.

Moreau, J. J. [1970], *C. R. Acad. Sci. Paris* **271A**, 608.

Moreau, J. J. [1976], in *Applications of Methods of Functional Analysis to Problems in Mechanics* (ed. P. Germain and B. Nayroles), Springer-Verlag, Vienna–New York, p. 56.

Mortelhand, I. M. [1987], *Heron* **32**, 115.

Mróz, Z. [1967], *J. Mech. Phys. Solids* **15**, 163.

Munro, J. [1979], *Engineering Plasticity by Mathematical Programming* (Proc. NATO Adv. Study Inst., Waterloo, Canada, 1977, ed. by M. Z. Cohn and G. Maier), Pergamon Press, New York, Ch. 7.

Nabarro, F. R. N. [1950], *Some Recent Developments in Rheology*, British Rheologists' Club, London.

Nabarro, F. R. N. [1975], in *The Physics of Metals*, Vol. 2: *Defects* (ed. P. B. Hirsch), Cambridge University Press, Cambridge, England, p. 152.

Nadai, A. [1923], *Z. angew. Math. Mech.* **5**, 1.

Nadai, A. [1950], *The Theory of Flow and Fracture of Solids*, McGraw-Hill, New York.

Naghdi, P. M., and J. A. Trapp [1975], *Int. J. Eng. Sci.* **13**, 785.

Nagtegaal, J. C. [1982], *Computer Methods Appl. Mech. Engrg.* **33**, 469.

Nagtegaal, J. C., D. M. Parks, and J. R. Rice [1974], *Computer Methods Appl. Mech. Engrg.* **4**, 153.

Neal, B. G. [1963], *The Plastic Methods of Structural Analysis*, Chapman & Hall, London.

Neal, B. G., and P. S. Symonds [1951], *J. Instn. Civ. Engrs.* **35**, 21.

Neal, B. G., and P. S. Symonds [1952], *Proc. Instn. Civ. Engrs.* **1**, 58.

Needleman, A. [1972], *J. Mech. Phys. Solids* **20**, 111.

Nguyen, Q. S., and H. D. Bui [1974], *J. Méc.* **13**, 321.

Niordson, F. I. [1985], *Shell Theory*, North-Holland, Amsterdam.

Noll, W. [1955], *J. Ration. Mech. Anal.* **4**, 3.

Nowacki, W. K. [1978], *Stress Waves in Non-Elastic Solids*, Pergamon Press, Oxford.

Oden, J. T. [1972], *Finite Elements of Nonlinear Continua*, McGraw-Hill, New York.

Odqvist, F. K. G. [1933], *Z. angew. Math. Phys.* **13**, 360.

Onat, E. T., and R. M. Haythornthwaite [1956], *J. Appl. Mech.* **23**, 49.

Onat, E. T., and W. Prager [1953], *J. Mech. Phys. Solids* **1**, 77.

Onat, E. T., and W. Prager [1954], in *Proc. 1st Midwestern Conf. Solid. Mech.* (Urbana, 1953), p. 40.

Onat, E. T., and W. Prager [1955], *Ingenieur* **67**, 0.46.

Onat, E. T., and R. T. Shield [1955], in *Proc. 2nd U.S. Natl. Cong. Appl. Mech.* (Ann Arbor, 1954), American Society of Mechanical Engineers, New York, p. 535.

Onat, E. T., W. Schumann, and R. T. Shield [1957], *Z. angew. Math. Phys.* **8**, 485.

Orowan, E. [1934], *Z. Phys.* **89**, 605.

Orowan, E. [1944], *Proc. Instn. Mech. Engrs.* **151**, 133 (discussion of paper by H. O'Neill).

Ortiz, M., and E. P. Popov [1985], *Int. J. Num. Methods Engrg.* **21**, 1561.

Owen, D. R. [1968], *Arch. Ration. Mech. Anal.* **31**, 91.

Owen, D. R. J., and E. J. Hinton [1980], *Finite Elements in Plasticity: Theory and Practice*, Pineridge Press, Swansea.

Pastor, J. [1976], *Application de l'analyse limite à l'étude de la stabilité des pentes et des talus*, Thèse, Université de Grenoble (France).

Parkes, E. W. [1955], *Proc. Roy. Soc. London* **A228**, 462.

Perzyna, P. [1958], *Arch. Mech. Stos.* **10**, 635.

Perzyna, P. [1963], *Q. Appl. Math.* **20**, 321.

Perzyna, P. [1971], in *Advances in Applied Mechanics*(ed. by C.-S. Yih), Vol. 11, Academic Press, New York, p. 313.

Phillips, A. [1972], *AIAA J.* **10**, 951.

Phillips, A. [1986], *Int. J. Plast.* **2**, 315.

Phillips, A., and H. Moon [1977], *Acta Mech.* **27**, 91.

Pinsky, P. M., M. Ortiz, and K. S. Pister [1983], *Computer Methods Appl. Mech. Engrg.* **40**, 137.

Pipkin, A. C., and R. S. Rivlin [1965], *Z. angew. Math. Mech.* **16**, 313.

Prager, W. [1942], *Duke Math. J.* **9**, 228.

Prager, W. [1947], *J. Appl. Phys.* **18**, 375.

Prager, W. [1948], in *Courant Anniversary Volume*, Interscience, New York, p. 289.

Prager, W. [1955a], *Proc. Instn. Mech. Engrs.* **169**, 41.

Prager, W. [1955b], *Ingenieur* **67**, 0.141.

Prager, W. [1956a], *J. Appl. Mech.* **23**, 493.

Prager, W. [1956b], in *Proc. 8th Int. Cong. Appl. Mech.* (Istanbul, 1952), Vol. 2, p. 65.

Prager, W. [1959], *An Introduction to Plasticity*, Addison-Wesley, Reading, Mass.

Prager, W. [1961], *An Introduction to Mechanics of Continua*, Ginn, Boston.

Prager, W., and P. G. Hodge, Jr. [1951], *Theory of Perfectly Plastic Solids*, Wiley, New York.

Prandtl, L. [1903], *Phys. Z.* **4**, 758.

Prandtl, L. [1920], *Göttinger Nachr. Math.-Phys. Kl.*, 74.

Prandtl, L. [1923], *Z. angew. Math. Mech.* **3**, 401.

Prandtl, L. [1924], in *Proc. 1st Int. Cong. Appl. Mech.*(Delft, 1924), p. 43.

Rabotnov, Yu. N., [1969], *Creep Problems in Structural Members*, North-Holland, Amsterdam.

Radenkovic, D. [1961], *C. R. Acad. Sci. Paris* **252**, 4103.

Radenkovic, D. [1962], in *Sém. de Plasticité* (Ecole Polytechnique, Paris, 1961), P.S.T. Ministère de l'Air No. N.T. 116.

Rakhmatulin, Kh. A. [1945], *Prikl. Mat. Mekh.* **9**, 91, 449.

Rakhmatulin, Kh. A. [1948], *Prikl. Mat. Mekh.* **12**, 39.

Rakhmatulin, Kh. A., and G. S. Shapiro [1948], *Prikl. Mat. Mekh.* **12**, 369.

Raniecki, B. [1979], *Bull. Acad. Pol. Sci. Sér. Sci. Tech.* **27**, 391 [721].

Read, H. E., and G. A. Hegemier [1984], *Mech. of Matls.* **3**, 271.

Read, W. T., Jr. [1953], *Dislocations in Crystals*, McGraw-Hill, New York.

Reed, K. W., and S. N. Atluri [1985], *Int. J. Plast.* **1**, 63.

Reuss, A. [1930], *Z. angew. Math. Mech.* **10**, 266.

Rice, J. R. [1970], *J. Appl. Mech.* **37**, 728.

Rice, J. R. [1971], *J. Mech. Phys. Solids* **19**, 433.

Roscoe, K. H. [1953], in *Proc. 3rd Int. Conf. Soil Mech. Found. Engrg.* (Switzerland, 1953), Vol. 1, p. 186.

Rozvany, G. I. N. [1976], *Optimal Design of Flexural Systems*, Pergamon Press, Oxford.

Rubin, M. B. [1986], *Int. J. Engrg. Sci.* **24**, 1083.

Rubinstein, R., and Atluri, S. N. [1983], *Computer Methods Appl. Mech. Engrg.* **36**, 277.

Sacchi, G., and M. A. Save [1968], *Meccanica* **3**, 43.

Salençon, J. [1972], *Ann. I.T.B.T.P*, No. 295-296, 90.

Salençon, J. [1973], in *Proc. Int. Symp. Foundations of Plasticity* (Warsaw, 1972, ed. A. Sawczuk), Noordhoff, Leyden.

Salençon, J. [1977], *Applications of the Theory of Plasticity in Soil Mechanics*, Wiley, Chichester, England.

Salvadori, M. G., and F. L. DiMaggio [1953], *Q. Appl. Math.* **11**, 223.

Save, M. A., and C. E. Massonnet [1972], *Plastic Analysis of Plates, Shells and Disks*, North-Holland, Amsterdam.

Save, M. A., and W. Prager [1963], *J. Mech. Phys. Solids* **11**, 255.

Schleicher, F. [1926], *Z. angew. Math. Mech.* **6**, 199.

Schmid, E. [1924], in *Proc. 1st Int. Cong. Appl. Mech.* (Delft, 1924), p. 342.

Seiler, J. A., B. A. Cotter, and P. S. Symonds [1956], *J. Appl. Mech.* **23**, 515.

Seiler, J. A., and P. S. Symonds [1954], *J. Appl. Phys.* **25**, 556.

Shaffer, B. W., and R. N. House [1955], *J. Appl. Mech.* **22**, 305.

Shaffer, B. W., and R. N. House [1957], *J. Appl. Mech.* **24**, 447.

Shakir-Khalil, H., and G. S. Tadros [1973], *Struct. Engrg.* **51**, 239.

Shanley, F. S. [1947], *J. Aeronaut. Sci.* **14**, 261.

Simmons, J. A., F. Hauser, and J. E. Dorn [1962], *Univ. Calif. Pubs. Engrg.* **5**, 177.

Simo, J. C. [1988a], *Computer Methods Appl. Mech. Engrg.* **66**, 199.

Simo, J. C. [1988b], *Computer Methods Appl. Mech. Engrg.* **68**, 1.

Simo, J. C., and T. J. R. Hughes [1988], *Elastoplasticity and Viscoplasticity, Computational Aspects*, Stanford Univ., Dept. of Appl. Mech.

Simo, J. C., and M. Ortiz [1985], *Computer Methods Appl. Mech. Engrg.* **49**, 221.

Simo, J. C., and K. S. Pister [1984], *Computer Methods Appl. Mech. Engrg.* **46**, 201.

Simo, J. C., and R. L. Taylor [1985], *Computer Methods Appl. Mech. Engrg.* **48**, 101.

Simo, J. C., and R. L. Taylor [1986], *Int. J. Num. Methods Engrg.* **22**, 649.

Sokolnikoff, I. S. [1956], *Mathematical Theory of Elasticity*, 2nd ed., McGraw-Hill, New York.

Sokolovskii, V. V. [1942], *Prikl. Mat. Mekh.* **6**, 241.

Sokolovskii, V. V. [1945], *Prikl. Mat. Mekh.* **9**, 343.

Sokolovskii, V. V. [1948], *Prikl. Mat. Mekh.* **12**, 261.

Southwell, R. V. [1915], *Philos. Mag.* **29**, 67.

Steele, M. C. [1954], *J. Mech. Phys. Solids* **3**, 156.

Sternglass, E. J., and D. A. Stuart [1953], *J. Appl. Mech.* **20**, 1953.

Stickforth, J. [1986], *Int. J. Plast.* **2**, 347.

Symonds, P. S., and W. T. Fleming, Jr. [1984], *Int. J. Impact Engrg.* **2**, 1.

Symonds, P. S., and C.-F. A. Leth [1954], *J. Mech. Phys. Solids* **2**, 92.

Symonds, P. S., and T. J. Mentel [1958], *J. Mech. Phys. Solids* **6**, 186.

Symonds, P. S., and B. G. Neal [1951], *J. Franklin Inst.* **252**, 383, 469.

Taylor, D. W. [1937], *J. Boston. Soc. Civ. Engrs.* **24**, 197.

Taylor, D. W. [1948], *Fundamentals of Soil Mechanics*, Wiley, New York.

Taylor, G. I. [1934], *Proc. Roy. Soc. London* **A145**, 362, 388.

Taylor, G. I. [1938], *J. Inst. Metals* **62**, 307.

Taylor, G. I. [1946], *J. Instn. Civ. Engrs.* **26**, 486.

Taylor, G. I. [1947], *Proc. Roy. Soc. London* **A191**, 441.

Taylor, G. I. [1948], *Proc. Roy. Soc. London* **A194**, 289.

Taylor, G. I. [1958], in *Scientific Papers* (ed. G. K. Batchelor), Vol. 2, Cambridge University Press, Cambridge, England, p. 467.

Taylor, G. I., and H. Quinney [1931], *Philos. Trans. Roy. Soc. London* **A230**, 323.

Telles, J. C. F., and C. A. Brebbia [1979], *Appl. Math. Modelling* **3**, 466.

Temam, R. [1985], *Mathematical Problems in Plasticity*, Gauthier-Villars, Paris.

Thomas, T. Y. [1958], *J. Ration. Mech. Anal.* **5**, 251.

Thomas, T. Y. [1961], *Plastic Flow and Fracture in Solids*, Academic Press,

New York.

Timoshenko, S., and J. M. Gere [1961], *Theory of Elastic Stability*, 2nd ed., McGraw-Hill, New York.

Timoshenko, S. P., and J. N. Goodier [1970], *Theory of Elasticity*, 3rd ed., McGraw-Hill, New York.

Timoshenko, S., and S. Woinowsky-Krieger [1959], *Theory of Plates and Shells*, 2nd ed., McGraw-Hill, New York.

Ting, T. C. T. [1969], *J. Appl. Mech.* **36**, 203.

Ting, T. C. T., and N. Nan [1969], *J. Appl. Mech.* **36**, 189.

Ting, T. W. [1966a], *J. Math. Mech.* **15**, 15.

Ting, T. W. [1966b], *Trans. Am. Math. Soc.* **123**, 1966.

Ting, T. W. [1967], *Arch. Ration. Mech. Anal.* **25**, 342.

Ting, T. W. [1969a], *Arch. Ration. Mech. Anal.* **34**, 228.

Ting, T. W. [1969b], *J. Math. Mech.* **19**, 531.

Ting, T. W. [1971], *Indiana Univ. Math. J.* **20**, 1047.

Tran-Cong, T. [1985], *Ing.-Arch.* **55**, 13.

Truesdell, C. [1984], *Rational Thermodynamics*, Springer-Verlag, New York.

Truesdell, C., and W. Noll [1965], *The Nonlinear Field Theories,* in *Handbuch der Physik* (ed. S. Flügge) Vol. III/3, Springer-Verlag, Berlin.

Truesdell, C., and R. A. Toupin [1960], *The Classical Field Theories,* in *Handbuch der Physik* (ed. S. Flügge) Vol. III/1, Springer-Verlag, Berlin.

Turner, L. B. [1909], *Trans. Cambridge Philos. Soc.* **21**, 377.

Valanis, K. C. [1971], *Arch. Mech.* **23**, 517.

Valanis, K. C. [1980], *Arch. Mech.* **32**, 171.

Valanis, K. C. [1985], *J. Appl. Mech.* **52**, 649.

Walker, K. P. [1981], *NASA Report* CR-165533.

Wang, A. J. [1955], *J. Appl. Mech.* **22**, 375.

Wang, A. J., and H. G. Hopkins [1954], *J. Mech. Phys. Solids* **3**, 22.

Washizu, K. [1958], *J. Math. Phys.* **36**, 306.

Washizu, K. [1975], *Variational Methods in Elasticity and Plasticity*, 2nd ed., Pergamon Press, Oxford.

Watanabe, O., and S. N. Atluri [1986], *Int. J. Plast.* **2**, 37, 107.

Wilkins, M. L. [1964], in *Methods of Computational Physics*, Vol. 3 (ed. B. Alder et al.), Academic Press, New York.

Winzer, A., and G. F. Carrier [1948], *J. Appl. Mech.* **15**, 261.

Winzer, A., and G. F. Carrier [1949], *J. Appl. Mech.* **16**, 346.

Woods, L. C. [1981], *Q. Appl. Math.* **39**, 119; *Bull. Inst. Math. Appl.* **17**, 98.

Yao, D., and E. Krempl [1985], *Int. J. Plast.* **1**, 259.

Zaid, M. [1959]. *J. Appl. Mech.* **26**, 462.

Ziegler, H. [1959], *Q. Appl. Math.* **17**, 55.

Zienkiewicz, O. C. [1977], *The Finite Element Method*, 3rd ed., McGraw-Hill, London.

Zienkiewicz, O. C., and I. Cormeau [1974], *Int. J. Num. Methods Engrg.* **8**, 821.

Zienkiewicz, O. C., and R. L. Taylor [1989], *The Finite Element Method*, 4th ed., Vol. 1, McGraw-Hill, London.

Zienkiewicz, O. C., S. Valliappan, and I. P. King [1969], *Int. J. Num. Methods Engrg.* **1**, 75.

Index